ALZHEIMER DISEASE

From Molecular Biology
to Therapy

Advances in Alzheimer Disease Therapy

Series Editors:
Ezio Giacobini
Robert Becker

Books in this series provide reviews of current advances in basic and clinical sciences that are relevant in understanding the etiology, pathogenesis, diagnosis and treatment of Alzheimer disease. Experts from the various fields - pharmacologists, neurochemists, neuroanatomists, neurophysiologists, neurologists, psychiatrists and geriatric medicine report and clinically appraise new findings and latest developments in the potential of Alzheimer disease therapies.

Books in the Series

Cholinergic Basis for Alzheimer Therapy
Edited by Robert Becker and Ezio Giacobini
ISBN 0-8176-3566-1

Alzheimer Disease: Therapeutic Strategies
Edited by Ezio Giacobini and Robert Becker
ISBN 0-8176-3757-5

Alzheimer Disease: From Molecular Biology to Therapy
Edited by Robert Becker and Ezio Giacobini
ISBN 0-8176-3879-2

ALZHEIMER DISEASE

From Molecular Biology to Therapy

Robert Becker
Ezio Giacobini
Editors

with the editorial assistance of
Joyce M. Barton
and production assistance of
Mona Brown

Birkhäuser

Robert E. Becker
Department of Psychiatry
Southern Illinois University
School of Medicine
P.O. Box 19230
Springfield, IL 62794-9230
USA

Ezio Giacobini
Institutions Universitaires
de Geriatrie de Geneve
Route de Mon-Idée
1226 Thonex, Geneva
Switzerland

Library of Congress Cataloging In-Publication Data
Alzheimer disease: from molecular biology to therapy / Robert Becker and Ezio Giacobini, editors.
p. cm. -- (Advances in Alzheimer disease therapy)
Includes bibliographical references and index.
ISBN-13: 978-1-4612-8660-8 e-ISBN-13: 978-1-4612-4116-4
DOI: 10.1007/978-1-4612-4116-4
1. Alzheimer's disease--Chemotherapy. 2. Cholinesterase inhibitors--Therapeutic use. 3. Alzheimer's disease--Molecular aspects. 4. Alzheimer's disease--Treatment. I. Becker, Robert E. II. Giacobini, Ezio. III. Series.
[DNLM: 1. Alzheimer's Disease--therapy. WT 155 A476 1996]
RC523.A3967 1996
618.97'6831--dc20
DNLM/DLC
for Library of Congress 96-30911
CIP

Printed on acid-free paper

Birkhäuser
Softcover reprint of the hardcover 1st edition 1997

Camera-ready text prepared by the editors using WordPerfect 6.0 on an IBM PS2
Printed and bound by Quinn-Woodbine, Woodbine, NJ.

9 8 7 6 5 4 3 2 1

CONTENTS

Part I. Pathological Basis of Therapy in Alzheimer's Disease

Criteria for the Diagnosis of Alzheimer's Disease

Neuropathology Of Alzheimer's Disease

Processing of Amyloid, Cytoskeletal Damage and Apolipoproteins

Cell Death: Excitotoxins, Apoptosis And Hypometabolism

Cholinesterase Inhibitors in Alzheimer Disease Treatment

Part I

PATHOLOGICAL BASIS OF THERAPY IN ALZHEIMER DISEASE

Criteria for the Diagnosis of Alzheimer's Disease
Neuropathology of Alzheimer's Disease
Processing of Amyloid, Cytoskeletal Damage and Apolipoproteins
Cell Death: Excitotoxins, Apoptosis and Hypometabolism
Neurotrophins, Growth Factors and Neuroprotectors
Animal Models

Alzheimer Disease: From Molecular Biology to Therapy
edited by R. Becker and E Giacobini

CRITERIA FOR THE CLINICAL DIAGNOSIS OF ALZHEIMER'S DISEASE: TRANSCULTURAL ASPECTS

Luigi Amaducci
Department of Neurology, University of Florence, Florence, Italy

Marzia Baldereschi
National Research Council of Italy, Targeted Project on Aging, Florence, Italy.

The clinical diagnosis of Alzheimer's Disease (AD) is now usually based on the use of structured diagnostic criteria. Although that is an essential step in the standardization of AD clinical diagnosis, differences in the observation and interpretation of findings may affect the reliability of the diagnostic instrument.

Cultural variables, such as languages, social customs and traditions, the quality and quantity of education, are of great relevance to the diagnostic process of AD. The threshold for making a diagnosis of dementia and public expectations about quality of life for the elderly differ in different cultural settings. However, the main issue in international research on dementia lays in the design of instruments that yield quantitative and replicable findings and that can be used in different cultures. Cognitive impairment is generally accepted as the primary and most essential criterion for the diagnosis and staging of dementia. Hence, a standardized approach to define cognitive impairment and dementia equitably and meaningfully in diverse cultures is needed.

Existing test batteries yield high false-negative rates of detection among the highly educated subjects, while among the very elderly and the illiterate subjects high false-positive rates have been reported (Anthony et al., 1982; Folstein et al., 1985; Escobar et al., 1986). When English language tests are translated for use in different cultures, sensitivity and specificity for detecting dementia are further compromised (Bird et al., 1987; Jin et al., 1989, Baldereschi et al., 1994a). The World Health Organization (WHO) undertook a cross-national epidemiological project on dementia prevalence and dementia determinants, the Age-Associated Dementia project (AAD).

RELIABILITY OF THE DIAGNOSIS OF DEMENTIA AND DEMENTING DISEASES

The use of standardized clinical criteria to achieve a reliable diagnosis of dementia across different cultures is an essential requirement for data comparison and generalizations of results.

The WHO-AAD project carried out an interrater reliability study (Baldereschi et al., 1994b) among 14 clinicians from Canada, Chile, Malta, Nigeria, Spain and the USA. The level of agreement was measured using both the percentage statistics and the extension of the Kappa index for multiple raters. The results are summarized in Table I and II.

These results show that clinicians from different medical traditions and cultures may reliably use DSM-III-R, ICD10 and NINCDS-ADRDA criteria. Nevertheless it has to be pointed out that only a mild agreement was achieved on the diagnosis of possible AD. The disagreement might have arisen from different levels of certainty among clinicians as to the significance of concomitant diseases, atypical presentations, or equivocal or very mild dementia as already reported by Lopez et al. (1990). In spite of those satisfying results, a specific training of the investigators in the correct and homogeneous use of the criteria sets can be surely helpful to improve the reproducibility of data collected in international and multicenter studies.

CROSS-NATIONAL VALIDITY OF SCREENING TESTS

The further step of the WHO-AAD project has been the selection of an instrument to screen for dementia across different cultures (Baldereschi et al., 1994a, 1995). The Mini-Mental State Examination (MMSE) (Folstein et al., 1975) and the Pfeffer Functional Activities Questionnaire (PFAQ) (Pfeffer et al., 1982) were adapted to the Chilean, Maltese and Spanish socio-cultural backgrounds and then validated against the previously standardized DSM III-R clinical diagnosis for dementia.

As shown in Table III, the selection of 90% and 100% sensitivity cut-off scores yielded to a wide range of specificity values for both the screening instruments. To circumvent the trade-off between sensitivity and specificity, as well as to obtain a comparable level of accuracy across the three cultures, it was decided to use the MMSE and the PFAQ in a serial testing procedure.

In conclusion, it has been outlined what must be done for adopting instruments and diagnostic criteria for use in cross-cultural researches. Though difficult, it represents an essential requirement for instrument development in this type of research that lacks clear culture-free standards.

	Agreement on positives (%)	Agreement on negatives (%)	Observed agreement (%)	K	Z
Dementia[1] DSM-III-R	96	86	89	0.67	46.0
Dementia[1] ICD-10	96	83	92	0.69	46.8
Depression[2] DSM-III-R	77	97	93	0.61	41.8
Cognitive[3] impairment	0	93	93	0.10	7.1

TABLE I. Interrater agreement for 14 raters on the DSM-III-R diagnoses of dementia and depression, on the ICD-10 diagnosis of dementia, and on the diagnosis of cognitive impairment. [1]Two diagnostic categories: dementia vs nondementia; [2]Two diagnostic categories: depression vs nondepression; [3]Two diagnostic categories: cognitive impairment vs non cognitive impairment.

	Agreement on positives (%)	Agreement on negatives (%)	Observed agreement (%)	K	Z
Probable AD[1]	77	95	88	0.58	39.4
Possible AD[1]	57	90	90	0.12	8.5
Prob + Poss AD[1]	91	95	90	0.72	49
Vascular dementia[1]	82	98	94	0.66	45.2
Other dementia[1]	93	76	91	0.40	27.1

TABLE II. Interrater agreement for 14 observers on the NINCDS-ADRDA clinical diagnoses and on the ICD-10 diagnoses of other dementing disorders [1]Two diagnostic categories: yes vs no.

	PFAQ and MMSEm		MMSEm only		Cut-off used	
	sensitivity	specificity	sensitivity	specificity	MMSEm	PFAQ
Chile	.94	.82	.94	.52	20/21	>5
Malta	.95	.78	.95	.47	23/24	>8
Spain	.88	.87	1.00	.55	23/24	>5

TABLE III. Serial testing procedure and cut-off used

ACKNOWLEDGMENTS

The study is supported by the World Health Organization - Program for Research on Aging. We thank M. E. Della Santa for preparing the manuscript.

REFERENCES

Anthony JC, Le Resche L and Niaz U (1982): Limits of the Mini-Mental-State as a screening test for dementia and delirium among hospital patients. *Psychol Med* 12:397-408.

Baldereschi M, Meneghini F, Quiroga P, Albala C, Mamo J, Muscat P, Gabriel R, Bermejo F, Amaducci L and Katzman R (1994a): Cognitive versus functional screening for dementia across different countries: cross-cultural validation of the Mini-Mental-State Examination (MMSE) and of the Pfeffer Functional Activities Questionnaire (PFAQ) against the standardized clinical diagnosis of dementia. *Neurology* 44 (suppl. 2):A365.

Baldereschi M, Amato MP, Nencini P, Pracucci G, Lippi A, Amaducci L, Gauthier S, Beatty L, Quiroga P, Klassen G, Galea A and Muscat P, Osuntokun B, Ogunniyi A, Portera-Sanchez A, Bermejo F, Hendrie H, Burdine V, Brashear A, Farlow M, Maggi S, Katzman R (1994b): Cross-national interrater agreement on the clinical diagnostic criteria for dementia. *Neurology* 44:239-242.

Baldereschi M, Grigoletto F, Mamo J, Muscat P, Gabriel R, Bermejo F, Quiroga P, Albala C, Amaducci L and Katzman R (1995): Cross-cultural validation of a modified version of the Mini-Mental-State Examination and of a serial testing procedure. *Neurology* 45 (suppl. 4):A326.

Bird HR, Canino G and Stipec MR (1987): Use of the Mini-Mental-State Examination in a probability sample of an hispanic population. *J New Ment Dis* 175:731-737.

Escobar JI, Burnam A and Karho M (1986): Use of the Mini-Mental-State Examination (MMSE) in a community population of mixed ethnicity. Cultural and linguistic artifacts. *J New Ment Dis* 174:607-614.

Folstein M, Anthony JC and Parhad I (1985): The meaning of cognitive impairment in the elderly. *J Am Geriatr Soc* 33:228-235.

Folstein MF, Folstein SE and McHugh PR (1975): Mini-Mental-State: a practical method for grading the cognitive state of patients for the clinicians. *J Psychiatr Res* 12:189-198.

Jin H, Zhang HY and Qu Oy (1989): Cross-cultural studies on dementia: use of a chinese version of the Blessed-Roth Information-Memory-Concentration test in a Shangai dementia survey. *Psychol Aging* 4:471-479

Lopez OL, Swihart AA, Becker JT, Reinmuth OM, Reynolds III CF, Rezek DL and Daly FL (1990): Reliability of NINCDS-ADRDA clinical criteria for the diagnosis of Alzheimer's disease. *Neurology* 40:1517-1522.

Pfeffer R, Kurosaki TT, Harrah CH, Chance JM and Filos S, (1982): Measurement of functional activities in older adults in the community. *J Gerontol* 37:323-329.

Alzheimer Disease: From Molecular Biology to Therapy
edited by R. Becker and E. Giacobini

CLINICAL AND NEUROPATHOLOGICAL FINDINGS FROM CERAD

John C. Morris
Department of Neurology, Washington University School of Medicine, St. Louis, MO

The Consortium to Establish a Registry for Alzheimer's Disease (CERAD) was established in 1986 to develop standard methods to evaluate persons with Alzheimer's disease (AD) and to gather clinical, neuropsychological, and neuropathological information about the illness (Morris et al., 1989). Rather than a true registry, CERAD utilized convenience samples from 24 U.S. AD research centers to accomplish its aims of standardizing and testing the reliability of brief assessment instruments for AD and thus provide tools for use in epidemiologic surveys, dementia registries, and AD research protocols, including clinical drug trials. Since the onset of patient recruitment in 1987, CERAD has enrolled 1,281 cases and 472 controls into its longitudinal studies. Analyses of the CERAD database have addressed clinical diagnostic accuracy, neuropsychological staging of dementia severity, rates of cognitive decline, and other issues in the clinico-pathological characterization of AD.

CERAD INSTRUMENTS

The CERAD protocols were designed for ease of administration, brevity, and conventionality to ensure their uniform application across multiple centers. The clinical and neuropsychological protocols are administered annually; annual return rates for CERAD subjects are between 65-75%.

The *clinical assessment protocol* was designed to provide clinicians with the minimum information necessary to make a confident diagnosis of AD (Morris et al., 1989). It contains semistructured interviews with both the patient and an informant, a general physical and neurological examination of the patient (including a structured assessment for extrapyramidal dysfunction), and brief cognitive scales. Global dementia severity is staged in accordance with the Clinical Dementia Rating (CDR). Diagnostic criteria for AD are adapted from those proposed by the National Institute of Neurological and

Communicative Disorders and Stroke/Alzheimer's Disease and Related Disorders Association (NINCDS/ADRDA) (McKhann et al., 1984).

The *neuropsychological assessment protocol* is administered independently of the clinical protocol (Morris et al., 1989). Measures were chosen to assess the primary cognitive manifestations of AD and to evaluate deficits over much of the course of the disease. Individual measures include Category Fluency, modified Boston Naming Test, Mini-Mental State, Word List Learning, Recall, and Recognition, and Constructional Praxis.

The neuropathology assessment protocol documents gross and microscopic central nervous system abnormalities relevant for dementia (Mirra et al., 1991). Silver (e.g., Bielschowsky) or thioflavine S methods are used to detect senile plaques (SP) and neurofibrillary tangles (NFT) in five anatomic regions: middle frontal gyrus, superior and middle temporal gyri, inferior parietal lobule, hippocampus and entorhinal cortex, and midbrain. Diagnostic criteria for AD are based on semiquantitative assessment of neuritic neocortical SP (i.e., those with thickened silver-positive neurites) in the most severely affected regions. An age-adjusted plaque score is integrated with clinical information about the presence or absence of dementia to determine the level of certainty for the diagnosis of AD.

Coronal T1-weighted spin echo sequences from magnetic resonance images (MRI) of the brain are used in the MRI assessment protocol to rate atrophy (focal and global), white matter abnormalities, and areas of cerebral infarction or hemorrhage (Davis et al., 1992). The family history assessment protocol collects demographic and cognitive information on family members of probands by means of a structured telephone interview by trained family historians (Silverman et al., 1994). The Behavior Rating Scale for Dementia samples the psychopathology encountered in demented patients, including mood disorders, agitation, aggression, and psychotic symptoms (Tariot et al., 1995), and is administered to an informant by a trained interviewer. Items are scaled by frequency of occurrence within the previous month.

STANDARDIZATION AND RELIABILITY

Instruction manuals have been developed for all CERAD protocols. The individual guidebooks, the training methods, and data management procedures are contained in the CERAD Manual of Operations.

Clinical and neuropsychological protocol standardization is accomplished in part through videotaped demonstrations. Each center is certified for use of the protocols after demonstrating standard administration and scoring procedures. Uniformity and completeness are maintained by monthly random review of subject protocols and regular communication with each center. Interrater agreement on the quantitative CERAD tests is high, with intraclass correlation coefficients ranging from 0.92 (Constructional Praxis) to 1.0

(Word List Recall). Test-retest reliability also is substantial (Morris et al., 1989).

The neuropathology protocol is standardized through the guidebook and a primer (Mirra et al., 1993). Individual CERAD neuropathologists vary in histologic techniques (e.g., Bielschowsky versus thioflavine S) and in interpretation of findings (e.g., consideration of diffuse as well as neuritic NPs). These methodological inconsistencies result in significant differences for quantitative measures of SP densities (interrater reliability = 0.32 for SP, 0.66 for NFT), whereas semiquantitative analysis improves interrater reliability for SP to 0.62 (Mirra et al., 1994). Rank order analysis yields an interrater reliability = 0.75, indicating that neuropathologists have good agreement for the overall severity of AD changes (Mirra et al., 1994).

The good to excellent interrater reliability of the standardized CERAD protocols contributes to their wide acceptance. The number of requests for the protocols exceeds 250. Another measure of their success is the translation of the CERAD protocols into 12 languages, including French, Spanish, Italian, Chinese, Japanese, Korean, Portuguese, Dutch, Czech, Bulgarian, Greek and Hebrew. CERAD procedures emphasize linguistic and cultural equivalence of the translated instruments (Demers et al., 1994).

DIAGNOSTIC ACCURACY

The CERAD autopsy rate is 51%. Ninety-two of the initial 106 cases diagnosed by CERAD clinicians with probable (n=83) or possible (n=23) AD by NINCDS/ADRDA criteria had AD verified histologically for a clinical diagnostic accuracy rate of 87% (Gearing et al., 1995). Although not considered to be the primary neuropathological diagnosis, AD also was present in 6 of the 14 remaining cases (7 clinically diagnosed as possible AD). These data support the validity of CERAD clinical criteria for AD.

STAGING OF DEMENTIA SEVERITY

CERAD neuropsychological measures effectively distinguish cases with mild probable AD from nondemented age- and education-matched control subjects; Word-List Delayed Recall is the best discriminator (Welsh et al., 1991). Control performance on the Delayed Recall measure is only minimally affected by age after adjustment for initial acquisition of material (Welsh et al., 1994). Perhaps because of "floor effects," however, this and other verbal memory tests are less effective in staging dementia severity than other measures (confrontation naming; fluency; constructional praxis) (Welsh et al., 1992).

RATES AND PREDICTORS OF DEMENTIA PROGRESSION

Rates of cognitive change are nonlinear as described by certain measures and are a function of the severity of cognitive impairment such that the less the dementia severity, the slower the rate of decline (Morris et al., 1993). Optimal reliability for change in cognitive status occurs when the observation period is greater than 1 year. Age affects the rate of cognitive decline such that younger patients decline more rapidly than older patients (Koss et al., 1996). Clinical milestones, including functional changes as measured by the CDR, are effective in charting the progression of AD and avoid issues of nonlinearity, "floor" effects, questionable clinical relevance, and other complications of cognitive tests (Galasko et al., 1995). Sex, age, and severity of dementia as measured by the CDR are powerful predictors of survival, with median duration of survival after entry into CERAD of about 6 years (Heyman et al., 1996).

CONCLUSIONS

CERAD has developed valid and reliable procedures and protocols for the standardized assessment of AD. Discrimination of mild AD from non-demented aging, staging of dementia severity throughout the course of disease, and description of rates of cognitive and functional decline have been reported from the large, longitudinally assessed CERAD sample recruited from multiple sites. The standardized CERAD clinical, neuropsychological, and neuropathological instruments fill a need for brief, easily administered methods for the characterization of AD.

ACKNOWLEDGMENTS

This report is based on the body of work produced by the collaborative CERAD team: Albert Heyman (Principal Investigator, CERAD); Leonard Berg, Steven Edland, Gerda Fillenbaum, Suzanne Mirra, Richard Mohs, Gerald van Belle, and Kathleen Welsh; and the investigators at participating CERAD centers. Supported by National Institute on Aging grant AGO6790.

REFERENCES

Davis PC, Gray L, Albert M, et al. (1992): The Consortium to Establish a Registry for Alzheimer's Disease (CERAD). Part III. Reliability of a standardized MRI evaluation of Alzheimer's disease. *Neurology* 42:1676-1680.

Demers P, Robillard A, Lafleche G, et al. (1994): Translation of clinical and neuropsychological instruments into French: the CERAD experience. *Age and Ageing* 23:449-451.

Galasko D, Edland, SD, Morris JC, et al. (1995): The Consortium to Establish a Registry for Alzheimer's Disease (CERAD). Part XI. Clinical milestones in patients with Alzheimer's disease followed over 3 years. *Neurology* 45:1451-1455

Gearing M, Mirra SS, Hedreen JC, et al. (1995): The Consortium to Establish a Registry for Alzheimer's Disease (CERAD). Part X. Neuropathology confirmation of the clinical diagnosis of Alzheimer's disease. *Neurology* 45:461-466.

Heyman A, Peterson B, Fillenbaum G and Pieper C. (1996): The Consortium to Establish a Registry for Alzheimer's Disease (CERAD). Part XIV: Demographic and clinical predictors of survival in patients with Alzheimer's disease. *Neurology* 46:656-660.

Koss E, Edland S, Fillenbaum G, et al. (1996): Clinical and neuropsychological differences between patients with earlier and later onset of Alzheimer's disease: A CERAD analysis, Part XII. *Neurology* 46:136-141.

McKhann G, Drachman D, Folstein M, et al. (1984): Clinical diagnosis of Alzheimer's disease: Report of the NINCDS-ADRDA work group under the auspices of Department of Health and Human Services Task Force on Alzheimer's Disease. *Neurology* 34:939-944.

Mirra SS, Gearing M, McKeel W Jr, et al. (1994): Interlaboratory comparison of neuropathology assessments in Alzheimer's disease: A study of the Consortium to Establish a Registry for Alzheimer's Disease (CERAD). *J Neuropathol Exp Neurol* 53:303-315.

Mirra SS, Heyman A, McKeel D, et al. (1991): The Consortium to Establish a Registry for Alzheimer's Disease (CERAD). Part II. Standardization of the neuropathologic assessment of Alzheimer's disease. *Neurology* 41:479-486.

Mirra S, Hart MN and Terry RD. (1993). Making the diagnosis of Alzheimer's disease. *Arch Pathol Lab Med* 117:132-144.

Morris JC, Heyman A, Mohs RC, et al. (1989): The Consortium to Establish a Registry for Alzheimer's Disease (CERAD). Part I. Clinical and neuropsychological assessment of Alzheimer's disease. *Neurology* 39:1159-1165.

Morris JC, Edland S, Clark C, et al. (1993): The Consortium to Establish a Registry for Alzheimer's Disease (CERAD). Part IV. Rates of cognitive change in the longitudinal assessment of probable Alzheimer's disease. *Neurology* 43:2457-2465.

Silverman JM, Raiford K, Edland S, et al. (1994): The Consortium to Establish a Registry for Alzheimer's Disease (CERAD). Part VI. Family history assessment: A multicenter study of first-degree relatives of Alzheimer's disease probands and nondemented spouse controls. *Neurology* 44:1253-1259.

Tariot PN, Mack JL, Patterson MB, et al. (1995): The behavior rating scale for dementia of the Consortium to Establish a Registry for Alzheimer's Disease. *Am J Psychiatry* 152:1349-1357.

Welsh K, Butters N, Hughes J, et al. (1991): Detection of abnormal memory decline in mild cases of Alzheimer's disease using CERAD neuropsychological measures. *Arch Neurol* 48:278-281.

Welsh KA, Butters N, Mohs RC, et al. (1994): The Consortium to Establish a Registry for Alzheimer's Disease (CERAD). Part V. A normative study of the neuropsychological battery. *Neurology* 44:609-614.

Welsh KA, Butters N, Hughes JP, Mohs RC and Heyman A (1992): Detection and staging of dementia in Alzheimer's disease. Use of the neuropsychological measures developed for the Consortium to Establish a Registry for Alzheimer's disease. *Arch Neurol* 49:448-452.

Alzheimer Disease: From Molecular Biology to Therapy
edited by R. Becker and E. Giacobini

DIFFERENTIATING ALZHEIMER DISEASE AND VASCULAR DEMENTIA: REFRAMING THE QUESTION

Helena Chui, Qian Zhang, Jeff Victoroff, Barbara Zaias
University of Southern California, Los Angeles, California

THE ESSENCE OF OUR TASK: REFRAMING THE QUESTION

The task at hand, in today's popular parlance, is to "differentiate Alzheimer disease from vascular dementia." Alzheimer disease (AD) refers to a slowly progressive neurodegenerative dementia associated with neurofibrillary tangles and neuritic plaques. Vascular dementia (VaD) refers to a heterogeneous group of dementias due to vascular disease. A comparative discussion of AD and VaD is important, since these disorders are recognized world-wide as the two most common causes of dementia.

Several problems arise, however, in differentiating AD and VaD. First, the comparison of AD vs. VaD is not a comparison of two homogeneous conditions or syndromes. There is heterogeneity in AD with regard to cause and clinical phenotype, even though this group of disorders shares a slowly progressive decline in cognition, characteristic histopathological features, and presumably a final common pathogenic pathway. The spectrum of heterogeneity in VaD is even greater, encompassing broad differences in cause, pathogenic mechanism, pathological features, clinical course and syndromes. This heterogeneity within each diagnosis makes it difficult to point to consistent differences between diagnoses. Second and more importantly, it can be argued that AD itself has an intrinsic vascular component. To a variable extent, the deposition of beta A4 is accelerated in blood vessels, as well as brain parenchyma. Furthermore, ultrastructural studies in AD cases demonstrate consistent morphological changes in the microvasculature (Perlmutter & Chui, 1990). It does not make sense to differentiate a part of the whole as something alien. Given these internal inconsistencies, it is worthwhile to consider the essence of the task at hand and to reframe our question.

The present discussion takes place in relation to the pharmacological treatment of AD. In this context, an important task is to sort out highly prevalent but distinct processes that might confound attempts to develop a treatment for AD. Although neurodegenerative and ischemic/hemorrhagic injury potentially lead to the common phenotype of dementia, they are not likely to respond to the same type of pharmacological intervention. Selection of a group of patients more homogeneous with respect to pathogenesis should enhance the likelihood of observing a significant treatment effect, and, conversely reduce the likelihood that treatment failure will result from treating a non-intended target. Our task can therefore be reframed as follows: "How can we improve the identification of vascular contributions to AD and other dementias."

EVIDENCE THAT DEMENTIA IS CAUSED BY CEREBROVASCULAR DISEASE

Cognitive impairment results from dysfunction of neurons subserving cognition. The relation between risk factors, cerebrovascular disease (CVD), cerebrovascular injury, and VaD can be likened to a series of nested boxes. Risk factors are not synonymous with CVD; CVD is not synonymous with cerebrovascular injury; and cerebrovascular injury is not synonymous with VaD. VaD results only when risk factors contribute to produce CVD that is sufficiently severe to produce cerebral injury. Further, that injury must be strategically located and must be sufficiently severe and lasting to meaningfully impair cognition.

It is worthwhile to note that the diagnosis of VaD and the prevention of VaD focus on opposite ends of this spectrum. The determination that dementia results from vascular disease requires, first, demonstration of cerebrovascular brain injury and, second, assessment of the lesion's strategic impact on cognition. In contrast, prevention of VaD does not depend upon either demonstration of injury or specific location of injury. Hopefully, reduction of risk factors will prevent the development of brain injury in the first place. Nonetheless, for the purpose of selecting AD patients for drug studies, the immediate concern is diagnosis. Thus, in the following sections, we consider the methods of recognizing and localizing cerebrovascular brain injury.

RECOGNITION OF CEREBROVASCULAR BRAIN INJURY: STATUS REPORT

CVD results in several types of parenchymal brain injury. Depending upon the size of the blood vessel involved, acute-focal ischemia may lead to brain infarction of variable size: involvement of major feeding arteries to large territorial infarcts; small penetrating arteries to small lacunes; and the capillary bed to microinfarcts. Chronic or global ischemia (e.g., from severe arteriolarsclerosis or systemic hypotension) may lead to various degrees of tissue injury, up to and including frank infarction in border or end-arteriole zones, such as the periventricular and deep white matter. Hemorrhage from rupture of the blood vessel wall damages brain tissue through a combination of mechanical and ischemic mechanisms. Differences can be expected in the sensitivity and specificity of current diagnostic tools to detect these various types of lesions. In reviewing present day ability to detect cerebrovascular lesions, we draw upon the experience at our Alzheimer center.

Clinical study

We reviewed 119 consecutive cases referred for assessment of memory loss to the Southern California Alzheimer Disease Diagnostic and Treatment Center (ADDTC) from January 1, 1994 to December 31, 1994. The sample comprised 42 males (35%) and 77 females (65%). The mean age and standard deviation was 69.9 ± 8.4 yrs for the males and 75.5 ± 10.1 yrs for the females.

The history was reviewed for the presence of risk factors, namely hypertension, diabetes mellitus, hyperlipidemia, and heart disease. Neuroimaging studies were obtained for all subjects: MRI in 66% of cases; CT in 49% cases; both MRI and CT in 15%. Neuroimaging studies were reviewed for evidence of stroke and severity of leukoaraiosis (4-point scale). In addition, laboratory studies including CBC, SMA-7, thyroid and liver function studies, B12 and folate levels, and syphilis serology were obtained to screen for reversible causes of dementia or medical co-morbidity. Diagnosis of AD was made by NINCDS-ADRDA criteria; diagnosis of ischemic vascular dementia (IVD) by ADDTC criteria (Chui et al., 1992).

History of stroke was ascertained in 15 (13%) of subjects, whereas neuroimaging evidence of stroke was observed in 26 (22%). Thus, evidence of unexpected stroke was obtained in 10 cases (9%).

After completed work-up, a diagnosis of probable AD was made in 53 (44.5%) and possible AD in 19 (16%) cases (total = 72 (60%). A diagnosis of probable IVD was made in 16 (13%), possible IVD in 7 (6%), and of CVD (not meeting criteria for IVD) in 4 (3%) (total = 27 (22%). Among all cases with a diagnosis of AD or VaD, 6 (5%) had a diagnosis of both AD and VaD (i.e., mixed dementia).

The frequency distribution of vascular risk factors and deep white matter changes differed for AD and VaD groups. Overall, significantly more patients with VaD than AD had at least one vascular risk factor (67% vs. 32%; Chi square $X^2 = 9.8$, $p < .002$). The frequency of each individual risk factor was also higher in VaD than AD: hypertension (44% vs. 26%), diabetes mellitus (30% vs. 11%), hyperlipidemia (7.4% vs. 2.8%), and heart disease (33% vs. 11%). Moderate to severe leukoaraiosis was observed in 32% of VaD and 14% of AD cases (Wilcoxon rank sum test, $p < .04$).

In summary, neuroimaging studies significantly enhanced our ability to detect evidence of stroke beyond the history and physical examination (additional 9%). A clinical diagnosis of mixed AD and VaD was made in 5% of cases. In patients with AD, vascular risk factors were present in 32% and moderate to severe leukoaraiosis was observed in 14% of cases.

Clinical-pathological study

Clinical and pathological diagnoses of dementia were compared in 196 cases contributed by seven ADDTC's in the State of California (Victoroff et al., 1995). The sample comprised 108 males and 88 females, who had died between 1985 and 1992. Most of these subjects had received a comprehensive and standardized diagnostic evaluation, including a neuroimaging study.

Using the pathological findings as the gold standard, the clinical diagnosis of AD showed a relatively high sensitivity of 89%, but a low specificity of 37%. The relatively low specificity reflects a high false positive rate. Of the 116 cases clinically diagnosed with AD, 100 (86%) had evidence of AD-type pathology. However, 20 (17%) revealed unexpected cerebral infarcts. (Note: this included large and small infarcts, but not microinfarcts). Of the 11 cases clinically diagnosed with VaD, 5 (46%) had CVD pathology (2 of these had Alzheimer-type pathology as well as infarcts). Six cases (54%) did not have CVD (4 had Alzheimer-type and 2 had Lewy body pathology). Finally, of the 19 cases clinically diagnosed as mixed AD/VaD, only 4 (21%) revealed mixed AD/VaD pathology; 9 had AD pathology without CVD, 2 had CVD without Alzheimer pathology, and 4 had neither AD nor CVD pathology.

This study calls attention to a large number of infarcts found in patients with AD at the time of autopsy, but not detected at the time of initial clinical evaluation. It is conceivable that some of these unsuspected infarcts transpired during the average 5 year interval between the initial clinical evaluation and death. Alternatively, one could hypothesize that neuropathological evaluation is more sensitive than clinical evaluation in detecting vascular lesions. In addition, the study indicates the difficulty of making accurate clinical diagnoses of VaD nor mixed AD/VaD.

Pathological study

The frequency of vascular pathology was systematically reviewed in 26 consecutive cases with neuropathologically-confirmed AD coming to autopsy at the Southern California ADDTC between 1991-1995. All of these cases had received a clinical diagnosis of probable AD and showed widespread neuritic plaques and neurofibrillary tangles in cerebral cortex. Moderate to severe atherosclerosis was found in 54%, arteriolosclerosis in 23%, and cerebral amyloid angiopathy in 81%. Lacunar infarcts were found in 15% and hemorrhages in 12%. Unsuspected microinfarcts, often multiple, were discovered in 14/26 (53%) cases. These data underscore the high frequency of vascular lesions in AD, which were not appreciated during the patient's life.

SUMMARY AND RECOMMENDATIONS

Taken together these studies suggest that: 1) vascular risk factors, infarcts (especially microinfarcts) and leukoaraiosis are quite prevalent in AD, as well as VaD; 2) while neuroimaging is more sensitive than the clinical exam, the neuropathological exam is more sensitive than neuroimaging for detecting cerebrovascular injury; 3) there is no sensitive clinical method for detecting microinfarcts; and 4) other factors such as lesion size, number, and location must be considered in assessing the relationship between cerebrovascular injury and cognitive impairment.

Several suggestions can be made for characterizing vascular contributions to dementia in candidates for pharmacological trials of AD.

1. Based on the neuroimaging study, the likelihood that vascular disease is contributing to dementia could be specified along several levels:
   a. Very likely: Patients with multiple infarcts/hemorrhages in areas likely to affect cognition or a single infarct/hemorrhage in a strategic location.
   b. Somewhat likely: Patients with a single infarct/hemorrhage in a non-strategic location or moderate to severe leukoaraiosis.
   c. Not likely: Patients with neither infarcts nor significant leukoaraiosis, recognizing that for the present time there is no sensitive clinical method for excluding the presence of microinfarcts.
2. Patients should meet criteria for probable AD (NINCDS-ADRDA) or dementia of the Alzheimer-type (DSM-IV).
3. Depending upon the targeted pathogenic mechanism, a decision to further include or exclude patients for clinical drug trials can be made based upon the relative likelihood that vascular disease is contributing significantly to the dementia. For example, in some pharmacological trials, a decision could be made to exclude all patients with 1a or 1b, while in another trial only patients with 1a would be excluded.

ACKNOWLEDGEMENTS

This work was supported by the National Institutes of Health (P01-AG12435, P50-AG05142, and T32-AG00093) and the State of California Department of Health Services.

REFERENCES

Chui HC, Victoroff J, Margolin D, Jagust W, Shankle W, Katzman R (1992): Criteria for the diagnosis of VaD proposed by the State of California Alzheimer Disease Diagnostic and Treatment Centers (ADDTC). *Neurology* 42:473-480.

Perlmutter LS, Chui HC (1990): Microangiopathy in the vascular basement membrane: A review. *Brain Res Bull* 24:677-686.

Victoroff JV, Mack WJ, Lyness SA, Chui HC (1995): Multicenter clinicopathological correlation in dementia. *Am J Psychiatry* 152: 1476-1484.

Alzheimer Disease: From Molecular Biology to Therapy
edited by R. Becker and E. Giacobini

BASIS OF STRUCTURAL ALZHEIMER DISEASE AND SOME PATHOGENIC CONCEPTS

Robert D. Terry
Departments of Neurosciences and Pathology, School of Medicine,
University of California San Diego, La Jolla, CA

INTRODUCTION

By now it may be fairly assumed that just about everyone concerned with the biology of this disorder is aware of its morphologic aspects. The positive findings are the senile plaques, the neurofibrillary tangles, activated microglia and the neuropil threads; while the negative findings are loss of neurons and of synapses (Table 1). The plaque is comprised of a cluster of dystrophic neurites and a few abnormal synapses surrounding a core of β-amyloid (Terry et al., 1964). Fibrous astrocytes are prominent on the periphery, while microglia hold a close relationship to the amyloid and are scattered in the neuropil. The tangles are found principally within the large and medium neurons of the hippocampal formation and entorhinal area, and in the neocortex. They are made up of paired helical filaments (PHF) (Kidd, 1963), which in turn are composed largely of abnormally phosphorylated tau protein (Brion et al., 1985). Neuropil threads contain PHF and neurofilaments (Kowall and Kosik, 1987). Tangles, especially in the entorhinal and hippocampal areas, may be left outside neurons as a result of their formative neurons having died. This is very rare in the neocortex despite extensive loss of neurons here.

The neurons that are lost have come from populations of large and medium sized pyramidal cells in entorhinal (Hyman et al., 1984) and hippocampal areas (Ball, 1977), and in the neocortex (Terry et al., 1981). Locus ceruleus (Bondareff et al., 1982), dorsal raphe (Aletrino et al., 1992) and certain other stem nuclei may also be involved. Tangles need not be involved in cell loss, especially from neocortex. Synaptic loss has been measured in cortex (Scheff et al., 1990; Masliah et al., 1989) and hippocampal (Samuel et al., 1994) areas and is significant in the neuropil between as well as within plaques. The loss of synapses is greater than that of neurons.

Given the prominence of the two major (and oldest) findings –plaques and tangles– it is not surprising that there are two major opposing concepts

POSITIVE	NEGATIVE
• Plaques & Amyloid	• Neuron Loss
• Tangles & PHF	• Synapse Loss
• Neuropil Threads	
• Activated Microglia	

TABLE I: Structural changes in AD

concerning the primary element of the pathogenesis. These two camps have been facetiously named the "baptists," because they favor the primacy of β-amyloid protein; and the "taoists," those favoring the tangle and its tau protein. It is, however, increasingly clear that there are logical problems in both positions. Table II cites several circumstances in which β-amyloid is present without tangles and therefore, this amyloid does not necessarily induce tangles. There are other instances where tangles are present without amyloid; and so tangles do not necessarily lead to the formation of amyloid. There must be other factors involved in Alzheimer disease (AD).

Some of the argument on pathogenesis has been based on which lesion came first. Diffuse plaques, that is immunoreactive amyloid largely without formed filaments and without neurites, are found very early in Down's syndrome (Rumble et al., 1989). The assumption is often made that these diffuse plaques go on to mature as typical neuritic senile plaques. But this has not been proven, and it is not inconceivable that the diffuse plaque and the mature plaque develop as two quite separate elements with only occasional conversion. Both are present in advanced cases of the disease. The aged primate displays clusters of dystrophic neurites without amyloid prior to the presence of fully formed typical plaques (rarely with PHF in the neurites) (Wisniewski et al., 1973). In electron microscopic studies of human brain biopsies, it was noted that the subtlest lesion was made up exclusively of dystrophic neurites without amyloid, and that amyloid appeared only when there were three or more such neurites clustered together (Terry and Wisniewski, 1970). The Athena mouse displays synapse loss preceding the appearance of amyloid (Masliah, 1996). Thus, amyloid may well not be the leading event.

PHF without significant amyloid	β-amyloid without PHF
Post-Encephalitic Parkinson's disease	Normal Aging
Guam Parkinson – Dementia	Plaque-only dementia
Tuberous Sclerosis	Ischemia
Subacute Sclerosing Pan Encephalitis	Lewy Body Variant
Dementia Pugilistica	Arterio-Venous Malformation
Niemann-Pick type C	Athena tg Mouse
	Aged Primate

TABLE II: Does the *one cause* the *other*?

In regard to etiology, many attempts have been made to correlate one or another structural/chemical finding with the measured clinical severity of the disease. In fact, none of the classical structural lesions correlated with great strength until techniques were developed to measure the population density of synapses (Scheff et al., 1990; Masliah et al., 1989). This synaptic density has been shown by several laboratories to correlate very closely with the severity of the disease as measured by common psychometric tests (Scheff et al., 1990; Terry et al., 1991). The strength of the correlation and that of the neurobiologic rationale is such that it is now widely accepted that the immediate cause of Alzheimer dementia is indeed loss of synapses with consequent cerebral disconnections (Hof and Morrison, 1994).

A reasonable cause of synaptic loss is deficient axoplasmic flow. There is no direct evidence that axoplasmic flow is diminished in AD, but the implication is clear since the neuronal cytoskeleton is abnormal (neurofibrillary tangles, neuropil threads, and where the tau is phosphorylated without yet forming PHF), while the process is dependent on the intact cytoskeleton. This was first suggested in 1967 (Suzuki and Terry), but measurement of axoplasmic flow in the human is not possible, and there are still no good animal models of deranged cytoskeleton. One might propose, however, that in the early stages of AD there is diminished axoplasmic flow and that as a result, synapses distant from the cell bodies suffer to the extent of dystrophic changes and degeneration. These changes would activate microglia. The loss of synaptic contacts would also result in lessened availability of neurotrophic factors from the target areas, so these factors would not get back to the cell bodies, which would ultimately die. The dystrophic neurites contain APP, and from them β-amyloid would be derived (in association with microglia) to form the nidus of the plaque. Ultimately, of course, there is such loss of synapses that their density falls below threshold, and dementia becomes clinically apparent.

A relatively recent hypothesis proposes that potentially treatable inflammation plays a major role in the degenerative process (Rogers, 1992). Although lymphocytes are not present, microglia and astrocytes are common in the afflicted brain parenchyma. Activated microglia and reactive astrocytes produce cytokines and complement both of which can damage cellular components of the CNS. Indeed, in AD there are great numbers of activated microglia (McGeer et al., 1994), especially in relation to the plaque and its amyloid, but also in the neuropil between plaques. The most prominent microglial change is to be found in the white matter where there is obvious proliferation of this cell type showing intense activation with short processes and granular cytoplasm. The relationship between microglia and amyloid was pointed out as long ago as 1964 when it was reported that these cells displayed deep cytoplasmic indentations filled with amyloid filaments (Terry et al., 1964). It was suggested that the microglia might be changing amyloid from nonfibrillar substrates to the extracellular filamentous form, as subsequently elaborated by Wisniewski and Wegiel (1994). A major normal function of microglia is to remodel synapses. Degenerative synapses might well activate the microglia and begin the cascade of events involving the cytokines and complement.

Of course, many other hypotheses have been suggested for etiology, primacy, and pathogenesis of AD in addition to amyloid, tau, axoplasmic flow and inflammation. One might elaborate on calcium and iron effects, mitochondria, free radicals, aluminum, etc. But for myself, at this time, I prefer the scheme in Figure 1.

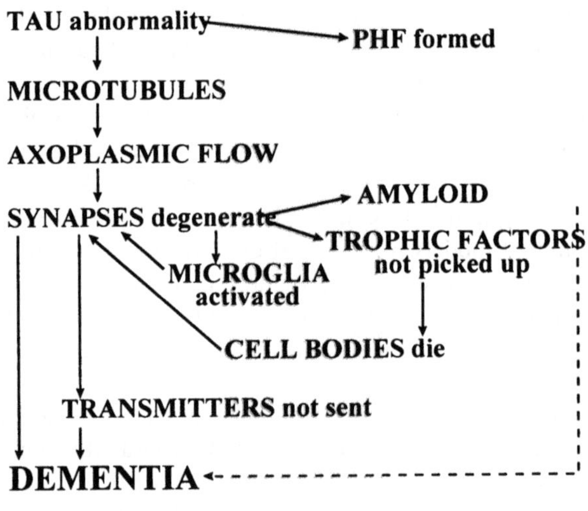

FIGURE 1.

REFERENCES

Aletrino MA, Vogels OJM, Van Domburg PHMF and Ten Donkelaar HJ (1992): Cell loss in the nucleus raphe dorsalis in Alzheimer's disease. *Neurobiol Aging* 13:461-468.

Ball MJ (1977): Neuronal loss, neurofibrillary tangles and granulovacuolar degeneration in the hippocampus with aging and dementia. A quantitative study. *Acta Neuropathol* 37:111-118.,

Bondareff W Mountjoy CQ and Roth M (1982): Loss of neurons of origin of the adrenergic projection to cerebral cortex (nucleus locus ceruleus) in senile dementia. *Neurology* 32:164-168.

Brion JP, Passareiro H, Nunez J, Flament-Durand J (1985): Immunological detection of tau protein in neurofibrillary tangles of Alzheimer's disease. *Arch Biol* 96:229-235.

Hof PR and Morrison JH (1994): The cellular basis of cortical disconnection in Alzheimer disease and related dementing conditions. In: *Alzheimer Disease*, Terry RD, Katzman R and Bick KL, eds. New York: Raven Press, pp. 197- 230.

Hyman BT, Van HGW, Damasio AR and Barnes CL (1984): Alzheimer's disease: cell-specific pathology isolates the hippocampal formation. *Science* 225:1168-1170.

Kidd M (1963): Paired helical filaments in electron microscopy in Alzheimer's disease. *Nature* 197:192-193.

Kowall NW and Kosik KS (1987): Axonal disruption and aberrant localization of tau protein characterize the neuropil pathology of Alzheimer's disease. *Ann Neurol* 22:639-643.

Masliah E (1996): Personal communication.

Masliah E, Terry RD, DeTeresa RM and Hansen LA (1989): Immunohistochemical quantification of the synapse-related protein synaptophysin in Alzheimer disease. *Neurosci Letts* 103:234-239.

McGeer PL, Walker DG, Akiyama H, Yasuhara O and McGeer EG (1994): Involvement of microglia in Alzheimer's disease. *Neuropath Appl Neurobiol* 20:191-192.

Rogers J (1992): Anti-inflammatory agents as a therapeutic approach to Alzheimer's disease. *Neurology* 42:447-449.

Rumble B, Retallack R, Hilbich C, Simms G, Multhaup G, Martins R, Hockey A, Montgomery P, Beyreuther K, Masters CL (1989): Amyloid A4 protein and its precursor in Down's syndrome and Alzheimer's disease. *N Engl J Med* 320:1446-1452.

Samuel W, Masliah E, Hill LR, Butters N, Terry R (1994): Hippocampal connectivity and Alzheimer's dementia: effects of synapse loss and tangle frequency in a two-component model. *Neurology* 44:2081-2088.

Scheff SW, DeKosky ST and Price DA (1990): Quantitative assessment of cortical synaptic density in Alzheimer's disease. *Neurobiol Aging* 11:29-37.

Suzuki K and Terry RD (1967): Fine structural localization of acid phosphatase in senile plaques in Alzheimer's presenile dementia. *Acta Neuropathol* 8:276-284.

Terry RD, Gonatas NK and Weiss M (1964): Ultrastructural studies in Alzheimer's presenile dementia. *Amer J Path* 44:269-297.

Terry RD, Peck A, DeTeresa R, Schechter R and Horoupian DS (1981): Some morphometric aspects of the brain in senile dementia of the Alzheimer type. *Ann Neurol* 10:184-192.

Terry RD, Masliah E, Salmon DP, Butters N, DeTeresa R, Hill R, Hansen LA and Katzman R (1991): Physical basis of cognitive alterations in Alzheimer disease: synapse loss is the major correlate of cognitive impairment. *Ann Neurol* 30:572-580.

Terry RD and Wisniewski HM (1970): The ultrastructure of the neurofibrillary tangle and the senile plaque. In: *CIBA Foundation Symposium on Alzheimer's Disease and Related Conditions*, Wolstenholme GEW and O'Connor M, eds. London: J & A Churchill, pp. 145-168.

Wisniewski HM, Ghetti B and Terry RD (1973): Neuritic (senile) plaques and filamentous changes in aged rhesus monkeys. *J Neuropath Exper Neurol* 32:566-584.

Wisniewski HM and Wegiel J (1994): The role of microglia in amyloid fibril formation. *Neuropath Appl Neurobiol* 20:192-194.

Alzheimer Disease: From Molecular Biology to Therapy
edited by R. Becker and E. Giacobini

REGIONAL DISTRIBUTION OF NEUROPATHOLOGICAL CHANGES IN ALZHEIMER'S DISEASE

Constantin Bouras, Pandelis Giannakopoulos, Philippe G. Vallet
Department of Psychiatry, HUG Belle-Idée,
University of Geneva School of Medicine, 1225 Geneva, Switzerland

INTRODUCTION

Alzheimer's disease (AD) is characterized neuropathologically by the presence of two classical histologic hallmarks, neurofibrillary tangles (NFT) and senile plaques (SP). Recent analyses of the distribution of the pathologic changes in AD suggest that this disease affects select populations of neurons within the cerebral cortex that are characterized by specific regional and laminar distribution and connectivity patterns, whereas other neuronal subgroups are remarkably resistant to the degenerative process. Based on the distribution of NFT and SP, it has been proposed that a global corticocortical disconnection leads to the loss of integrated functions exhibited by AD patients. Furthermore, it appears that lesion distribution follows certain organization schemes and that these regional and laminar patterns may relate cortical circuits to clinical symptoms. Several neuropathologic studies have documented the presence of NFT and SP in the cerebral cortex of non-demented elderly individuals (Mountjoy et al., 1983; Tomlinson et al., 1968). More recently, it has been shown that amyloid deposition and NFT formation are not only common features of brain aging, but that they are present within certain regions of the cerebral cortex with densities comparable to those observed in demented patients (Bouras et al., 1993; Hof et al., 1992). Thus, high NFT densities have been demonstrated in the hippocampal formation and temporal neocortex, but sparse NFT in the other neocortical regions in cases presenting with very mild signs of cognitive decline (Bouras et al., 1993; Hof et al., 1992). Non-demented cases have been described with NFT preferentially distributed in the entorhinal cortex, and lower NFT counts in the CA1 and temporal neocortex, and numerous SP in the neocortex (Braak and Braak, 1991). Also, Braak and Braak (1990, 1991) have demonstrated that the transition zone between the allo- and neocortex was consistently involved early in the degenerating process, and have proposed a neuropathologic staging of dementia based on the progressive alteration of the hippocampal formation. In this article we summarize the results of our comparative studies of over 1250 brains obtained from patients with AD, related forms of dementia, and from elderly non-demented individuals.

MATERIAL AND METHODS

The human brains used for these studies were obtained from patients autopsied in the Department of Psychiatry and the Geriatric Hospital, University of Geneva School of Medicine, Switzerland between 1982 and 1992. There were 540 men (81.0 ± 7.1 years old; 43%) and 718 women (82.8 ± 7.3 years old; 57%). Clinical diagnoses were obtained from the medical records and were subsequently confirmed neuropathologically for all of the demented cases. All of these brains (post-mortem delay: 3-20 hours) were fixed in a 10% formalin solution for at least 6 weeks, and cut into 1 cm thick coronal slices. After macroscopic examination, tissue blocks were taken from the mid portion of the hippocampal formation, the inferior temporal, superior frontal and occipital cortex in the left hemisphere. For routine neuropathological evaluation 20 µm thick sections were prepared from the selected areas and were stained with hematoxylin-eosin and cresyl violet. Modified Gallyas and Globus silver impregnation techniques or modified thioflavine S were used for the quantitative assessment of the distribution of NFT and SP in all of these cases (Vallet et al., 1992). For immunohistochemical assessment of lesion distribution, additional sections were processed with highly specific and previously fully characterized antibodies to the microtubule-associated protein tau and to the amyloid protein (Giannakopoulos et al., 1994a). All quantitative analyses were performed on computer-assisted image analysis system.

RESULTS

The neuropathological evaluation and immunohisto-chemical analysis of one-year autopsy cases reveal that although the principal lesions observed in AD brains are found in non-demented cases, their density and distribution differs significantly between demented and non-demented patients (Giannakopoulos et al., 1994a). They also indicate that the quantitative distribution of NFT and SP shows two distinct profiles in elderly individuals regardless of the clinical diagnosis.

In both non-demented cases and cases with mild cognitive impairment, NFT first developed in layer II of the entorhinal cortex in all of the cases. The CA1 field of the hippocampus and inferior temporal cortex were frequently affected by NFT formation. Numerous cases had moderate to high NFT densities in layer III of the inferior temporal cortex, whereas the superior frontal and occipital cortex were rather preserved, showing very low NFT densities in a few cases only . Similarly, in AD cases, NFT were widespread in the hippocampal formation and predominated in the CA1 field, subiculum, and layers II and V of the entorhinal cortex. The inferior temporal cortex was the principal site of NFT development in the neocortex whereas the superior frontal and occipital cortex displayed low NFT densities in the majority of AD cases. Quantitatively, the onset of cognitive decline was correlated with the NFT density in the entorhinal cortex in the general elderly population, while the onset of AD was with the NFT density in the inferior temporal cortex (Bouras et al., 1993). Thus, the massive involvement of the inferior temporal cortex appears to be a crucial factor for the clinical expression of dementia. Only neocortical, but not hippocampal, NFT densities represent a reliable index of dementia, suggesting that involvement of cortical structures outside the hippocampus is a prerequisite for the development of dementia. In this perspective, some of the non-demented cases with relatively high NFT densities in the inferior temporal cortex could represent a preclinical stage of AD, as suggested (Hubbard et al., 1990; Bouras et al., 1993, 1994; Giannakopoulos et al., 1993, 1994a, b; Hof and Morrison 1990; Hof et al., 1992).

A separate set of observations revealed that correlations between the topography of senile lesions and cognitive impairment change with age. Furthermore, the comparisons between younger and very old AD patients suggest that the degenerative process involves different cortical structures in the oldest-old than in younger people (Bouras et al., 1993, 1994; Giannakopoulos et al., 1993, 1994a,b). Quantitative analyses of NFT and SP distribution have been so far directed at populations below 90 years of age. A few classical histological studies of nonagenarians and centenarians have shown significant differences in the distribution of NFT and SP in this age group in comparison with younger demented and non-demented subjects (Hauw et al., 1986; Mitzutani and Shimada, 1992). In the present series, the study of centenarians was performed using immunohistochemistry for a better visualization of the neuropathological lesions for quantitative purposes (Giannakopoulos et al., 1993, 1994b). NFT were present in all of the cases. The hippocampal and parahippocampal areas were the most affected in both centenarians with AD and those with no or minimal cognitive impairment. The inferior temporal cortex was involved in 90% of non-demented cases and in 100% of demented cases and displayed moderate NFT densities regardless of the clinical diagnosis. The involvement of the superior frontal cortex was more frequent among AD centenarians but only few NFT were found in this area in demented and intellectually preserved patients in this age group. Quantitatively, the only statistically significant difference between AD and non-demented cases was observed in the CA1 hippocampal field. Interestingly, demented centenarians showed lower NFT counts in the neocortex than younger cases with the same clinical diagnosis. Moreover, the laminar distribution of pathological changes in very old AD cases differed from those previously reported in younger demented subjects since the NFT predominated in layers II and III but not in layers V and VI of the inferior temporal and superior frontal cortex. SP distribution in centenarians was comparable to that seen in younger age groups in that neocortical regions were more frequently involved and exhibited higher SP densities than limbic structures. There was no relationship between the laminar distribution of SP and NFT and no correlation was found between SP densities and the severity of cognitive changes in these very old people.

CONCLUSIONS

In conclusion, the extensive study of large autopsy populations demonstrate that the progression of the degenerative process within the cerebral cortex is much more complex than originally thought in that it involves different cortical fields and cellular populations at different points in time. With respect to clinicopathological correlations, these observations suggest that the final clinical symptomatology of AD may correspond to several distinct patterns of lesion distribution depending on the age and the morphologic and neurochemical characteristics specific to susceptible neuronal subpopulations. Altogether, these observations suggest that the pathologic changes found at a regional level in the cerebral cortex reflect the disconnection of specific cortical pathways subserved by select neuronal populations. The early pathologic alterations restricted to the hippocampal formation are strongly correlated with incipient stages of the dementing process, that may correspond to a clinical condition referred to as benign senescent forgetfulness. The involvement of the inferior temporal cortex is a crucial step to the development of overt clinical signs of AD. In very old demented patients the neuronal degeneration involves cortical areas usually preserved at the early stages of the dementing process such as the superior parietal, anterior and posterior cingulate cortex. SP formation in certain neocortical areas may be a pathologic hallmark of the severity of dementia in this particular age group. Thus, the key to dementia may reside in the mechanisms by which pathological changes that are limited to the hippocampal formation at early stages of the disease, are extending to neocortical circuits. In this context, if the neurochemical and anatomical features that are most clearly linked to differential cellular vulnerability in AD can be isolated, strategies to protect those particularly vulnerable neurons could be developed.

REFERENCES

Bouras C, Hof PR, Morrison JH (1993): Neurofibrillary tangle densities in the hippocampal formation in a non-demented population define subgroups of patients with differential early pathologic changes. *Neurosci Lett* 153:131-135.

Bouras C, Hof PR, Giannakopoulos P, Michel JP, Morrison JH (1994): Regional distribution of neurofibrillary tangles and senile plaques in the cerebral cortex of elderly patients: a quantitative evaluation of a one-year autopsy population from a geriatric hospital. *Cereb Cortex* 4:138-150.

Braak H, Braak E (1990): Neurofibrillary changes confined to the entorhinal region and an abundance of cortical amyloid in cases of presenile and senile dementia. *Acta Neuropathol* 80:479-486.

Braak H, Braak E (1991): Neuropathological stageing of Alzheimer-related changes. *Acta Neuropathol* 82:239-259.

Giannakopoulos P, Hof PR, Surini M, Michel JP, Bouras C (1993): Quantitative immunohistochemical analysis of the distribution of neurofibrillary tangles and senile plaques in the cerebral cortex of nonagenarians and centenarians. *Acta Neuropathol* 85:602-610

Giannakopoulos P, Hof PR, Mottier S, Michel JP, Bouras C (1994a): Neuropathologic changes in the neocortex of 1258 cases from a geriatric hospital: retrospective clinicopathologic evaluation of a ten year autopsy population. *Acta Neuropathol* 87:456-468.

Giannakopoulos P, Hof PR, Giannakopoulos AS, Buée-Scherrer V, Delacourte A, Bouras C (1994b): Dementia in the oldest-old: quantitative analysis of 12 cases from a psychiatric hospital. *Dementia* 5:348-356.

Hauw JJ, Vignolo P, Duyckaerts C, Beck H, Forette F, Henry JF, Laurent M, Piette F, Sachet A, Berthaux P (1986): Etude neuropathologique de 12 centenaires: la fréquence de la démence sénile de type Alzheimer n'est pas particulièrement élevée dans ce groupe de personne très agées. *Rev Neurol* 142:107-115.

Hof PR, Morrison JH (1990): Quantitative analysis of a vulnerable subset of pyramidal neurons in Alzheimer's disease: II. Primary and secondary visual cortex. *J Comp Neurol* 301:55-64.

Hof PR, Bierer LM, Perl DP, Delacourte A, Buée L, Bouras C, Morrison JH (1992): Evidence for early vulnerability of the medial and inferior aspects of the temporal lobe in an 82-year-old patient with preclinical signs of dementia - Regional and laminar distribution of neurofibrillary tangles and senile plaques. *Arch Neurol* 49:946-953.

Hubbard BM, Fenton GW, Anderson JM (1990): A quantitative histological study of early clinical and preclinical Alzheimer's disease. *Neuropathol Appl Neurobiol* 16:111-121.

Mizutani T, Shimada H (1992): Neuropathological background of twenty-seven centenarian brains. *J Neurol Sci* 108:168-177.

Mountjoy CQ, Roth M, Evans NJR, Evans HM (1983): Cortical neuronal counts in normal elderly controls and demented patients. *Neurobiol Aging* 4:1-11.

Tomlinson BE, Blessed G, Roth M (1968): Observations on the brains of non-demented old people. *J Neurol Sci* 7:331-356.

Vallet PG, Guntern R, Hof PR, Golaz J, Delacourte A, Robakis NK, Bouras C (1992): A comparative study of histological and immunohistochemical methods for neurofibrillary tangles and senile plaques in Alzheimer's disease. *Acta Neuropathol* 83:170-178.

Alzheimer Disease: From Molecular Biology to Therapy
edited by R. Becker and E. Giacobini

ALZHEIMER NEUROFIBRILLARY DEGENERATION: A FEASIBLE AND KEY TARGET FOR THERAPEUTICS

Khalid Iqbal, Alejandra del C. Alonso, Cheng-Xin Gong, Niloufar Haque, Sabiha Khatoon, Takashi Kudo, Jin-Jing Pei, Toolsee J. Singh, Toshihisa Tanaka, Jian-Zhi Wang and Inge Grundke-Iqbal
New York State Institute for Basic Research in Developmental Disabilities, Staten Island, NY 10314, USA

INTRODUCTION

Independent of the etiology, i.e. whether genetic or non-genetic, Alzheimer disease (AD) is characterized histopathologically by the intraneuronal accumulation of paired helical filaments (PHF), forming neurofibrillary tangles, neuropil threads and dystrophic neurites surrounding the extracellular deposits of β-amyloid, the second major lesion. The clinical expression of AD correlates with the presence of neurofibrillary degeneration; β-amyloidosis alone does not produce the disease clinically. Thus, arresting neurofibrillary degeneration offers a promising key target for therapeutic intervention of AD. In this chapter the molecular pathology, the mechanisms of neurofibrillary degeneration and therapeutic strategies to arrest this brain lesion in AD are briefly reviewed.

Molecular Pathology of Alzheimer Neurofibrillary Degeneration

For a neuron to function, it must be able to transport materials between its cell body and synapses, and integrity of the microtubule system is essential for this axonal transport. In certain selected neurons in AD brain, especially hippocampus and neocortex, the microtubule system is disrupted and replaced by neurofibrillary tangles of PHF. Some of the tangles also contain 2.1 nm thin filaments structurally identical to the tau filaments (Ruben et al., 1991).

Based on solubility in detergents, there are two general populations of Alzheimer neurofibrillary tangles (ANT), the readily soluble (ANT/PHF I) and the sparingly soluble (ANT/PHF II) types (Iqbal et al., 1984). Microtubule associated protein tau, which promotes the assembly and maintains the structure of microtubules in a normal neuron (Weingarten et al., 1975), is the major protein subunit of PHF (Grundke-Iqbal et al., 1986a,b; Iqbal et al., 1989). Tau in PHF is present in abnormally hyperphosphorylated forms (Grundke-Iqbal et al., 1986b; Iqbal et al., 1989). In addition to the PHF, there is a significant pool of the abnormal tau as yet unpolymerized into PHF present in the AD brain (Iqbal et al., 1986; Köpke et al., 1993). Small amounts of ubiquitin (<5%) are associated with PHF II but neither with PHF I nor with the non-PHF abnormal tau (AD P-tau) in AD brain (see Iqbal et al., 1993). Furthermore, the pretangle neurons, i.e. neurons containing accumulation of AD P-tau but no tangles, are readily immunolabeled for the abnormally phosphorylated tau but not for ubiquitin (Bancher et al., 1989).

Mechanisms of Alzheimer Neurofibrillary Degeneration

The abnormal tau (AD P-tau) isolated from AD brain contains 5 to 9 moles of phosphate/mole of the protein, whereas the normal tau contains only 2 to 3 moles of phosphate/mole of the protein (Köpke et al., 1993). Unlike normal tau, the AD P-tau because of hyperphosphorylation does not stimulate assembly of microtubules *in vitro*, unless dephosphorylated prior to incubation with tubulin (Alonso et al., 1994; Iqbal et al., 1994). Furthermore, microtubules are assembled *in vitro* from brain cytosol of normal aged control cases but not from identically-treated AD tissue obtained within 5 hours postmortem; the assembly of microtubules is, however, achieved from AD brain cytosol in the presence of DEAE-dextran, a polycation that mimics the action of tau in stimulating the assembly (Iqbal et al., 1986). The lack of microtubule assembly from AD brain cytosol is not due to lack of normal tau because the levels of normal tau in the diseased brains are similar to that in the normal aged control brains. In addition to normal levels of non-hyperphosphorylated tau, there are several-fold amounts of the abnormally phosphorylated tau in AD brain (Khatoon et al., 1992, 1994). The failure to achieve microtubule assembly from AD brain cytosol is most likely due to the sequestration of normal tau by the AD P-tau (Alonso et al., 1994). AD P-tau readily associates with normal tau under physiological conditions and inhibits both tau-assembled (Alonso et al., 1994) and high molecular weight associated proteins MAP 1- and MAP 2-assembled microtubules (Alonso et al., in preparation). AD P-tau also disassembles preassembled microtubules *in vitro*. This association of AD P-tau and normal tau results in formation of tangles of thin tau filaments (Alonso et al., in preparation). This association between the AD P-tau and normal tau not only is observed with isolated proteins but is also found in AD brain cytosol. The interaction between the AD P-tau and normal tau appears to be due to the hyperphosphorylation of tau, because on enzymatic dephosphorylation the AD P-tau loses its ability to sequester normal tau and to inhibit microtubule assembly. Thus, the abnormal hyperphosphorylation of tau appears to lead to the sequestration of normal tau by the aberrant tau, causing inhibition of assembly and disruption of preformed microtubules, and the accumulation of the tau as tangles of filaments. The latter subsequently, by as yet not fully understood mechanisms, are converted into PHF tangles.

The disruption of the microtubule system in the affected neurons probably leads to compromised axonal transport and eventually a retrograde degeneration of the affected neurons. The retrograde degeneration in AD brain has been confirmed both by Golgi stain studies, showing a massive loss of the neuronal arborization (Scheibel et al., 1976), and by the presence of thickening of the distal ends of the neurites, in pretangle neurons immunostained for the abnormally phosphorylated tau (Braak et al., 1994).

While the affected neurons are undergoing retrograde degeneration as a result of the breakdown of the microtubule system, the abnormally phosphorylated tau continues forming accumulation of PHF. This second insult, a space occupying lesion beyond a certain size, also probably starts interfering with cellular functions and eventually, though most likely over a period of several months to years, leads to cell death, leaving behind the "tombstones," also called ghost tangles. Although the ubiquitin levels are increased several-fold in AD brain (Kudo et al., 1994), the neurofibrillary tangles appear to persist, suggesting that the ubiquitin system might be minimally effective in proteolyzing PHF.

Strategies to Treat Alzheimer Disease by Arresting Neurofibrillary Degeneration

From the foregoing it is clear that inhibiting abnormal hyperphosphorylation should arrest neurofibrillary degeneration. The phosphorylation state of a protein is the function of both the activities of protein kinases as well as the protein phosphatases. To date, 21 sites at which tau is abnormally phosphorylated in AD brain have been identified (see Iqbal et al., 1995). Ten of the twenty-one sites are canonical sites for the proline-directed protein kinases (PDPKs) and the remaining eleven are the non-PDPK sites. Tau can be phosphorylated at some of the same sites as the AD P-tau by several protein kinases, and different tau isoforms are phosphorylated at different efficiency by different protein kinases (Singh et al., 1996). Thus, probably several protein kinases are involved in the phosphorylation of tau. However, to date, neither the exact identification of the protein kinases nor the sequence in which they phosphorylate tau to an Alzheimer-like state is fully understood. Furthermore, currently there is no clear evidence for the upregulation of the activity of any tau kinase(s) in AD brain. On the other hand, phosphoseryl/phosphothreonyl protein phosphatases (PP)-2A, -2B and, to a lesser degree, PP-1 have been demonstrated to dephosphorylate the abnormally phosphorylated tau, both AD P-tau and tau in PHF, at several abnormal sites and restore the microtubule assembly promoting activity of the aberrant protein (Gong et al., 1994a,b;c; Wang et al., 1995, 1996). Furthermore, the activities of these tau phosphatases, all of which have been immunocytochemically shown to be present in neurons (Pei et al., 1994), are reduced by approximately 30% in AD brain (Gong et al., 1993, 1995). *In vitro* dephosphorylation of neurofibrillary tangles isolated from AD brain causes their dissociation, releasing tau that can stimulate assembly of tubulin into microtubules (Wang et al., 1995).

The abnormal hyperphosphorylation of tau, thus, appears to be an imbalance of the protein phosphorylation and dephosphorylation in the affected neurons. A compound or compounds that can increase the tau phosphatase activity(s) in AD brain to correct the imbalance should arrest neurofibrillary degeneration and thereby AD. The activities of all the three tau phosphatases, PP-2A, PP-2B and PP-1 are considerably increased *in vitro* in the presence of 50-100 μM Mn^{2+} (Gong et al., 1994a). The future studies should develop such compounds that can inhibit the abnormal hyperphosphorylation of tau.

CONCLUSIONS

Tau is abnormally hyperphosphorylated in AD brain and this altered tau is the major protein subunit of PHF. The abnormal phosphorylation of tau probably precedes its polymerization into PHF. The abnormally phosphorylated tau does not bind to tubulin but competes with tubulin in binding to normal tau. This sequestration of normal tau by the abnormal tau inhibits the assembly and disrupts even the preformed microtubules. The abnormally hyperphosphorylated tau can, however, be converted into normal functional protein by dephosphorylation by protein phosphatases PP-2A, PP-2B and to a lesser extent by PP-1. The abnormally phosphorylated tau-phosphatase activity is decreased in AD brain which probably is the cause of the hyperphosphorylation of tau. Alzheimer neurofibrillary degeneration can probably be inhibited by increasing the tau-phosphatase activity in the brain of patients with AD.

ACKNOWLEDGMENTS

Secretarial support was provided by Joanne Lopez and Maritza Kaufmann. This work was supported in part by funds from the New York State Office of Mental Retardation and Developmental Disabilities, National Institutes of Health Grants AG05892, AG08076, NS18105, TW00507, and Zenith Award from the Alzheimer's Association, USA.

REFERENCES

Alonso A del C, Zaidi T, Grundke-Iqbal I and Iqbal K (1994): Role of abnormally phosphorylated tau in the breakdown of microtubules in Alzheimer disease. *Proc Natl Acad Sci (USA)* 91:5562-5566.

Bancher C, Brunner C, Lassmann H, Budka H, Jellinger K, Wiche G, Seitelberger F, Grundke-Iqbal I, Iqbal K. and Wisniewski HM (1989): Accumulation of abnormally phosphorylated tau precedes the formation of neurofibrillary tangles in Alzheimer's disease. *Brain Res* 477:90-99.

Braak E, Braak H and Mandelkow EM (1994): A sequence of cytoskeleton changes related to the formation of neurofibrillary tangles and neuropil threads. *Acta Neuropathol (Berl)* 87:554-567.

Gong C-X, Singh TJ, Grundke-Iqbal I and Iqbal K (1994a): Alzheimer disease abnormally phosphorylated tau is dephosphorylated by protein phosphatase 2B (calcineurin). *J Neurochem* 62:803-806.

Gong C-X, Grundke-Iqbal I and Iqbal K (1994b): Dephosphorylation of Alzheimer disease abnormally phosphorylated tau by protein phosphatase-2A. *Neurosci* 61:765-772, 1994.

Gong C-X, Grundke-Iqbal I, Damuni Z and Iqbal K (1994c): Dephosphorylation of microtubule-associated protein tau by protein phosphatase-1 and -2C and its implication in Alzheimer disease. *FEBS Lett* 341:94-98.

Gong C-X, Shaikh S, Wang J-Z, Zaidi T, Grundke-Iqbal I and Iqbal K (1995): Phosphatase activity toward abnormally phosphorylated τ: decrease in Alzheimer disease brain. *J Neurochem* 65:732-738.

Gong C-X, Singh TJ, Grundke-Iqbal I and Iqbal K (1993): Phosphoprotein phosphatase activities in Alzheimer disease. *J Neurochem* 61:921-927.

Grundke-Iqbal I, Iqbal K, Quinlan M, Tung Y-C, Zaidi MS and Wisniewski HM (1986a): Microtubule-associated protein tau: A component of Alzheimer paired helical filaments. *J Biol Chem* 261:6084-6089.

Grundke-Iqbal I, Iqbal K, Tung Y-C, Quinlan M, Wisniewski HM and Binder LI (1986b): Abnormal phosphorylation of the microtubule associated protein τ (tau) in Alzheimer cytoskeletal pathology. *Proc Natl Acad Sci (USA)* 83:4913-4917.

Iqbal K, Alonso A, Gong C-X, Khatoon S, Kudo T, Singh TJ, Grundke-Iqbal I (1993): Molecular pathology of Alzheimer neurofibrillary degeneration. *Acta Neurobiol Exp* 53:325-335.

Iqbal K and Grundke-Iqbal I (1995): Alzheimer abnormally phosphorylated tau is more hyperphosphorylated than the fetal tau and causes the disruption of microtubules. *Neurobiol Aging* 16:375-379.

Iqbal K, Grundke-Iqbal I, Smith AJ, George L, Tung Y-C, Zaidi T (1989): Identification and localization of a tau peptide to paired helical filaments of Alzheimer disease. *Proc Natl Acad Sci (USA)* 86:5646-5650.

Iqbal K, Grundke-Iqbal I, Zaidi T, Merz PA, Wen GY, Shaikh SS, Wisniewski HM, Alafuzoff I, Winblad B (1986): Defective brain microtubule assembly in Alzheimer's disease. *Lancet* 2:421-426.

Iqbal K, Zaidi T, Bancher C and Grundke-Iqbal I (1994): Alzheimer paired helical filaments: Restoration of the biological activity by dephosphorylation. *FEBS Lett* 349:104-108.

Iqbal K, Zaidi T, Thompson CH, Merz PA, Wisniewski HM (1984): Alzheimer paired helical filaments: bulk isolation, solubility and protein composition. *Acta Neuropathol (Berl)* 62:167-177.

Khatoon S, Grundke-Iqbal I and Iqbal K (1992): Brain levels of microtubule-associated protein tau are elevated in Alzheimer's disease: a radioimmuno-slot-blot assay for nanograms of the protein. *J Neurochem* 59:750-753.

Khatoon S, Grundke-Iqbal I and Iqbal K (1994): Levels of normal and abnormally phosphorylated tau in different cellular and regional compartments of Alzheimer disease and control brains. *FEBS Lett* 351:80-84.

Köpke E, Tung Y-C, Shaikh S, Alonso A del C, Iqbal K, Grundke-Iqbal I (1993): Microbutule associated protein tau: abnormal phosphorylation of a non-paired helical filament pool in Alzheimer disease. *J Biol Chem* 268:24374-24384.

Kudo T, Iqbal K, Ravid R, Swaab DF and Grundke-Iqbal I (1994): Alzheimer disease: correlation of cerebrospinal fluid and brain ubiquitin levels. *Brain Res* 639:1-7.

Pei J-J, Sersen E, Iqbal K and Grundke-Iqbal I (1994): Expression of protein phosphatases PP-1, PP-2A, PP-2B and PTP-1B and protein kinases MAP kinase and $P34^{cdc2}$ in the hippocampus of patients with Alzheimer disease and normal aged individuals. *Brain Res* 655:70-76.

Ruben GC, Iqbal K, Grundke-Iqbal I, Wisniewski HM, Ciardelli TL and Johnson JE (1991): The microtubule associated protein tau forms a triple-stranded left-hand helical polymer. *J Biol Chem* 266:22019-22027.

Scheibel ME, Lindsay RD, Tomiyasu U and Scheibel AB (1976): Progressive dendritic changes in the aging human limbic system. *Exp Neurol* 53:420-430.

Singh TJ, Grundke-Iqbal I and Iqbal K (1996): Differential phosphorylation of human tau isoforms containing three repeats by several protein kinases. *Arch Biochem Biophys*. In Press.

Wang J-Z, Gong C-X, Zaidi T, Grundke-Iqbal I and Iqbal K (1995): Dephosphorylation of Alzheimer paired helical filaments by protein phosphatase-2A and -2B. *J Biol Chem* 270:4854-4860.

Wang J-Z, Grundke-Iqbal I and Iqbal K (1996): Restoration of biological activity of Alzheimer abnormally phosphorylated τ by dephosphorylation with protein phosphatase-2A, -2B and -1. *Molec Brain Res* In Press.

Weingarten MD, Lockwood AH, Hwo S-Y and Kirschner MW (1975): A protein factor essential for microtubule assembly. *Proc Natl Acad Sci (USA)* 72:1858-1862.

Alzheimer Disease: From Molecular Biology to Therapy
edited by R. Becker and E. Giacobini

DIABETES AND DEMENTIA: A RETROSPECTIVE NEUROPATHOLOGIC STUDY OF OLD DIABETICS COMPARED TO NON-DIABETIC CONTROLS

Jean-Pierre Michel[1], Pius Muller[1], Gabriel Gold[1], Willliam Mac Gee[1], Reihnild Mulligan[1], Constantin Bouras[2]
[1]Department of Geriatrics
[2]Neuropathology Division, Department of Psychiatry
University Hospitals of Geneva, 1226 - Thônex-Geneva, Switzerland

INTRODUCTION

Non-insulin-dependent diabetes mellitus (NIDDM) and dementia are both characterized by high prevalence in later life (Muller, 1993). Differences in prevalence of NIDDM according to regions and races are well-known. In an industrialized society, with a predominantly white population, the rate of NIDDM increases linearly starting from 35 years old, reaching 15% at the age of 75 (Wilson et al., 1986) and 20% at the age of 80 (Morley and Perry, 1991). As for Alzheimer's Disease (AD), it follows an exponential curve starting from the age of 50, reaching 30% for the 85 year-old population (Skoog et al., 1993).

			Percentage of Diabetes	
Year	Author	n of Patients with Dementia	AD(*)	Vascular Dementia
1983	Bucht et al.	839	0.0%	7.5%
1988	Wolf-Klein et al.	206	3.9%	7.7%
1992	Small et al.	69	5.9%	-
1993	Lanvin et al.	71	0.0%	36.0%
1993	Mortel et al.	240	6.0%	23.0%
1996	Nielson et al.	265	0.8%	11.8%

TABLE I: Frequency of diabetes according to the etiology of dementia (*)AD: dementia of Alzheimer type.

Yet, there has been little research on the interaction between these two diseases. The frequency of diabetes in patients with dementia seems different according to the etiology of the dementia (Table I). In vascular dementia patients, the rate of diabetes mellitus is estimated to be between 7.7% (Bucht 1983; Nielson et al., 1996) and 36% (Lanvin et al., 1993). On the contrary, in AD patients, the rate of diabetes seems to be very low. Neither Bucht et al. (1983), nor Lanvin et al. (1993) found an association between diabetes and AD. Nielson et al. (1996) found only 0.8% of diabetics among AD patients. These estimations are confirmed by data from Wolf-Klein et al. (1988) and Mortel et al. (1993), who found a frequency of 3.9% and 6.0% of diabetes among AD patients. These data allowed Croxon and Jagger (1995) to write that "diabetics are less likely to develop AD and much more likely to develop vascular dementia."

In order to validate previous clinical studies, we designed a clinical neuropathologic comparison with the intention of determining whether or not diabetes "protects" from the onset of AD. The demonstration of such relationship would imply physiopathologic consequences related to etiology and treatment of the disease.

PATIENTS AND METHODOLOGY

The University Hospitals of Geneva Geriatrics Department admitted approximately 3,000 elderly patients per year (mean age 83 years), most of whom were community subjects. The mean length of stay was 40 days. The annual death rate approximated 15% and the autopsy rate averaged 30%. Clinical care of patients was handled by medical teams sharing the same philosophy of care. All cases were examined by the same pathologist (WMG) and neuropathologist (CB).

First, among all patients autopsied during the study period (5 years), 96 diabetics were identified of whom 52 met WHO criteria for NIDDM and were included in the study (Muller, 1993) (Table II).

In a second step, records of 104 patients autopsied, but not diagnosed as diabetics, were selected. Selection criteria were: same sex and age ± 5 years, same month of death, no history of diabetes and fasting blood glucose below 7.8 mmol/l in the course of hospitalization. After careful scrutiny of the selected records these patients were collected into two groups (Table II):

- 54 strictly normoglycemic subjects (fasting blood glucose below 6 mmol/l and postprandial glycemia below 7.8 mmol/l), called "controls".
- 50 patients with "impaired glucose tolerance" (IGT) (fasting blood glucose between 6 and 7.8 mmol/l and postprandial glycemia below 11.1 mmol/l)

	Non-insulin-dependent diabetes (n=52)	Impaired glucose tolerance (n=50)	Normo-glycemic (n=54)	P
Males /females	0.58	0.61	0.50	NS
Mean age (years)	80.7	80.1	83.0	<.05
Alone at home before hospitalization (%)	36.5	50.0	48.1	NS
Home help (%)	82.7	72.0	68.5	NS
Medical ground for hospitalization (%)	71.7	70.0	64.9	NS
Mean duration of hospitalization (weeks)	10.0	13.5	17.0	NS
Normal cognitive status at admission (%)	53.8	48.0	53.7	NS
Clinical dementia at admission (%)	21.2	26.0	18.5	NS
Functional independence at admission (%)	17.3	16.0	35.2	NS
Total functional dependence at admission (%)	25.0	38.0	20.4	NS
Mean weight at admission (kg)	59.2 ± 7.4	60.8 ± 14.6	49.1 ± 12.7	<.001
Mean body mass index (Kg/m2)	23.8 ± 3.8	24.0 ± 4.0	20.6 ± 4.9	<.05
Mean glycemia at admission(µmol/l)	17.4 ± 7.4	7.0 ± 1.4	5.1 ± 0.6	<.001
Mean creatinine at admission (µmol/l)	109.5 ± 53.3	135.7 ± 153.1	101.8 ± 44.1	NS
Mean creatinine clearance (ml/mn)	42.5 ± 14.5	44.5 ± 15.5	34.8 ± 9.4	<.05
Death by pulmonary infection (%)	13.5	14.0	18.0	NS
Death by cancer (%)	15.4	32.0	20.4	NS
Death by myocardial infarction (%)	38.5	16.0	14.8	<.01

TABLE II: Main characteristics of the population of the three study groups.

Accordingly, the population surveyed in this study comprises three groups of autopsied patients (Table II):

- group 1, "non-insulin-dependent diabetics," composed of 52 patients;
- group 2, "impaired glucose tolerance group," composed of 50 patients with not strictly normal glycoregulation;
- group 3, "non-diabetic controls," composed of 54 subjects with no glycoregulation disorder.

The 3 groups are homogeneous with regard to sex (males-females ratio 0.58, 0.61 and 0.50, respectively). There exists an age difference among the 3 groups of patients (80.7 ± 5.5 years, 80.1 ± 5.4 and 83.0 ± 4.9 years, respectively, $p<.05$). There is no significant difference among the 3 groups with respect to marital status, home help prior to hospitalization, grounds for hospitalization and mean duration of stay. There is no significant difference among the 3 groups with respect to cognitive status at admission as measured by Folstein's MMSE (Mini Mental State Examination) or degree of functional dependence according to Katz's scale for ADL (Activities of Daily Living). Weight, body mass index and creatinine clearance of the controls were significantly lower than those of the 2 other groups (Table II).

A list of the 156 patients with no indication of group relationship was passed on to the neuropathologist for a "blind" review of the brains. Criteria used were consistent with those recognized as referenced in the literature (Blessed, 1990; Bouras and Hof, 1992; Ulrich et al., 1986). The brains studied were classified into four categories as follows (Table III):

- Alzheimer's Disease (AD) with typical histologic lesions such as senile plaques, neurofibrillary tangles, granulovacuolar degeneration and cerebral amyloid angiopathy (Bouras and Hof, 1992)
- Vascular Dementia (VD) with lesions of cortical and/or severe infarctual type, subcortical atrophy, demyelinization, cerebral amyloid angiopathy, or granular atrophy (Wade and Hachinski, 1987). All primary hematomas and traumatic lesions were excluded.
- Mixed-type Dementia (MD) with both degenerative and vascular lesions. The predominance of one lesional type as opposed to another allows for inclusion into this category of lesions that were essentially, though not exclusively, degenerative or vascular.

Groups	Alzheimer	Vascular dementia	Mixed-type dementia	Normal	Alz/Vx/Mix/Nl
Diabetic (n=52)	40.4	32.7	15.4	11.5	NS
Impaired glucose tolerance (n=50)	32.0	32.0	20.0	16.0	NS
Normoglycemic-(n=54)	53.7	22.2	11.1	13.0	NS

TABLE III: Percent of neuropathologic lesions in the three study groups.

Three stages of degenerative lesion were identified:

- mild or isolated hippocampal and parahippocampal lesions,
- severe lesions of these structures and lastly,
- diffuse lesions affecting mainly associative regions of the cortex and certain subcortical structures.
- Vascular lesions were defined as: focal when lesions did not support the diagnosis of multiple infarct dementia (MID). Conversely, multifocal vascular lesions with major destruction of cerebral parenchyma accounted for most of the clinical manifestations (Muller, 1993).

- Normal Brain without any macroscopic and/or microscopic lesion (Muller, 1993).

After separate clinical assessment of diabetes and neuropathologic classification, data were analyzed using the Statistical Package for Social Science software. The $\chi 2$ test was used to evaluate the heterogeneity of contingence tables and F test was used to evaluate the differences between mean scores. Statistical significance was set at 5 % error probability ($P<0.05$)

RESULTS

In the NIDDM group (N=52), neuropathologic lesions of AD type were found in 40.4%, lesions of VD type in 32.7%, MD lesions in 15.4%. In this group, only 11.5% of the brains were classified as devoid of pathologic lesions (Table III).

In the impaired glucose tolerance group, neuropathologic lesions of AD, VD and MD types were found in 32.0%, 32.0% and 20.0%, respectively. In this group, there were 16.0% of brains devoid of macroscopic and microscopic lesions.

In the control group of non-diabetics, neuropathologic lesions of AD, VD and MD types were found in 53.7%, 22.2% et 11.1%, respectively. In this control group, there were 13.0% of brains devoid of macroscopic and microscopic lesions.

Matching of the above neuropathologic results against the cognitive status of patients from the 3 groups considered (NIDDM, IGT and normoglycemics) produces highly interesting results. Among the 28 diabetic patients presenting with a "normal" cognitive status at admission (MMSE> 28/30), only 17.9% had a brain devoid of macroscopic and microscopic lesion, whereas 42.9% had lesions of AD type, 25.0% had VD lesions, and 14.2% had MD lesions (Table IV). Among the 11 diabetic patients considered demented at admission (MMSE <21/30), 9.1% had no cerebral neuropathologic lesions, 36.4% had lesions of the Alzheimer type, 18.1% had lesions of the VD type, and 36.4% had MD lesions (Table IV).

Neuropathological diagnosis	Cognitive status (MMSE) at admission (n=52)		
	MMSE>28 (n=28)	MMSE>21 (n=13)	MMSE<21 (n=11)
Dementia of Alzheimer type	42.9%	38.5%	36.4%
Vascular dementia	14.2%	15.4%	18.1%
Mixed-type dementia	25.0%	46.1%	36.4%
Absence of neuropathologic lesion	17.9%	0.0	9.1%

TABLE IV: Relation of cognitive status at admission and neuropathological diagnosis in 52 autopsied diabetic patients.

It appears, therefore, that diabetics presenting with lesions of Alzheimer type display clinically cognitive dysfunction in the form of poor intellectual performance (MMSE<21) in only 43% of cases.

DISCUSSION

This study is unique insofar as, to our knowledge, it is the first to attempt to establish a correlation between diabetes and dementia based upon neuropathology. However, it suffers of the limitation of being retrospective, that is, for part of the patients, comprehensive neuropsychological examination, systematic CT scan, routine glycosylated hemoglobin at admission, and complete genotype are missing. The neuropsychological assessment is based on MMSE at the time of hospitalization, complimented by a clinical examination non-targeted to the geriatric psychiatric diagnosis. Nevertheless, it presents the advantage of being clinically (single geriatric service), pathologically and neuropathologically (same pathologist and same neuropathologist) homogeneous. Choice of cases is based on specific criteria for diabetes mellitus, glycoregulation disorder or normoglycemia (WHO, 1985).

Most clinical and neuropsychological studies have been performed on elderly people with no patent cognitive impairment, but presenting with diabetes mellitus. Only two studies (Mattlar et al., 1985, Soininen et al., 1992), report no difference between diabetics and controls from the neuropsychological viewpoint. But in both such studies, outcomes are limited by selection expedients (age, association with hypertension, types of diet and of treatment). In all the other studies, comparison of NIDDM with a control group of age- and sex-matched non-diabetic subjects (between 65 and 70 years of age) demonstrates heterogeneous cognitive functions in diabetics. Attention and memory retention abilities and performance of complex tasks are significantly impaired in older diabetic subjects (Perlmuter et al., 1984; Mooradian et al., 1988; Tun et al., 1990; U'Ren et al., 1990). In contrast, immediate recall, semantic memory, verbal fluency, speed of comprehension and performance of simple tasks are not affected in elderly diabetics as compared with the control group (Perlmuter, 1984; Reaven et al., 1990; U'Ren et al., 1990). Recent comparison of the modified MMSE and delayed word recall in 90 community-dwelling NIDDM over 50 years of ages and 90 non-diabetic control persons matched for sex and age, confirms a significant decline in performances in the diabetics (Worall et al., 1993).

The recent discovery of insulin receptors in such areas as hypothalamus, olfactory bulb, CA1, CA3 brain areas and the dental gyrus region of the hippocampus (Schwartz et al., 1989; Unger et al., 1989) helps to better understand glucose utilization by the brain, since the latter is able neither to synthesize nor to store glucose (Craft et al., 1996). Such a finding may help to explain:

- the vulnerability of the hippocampus, a structure critical to memory function, since severe hypoglycemia may impair declarative memory (Craft et al., 1996).
- the variations of memory functions in non-diabetic elderly subjects with glycemia levels, since mild glycemias enhance memory whereas severe hyperglycemias impair memory abilities (Tun et al., 1990).

It is also possible that the U-shaped effect of glycemia on memory functions is more linked to the specific action of insulin on the brain than to that of glucose (Tun et al., 1990). Moreover, oral antidiabetics used for the treatment of NIDDM significantly enhance mental performances in the fields of attention, learning and problem solving (Gradman et al., 1993; Meneilly et al., 1993).

It has been suggested that a perturbation of glycemic regulation can accompany dementia of Alzheimer type, thus contributing to the severity of memory disorders (Craft et al., 1996). A transient hyperinsulinism has been noted in the early stages of AD (Bucht et al., 1983 and Craft et al., 1993), in contrast to a hypoinsulinism that is present in the later stages of the disease (Craft et al., 1996). Furthermore, positron-emission tomography (PET) has identified specific regions of brain in which the rate of glucose metabolism declines progressively in patients with probable AD (Reiman et al., 1996). These data strengthen the hypometabolic hypothesis or the inversed metabolic theory of AD (Lanvin et al., 1993). For that matter, at a mild stage of AD, an induced hyperinsulinemia with normoglycemia enhances patients' memory performances by promoting glucose utilization in the hippocampal formation, hence a facilitation of the glycolytic output of the precursor of acetyl coenzyme A, resulting in an increase in the level of cholinergic neurotransmitters (Craft et al., 1996). This enhanced understanding of interrelations between glycoregulation and memory abilities could suggest that elderly diabetics that are properly treated with an oral antidiabetic or insulin, have "relatively preserved or stimulated" memory functions.

This physiopathologic explanation may account for the discrepancy between the apparent low frequency (from 0 to 6%) of AD in older diabetics in clinical studies (Bucht et al., 1983; Wolf-Klein et al., 1988; Small et al., 1992; Lanvin et al., 1993; Mortel et al., 1993 and Nielson et al., 1996) and the striking neuropathologic findings of the present study, attesting that 40.4% of the autopsied elderly diabetics present with the pathologic marks of AD (Muller et al., 1993). This may suggest that diabetics are not more protected from the onset of Alzheimer-type cerebral lesions than non-diabetic controls. Although from the few clinical surveys it appears that AD symptomatic expression may be less pronounced in diabetics than in non-diabetics, this explanation is too simplistic insofar as it does not take into account the genetic prevalences observed of late in clinically diagnosed AD patients (Nielson et al., 1996). According to this study:

- 71.5 % were holders of at least one ApoE4 allele; and
- only 1 out of 123 patients (0.8%) was diabetic, and holder of only one ApoE4 allele (i.e. 0.8% of the cohort!).

Similarly, the pattern of inheritance of classic late-onset NIDDM is complex, suggesting that both genetic and non-genetic factors are involved (Polonski et al., 1996). We might conclude that more complete longitudinal studies are necessary before defining a definite link between the two diseases.

CONCLUSION

Although it is difficult to draw a definite conclusion concerning the type of relationship that may exist between dementia of the Alzheimer type and diabetes, our results support the view that in elderly diabetics, frequency of neuropathologic lesions of AD type is the same as in non-diabetic controls. Therefore, diabetes does not seem to protect against AD. Nevertheless, it appears that in the presence of hyperglycemia and drug-induced hyperinsulinism, diagnosis of AD is harder to determine in diabetics than in non-diabetics. Only longitudinal studies of diabetic populations and controls can determine what kind of links exist between diabetes mellitus and AD.

REFERENCES

Blessed G (1990): Definition and classification of the dementias: the point of view of a clinician. In: *Innovative Trends in Psychogeriatrics, Interdiscipl Top in Gerontol.* Wertheimer J, Bauman P, Gaillard M, Schwed P eds. Basel: Karger pp 63-9.

Bouras C, Hof P (1992): Neuropathologie de la maladie d'Alzheimer In: Editions techniques. *Enc Med Chir Psychiatr* 37280 A20.

Bucht G, Adolfsson R, Lithner et al. (1983): Changes in blood glucose and insulin secretion in patients with senile dementia of Alzheimer type. *Acta Med Scand* 213: 387-92.

Craft S, Dagogo-Jack SE, Wiethop B et al. (1993): The effects of hyperglycemia on memory hormone levels in dementia of the Alzheimer type: a longitudinal study. *Behav Neurosci* 107:223-7.

Craft S, Newcomer J, Kanne S et al. (1996): Memory improvement following induced hyperinsulinemia in AD. *Neurobiol Aging* 17:123-30.

Croxon SCM, Jagger C (1995): Diabetes and cognitive impairment: a community-based study of elderly subjects *Age Aging* 24:421-4.

Gradman TJ, Laws A, Thompson LW et al. (1993): Verbal learning and/or memory improves with glycemic control in older subjects with non-insulin-dependent diabetes mellitus *J Am Geriatr Soc* 41:1305-12.

Lanvin K, Blennow K, Wallin A et al. (1993): Low blood pressure and blood glucose levels in AD: Evidence for a hypometabolic disorder? *J Internal Med* 233: 357-63.

Mattlar CE, Falck B, Rönnema T et al. (1985): Neuropsychological cognitive performance of patients with type II diabetes. *Scand J Rehab Med* 17:101-5.

Meneilly GS, Cheung E, Tessier et al. (1993): The effects of improved glycemic control on cognitive functions in the elderly patient with diabetes. *J Gerontol* 48: M117-21.

Morley J, Perry HM (1991): The management of diabetes mellitus in older individuals. *Drugs* 41:548-65.

Mortel KF, Wood S, Pavol MA et al. (1993): Analysis of familial and individual risk factors among patients with ischemic vascular dementia and AD. *Angiology* 44:599-605.

Mooradian AD, Perryman K, Fitten J et al. (1988): Cortical function in elderly non insulin dependent diabetic patients. *Arch Intern Med* 148:2369-72.

Muller PJ (1993): Neuropathologie cérébrale des diabétiques âgés *Thèse doctorat en médecine*, Faculté de Médecine de Genève N°9477.

Nielson KA, Nolan JH, Berchtold NC et al. (1996): Apolipoprotein-E genotyping of diabetic dementia patients: is diabetes rare in Alzheimer disease? *J Am Geriatr Soc* (in press).

Perlmutter LC, Hakami MK, Hodgson-Harrington C et al. (1984): Decreased cognitive function in aging non-insulin dependent diabetic patients. *Am J Med* 77:1043-8.

Polonsky KS, Sturis J, Bell GI (1996): Non-insulin-dependent diabetes: a genetically programmed failure of the beta cell to compensate for insulin resistance. *New Engl J Med* 334:777-83.

Reaven GM, Thompson LW, Nahum D et al. (1990): Relationship between hyperglycemia and cognitive function in older NIDDM patients. *Diabetes Care* 13:16-21.

Rieman EM, Caselli RJ, Yun LS et al. (1996): Preclinical evidence of AD in persons homozygous for the ε4 allele for apolipoprotein E. *New Engl J Med*; 334:752-8.

Schwartz MW, Figlewicz DP, Baskin DG (1989): Insulin in the brain: a hormonal regulator of energy balance. *Mol Neurobiol* 3:71-5.

Skoog I, Nilsson L, Palmertz B et al. (1993): A population-based study of dementia in 85-year-olds. *New Engl J Med* 328:153-8.

Small GW, Rosenthal MJ (1992): Coexistence of AD and diabetes mellitus. *J Am Geriatr Soc* 40:1075-6.

Soininen H, Puranen M, Helkala EL et al. (1992): Diabetes mellitus and brain atrophy: a computed tomography study in an elderly population. *Neurobiol Aging* 13:717-21.

Tun PA, Nathan DM, Perlmuter LC (1990): Diabetes mellitus in the elderly: cognitive and affective disorders in elderly diabetics. *Clinics Geriatr* 6:731-46.

Ulrich J, Probst A, Wüest M (1986): The brain diseases causing senile dementia. A morphological study on 54 consecutive autopsy cases. *J Neurol* 233:118-22.

Unger J, McNeill T, Moxley RT III et al. (1989): Distribution of insulin receptor-like immunoreactivity in the rat forebrain. *Neurosci* 3:143-5.

U'Ren RC, Riddle MC, Lezack MD et al. (1990): The mental efficiency of the elderly person with type II diabetes mellitus. *J Am Geriatr Soc* 38:505-10.

Wade JPH, Hachinski VC (1987): Multi-infarct dementia In: *Dementia.* Churchill, Livingstone, Edinburgh, eds. PITT B 209-28.

WHO (1985): Diabetes mellitus. *Technical report 727*, WHO Geneva.

Wilson PW, Keaven M, Anderson PD (1986): Epidemiology of diabetes mellitus in the elderly. The Framingham study. *Am J Med* 80 suppl 5A:3-9.

Wolf-Klein GP, Silverstone FA, Brod MS et al. (1988): Are Alzheimer patients healthier? *J Am Geriatr Soc* 36:219-24.

Worrall G, Moulton N, Briffett E (1993): Effect of type II diabetes mellitus on cognitive function. *J Family Practice* 36:639-43.

Alzheimer Disease: From Molecular Biology to Therapy
edited by R. Becker and E. Giacobini

ISOFORM-SPECIFIC METABOLISM OF APOLIPOPROTEIN E: IMPLICATIONS FOR ALZHEIMER 'S DISEASE

Warren J. Strittmatter
Departments of Medicine (Neurology) and Neurobiology
Duke University Medical Center, Durham, NC, 27710, USA

INTRODUCTION

Inheritance of specific apolipoprotein E alleles determines, in large part, the risk and mean age of onset of late-onset familial and sporadic Alzheimer disease (AD) (see Strittmatter and Roses, 1995). The mechanism by which the apolipoprotein E isoforms differentially contribute to disease expression is, however, unknown. Many isoform-specific interactions of apolipoprotein E have been identified. Identification of isoform-specific interactions of apolipoprotein E give rise to testable hypotheses for the mechanism of pathogenesis of AD. An unresolved issue of increasing importance is the relationship between the structural pathological lesions and the cellular pathogenesis responsible for the clinical disease phenotype, progressive dementia.

ApoE IN THE CENTRAL NERVOUS SYSTEM

ApoE may perform multiple metabolic functions in the brain. ApoE is in the extracellular space, as a free and as a bound protein. It is also found within some neurons, both in the cytoplasmic space and in the intravesicular space of endosomes, lysosomes and peroxisomes. This multitude of sites suggest multiple metabolic functions. Not all of the functions of ApoE may be relevant to its role in AD. Interactions which are qualitatively or quantitatively different involving the ApoE isoforms are candidates for developing hypotheses of how the various ApoE alleles differentially regulate disease expression.

In situ hybridization studies on brain have indicated that ApoE mRNA is expressed in astrocytes and microglia, and not in neurons. However, recent immunohistochemical studies on AD brain have shown ApoE in both tangle-bearing and tangle-free nerve cells, indicating that ApoE may play a role in neuronal metabolism and in neuronal degeneration or regeneration in the central nervous system (Han et al., 1994a;b; Strittmatter et al., 1993a). ApoE immunoreactive cortical neurons can be detected with certain ApoE antibodies at both the light and electron microscopic levels. ApoE immunoreactivity defines specific subsets of neurons, with punctate staining around soma, diffuse cytoplasmic staining, and granular staining.

The mechanism of ApoE entry into the cytoplasmic compartment of nerve cells is not yet understood. Other proteins, including the bacterial toxins pertussis and botulinum, and lactoferrin, are also capable of crossing cellular membranes, gaining access to the cytoplasm from the extracellular space.

THE ApoE ISOFORMS

The most common isoform of ApoE in the general population, ApoE3, is secreted as a 299 amino acid protein containing a single cysteine residue at position 112. The other two common isoforms, ApoE2 and ApoE4, differ at one of two positions (residues 112 and 158) from ApoE3 by cysteine-arginine interchanges: ApoE2 contains a cysteine at position 158 and ApoE4 contains an arginine at 112. The single amino acid differences among the common isoforms of ApoE result in significant differences in the biochemistry and cellular metabolism of these proteins, and in marked differences in the risk of AD. The single cysteine in ApoE3 permits disulfide bond formation with other molecules, including itself. ApoE3 forms heterodimers with apolipoprotein A-II and homodimers with another ApoE molecule. ApoE4 lacks a cysteine and cannot form these disulfide-mediated complexes (Weisgraber, 1994).

ISOFORM SPECIFIC INTERACTIONS OF ApoE IN CHOLESTEROL TRANSPORT AND RECEPTOR BINDING

The ApoE isoforms differ in their interactions with the low density lipoprotein (LDL) receptor and in binding cholesterol containing lipid particles (Weisgraber, 1994). ApoE3 and ApoE4 bind the LDL receptor with high affinity, while ApoE2 binds with very low affinity (approximately 1% of ApoE3). ApoE binds other cell surface receptors, including the low density lipoprotein receptor related protein, the very-low density lipoprotein receptor (VLDL-receptor), and scavenger receptors, but the interactions of the various isoforms of ApoE with these receptors have not been extensively investigated. The ability of ApoE to bind phospholipid particles is isoform-specific, with ApoE2 and ApoE3 preferentially associating with high-density lipoprotein (HDL) particles and ApoE4 preferentially associating with VLDL particles.

ApoE contains two functionally important domains, one that binds the LDL receptor and the other that binds lipoprotein particles (VLDL or HDL).

The region of ApoE that binds the LDL receptor is contained in the amino-terminal domain (residues 1-191); the major lipid binding region of ApoE resides in the carboxyl-terminal domain, with residues carboxyl-terminal to 244 playing a major role. The cysteine:arginine interchange at position 112, which distinguishes ApoE3 from ApoE4, is not contained in the major lipid binding region of ApoE, yet influences the distribution of these isoforms among the various lipoprotein particle classes (Weisgraber, 1994).

ApoE AND THE MICROTUBULE BINDING PROTEINS TAU AND MAP-2

Neurofibrillary tangles (NFT) are dense bundles of long unbranched filaments in the cytoplasm of some neurons. These filamentous structures are paired helical filaments. Each filament is 10 nm in width, and two filaments are helically twisted about each other with a periodic full twist every 160 nm. Paired helical filaments may also be found in neurites undergoing degeneration. The filaments consists primarily, and probably exclusively, of the microtubule-associated protein tau.

Tau normally binds and stabilizes microtubules, and promotes the assembly of microtubules by polymerizing tubulin. Microtubules are necessary for neurite extension and maintenance and for the transport of materials along the axon and dendrites in both orthograde and retrograde directions. In AD, tau becomes abnormally phosphorylated, and self-assembles into the pathological paired-helical filaments forming neurofibrillary tangles.

Dementia in AD is generally accepted to be better correlated with NFT pathology than with the extent of Aβ deposition. Because of the genetic relevance of ApoE4 and the presence of immunoreactive ApoE in neurons containing neurofibrillary tangles, isoform-specific interactions of ApoE with tau are being studied. *In vitro*, human ApoE3 binds tau, forming a molecular complex which resists dissociation by boiling in 2% sodium dodecyl sulfate (Strittmatter et al., 1994a;b) ApoE4 does not irreversibly bind tau under these conditions.

Paired helical filament tau is phosphorylated at a number of serine/threonine - proline sites. At least some of these sites are phosphorylated by incubating recombinant tau with a crude rat brain extract and ATP. Recombinant tau-40 phosphorylated in this manner does not bind either ApoE3 or ApoE4. In AD, hyperphosphorylated tau is believed to self-assemble into the paired helical filament by formation of anti-parallel dimers of the microtubule-binding repeat region. *In vitro*, these microtubule-binding repeats of tau self-assemble into paired helical-like filaments. ApoE3 irreversibly binds these microtubule-binding repeat regions of tau, whereas ApoE4 does not bind with this avidity (Strittmatter et al., 1994a).

We have used synthetic peptides corresponding to the 18 amino acid microtubule-binding region from each of the four tau repeats (designated repeat peptides I - IV) to study the interactions between tau and ApoE. Each microtubule-binding repeat is capable of binding ApoE3, the complete microtubule-binding domain is not required. Of the four peptides, only those from repeats II and III contain a cysteine residue, indicating that disulfide bond formation between tau and ApoE3 cannot be the sole reason for these *in vitro* interactions. Serine262 is the only residue within the repeat region known to be phosphorylated in PHF-tau, and its phosphorylation reduces the ability of tau to bind to microtubules. Phosphorylation of repeat peptide I at a position equivalent to Serine262 completely abolishes its ability to bind ApoE3 (Huang et al., 1995).

MAP2c is another member of a group of microtubule-associated proteins (MAPs) that bind microtubules and promote their assembly, and stabilize their polymerized structure. *In vitro,* ApoE3 avidly binds MAP2c, similar to its isoform-specific interaction with tau (Huang et al., 1994). ApoE4 does not bind MAP2c under identical conditions. Tau and MAP2c contain a highly conserved microtubule-binding repeat region. This region consists of three or four conserved amino acid sequences (depending on the isoform), each 31 or 32 amino acids. MAP2c contains three copies of these microtubule-binding repeats, homologous to the microtubule-binding repeats of tau.

Isoform-specific interactions of ApoE with tau and MAP2c may regulate intraneuronal metabolism in AD and alter the rate of formation of paired helical filaments and neurofibrillary tangles. Such isoform-specific interactions of ApoE with tau, MAP2c, and other cytoskeletal proteins (Fleming et al., 1996) might alter their metabolism, their function in microtubule assembly and stabilization, and may be central to disease mechanism. Such isoform-specific effects of apolipoprotein E on neurite outgrowth and microtubule organization are observed in cultured Neuro-2a cells (Nathan et al., 1995).

ApoE AND βA PEPTIDE

Neuritic plaques are extracellular structures with complex molecular and cellular constituents. Plaques contain Aβ, a peptide of 39-43 amino acids, which is produced by proteolytic cleavage of the amyloid precursor protein (APP) in its normal metabolism. The complete molecular composition of the fibrillar structures in the plaque, the mechanism of assembly, and their role in the disease is unknown.

ApoE accumulates extracellularly in the senile plaque (Namba et al., 1991; Strittmatter et al., 1993a). *In vitro,* ApoE avidly binds to synthetic βA peptide, the primary constituent of the senile plaque (Strittmatter et al., 1993a,b). ApoE 4 binds βA peptide more rapidly, and with a different pH dependence than does ApoE3. LaDu et al. (1994) have demonstrated that such isoform-dependent interactions depend on the state of the protein, including the presence of bound lipid particles.

The domain of ApoE that binds BA peptide was determined by examining various recombinantly expressed ApoE fragments (Strittmatter et al., 1993b). βA binding to ApoE appears to require the domain of ApoE between amino acids 244 to 272, within the region of ApoE previously demonstrated to mediate binding to lipoprotein particles.

After days of *in vitro* incubation, ApoE and βA peptide form unique fibrils, with ApoE4 forming more abundant fibrils than ApoE3 (Sanan et al., 1994). These 10 nm fibrils are distinctly different than the twisted ribbons formed by βA alone. Immunogold labeling shows that ApoE associates with these fibrils along their full length, suggesting that ApoE may be adsorbed to the outside of the simple fibrils or intercalated between the βA peptide monomers, perhaps forming a hybrid fibril.

In vitro differences in the binding of these ApoE isoforms with βA peptide are paralleled by differences *in vivo.* In a study of the comparative neuropathology of brain tissue from AD patients homozygous for ApoE4 or ApoE3, congophilic staining of amyloid in senile plaques was greatly increased in ApoE4 patients (Schmechel et al., 1993).

ApoE AND PRODUCTS OF LIPID PEROXIDATION

One of the consequences of oxidative damage to cells is lipid peroxidation, an autocatalytic process that damages lipid-containing structures and yields reactive by-products, primarily malondialdehyde (MDA) and 4-hydroxy-2-nonenal (HNE). MDA and HNE are electrophilic species that can covalently modify cellular macromolecules; each react with nucleophilic groups in amino acid residues to form monoadducts as well as intra- and intermolecular crosslinks. These two products of lipid peroxidation, MDA and HNE, covalently modify ApoE and may alter its metabolism.

In vitro, both HNE and MDA crosslink purified ApoE3 and ApoE4 (Montine et al., 1996a). HNE is a more potent crosslinker than MDA, and ApoE3 is more susceptible to crosslinking by HNE than ApoE4. In P19 neuroglial cultures, oxidative stress with lipid peroxidation leads to increased intracellular accumulation of an anti-HNE and anti-ApoE immunoreactive proteins of approximately 50 kDa (Montine et al. 1996b). In CSF, a 50 kDa ApoE-immunoreactive protein co-migrated with proteins most immuno-reactive for HNE and MDA adducts. These proteins are in CSF from adult subjects (with or without dementia), and in AD patients homozygous for ApoE3 or ApoE4 alleles. These data suggest that HNE covalently crosslinks ApoE in P19 neuroglial cultures to form a 50 kDa protein, and that similar modifications of ApoE appear to occur *in vivo.*

Lipid peroxidation, an important consequence of oxidative damage, increases with age in brain and is further increased in patients with AD. ApoE may be especially vulnerable to covalent modification by electrophilic species generated from lipid peroxidation. The resulting altered metabolism of ApoE provides a potential link between processes of aging, brain lipid peroxidation, and a protein critical to the pathogenesis of late-onset AD.

FUTURE DIRECTIONS

Since the alleles of ApoE determine the major risk of developing AD, as well as the average age of onset, identification of the isoform-specific metabolism of ApoE in disease pathogenesis is of central importance. Several isoform-specific interactions of ApoE with other molecules have been identified, which include interactions with molecules responsible for disease pathology; tau, which forms the neurofibrillary tangle, and βA peptide, the primary constituent of the neuritic plaque. Adduction and crosslinking of the ApoE isoforms by peroxidized lipids may contribute to the relevant isoform-specific metabolism in AD. Identification of additional isoform-specific protein interactions, and other isoform-specific post-translational modifications will produce new hypotheses of the role of ApoE in AD.

REFERENCES

Fleming L, Weisgraber KH, Strittmatter WJ, Troncoso JC, and Johnson GVW (1996): Differential binding of apoE isoforms to tau and other cytoskeletal proteins. *Experimental Neurology* (In Press).

Han SH, Hulette C, Saunders AM, Einstein G, Pericak-Vance M, Strittmatter WJ, Roses A D, Schmechel DE, (1994a): Apolipoprotein E is present in hippocampal neurons without neurofibrillary tangles in AD and in age-matched controls. *Exptl Neurol* 128:13-26.

Han SH, Einstein G, Weisgraber KH, Strittmatter WJ, Saunders A, Pericak-Vance M, Roses AD and Schmechel DE (1994b): Apolipoprotein E is localized to the cytoplasm of human cortical neurons: A light and electron microscopic study. J *Neuropath Exp Neurol* 53: 535-544.

Huang DY, Goedert M, Jakes R, Weisgraber KH, Garner C, Saunders AM, Pericak-Vance M., Schmechel D, Roses AD, Strittmatter WJ (1994): Isoform-specific interactions of apolipoprotein E with the microtubule-associated protein MAP 2c: Implications for Alzheimer's disease, *Neuroscience Letters* 182: 55-58.

Huang DY, Weisgraber KH, Goedert M, Saunders AM, Roses AD, and Strittmatter WJ (1995): ApoE3 binding to tau tandem repeat I is abolished by tau serine 262 phosphorylation. *Neuroscience Letters* 192: 209-212.

LaDu MJ, Falduto MT, Manelli AM, Reardon CA, Getz GS, and Frail DE (1994): Isoform-specific binding of apolipoprotein E to 13-amyloid. *J Biol Chem* 269: 23403-23406.

Montine TJ, Huang DY, Valentine WM, Amarnath V, Saunders AM, Weisgraber KH, Graham DG, and Strittmatter WJ (1996a):, Crosslinking of apolipoprotein E by products of lipid peroxidation. *J Neuropath Expt Neurology* (In Press).

Montine TJ, Amarnath V, Martin ME, Strittmatter WJ, and Graham DG (1996b), E-4-hydroxy-2-nonenal, is cytotoxic and covalently crosslinks cytoskeletal proteins in P19 neuroglial cells. *Amer J Path* (In Press).

Namba Y, Tomonaga M, Kawasaki H, Otomo E, and Ikeda K (1991): Apolipoprotein E immunoreactivity in cerebral amyloid deposits and neurofibrillary tangles in Alzheimer's disease and kuru plaque amyloid in Creutzfelt-Jakob disease. *Brain Res* 541:163-166.

Nathan BP, Chang KC, Bellosta S, Brisch E, Ge N, Mahley RW and Pitas RE (1995): The inhibitory effect of apolipoprotein M on neurite outgrowth is associated with microttubule depolymerization *J Biol Chem* 270: 19791-19799.

Sanan D, Weisgraber KH, Mahley RW, Huang DY, Russell SJ, Saunders A, Schmechel D, Wisniewski T, Frangione B, Roses AD and Strittmatter WJ (1994): Apolipoprotein E associates with AB amyloid peptide to form novel monofibrils in vitro: Isoform ApoE4 associates more efficiently than ApoE3. *J Clin Invest* 94: 860-869.

Schmechel DE, Saunders AM, Strittmatter WJ, JOO SH, Hulette C, Crain B, Goldgaber D, and Roses AD (1993): Increased vascular and plaque 13A4 amyloid deposits in sporadic Alzheimer disease patients with apolipoprotein E-e4 allele. *Proc Natl Acad Sci* 90:9649-9653.

Strittmatter WJ, Saunders AM, Pericak-Vance M, Salvesen GS, Enghild J and Roses AD (1993a): Apolipoprotein E: High avidity binding to βA amyloid and increased frequency of type 4 isoform in familial Alzheimer Disease. *Proc Natl Acad Sci* 90:1977-1981.

Strittmatter WJ, Weisgraber KH, Huang D, Dong LM, Salvesen GS, Pericak-Vance M Schmechel D, Saunders AM, Goldgaber DM, and Roses AD (1993b): Binding of human apolipoprotein E to βA4 peptide: Isoform-specific effects and implications for late-onset Alzheimer disease. *Proc Natl Acad Sci* 90: 8098-8102.

Strittmatter WJ, Weisgraber KH, Goedert M, Saunders AM, Huang D, Corder, EH, Dong LM, Jakes R, Alberts MJ, Gilbert JR, Schmechel DE, Pericak-Vance MA, and Roses AD, (1994a) Hypothesis: Microtubule instability and paired helical filament formation in the Alzheimer disease brain as a function of apolipoprotein E genotype. *Exptl Neurol* 125:163-171.

Strittmatter WJ, Saunders AM, Goedert M, Weisgraber K, Dong LM, Jakes R, Huang DY, Pericak-Vance M, Schmechel D, and Roses AD (1994b) Isoform-specific interactions of apolipoprotein E with microtubule-associated protein tau: Implications for Alzheimer disease. *Proc Natl Acad Sci* 91:11183-11186.

Strittmatter WJ and Roses AD (1995): Apolipoprotein E and Alzheimer disease, *Proc Natl Acad Sci* 92: 4725-4727.

Weisgraber KH (1994): Apolipoprotein E: Structure-Function Relationships, in *Advances in Protein Chemistry,* Vol 45, pp.249-301.

Alzheimer Disease: From Molecular Biology to Therapy
edited by R. Becker and E. Giacobini

APOLIPOPROTEIN E4 AND CHOLINERGIC ACTIVITY IN ALZHEIMER'S DISEASE

Judes Poirier, Isabelle Aubert, Rémi Quirion, Serge Gauthier, Martin Farlow, and Josephine Nalbantoglu
McGill Centre for Studies in Aging, Montreal Neurological Institute,
and Douglas Hospital Research Centre, McGill University,
Montréal, Québec, Canada

Maire–Claude Delisle
Department of Neurology, School of Medicine,
University of Indiana, Indianapolis, Indiana

Steve Gracon
Parke-Davis Corporation, Ann Arbor, Michigan, US

INTRODUCTION

Apolipoprotein E (apoE) is a well characterized lipophilic protein associated with plasma and CSF lipoproteins. ApoE is synthesized primarily by the liver, but also at other sites including brain, macrophages and adrenals. Furthermore, apoE is unique among apolipoproteins in that it has a special relevance to the central and peripheral nervous systems. It is a key determinant in the cellular recognition and internalization of cholesterol- and phospholipid-rich lipoproteins in the developing brain and in the response to neuronal injury (Boyles et al., 1989, Poirier et al., 1991, 1991a, 1993). It was shown to play a fundamental role in the CNS during hippocampal synaptic plasticity induced by entorhinal cortex lesions in the rat (Poirier et al., 1991, 1991a and 1993).

ApoE is synthesized and secreted by astrocytes in the deafferented zone of the hippocampus following lesions of the entorhinal cortex (Poirier et al., 1991). Experimental evidences suggest that apoE and its main receptor, the so-called LDL receptor, are directly involved in cholinergic synaptic remodelling caused by the loss of entorhinal cortex neurons; a pathophysiological process also observed in AD (Hyman et al., 1984).

The human apoE gene on chromosome 19 has three common alleles (E2, E3, E4), which encode three major apoE isoforms. Recently, the frequency of the apoε4 allele was shown to be markedly increased in sporadic (Poirier et al., 1993a, Saunders et al., 1993, Noguchi et al., 1993) and late onset familial AD (Corder et al., 1993, Payami et al., 1993). Most interestingly, a gene dosage effect was observed in both familial (Corder et al., 1993) and sporadic (Poirier et al. 1993a) cases (i.e. as age of onset increases, E4 allele copy number decreases). The discovery that the apoE4 allele is strongly linked to both sporadic and familial late onset AD raises the possibility that a dysfunction of the lipid transport system associated with compensatory plasticity and synaptic remodelling could be central to the pathophysiology of AD.

The role of apoE in the CNS is particularly important in relation to the function of the cholinergic system which relies heavily on the integrity of lipid homeostasis to synthesize acetylcholine in neurons. Brain membrane phospholipids, especially phosphatidylcholine (PC) and phosphatidylethanolamine (PE), have been shown to be involved in the availability of choline, a rate-limiting precursor of acetylcholine (ACh) (Blusztajn et al., 1987). The release from PC of free choline precursor for ACh synthesis is accomplished in a one step process through a phospholipase-D type enzyme in cholinergic neurons. Brain levels of choline are decreased by up to 40-50% in frontal and parietal cortices (Nitsch et al., 1992) of AD patients whereas cholesterol, which is required for the proper functioning of nicotinic receptor sub-type (Jones and McNamee, 1988), was shown to be markedly reduced in AD versus control subjects (Svennerholm and Gottfries, 1994). It was recently proposed that the low levels of apoE reported in the brain and CSF of apoε4 AD subjects may compromise cholesterol and phospholipid transport in the CNS and selectively damage the cholinergic system which relies heavily on lipid homeostasis (Poirier, 1994). The relationship between the apoE4 genotype and cholinergic deficits is highly relevant to investigate in genetically distinct individuals. Preliminary studies indicated that ChAT activity is markedly reduced in the hippocampus and cortex of apoE4-AD subjects (Poirier, 1994, Sioninen et al., 1995, Poirier et al., 1995). The major aims of the present study were thus a) to investigate the apparent status and integrity of the cholinergic system in the brain of AD and control subjects with different apoE genotypes; and b) to examine the effect of apoE genotype on the therapeutic response to cholinomimetic treatments.

RESULTS

Pathological Studies

On the basis of its well-recognized post-mortem stability, ChAT activity was examined in relation to apoE genotype in control and AD post-mortem subjects. Figure 1A illustrates that the loss in ChAT activity (Poirier et al., 1995) in the hippocampus of control and AD cases as a function of apoE4 allele copy number (i.e. as apoE4 allele copy number increased, ChAT activity decreased). These results indicate the existence of distinct genetic entities in sporadic AD which show differential degree of alterations of cholinergic innervation, at least as revealed by ChAT activity. Muscarinic M1 receptors, which are mostly post-synaptic in the hippocampus and temporal cortex of human, were found to be relatively spared in AD subjects (Figure 1B). Genotype had little or no impact of M1 binding sites density.

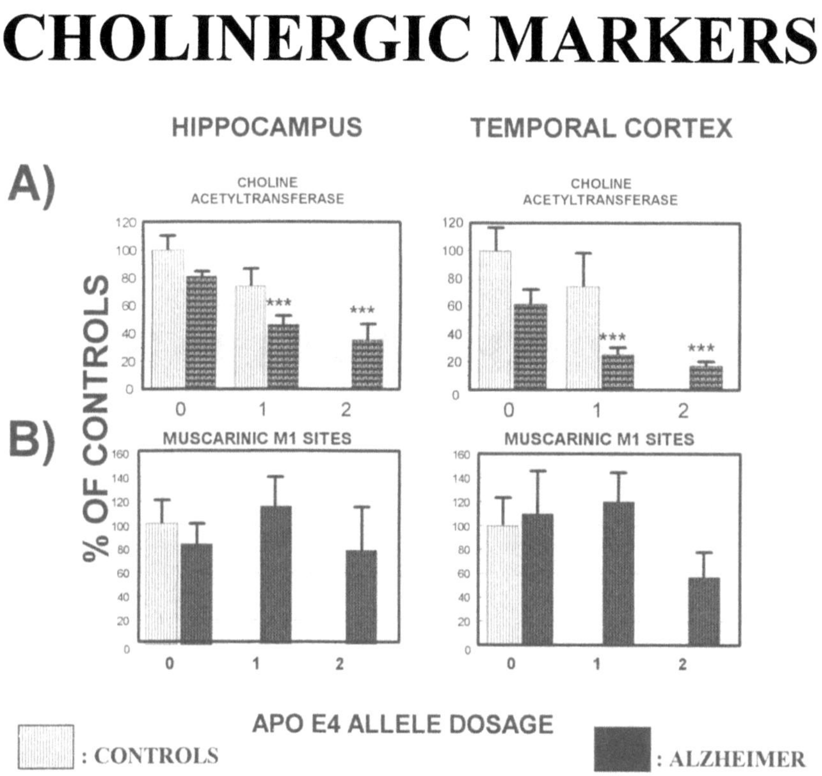

Figure 1: ChAT activity (A) and ^{3}H–pirenzepine muscarinic M_1 (B) binding densities in the hippocampus and temporal cortex of AD and control subjects as a function of apoE4 allele copy number. Protocols for all cholinergic markers shown here have been described in Poirier et al.,(1995). Data are mean ± S.E.M. (bars) values from 2 to 27 brains. Statistically significant differences from normal control values are indicated: *: p<0.05, ***: p<0.001.

Clinical Studies
A small group of patients (n=40) were selected from a large cohort of AD patients (n=572), subjects completing a 30 week tacrine trial (Knapp et al., 1994). The patients were selected if they showed maximal worsening or improvement on the AD Assessement Scale–cognitive component (ADAS-Cog). All patients received the maximum dose of 160mg/day. The ADAS-Cog is a 70-point scale where decline in scores indicates improvement. Differences were judged between baseline and 30 week evaluations. Genotype determination for apoE was then performed as described previously (Poirier et al., 1993a) and cognitive performance was contrasted with the presence or absence of the apoE4 allele. Figure 2 summarizes the results obtained in this preliminary study.

Overall more than 80% of the non-apoE4 carrying patients had improvement in response to tacrine. In contrast, 60% of E4 patients were unchanged or worse after 30 weeks. These differences were analyzed by t-test

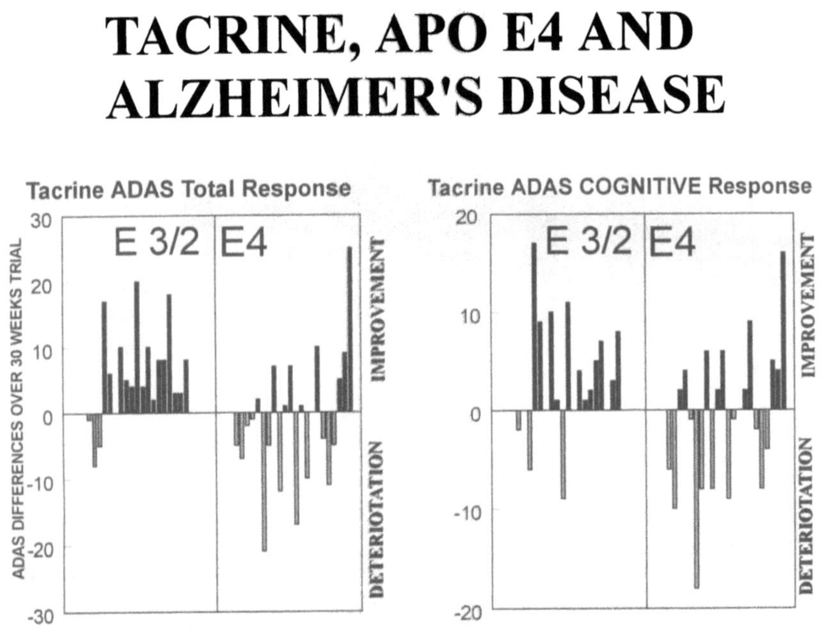

Figure 2: Effect of apoε4 allele copy numbers on tacrine drug responsiveness in AD subjects. Fourty subjects enrolled in the 30-weeks randomized controlled trial of high-dose tacrine in patients with AD were selected out of the original 572 subjects cohort (Knapp et al., 1994). Patients were selected prior to apoE phenotype determination and blind to genotype. Graph represents individual variation (ADAS score before and after drug trial) in cognitive performances as a function of apoε4 allele incidence.

procedure. These results indicate the patients not carrying apoE4 show a more solid response to tacrine, as a group, than those carrying the E4 allele (ADAS-TOTAL: $p < 0.04$). A significant difference in favor of non-apoE4 patients being better responders was also seen in the ADAS-Cog.

This result should not be interpreted as evidence of the presence or absence of drug responsiveness in tacrine treated patients but instead, it should be seen as a genotype-dependent difference in the quality (and size) of the response. Furthermore, recent evidence indicate that genotype, in combination with tacrine administration, can markedly affect the course of AD to the point of delaying nursing home placement by more than 400 days.

In summary, both pathological and clinical data clearly suggest that apoE4 genotype influence the function and integrity of the cholinergic system in the brain. This observation should have some significant impact on the design of future cholinomimetic-based trials in AD in addition to focusing efforts to understand mechanisms involving apoE4 in the cholinergic deficit in AD.

ACKNOWLEDGEMENTS

The research from our laboratory reviewed here was supported by grants from the National Institute on Aging, Alzheimer Society of Canada and the "Fonds de la Recherche en Santé du Québec".

REFERENCES

Blusztajn JK Liscovitch M and Richardson UI (1987): Synthesis of acetylcholine from choline derived from phosphatidylcholine in a human neuronal cell line. *Proc. Natl Acad Sci* 84: 5474–5477.

Boyles JK, Zoellner CD, Anderson LJ, Kosick LM, Pitas RE, Weisgraber KH, Hui DY, Mahley RW, Gebicke-Haeter PJ, Ignatius MJ and Shooter EM (1989): A role for apolipoprotein E, apolipoprotein A-1, and low density lipoprotein receptors in cholesterol transport during regeneration and remyelination of the rat sciatic nerve. *J Clin Invest* 83:1015–1031.

Corder EH, Saunders AM, Strittmatter WJ, Schmechel DE, Gaskell PC, Small GW, Roses AD and Pericak-Vance MA (1993): Gene dose of apolipoprotein E type 4 and risk of Alzheimer's disease in late onset families. *Science* 261:921–923.

Hyman BT, Van Hoesen GW, Damasion AR and Barnes CL (1984): Alzheimer's Disease: Cell-specific pathology isolates the hippocampal formation. *Science* 225:1168–1170.

Jones OT and McNamee MG (1988): Annular and non-annular binding sites for cholesterol associated with nicotinic acetylcholine receptor. *Biochemistry* 27: 2364–2374.

Knapp MJ, Knopman DS, Solomon PR, Davis CS and Gracon SI (1994): A 30-week randomized controlled trial of high dose tacrine in patients with AD. *J Am Med Assn* 271: 985–991.

Nitsch RM, Blusztajn JK, Pitas AG, Slack BE and Wurtman RJ (1992): Evidence for a membrane defect in Alzheimer's disease. *Proc Natl Acad Sci* 89:1671–1675.

Noguchi S, Murakami K and Yamada N (1993): Apolipoprotein E and Alzheimer's disease. *Lancet* (Letter) 342:737.

Payami H, Kaye J, Heston LL and Schellenberg GD (1993): Apolipo-protein E and Alzheimer's disease. *Lancet* (letter) 342:738

Poirier J, Hess M, May PC and Finch CE (1991): Cloning of hippocampal poly(A+) RNA sequences that increase after entorhinal cortex lesion in adult rat. *Mol Br Res* 9:191–195.

Poirier J, Hess M, May PC and Finch CE (1991a): Apolipoprotein E-and GFAP–RNA in hippocampus during reactive synaptogenesis and terminal proliferation. *Mol Brain Res* 11:97–106

Poirier J, Baccichet A, Dea D and Gauthier S (1993): Role of hippocampal cholesterol synthesis and uptake during reactive synaptogenesis in adult rats. *Neuroscience* 55:81–90.

Poirier J, Davignon J, Bouthillier D, Bertrand P and Gauthier S (1993a): Apolipoprotein E and Alzheimer's disease. *Lancet* 342: 697–699.

Poirier J. (1994): Apolipoprotein E in animal models of brain injury and in Alzheimer's disease. *Trends Neurosci* 12:525–530.

Poirier J, Aubert I, Quirion R, Farlow M, Nalbantoglu J, Gilfix BG and Gauthier S (1995): Apolipoprotein E4 allele as a predictor of cholinergic deficits and treatment outcome in Alzheimer's disease. *Proc Natl Acad Sci* 92:12260–12264.

Sioninen H, Kosunen O, Helisalmi S, Mannermaa A, Paljarvi L and Riekkinen O (1995): A severe loss of choline acetyltransferase in the frontal cortex of patients carrying apolipoprotein E4 allele. *Neurosci Lett* 187:79–82.

Saunders AM, Strittmatter WJ, Schmechel D, St. George-Hyslop PH, Pericak-Vance MA, Joo SH, Rosi BA, Gusella JF, McClaclans DR, Alberts MJ and Roses AD (1993): Association of apolipoprotein E4 allele with late onset familial and sporadic Alzheimer's disease. *Neurology* 43:1467–1472.

Svennerholm L and Gottfries CG (1994): Membrane lipids, selectively diminished in Alzheimer's disease brain, suggest a synapse loss as a primary event in early onset form. *J Neurochem* 62:1039–1047.

Alzheimer Disease: From Molecular Biology to Therapy
edited by R. Becker and E. Giacobini

A NEW FOCUS ON CYTOSKELETAL THERAPY IN ALZHEIMER'S DISEASE

Hugo Geerts, Rony Nuydens, Mirjam de Jong, Gerd van de Kieboom
Dept Cellular Physiology, Janssen Research Foundation, Beerse Belgium

TOMBSTONE REMNANTS AND NEURONAL AGONY

The classical approach to Alzheimer's Disease (AD) starts from the neuropathological findings and assumes any reasonable model should at least incorporate either β-amyloid plaques or neurofibrillary tangles. However, these neuropathological features are just tombstones of a disease process going on for several years. In the absence of specific and sensitive imaging markers for these aberrant protein forms, we will not be able to judge properly on their relevance to the course of the pathology per se.

APP knockout mice display altered behaviour and reactive gliosis (Zheng et al., 1995). Similarly, it has been documented that during stage I and II of the Braak and Braak (1992) staging, aberrant tau phosphorylation is significantly increased long before any neuropathological evidence. Furthermore, rats chronically infused intraventricularly with okadaic acid during three weeks, show deficient behaviour in the Morris water maze, related to aberrant tau phosphorylation in the hippocampus (Arendt et al., 1995).

Cellular congestion in some transgenic animals

A number of observations in transgenic animals have suggested that either overexpression of the βA4 peptide (La Ferla, 1995), or of an APP isoform carrying a mutated α-secretase binding site (Moechars et al., 1996) in the brain give rise to intraneuronal cytotoxicity. Overexpressing the inappropriate APP isoform resulted in impaired functional behaviour (Moran et al., 1995). This suggested that inappropriate intracellular localization, possibly due to "congestion" of the neuronal cells (as a consequence of forced high protein levels), may lead to loss of control over normally regulated metabolism. We hypothesized that during the clinical course of AD pathology, similar mechanisms may act, and propose reduction of fast axonal transport, as a cause for such an inappropriate localization.

Decreased fast axonal transport in Alzheimer's disease

Past studies have suggested that fast axonal transport may be reduced in AD (Richard et al., 1985, Burke et al., 1990). Recent studies on the level of NGF have suggested that retrograde axonal transport of the receptor may be significantly reduced despite an overall small increase in NGF levels (Mufson et al., 1995)

Fast axonal transport in neurons necessarily needs a functional microtubule network. Aberrant tau phosphorylation leads to decreased association between tau and microtubules (Iqbal et al., 1986) and to a subsequent loss of microtubule stability (see Prapotnik et al., 1996).

Recently we have shown that *in vitro*, increased tau phosphorylation in neuronal cell cultures leads to decreased fast axonal transport (Nuydens et al., 1996). In both primary cultures from neonatal rat hippocampus and human neuroblastoma, the level of aberrantly phosphorylated elevated phosphorylation of tau protein was elevated by a mixture of kinase stimulators such as db-cAMP, gangliosides, sodium butyrate and nerve growth factor. Video-enhanced contrast microscopy suggested that at certain positions along the axons, trains of vesicles were immobilized for longer time. These "traffic jams" were confirmed by quantitative fluorescence immunocytochemistry of markers for anterograde fast axonal transport, and by ultrastructural observations we confirmed the relationship with aberrantly phosphorylated tau protein.

Consequences of impaired fast axonal transport

One of the immediate consequences of such an early decrease in axonal transport is a lowered glucose metabolism. Indeed, fast axonal transport takes a large part of total energy consumption of the neuron (Hammerschlag et al., 1992). On the other hand, decreased glucose metabolism in Alzheimer patients has been observed at an early diagnostic state by both SPECT and PET tomography (Wolfe et al., 1995). Therefore, we suggest that part of this metabolic decrease is a direct consequence of impaired fast axonal transport.

Impaired transport may lead to inappropriate localization of proteins or enzymatic activities within intracellular compartments. Such a situation may be simulated in a number of transgenic animals, where over-expression of APP may lead to cellular congestion as suggested above. Furthermore, impaired axonal transport may have the following consequences.

1) Anterograde transport of key-proteins (such as GAP-43) and lipid material needed for synaptic remodelling will be decreased. High NFT density has been correlated with reduced axonal plasticity, lower GAP-43 expression and decreased synapse density (Callahan et al., 1995).

2) Normal neurotransmitter functioning will be impaired by a reduced turnover of proteins for normal neurotransmitter functioning due to lower transport of newly synthesized and ubiquinated proteins respectively (Lassman et al., 1992).

3) The retrograde transport of neurotrophic factors (Mufson et al., 1995, Scott et al., 1995) and related second messenger pathway molecules from the (distant) peripheral nerve ending will be reduced, leading to a deficient signal transduction.

4) Reduced transport capability of APP may lead to accumulation in swollen neurites (Kawarabayashi et al., 1993). This inappropriate location of APP may enhance amyloidogenesis by defectively regulated metabolism.

5) Accumulation of toxic components at inappropriate localizations; for example, a decreased anterograde transport of tryptophan hydroxylase in dorsal raphe, leads to the inappropriate perikaryal accumulation of serotonin and its oxidative metabolite 5-hydroxyindoleacetic acid in the dorse raphe perikarya (Burke et al., 1990).

What is the origin of aberrant phosphorylation?

We must explain which processes lead to aberrant tau phosphorylation early during the pathology. There is evidence in the Alzheimer brain that a wave of neuronal plasticity takes place before overt neuronal cell death. Coleman and Flood (1987) have shown that neurons of aged individuals do show an increasingly complex dendritic arborization, probably triggered by a number of neurons who die simply as a cause of aging. Increased neuronal plasticity requires the timely and spatial activation of kinases, such as cyclin-dependent kinase 5 (cdk5) (Nikolic et al., 1995), leading to a transient hyperphosphorylation of microtubule-associated proteins (MAP) during this process. Therefore, all premises are in place to activate a number of kinases with tau as a substrate.

Secondarily, increasing evidence suggests an apoptotic pathway in AD brain (Bredesen et al, 1995). There is increasing evidence from different labs, that during apoptosis, kinase activity is increased. In addition, apoptosis has often been described as a frustrated mitotic attempt. The recent similarity found between a large fragment of Presenilin-2 and Apoptosis-Related Gene (ALG-3) (Vito et al., 1996) suggests that Presenilin-2 could inhibit apoptosis and that mutations in this gene actually could result in a higher apoptotic load.

THERAPEUTIC APPROACHES

An obvious rationale is to inhibit the activity of the kinases responsible for aberrant phosphorylation. Altough a number of candidate kinases have been described, such as MAP kinase, GSK3, cdk5, cdc2-p34 and MARCKS 110 kD kinase (see Kawahima et al., 1995), it is still unclear which should be the preferential target *in vivo* and if possibly a synergistic action of kinases is necessary. In addition, brain-specific kinases should be preferentially targeted.

Another approach is to increase the activity of the phosphatases (Gong et al., 1993). This is still a largely uncovered area and the intracellular regulation seems to be very complex. Here, also, systemic side-effects may be limiting.

A third approach consists to inhibit apoptosis, as this may lead to aberrant phosphorylation. The aim is again to define neuron-specific mechanisms in order to avoid the risk of malignant tumour formation.

Finally, stabilizing the neuronal cytoskeleton may be a valuable strategy. The microtubule-bundling compound taxol has been proposed by Lee et al., (1994) as a therapeutic option. This approach, when focussed at the level of tau-MT interaction, may lead to very brain-specific interventions.

SUMMARY

This article presents a working hypothesis with a completely different angle into the pathology of AD. It starts from the observation that a trigger for neuronal plasticity and an inappropriate regulation of apoptosis may lead to aberrant tau phosphorylation. This in turn leads to deficient fast axonal transport (especially in long fibers), a process which may be possibly mirrored in deficient energy metabolism in well-defined regions of the AD brain. Consequently, inappropriate regulation of protein metabolism may lead to build-up of toxins. It also explains deficient neurotransmitter functioning, impaired neuronal plasticity and reduced neurotrophic factor sensitivy.

REFERENCES

Arendt T et al. (1995): PHF-like phosphorylation of tau, deposition of βA4 amyloid and memory impairment in rat induced by chronic inhibition of phosphatase 1 and 2A. *Neurosci* 69, 691-698.

Braak H and Braak E (1992): Neuropathological staging of Alzheimer-related changes. *Acta Neuropath.* 82: 239-259.

Bredesen D (1995): Neural apoptosis. *Ann Neurol* 38, 839-851.

Burke W, Park D, Chung H, Marshall G , Haring J, Joh T (1990) : Evidence of decreased transport of tryptophan hydroxylase in Alzheimer's disease. *Brain Res* 537: 83-87.

Callahan L, Coleman P (1995): Neurons bearing neurofibrillary tangles are responsible for selected synaptic deficits in Alzheimer's Disease. *Neurobiol of Aging* 16:311-314.

Coleman P, Flood D (1987): Neuron numbers and dendritic extent in normal aging and Alzheimer's disease. *Neurobiol of Aging*, 8:521-545.

Gong C, Singh T, Grundke I, Iqbal K (1993): Phosphoprotein phosphatase activities in Alzheimer Disease Brain. *J. of Neurochemistry* 61: 921-927.

Hammerschlag R. and Bobinski J. (1992): Does nerve impulse activity modulate fast axonal transport? *Molec Neurobiol* 6:191-201.

Iqbal K, Iqbal I, Zaidi T, Merz PA, Wen GY, Shaikh SS, and Wisniewski HM (1986) : Defective brain microtubule assembly in Alzheimer's disease. *Lancet. 2:*421-426.

Kawahima M Morishima-, Hasegawa M, Takio K, Suzuki M, Yoshida H, Titani K, Ihara Y (1995): Proline-directed and non-proline directed phosphorylation of PHF-tau. *Journ Biol Chem* 270:823-829.

Kawarabayashi T , Shojii M, Yamaguchi H, Yanaka M, Harigaya Y, Ishiguro K, Hirai S (1993): Amyloid β-protein precursor accumulates in swollen neurites throughout rat brain with aging. *Neurosci Lett* 153 73-76.

La Ferla F, Tinkle B, Bieberich C, Huadenschild C and Jay G (1995): The Alzheimer Aβ peptide induces neurodegeneration and apoptotic cell death in transgenic mice.*Nature Genet* 9:21-29.

Lassmann H, Weiler R, Fischerm P, Bancher C, Jellinger K, Floor E, Danielczyk W, Seitelberger F, Winkler H (1992): Synaptic pathology in Alzheimer's disease: Immunological data for markers of synaptic and large dense-core vesicles. *Neuroscience* 46:1-8.

Lee V, Daughenbaugh R, Trojanowski J (1994): Microtubule stabilizing drugs for the treatment of Alzheimer's disease. *Neurobiol Aging* 15:s2, S87-S89.

Moechars D, Lorent K, De Strooper B, Dewachter I, van Leuven F (1996): Expression in brain of APP mutated in the α-secretase site, causes disturbed behavior, neuronal degeneration and premature death in transgenic mice. *EMBO J*, in press.

Moran P, Higgins L, Cordell B, Moser P. (1995): Age-related learnig deficits in transgenic mice expressing the 751-amino acid isoform of human β-amyloid precursor protein. *Proc Natl Acad Sci USA* 92:5341-5345.

Mufson E, Conner J, and Kordower J (1995): Nerve growth factor in Alzheimer's disease: defective retrograde transport to nucleus basalis. *NeuroReport* 6:1063-1066

Nikolic M, Delalle I, Tsai L (1995): The role of p35/cdk5 kinase in neuronal differentiation and neurite outgrowth. *Soc Neurosci Abst*, 21. part 3, p2001

Nuydens R., De Jong M., Nuyens R., Cornelissen F, Geerts H (1996): Aberrant tau phosphorylation decreases fast axonal transport. (Submitted).

Prapotnik D, Smith M, Richey P, Vinters H, Perry G (1996): Filament heterogeneity within the dystrophic neurites of senile plaques suggests blockage of fast axonal transport in Alzheimer's disease. *Acta Neuropathol* 91:226-235.

Richard S, Brion JP, Couck AM, Flament-Durand J (1985): Accumulation of smooth endoplasmic reticulum in Alzheimer's disease: new morphological evidence of axoplasmic flow disturbances. *J. Submicr. Cytol Pathol* 21:461-465.

Scott S, Mufson E, Weingartner J, Skau K, Crutcher K (1995): Nerve Growth factor in Alzheimer's disease : increased levels throughout the brain coupled with declines in Nucleus basalis. *Journ Neurosci* 15:6213-6221.

Vito P, Lacana E, d'Adamio L (1996): Interfering with apoptosis: Ca-binding protein ALG-3 and Alzheimer's disease gene ALG-3. *Science* 271:521-524.

Wolfe N, Reed B, Eberling J, Jagust W (1995): Temporal lobe perfusion on SPECT predicts the rate of cognitive decline in Alzheimer's disease. *Arch Neurol* 52:257-262.

Zheng Y, Jiang M, Trumbauer M, Sirinathsinghji D, Hopkins R, Smith D, Heavens R, Dawson G, Boyce S, Connor M, Stevens K, Slunt H, Sisodia S, Chen H, Van der Ploeg L (1995): β-amyloid precursor protein deficient mice show reactive gliosis and decreased locomotor activity. *Cell* 81:525-531.

Alzheimer Disease: From Molecular Biology to Therapy
edited by R. Becker and E. Giacobini

NOVEL CATHEPSIN D INHIBITORS PREVENT THE β-SECRETASE-DERIVED INTRACELLULAR FORMATION OF A 12 kDa POTENTIALLY AMYLOIDOGENIC PRODUCT IN HUMAN CELLS

Nathalie Chevallier, Philippe Marambaud, Jean-Pierre Vincent and Frédéric Checler
Institut de Pharmacologie Moléculaire et Cellulaire, CNRS UPR411, 660 route des Lucioles, Sophia-Antipolis, 06560 Valbonne (France)

Jean Vizzavona, Pierre Fulcrand and Jean Martinez
URA CNRS 1845, Av. Charles Flahault, 34060 Montpellier (France)

Claus-Peters Baur, Maria Spillantini and Michel Goedert
MRC Laboratory of Molecular Biology, Hills Road, Cambridge, CB2 2QH (U.K.)

INTRODUCTION

The β-amyloid precursor protein (βAPP) is a transmembrane protein that can exist in several isoforms resulting from the alternative splicing of a single gene (Mullan and Crawford, 1993). This protein undergoes complex processing events, the alterations of which have been postulated to be likely linked to the Alzheimer's disease (AD) neuropathology. In physiological conditions, βAPP is cleaved near the membrane by a proteolytic activity, namely α–secretase, that releases an N-terminal fragment (called APPα) that was demonstrated to participate in coagulation and wound repair (Oltersdorf et al., 1989; Smith et al., 1990) as well as in neurotrophic and cytoprotective processes (Qiu et al., 1995; Saitoh et al., 1989; Mattson et al., 1993). In normal conditions, a 40 amino-acid-long fragment called Aβ peptide can also be generated upon βAPP proteolysis by the action of two enzymes called β–and γ–secretases that liberate the N- and C-terminal moieties of Aβ peptide, respectively (see Haass and Selkoe, 1993; Checler, 1995).

In AD, the Aβ peptide accumulates and represents the main component of the senile plaques that are one of the major histopathological hallmarks detected in the brain of affected patients (Glenner and Wong, 1984; Masters et al., 1985). It is therefore of major importance to understand the molecular events that trigger the shift from the physiological to the pathological situation. Several lines of evidence indicate that both the increase in the production of the Aβ peptide or the production of a longer 42-amino-acid Aβ species, could lead to exacerbated aggregation processes (Citron et al., 1992; Cai et al., 1993; Felsenstein et al., 1994; Suzuki et al., 1994). Additional events involved in Aβ chaperoning (Strittmatter et al., 1993) or catabolic breakdown could also contribute to the alteration of the physiological Aβ concentration. Altogether, these data emphasized the fact that, even if the production of Aβ peptide is not the etiological cause of the disease, it remains that the conversion of Aβ from its precursor necessitates proteolytic activities, the blockade of which could ultimately prevent Aβ formation. It is, therefore, of interest to characterize the various secretases, the nature of which remains largely unknown.

RESULTS

The present study particularly focused on the identification of possible β-secretases candidates in human cells overexpressing the Swedish mutated form of βAPP751. This double mutation substitutes the Lys-Met dipeptide (located adjacently to the N-terminus of Aβ) for Asn-Leu and has been shown to co-segregate with an early onset clinically diagnosed form of AD described

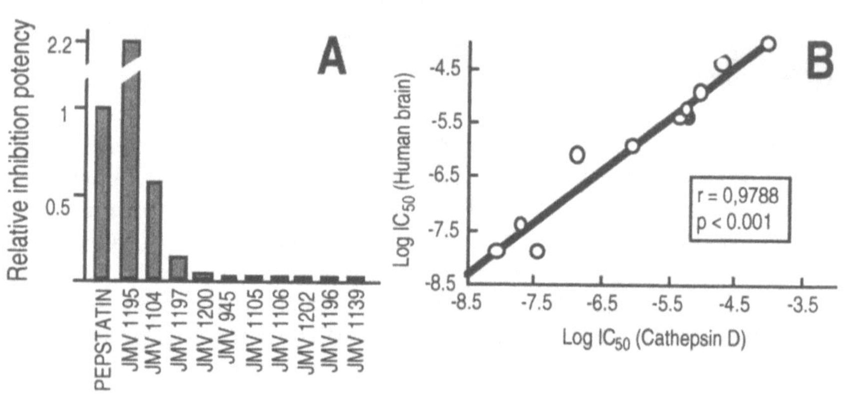

FIGURE 1: ΔNL peptide (10 nmol) was incubated with cathepsin D or with human brain membranes (0,6 to 1 mg/ml) in a final volume of 200 μl of sodium acetate, pH 3,5 in the absence or in the presence of 5 to 7 concentrations of the indicated inhibitors. IC_{50}s displayed on cathepsin D are expressed relative to that of pepstatin (IC_{50} = 0,02 μM) taken as 1 (A). Correlation between IC_{50}s on cathepsin D and human brain activity is shown in B.

in a Swedish kindred (Mullan and Crawford, 1993). This "virulent" form of the disease is likely due to the exacerbation of the production of Aβ and Aβ-containing products as suggested by transfection analysis experiments performed with neuroblastoma and human kidney cells (see Checler, 1995).

Our preliminary experiments indicated that human brain contained an acidic protease displaying preferred catalytic activity towards synthetic peptides encompassing the sequence targeted by β-secretase and bearing the double Swedish mutations (Ac-I-S-E-V-N-L-D-A-E-F-R-H-D-NH2, referred to as ΔNL peptides), than for peptides mimicking the wild type sequence. This, and other biochemical characteristics reminiscent of the acidic protease cathepsin D prompted us to examine such enzyme as putative β-secretase candidate. First, we have developed a series of novel inhibitors that affect cathepsin D activity with affinities ranging from 9 nM (JMV 1195) to more than 100 μM (compound JMV 1245, Fig. 1A). Interestingly, these inhibitors prevent the hydrolysis of the ΔNL peptide by human brain membranes with relative affinities that fully matched those observed with cathepsin D as indicated by the correlation factor observed in Fig. 1B.

In order to examine the ability of cathepsin D to trigger β-secretase-like cleavages on the natural βAPP substrate, we produced and purified recombinant wild type βAPP751 and its Swedish mutated counterpart (ΔNL βAPP751) in baculovirus (Fig. 2A). The ΔNL βAPP751 protein is cleaved by purified cathepsin D as shown in Fig. 2B. Furthermore, we established that one of the breakdown products eluted as a C-terminal 12 kDa fragment that was recognized by polyclonal antibodies fully specific of the N-terminal end of Aβ (Fig. 2C).

Therefore, cathepsin D triggers the release of the C-terminal fragment that is expected to derive from a β-secretase cleavage. Interestingly, ΔNL βAPP751 appeared to undergo a 4 to 6-fold faster cleavage rate by cathepsin D than that observed with the wild type βAPP751 (Fig. 2D).

In order to further examine the possible involvement of cathepsin D as β-secretase, human embryonic kidney cells were stably transfected with wild type and ΔNL βAPP751 cDNAs (in the $pcDNA_3$ eucaryotic vector). These cells secrete Aβ (Fig. 3A) and produce an intracellular 12 kDa product that is immunoprecipitated by our anti- Aβ antibodies (Fig. 3B, lane 1). The formation of this potentially amyloidogenic C-terminal fragment is strongly reduced by treatment of the cells with the phorbol ester PDBu (not shown). This agrees well with previous studies reporting on the decrease of the β-secretase-derived products after stimulation of the protein kinase C pathway (Gabuzda et al., 1993; Buxbaum et al., 1993; Checler, 1995). Interestingly, treatment of the cells with one of our potent cathepsin D inhibitors (JMV 1195) drastically reduced the 12 kDa product formation while the inactive inhibitor (JMV 1245) did not (Fig. 3B, lanes 2 and 3). Altogether, our data confirm the fact that cathepsin D participates to the formation of the 12 kDa product and support the hypothesis that, at least in cells expressing the Swedish mutated form of βAPP, cathepsin D could be considered as a putative β-secretase.

DISCUSSION

It has been recently described that the amyloidogenic processing of the human βAPP in hippocampal neurons led to the intracellular formation

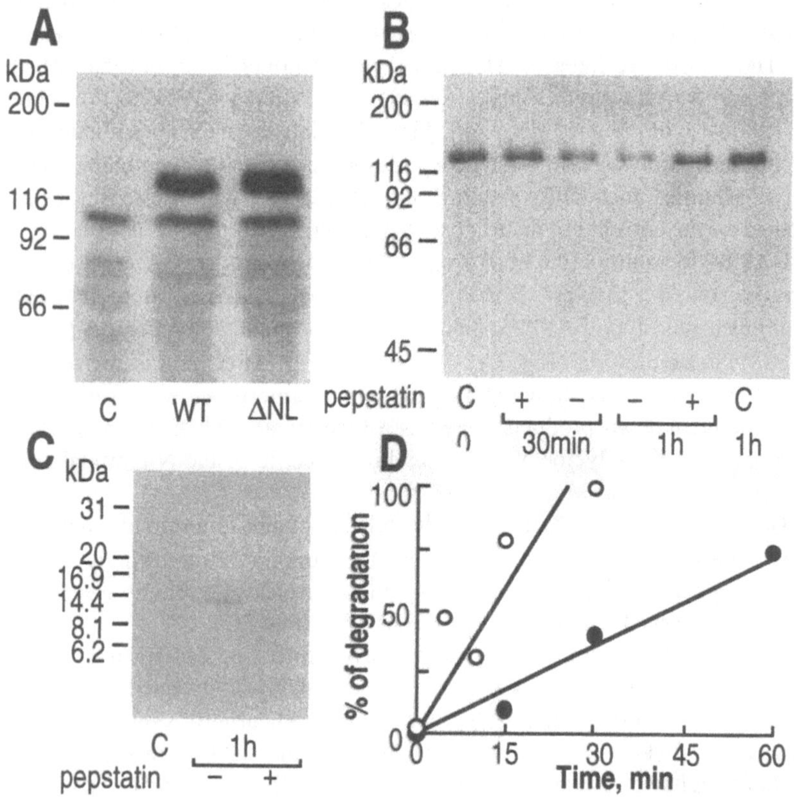

FIGURE 2: A: Production of recombinant wild-type and ΔNLβAPP751 proteins. B: Purified recombinantΔNLβAPP751 was incubated without or with cathepsin D (1 μg), in absence or in the presence of 10 μM pepstatin. Proteins were electrophoresed, Western blotted and immunostained with C-terminal directed BR188 antiserum. C: Tris-Tricine SDS PAGE analysis of the C-terminal degradation product of ΔNLβAPP751 by means of anti Aβ antibodies (FCA18). D: Compared degradation of ΔNLβAPP751 (○) or wild-type βAPP751 (●) was estimated by densitometric analysis.

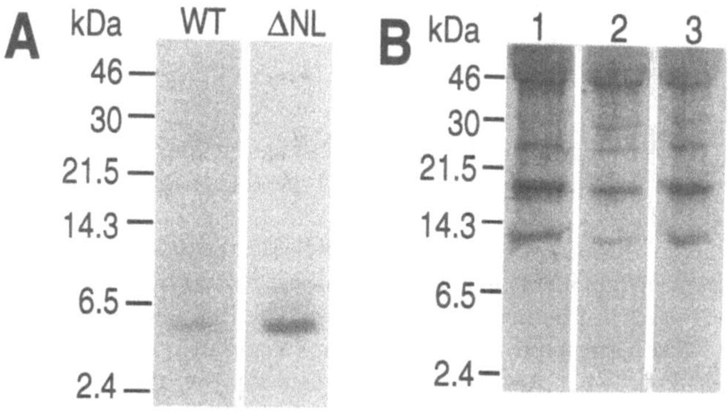

FIGURE 3: Transfected HK293 cells overexpressing wild-type βAPP751 (WT) or ΔNLβAPP751 (ΔNL) were metabolically labelled then extracellular medium was immunoprecipitated with FCA18 antibodies (A). ΔNLβAPP751 transfected cells were treated without (1) or with JMV1195 (2) or JMV1245 (3) then washed, lysed and extracts were immunoprecipitated as above with FCA18 antibodies (B).

of 10 to 12 kDa Aβ-containing C-terminal fragments that likely correspond to that produced by our cells (Simons et al., 1996). Furthermore, a recent study clearly reported on the toxicity of such a C-terminal fragment. Thus, stably transfected PC12 cells overexpressing the sequence corresponding to the last 100 amino-acids of the βAPP C-terminus (the theoretical 12 kDa product of a single β-secretase cleavage on βAPP) displayed altered response to growth factors as indicated by the abolishment of the nerve growth factor-induced differentiation processes (Sandhu et al., 1996). Therefore, the blockade of the β-secretase-derived production of this C-terminal fragment by our cathepsin D inhibitors is of potential interest.

It is important to note that, although the bulk of studies dealing with cathepsin D indicated that this protease was essentially located in lysosomes, it is now clear that the enzyme is also present in early and late endosomes (see Nixon and Cataldo, 1995), two intracellular compartments reminiscent of those suspected to contain β-secretase(s) (see Haass and Selkoe, 1993; Checler, 1995). Furthermore, it has been described that elevated levels of cathepsin D could be recovered in senile plaques (Cataldo and Nixon, 1990) as well as in the cerebrospinal fluids of AD patients (Schwagerl et al., 1995). Finally, a recent study established that AD brains displayed increased expression of the cathepsin D mRNA and immunoreactive proteins (Cataldo et al., 1995). Altogether, these data argue in favor of our current proposal that, at least in the case of the pathology linked to the βAPP Swedish mutations, cathepsin D could be considered as a potential β-secretase candidate. Therefore, this enzyme could be considered as a putative target of a therapeutic strategy aimed at blocking β-secretase and therefore, slowing down the neurodegenerative process responsible for some genetic forms of AD pathology.

REFERENCES

Buxbaum JD, Koo EH and Greengard P (1993): Protein phosphorylation inhibits production of Alzheimer amyloid β/A4 peptide. *Proc Natl Acad Sci USA* 90:9195-9198.

Cai X-D, Golde TE and Younkin SG (1993): Release of excess amyloid β protein from a mutant amyloid β protein precursor. *Science* 259: 514-516.

Cataldo A, Barnett JL, Berman SA, Li J, Quarless S, Bursztajn S, Lippa C and Nixon RA (1995): Gene expression and cellular content of cathepsin D in Alzheimer's disease brain: evidence for early up-regulation of the endosomal-lysosomal system. *Neuron* 14: 671-680.

Cataldo AM and Nixon RA (1990): Enzymatically active lysosomal proteases are associated with amyloid deposits in Alzheimer brain. *Proc Natl Acad Sci USA* 87:3861-3865.

Checler F (1995): Processing of the β-amyloid precursor protein and its regulation in Alzheimer's disease. *J Neurochem* 65:1431-1444.

Citron M, Oltersdorf T, Haass C, McConlogue L, Hung AY, Seubert P, Vigo-Pelfrey C, Lieberburg I and Selkoe DJ (1992): Mutation of the β-amyloid precursor protein in familial Alzheimer's disease increases β-protein production. *Nature* 360:672-674.

Felsenstein KM, Hunihan LW and Roberts SB (1994): Altered cleavage and secretion of a recombinant β-APP bearing the swedish familial Alzheimer's disease mutation. *Nature Genetics* 6:251-256.

Gabuzda D, Busciglio J and Yankner BA (1993): Inhibition of β-amyloid production by activation of protein kinase C. *J Neurochem* 61: 326-2329.

Glenner GG and Wong CW (1984): Alzheimer's disease: initial report of the purification and characterization of a novel cerebrovascular amyloid protein. *Biochem Biophys Res Commun* 120 (3):885-890.

Haass C and Selkoe DJ (1993): Cellular processing of β-amyloid precursor protein and the genesis of amyloid β-peptide. *Cell* 75:1039-1042.

Masters CL, Simons G, Weinman NA, Multhaup G, Mc Donald BL and Beyreuther K (1985): Amyloid plaque core protein in Alzheimer's Disease and Down Syndrome. *Proc Natl Acad Sci USA* 82:4245-4249.

Mattson MP, Cheng B, Culwell AR, Esch FS, Lieberburg I and Rydel RE (1993): Evidence for excitoprotective and intraveuronal calcium-regulating roles for secreted forms of the beta-amyloid precursor protein. *Neuron* 10:243-254.

Mullan M and Crawford F (1993): Genetic and molecular advances in Alzheimer's disease. *Trends in Neurosci* 16:398-403.

Nixon RA and Cataldo AM (1995): The endosomal-lysosomal system of neurons: new roles. *Trends in Neurosci* 18:489-496.

Oltersdorf T, Fritz LC, Schen DB, Lieberburg I, Johnson-Wood KL, Beattie EC, Ward PJ, Blachen RW, Dovey HF and Sinha S (1989): The secreted form of the Alzheimer's amyloid precursor protein with the Kunitz domain is protease nexin-II. *Nature* 341:144-147.

Qiu WQ, Ferreira A, Miller C, Koo EH and Selkoe DJ (1995): Cell-surface β-amyloid precursor protein stimulates neurite outgrowth of hippocampal neurons in an isoform-dependent manner. *J Neurosci* 15: 2157-2167.

Saitoh T, Sundsmo M, Roch JM, Kimura N, Cole G, Schubert D, Olsterdorf T and Schenk DB (1989): Secreted form of amyloid β protein precursor is involved in the growth regulation of fibroblasts. *Cell* 58:615-622.

Sandhu FA, Kim Y, Lapan KA, Salim M, Aliuddin V and Zain SB (1996): Expression of the C terminus of the amyloid precursor protein alters growth factor responsiveness in stably transfected PC12 cells. *Proc Natl Acad Sci USA* 93:2180-2185.

Schwagerl AL, Mohan PS, Cataldo AM, Vonsattel JP, Kowall NW and Nixon RA (1995): Elevated levels of the endosomal-lysosomal proteinase cathepsin D in cerebrospinal fluid in Alzheimer disease. *J Neurochem* 64:443-446.

Simons M, De Strooper B, Multhaup G, Tienary PJ, Dotti CG and Beyreuther K (1996): Amyloidogenic processing of the human amyloid precursor protein in primary cultures of rat hippocampal neurons. *J Neurosci* 16:899-908.

Smith RP, Higuchi DA and Broze GJ (1990): Platelet coagulation factor XIa-inhibitor, a form of Alzheimer amyloid precursor protein. *Science* 248:1126-1128.

Strittmatter WJ, Saunders AM, Schmechel D, Pericak-Vance M, Enghild J, Salvesen GS and Roses AD (1993): Apolipoprotein E: high avidity binding to β-amyloid and increased frequency of type 4 allele in late-onset familial Alzheimer disease. *Proc Natl Acad Sci USA* 90:1977-1981.

Suzuki N, Cheung TT, Cai X-D, Odaka A, Otvos Jr. L, Eckman C, Golde TE and Younkin SG (1994): An increased percentage of long amyloid β protein secreted by familial amyloid β protein precursor (βAPP717) mutants. *Science* 264:1336-1340.

Alzheimer Disease: From Molecular Biology to Therapy
edited by R. Becker and E. Giacobini

CELLULAR AND TEST TUBE MODELS OF AMYLOID-β FORMATION

Henryk M. Wisniewski[1], Janusz Frackowiak[1], Bozena Mazur-Kolecka[1,2], Jerzy Wegiel[1], Abha Chauhan[1], Ved P.S. Chauhan[1]
[1]NYS Institute for Basic Research in Developmental Disabilities, Staten Island, NY
[2]Institute of Pharmacology, Department of Immunobiology, Polish Academy of Sciences, Crakow, Poland

INTRODUCTION

Aggregation of Alzheimer's amyloid-β peptide (Aβ) and formation of amyloid plaques and vascular amyloid are considered a cause of dementia of the Alzheimer's type. The mechanisms and/or agents that trigger fibrillization of Aβ, which could be the targets of rational therapy for Alzheimer's disease (AD), are still unknown. The lack of appropriate experimental models has restricted the search for potential anti-β-amyloidogenic drugs and the development of treatment procedures. In the present study, we demonstrated the cellular model of β-amyloidogenesis, which appears to possess various features of the *in vivo* pathology and could be useful for drug evaluation. We also used cell-free *in vitro* methods to examine the ability of various factors to interfere with aggregation and fibrillization of synthetic Aβ.

CELLULAR MODEL OF AMYLOID-β FORMATION

Amyloid-β in blood vessels is produced by smooth muscle cells (SMCs), as demonstrated in AD, Down's syndrome (Frackowiak et al., 1993, 1994; Wisniewski, 1994a; Wisniewski and Wegiel, 1994b), and aged dogs (Dahme and Sprengler, 1976; Wegiel, 1995). Recently, we have shown that we can transfer the cellular pathology associated with amyloidogenesis into cell culture by using SMCs isolated from dogs' brain blood vessels with amyloid angiopathy (Frackowiak et al., 1995a,b; Mazur-Kolecka et al., 1994; Wisniewski, 1995). Characterization of cultured SMCs demonstrates that the system we created exhibits various features of *in vivo* β-amyloidogenesis and thus can be used to test risk factors for AD and potential anti-β-amyloidogenic drugs.

RESULTS

The remarkable variability of the involvement of vascular SMCs in amyloidogenesis *in vivo* in AD and in old dogs --indicated by the focal deposition of amyloid-β-- was confirmed by studies of isolated cells in culture. We were able to separate and subculture subpopulations of SMCs at different stages of engagement in amyloidogenesis from cases with amyloid angiopathy. The following parameters were evaluated to assess the cell pathology associated with amyloidosis.

Accumulation of Aβ in the form of intracellular deposits
Intracellular Aβ accumulation in culture is similar to what can be found in SMCs in some blood vessels in AD *in vivo.* Immunocytochemical staining with a panel of Aβ-specific antibodies (Kim et al., 1990; Hyman et al., 1991) revealed reactivity for Aβ sequence 1-16 (mAbs 6E10 and lOD5) and aa 17-24 (mAb 4G8) (Fig. 1), but not aa 36-42 (antiserum 277.2). These data indicate that Aβ 1-40, but not 1-42, is accumulated. The accumulation of Aβ was confirmed by immunoblotting, which revealed an Aβ-immunoreactive 4kD band in cell lysates. Some of these deposits contained fibrillar material, as demonstrated by thioflavin S staining. Aβ-immunoreactive amyloid fibrils were also detected by electron microscopy. Extracellular deposition of Aβ can be detected in SMCs cultured in exogenous extracellular matrix proteins. Aβ

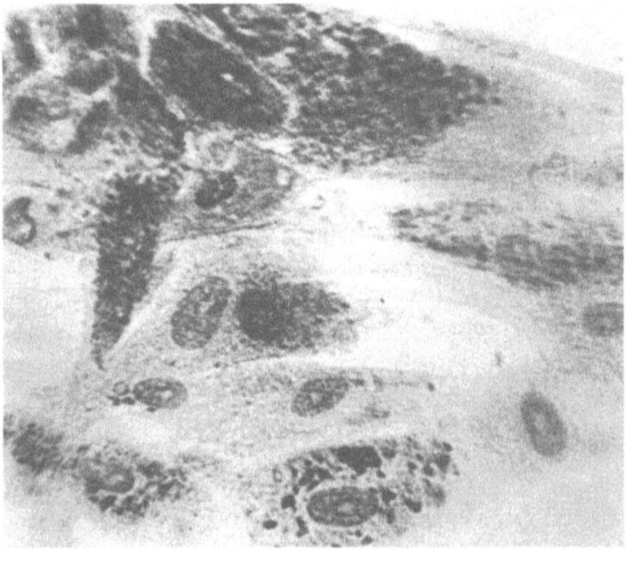

FIGURE 1. Vascular smooth muscle cells isolated from 16-yr-old dog. SMCs contain cytoplasmic deposits immunoreactive for Aβ with mAb 4G8. Magn. 360X.

is accumulated intracellularly and extracellularly in the form of diffuse deposits, similar to what is observed in blood vessels *in vivo*. These results confirm participation of extracellular matrix proteins in extracellular aggregation of Aβ and amyloid formation.

Altered processing of amyloid-β precursor protein (AβPP)
The SMCs in culture produce AβPP 751 and 770, similarly as in blood vessels *in vivo*. SMCs that accumulate Aβ show decreased secretion of sAβPP. These findings were correlated with the appearance of intracytoplasmic C-terminal fragments of AβPP that contain the full sequence of Aβ.

Increased secretion of Aβ as detected by ELISA and immunoprecipitation
Control SMCs secrete 93 ± 7.6 pg of Aβ/10^5 cells/24 hr, whereas in cultures that contain more than 50% of Aβ-immunopositive cells, the figure reaches 525±161 pg/10^5 cells 24 hr.

Features of cell senescence are correlated with accumulation of Aβ
We demonstrated decreased cell adherence to plastic, reduced metabolic activity (measured by the ability to reduce 3-[4,5-dimethylthiazol-2-yl]-2,5-diphenyltetrazolium bromide (MTT), and decreased proliferation and viability.

This characterization of SMCs demonstrates that our culture system mimics basic features of *in vivo* β-amyloidogenesis. Isolation of SMCs in different levels of involvement in β-amyloidogenesis allowed us not only to study production and metabolism of AβPP and Aβ in the context of amyloidogenesis but also to study various factors for their ability to induce and to modify amyloidogenesis. We applied this model to test factors postulated to be involved in AD pathology, including apolipoprotein E4 (apoE4). Intracellular accumulation of Aβ can be induced by treatment with apoE3 and apoE4; the higher stability and reduced metabolic activity are the first demonstration of differential effects of apoE isoforms on amyloidogenesis (Mazur-Kolecka et al., 1995). We also found that the spontaneous and apoE-induced accumulation of Aβ could be abolished by supplementing culture medium with a physiological concentration of transthyretin.

The unique SMC strains we collected appear to be suitable for quick and easily reproducible screening studies of possible AD risk factors. This system can also be used to search for potential drugs able to inhibit Aβ aggregation and fibrillization or to reduce the toxic effects of Aβ on cells.

CELL-FREE MODEL OF Aβ FIBRIL FORMATION

Synthetic Aβ has been used widely for *in vitro* studies of Aβ fibril formation (Jarett and Lansbury, 1993; Wisniewski et al., 1994). Synthetic peptides comprising the complete Aβ sequence (Aβ 1-40 and Aβ 1-42) behave essentially similarly to naturally occurring amyloid Aβ. They both form fibrils *in vitro* and show characteristic green birefringence after staining with Congo red. Various biophysical techniques such as X-ray diffraction, Fourier transform infrared (FTIR) spectroscopy, circular dichroism (CD) spectroscopy, and nuclear magnetic resonance (NMR) have been employed by different investigators to study Aβ fibril formation. However, these techniques are time-consuming and therefore impractical for screening drugs with anti-amyloidogenic activity. Thioflavin T (ThT) fluorescence spectroscopy and Congo red absorbance spectroscopy are rapid and sensitive methods to assess cross β-pleated sheet structures of Aβ, a defining characteristic of amyloid fibrils. ThT (LeVine, 1993, 1995) and Congo red (Shen, 1993) do not bind to dimers or tetramers, but only to higher-order aggregates, so-called protofibrils, and amyloid-like fibrils. The effects of different factors on the morphological characteristics of Aβ fibrils are visualized by electron microscopy to assess Aβ fibril morphology, their formation, elongation, and defibrillization.

RESULTS

We employed the techniques above to study (a) the effect of cerebrospinal fluid (CSF) from normal and AD patients with different apolipoprotein E phenotypes on *in vitro* fibrillogenesis of synthetic Aβ 1-40 (Chauhan, 1996); and, (b) the effect of different culture media and of normal and heat-inactivated sera on fibrillization of Aβ 1-40 and Aβ 1-42 (Wegiel, 1996). We found that fibrillization of synthetic Aβ 1-42 was much faster in all examined buffers (PBS, TRIS, pH 7.5) and media (DMEM, RPMl, OPTI-MEM) as compared with that of Aβ 1-40. The culture media promoted fibrillization of Aβ 1-40 and Aβ 1-42 faster than buffers (Wegiel,1996). The serum, on the other hand, inhibited Aβ fibril formation and elongation in buffers and culture media, and it also defibrillized preformed Aβ fibrils (Wegiel, 1996) as compared to buffer (Chauhan, 1996), in agreement with an earlier report (Wisniewski, 1993). Significant differences were observed in Aβ aggregation between CSF from AD patients and from control groups, and between CSF from different apoE phenotypes (Chauhan, 1996). There was a positive correlation between CSF apoE concentrations and Aβ aggregation, but no correlation between CSF-soluble Aβ levels or severity of dementia and Aβ aggregation (Chauhan, 1996).

The combination of ThT fluorescence spectroscopy and EM analysis in negative staining can be utilized to identify drugs that can prevent or inhibit Aβ fibril formation in test tube and possibly amyloid formation in AD.

ACKNOWLEDGMENTS

The authors want to thank Dr. Dale B. Schenk (Athena Neurosciences, Inc., San Francisco, CA) for antibody lOD5 and antiserum 277.2, and Dr. Kwang Soo Kim (IBR, Staten Island, NY) for antibodies 4G8 and 6E10. Supported in part by funds from the NYS/OMRDD and a grant from the National Institute on Aging, P01 AGO4220.

REFERENCES

Chauhan A, Pirttila T, Mehta P, Chauhan VPS and Wisniewski HM (1996): Effect of cerebrospinal fluid from normal and Alzheimer's patients with different apolipoprotein E phenotypes on in vitro aggregation of amyloid beta protein. *J Neurol* Sci (in press).

Dahme Von E and Sprengler B (1976): Weitere Überlegungen zur Pathogenese des Altersamyloids in Meninx- und Hirnrindenarterien beim Hund. *Ben Münch Tierärztl Wschr* 89:64-67.

Frackowiak J, Zoltowska A and Wisniewski HM (1993): Presence of non-fibrillar β-amyloid protein in smooth muscle cells of vessel walls in Alzheimer disease. *J Neuropathol Exp Neurol* 52:306.

Frackowiak J, Zoltowska A and Wisniewski HM (1994): Non-fibrillar β-amyloid protein *is* associated with smooth muscle cells of vessel walls in Alzheimer disease. *J Neuropathol Exp Neurol* 53:637-645.

Frackowiak J, Mazur-Kolecka B, Wisniewski HM, Potempska A, Carroll RT, Emmerling MR and Kim KS (19995a): Secretion and accumulation of Alzheimer's β-protein by cultured vascular smooth muscle cells from old and young dogs. *Brain Res* 676:225-230.

Frackowiak J, Mazur-Kolecka B, Wegiel J, Kim KS and Wisniewski HM (1995b): Culture of canine vascular myocytes as a model to study production and accumulation of β-protein by cells involved in amyloidogenesis. In: *Research Advances in Alzheimer's Disease and Related Disorders,* Iqbal K, Mortimer JA, Winblad B and Wisniewski HM, eds. Chichester: John Wiley & Sons, pp.747-754.

Hyman BT, Tanzi RB, Marzloff K, Barbour R and Schenk DB (1991): Kunitz protease inhibitor containing amyloid beta protein precursor protein immuno-reactivity in Alzheimer's disease. J *Neuropath Exp Neurol* 51:76-83.

Jarett JT and Lansbury PT Jr. (1993): Seeding "One-Dimensional Crystallization" of Amyloid: A pathogenic mechanism in Alzheimer's disease and scrapie? *Cell* 73: 1055-1058.

Kim KS, Wen GY, Bancher C, Chen CMJ, Sapienza VJ, Hong H and Wisniewski HM (1990): Detection and quantitation of amyloid β-peptide with two monoclonal antibodies. *Neurosci Res Commun* 7:113-122.

LeVine H III(1993): Thioflavin T interaction with synthetic Alzheimer's disease β-amyloid peptides: Detection of amyloid aggregation in solution. *Protein Sci* 2:404-410.

LeVine H III (1995): Soluble multimeric Alzheimer β (1-40) preamyloid complexes in dilute solution. *Neurobiol Aging* 16:755-764.

Mazur-Kolecka B, Frackowiak J, Wisniewski HM, Carroll RT, Emmerling MR and Kim KS (1994): Production of β-protein in canine vascular myocytes in long-term cultures. *Neurobiol Aging* 15:S52.

Mazur-Kolecka B, Frackowiak J and Wisniewski HM (1995): Apolipoproteins E3 and E4 induce, and transthyretin prevents accumulation of the Alzheimer's β-amyloid peptide in cultured vascular smooth muscle cells. *Brain Res* 698:217-222.

Shen CL, Scott CL, Merchant F and Murphy RM (1993): Light scattering analysis of fibril growth from the amino-terminal fragment β (1-28) of β-amyloid peptide. *Biophys J* 65:2383-2395.

Wegiel J, Wisniewski H, Dziewiatkowski J, Tarnawski M, Nowakowski J, Dziewiatkowska A and Soltysiak Z (1995): The origin of amyloid in cerebral vessels of aged dogs. *Brain Res* 705:225-234.

Wegiel J, Chauhan A, Wisniewski HM, Nowakowski J and LeVine H (1996): Promotion of synthetic amyloid β peptide fibrillization by culture media and cessation of fibrillization by serum. *Neurosci Lett* (in press).

Wisniewski HM, Frackowiak J, Zoltowska A and Kim KS (1994a): Vascular β-amyloid in Alzheimer disease angiopathy is produced by proliferating and degenerating smooth muscle cells. *Amyloid lnt J ClinExp Invest* 1:8-16.

Wisniewski HM and Wegiel J (1994b): β-amyloid formation by myocytes of leptomeningeal vessels. *Acta Neuropathol* 87:233-241.

Wisniewski HM, Frackowiak J and Mazur-Kolecka B (1995): In vitro production of β-amyloid in smooth muscle cells isolated from amyloid angiopathy-affected vessels. *Neurosci Letters* 183:120-123.

Wisniewski T, Castano E, Chiso J and Frangione B (1993): Cerebrospinal fluid inhibits Alzheimer β-amyloid fibril formation in vitro. *Ann Neurol* 34:631-633.

Wisniewski T, Castano EM, Golabek A, Vogel T and Frangione B (1994): Acceleration of Alzheimer's fibril formation by apolipoprotein E in vitro. *Am J Pathol* 145:1030-1035.

Alzheimer Disease: From Molecular Biology to Therapy
edited by R. Becker and E. Giacobini

EFFECTS OF β-AMYLOID (1-40) PEPTIDE INJECTION IN THE NUCLEUS BASALIS

Giancarlo Pepeu, Lisa Giovannelli, Fiorella Casamenti and Carla Scali
Department of Pharmacology, Florence University, Florence, Italy

INTRODUCTION

Alzheimer's disease (AD), the main cause of senile dementia, is a neuro-degenerative disorder whose incidence has increased steadily during the last decades. Mounting awareness of the social burden represented by this disease has generated a large research effort aimed to clarify its etiology. At present, a definite treatment or prevention for AD is still not available. However, much progress has been made in identifying the molecular alterations which accompany this disorder. Investigations have focused on the most prominent neuropathological hallmarks of AD, senile plaques and neurofibrillary tangles, in an attempt to dissect the molecular mechanisms of their formation and their relationship to the pathogenesis of AD. Much is known about the biology of β -amyloid protein (Aβ), the main component of senile plaques, and its precursor protein APP (Selkoe, 1994). The hypothesis that the deposition of Aβ is a critical event in the pathogenesis of AD has gained support from human genetic studies demonstrating a linkage between mutations in the APP gene and some familial forms of AD (Hardy, 1992), and from *in vitro* studies demonstrating that Aβ is neurotoxic to cultured neuronal cells (Yankner et al., 1990). *In vivo* studies aimed to define whether Aβ is neurotoxic when injected in the brain have provided conflicting results (Frautschy et al., 1991; Games et al., 1992). As AD is characterized by a highly regional-specific neuronal loss, particularly prominent in the basal forebrain and its target areas, cortex and hippocampus, we have focused on the forebrain nucleus basalis (NB) and studied the neurotoxicity of local injections of β-amyloid peptide (1-40) in the rat.

EXPERIMENTAL PROCEDURES

Male Wistar rats of 3 months of age were used. All experiments were carried out according to the guidelines of the European Community's Council for

Animal Experiments. ß-amyloid peptide (1-40) and the scrambled (25-35) peptide were dissolved in bidistilled water at the concentration of 10 µg/µl and the solutions were incubated at 37°C for 1 week before use. A stereotaxic unilateral injection into the right NB was performed under anesthesia. One µl of either peptide or saline solution was injected. The animals were sacrificed 7, 14, 21, 30, 60, 120 and 180 days later. ACh and GABA release from the parietal cortex was measured *in vivo* by means of the transversal microdialysis technique and assayed by high-performance liquid chromatography (Giovannini et al., 1994). At the end of the microdialysis experiments, the rats were deeply anesthetized and perfused through the ascending aorta with 4% paraformaldehyde. The brains were coronally cut in 20 µm-thick coronal sections throughout the extension of the injected area. For Congo Red staining of the amyloid deposits, the procedure described by Giovannelli et al. (1995) was used. Choline acetyltransferase (ChAT), parvalbumin (PV) and Aβ immunoreactivity (IR) were analysed using polyclonal rabbit antibodies visualized by means of the avidin-biotin-peroxidase method with diaminobenzidine (DAB).

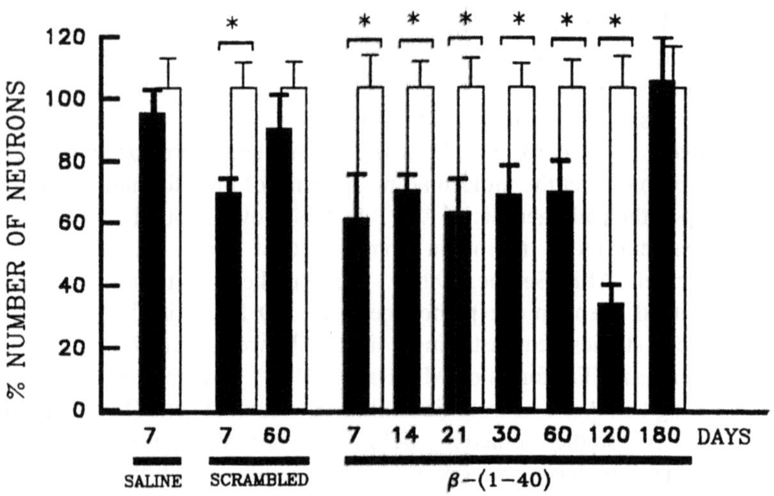

FIGURE 1. Effect of ß-peptide (1-40) injection on the number of ChAT-positive neurons in the NB at different times post-lesion. The percentage of the contralateral unlesioned NB was calculated per each histological section. White bars: contralateral unlesioned side values; solid bars: lesioned side values. Data represent the mean percentage ± S.E.M. of 3-5 animals (5 sections per each animal). Student's t test: * $p<0.02$; ** $p<0.01$ statistically significant difference from the respective control.

RESULTS

Injection of the β (1-40) peptide, but not of the scrambled peptide, into the NB induced the formation of a deposit which persisted at the injection site for at least 6 months, as demonstrated by Nissl staining and Aβ immunoreactivity (not shown). From 1 week up to 4 months post-injection, the deposit maintained a fibrillary conformation typical of Aβ in senile plaques. A transition from fibrillary to amorphous aggregation state took place between 4 and 6 months, when the deposit exhibited faint or no birefringency upon Congo Red staining (not shown). The persistence of a fibrillary deposit in the NB was accompanied by cholinergic hypofunction, as indicated by the decrease in the number of ChAT-positive neurons in the NB (Fig. 1), and basal ACh output from the cortex ipsilateral to the injection (Fig. 2). A recovery of both ChAT IR and ACh release was found at 6 months, concomitantly with the loss of the deposit's fibrillary conformation. Conversely, cortical GABA release was considerably increased in the β (1-40)-injected animals as compared to controls both at 21 and 180 days post injection (Fig. 3). The increase was accompanied by no change in PV IR, a marker of GABAergic neurons, in the NB complex (not shown).

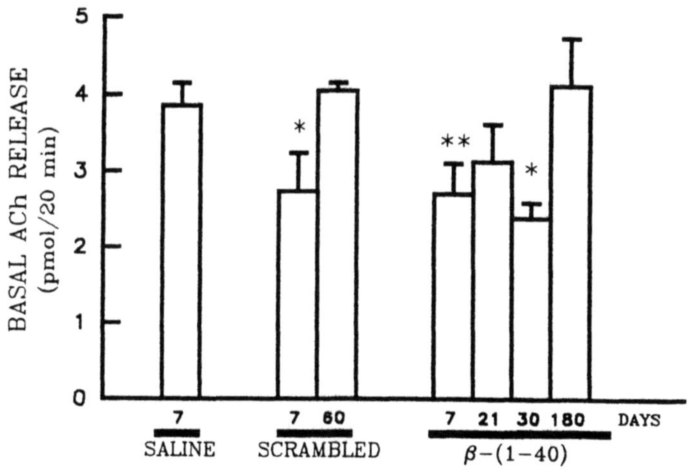

FIGURE 2. Effect of ß-peptide (1-40) injection into the NB on basal ACh release from the ipsilateral parietal cortex at different times post-lesion. Data represent the mean ± S.E.M. ACh efflux of six collection periods for each rat (n= 4-5). ANOVA $p<0.02$) followed by Fisher's LSD post-comparison test. * $p<0.05$; ** $p<0.02$ versus saline-injected rats.

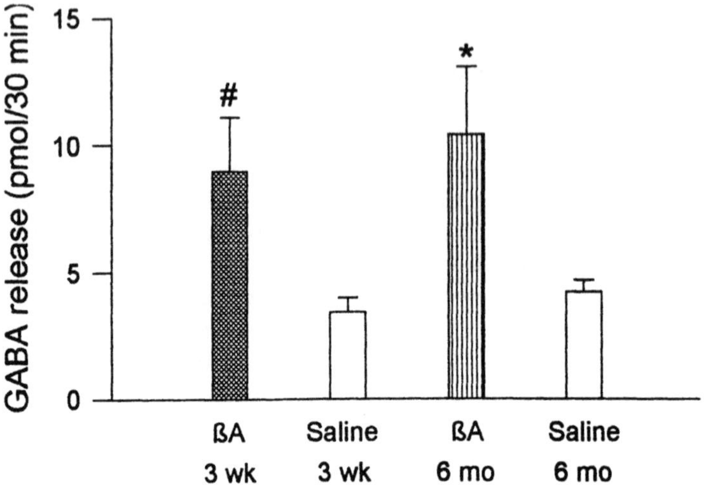

FIGURE 3. Effect of ß-peptide (1-40) injection into the NB on basal GABA release from the ipsilateral parietal cortex at 3 weeks and 6 months post-lesion. Data represent the mean ± S.E.M. GABA efflux of six collection periods for each rat (n= 5-7). Student's t test: * and # $p<0.05$ versus respective saline-injected rats.

DISCUSSION

The present work shows that injections of the β-amyloid peptide (1-40) in the NB exert different effects on cholinergic and GABAergic cortical extracellular levels. The decrease in ACh release is long-lasting but reversible, indicating that hypofunction but not cell death has been induced by the presence of amyloid deposits in proximity of the NB cholinergic cell bodies. Furthermore, the toxic effect on the cortically-projecting cholinergic neurons seems to be related to the fibrillary conformation of the peptide deposit, in agreement with previous *in vitro* studies (Lorenzo and Yankner, 1994). It is more difficult to explain the effect on cortical GABA levels. One possibility is that the amyloid deposit induces a hypofunction of the basal forebrain-cortical GABAergic pathway as well as of the cholinergic one, being the former longer-lasting. As the GABAergic cortical afferents from the forebrain have been shown to terminate mainly on GABAergic cortical interneurons (Freund and Gulyas, 1991), it is feasible that an impairment of the GABAergic forebrain function would increase cortical GABA output, leading to an enhanced local inhibition in the cortex. The use of markers of GABAergic neurons other than PV will help clarifying whether Aβ affects the GABAergic forebrain neurons. Furthermore, whether the increased cortical inhibition is related to the disruption of some memory processes observed after Aβ injection (Giovannelli et al., 1995) needs to be investigated.

ACKNOWLEDGMENTS

This work was supported by grant no. 93 00436 PF 40 from CNR, Target Project on Aging.

REFERENCES

Frautschy SA, Baird A and Cole GM (1991): Effects of injected Alzheimer ß-amyloid cores in rat brain. *Proc Natl Acad Sci USA* 88:8362-8366.

Freund TF and Gulyas AI (1991): GABAergic interneurons containing calbindin D28k or somatostatin are major targets of GABAergic basal forebrain afferents in the rat neocortex. *J Comp Neurol* 314:187-199.

Games D, Khan KM, Soriano FG, Davis, DL, Bryant K and Lieberburg I (1992): Lack of Alzheimer pathology after β-amyloid protein injections in the rat. *Neurobiol Aging* 13:569-576.

Giovannini MG, Mutolo D, Bianchi L, Michelassi A and Pepeu G (1994): NMDA receptor antagonists decrease GABA outflow from the septum and increase acetylcholine outflow from the hippocampus: a microdialysis study. *J Neurosci* 14:1358-1365.

Giovannelli L, Casamenti F, Scali C, Bartolini L and Pepeu G (1995): Differential effects of amyloid peptides β-(1-40) and β-(25-35) injections into the rat nucleus basalis. *Neuroscience* 66:781-792.

Hardy J (1992): Framing β-amyloid. *Nature Genet* 1:233-234.

Lorenzo A and Yankner BA (1994): β-amyloid neurotoxicity requires fibril formation and is inhibited by Congo Red. *Proc Natl Acad Sci USA* 91:12243-12247.

Selkoe DJ (1994): Normal and abnormal biology of the β-amyloid precursor protein. *Ann Rev Neurosci* 17:489-517.

Yankner BA, Duffy LK and Kirschner DA (1990): Neurotrophic and neurotoxic effects of amyloid β-protein: reversal by tachykinin neuropeptides. *Science* 250:279-282.

ACKNOWLEDGMENTS

This work was supported by grant no. [illegible]

REFERENCES

[illegible]

[illegible]

[illegible]

[illegible]

[illegible]

[illegible]

[illegible]

[illegible]

Alzheimer Disease: From Molecular Biology to Therapy
edited by R. Becker and E. Giacobini

ß-AMYLOID PRECURSOR PROTEIN - ROLE IN COGNITIVE BRAIN FUNCTION?

Gerda S. Huber, Jean-Luc Moreau and James R. Martin
Pharma Division, Preclinical CNS Research, F. Hoffmann-La Roche Ltd, 4070 Basel, Switzerland

Yannick Bailly
Laboratoire de Neurobiologie Cellulaire CNRS, Strasbourg, France

Jean Mariani
Université P. & M. Curie, CNRS URA 1488, Paris, France

Bernard Brugg
Hôpital de la Salpetrière, INSERM U289, Paris, France

INTRODUCTION

Alzheimer's disease (AD) is characterized by loss of higher brain functions and appearance of specific pathological lesions such as neurofibrillary tangles, ß-amyloid plaques and neuronal degeneration. As ß-amyloid generation and accumulation occur very early in the disease process the fate and role of its precursor proteins (ß-APPs) is important for understanding AD pathogenesis. The primary structure of ß-APP resembles glycosylated cell surface proteins (Kang et al., 1987) and they can be proteolyzed in two different ways giving rise either to extracellular secreted C-terminal truncated molecules or to the ß-amyloid peptide (Haass and Selkoe, 1993). Different length isoforms of ß-APP are derived from alternative mRNA splicing (ß-APP_{695}, ß-APP_{751}, ß-APP_{770}) all of which may be N- and O-glycosylated (Weidemann et al., 1989). The ß-APP gene products are expressed in different tissues and show highest abundance in neurons in the brain (Shivers et al., 1988). Co-localization of ß-APP with synaptophysin was found at synaptic sites (Schubert et al., 1991) and ß-APP levels, in particular ß-APP_{695}, increase transiently during synaptogenesis in rats (Löffler and Huber, 1992). The potential roles of ß-APPs in higher brain function were explored by intracerebroventricular injection of anti-ß-APP antibodies in rats followed by behavioural analysis

(Huber et al., 1993) and by comparison of ß-APP levels in brains of rats with different "cognitive" capabilities.

MATERIALS AND METHODS

Antibody injections into rat brain
20μl of anti-ß-APP-specific antibody (affinity-purified, 1.3 mg/ml, anti-N-terminal epitope aa 16-32) or control IgG (1.5 mg/ml) were injected intracerebroventricularly (icv) into adult rats. The rats were then trained in a passive avoidance task and were retested 24 hours later for retention (N's=14).

Environmental enrichment during early life
Pregnant rats were housed either individually in standard cages under standard lab conditions or in small groups in large boxes with a complex sensory-motor environment. This consisted of daily alterations in foods and fluids, bedding, intermittent music, olfactory stimulation, wooden sticks/branches for climbing, handling for about 5 min and periodic introduction of an intruder rat for half a day. At 19 days of age, the mother rats and female pups were removed and the male rats were continuously housed as described ("standard/impoverished" or "enriched"). Five weeks post-weaning the rats were tested in diverse behavioral tasks (N's = 20).

Brain protein analysis
For quantitative ß-APP analysis, brain extracts of 4 rats from enriched and from impoverished groups were prepared. Brains were dissected into higher cortical regions or cerebellum, homogenized in 4 volumes of 10% sucrose containing Hepes buffer(pH 7.5)/50 μM $CaCl_2$ and 10x SDS-PAGE sample buffer was added before boiling. Equal amounts of protein were analyzed by immunoblotting using antibodies against ß-APP, (anti-KPI, anti-$ß_{1\text{-}16}$, described in Löffler and Huber, 1992; 22C11 Boehringer Mannheim) or synaptophysin (Boehringer Mannheim). Immunoreactive bands were quantified using a video densitometer.

RESULTS AND DISCUSSION

In a passive avoidance retention test done 24 hr after training, anti-ß-APP-injected animals showed significantly shorter step-down latencies than the control IgG animals suggesting impaired cognitive function in the former group (Fig. 1A).

Furthermore, rats which had been exposed to an enriched environment during early life compared to rats raised under standard/impoverished conditions exhibited significantly better initial learning and long-term retention in a Morris water maze task and improved initial hurdle jump active

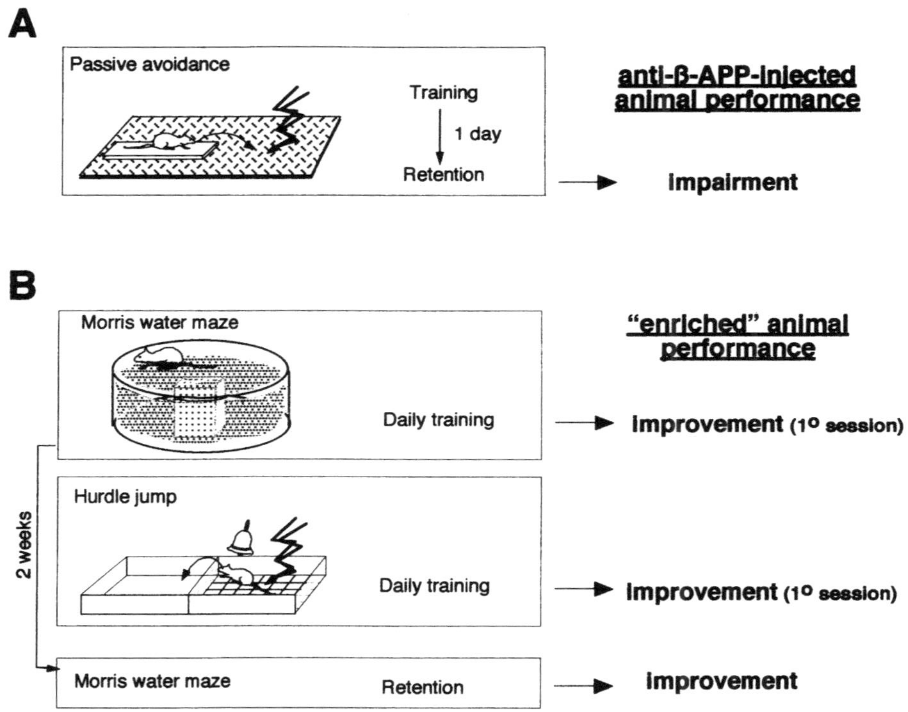

FIGURE 1. A: anti-β-APP injected rats showed impaired 24-hour memory in a step-down passive avoidance test compared to control IgG injected rats. B: "enriched" rats showed improved cognitive behavior in different tests as compared to "impoverished" animals.

avoidance learning. (Fig. 1B). The "enriched" rats also exhibited superior motor coordination determined in a rotarod task. Brain protein analysis revealed higher levels of synaptic marker and of ß-APP isoforms in cortical regions of "enriched" rats. Increased levels were associated with the increased immediate ability to learn new tasks and suggest an involvement of ß-APPs in higher brain function.

CONCLUSION

In conclusion, accumulating evidence suggests that ß-APPs may participate in synaptic plasticity possibly by modulation of cell-cell interaction at synaptic sites (Schubert et al., 1989; Moya et al, 1994). Loss of neuronal connectivity, as indexed by loss of synapses, is known to predict the degree of cognitive impairment in AD patients (DeKosky and Scheff, 1990). Thus, it is hypothesized that proper expression and functioning of ß-APP at synapses is one of the prerequisites for higher brain function. Disturbed proteolytic processing of ß-APPs, as may occur at AD synapses, could thus be a fundamental problem in the disease. ß-Amyloidosis represents a molecular cascade which offers targets for therapeutic intervention at different levels such as alternative proteolysis, Aß accumulation/aggregation/toxicity and in the light of these results perhaps even ß-APP functional substitution.

REFERENCES

DeKosky ST and Scheff SW (1990): Synapse loss in frontal cortex biopsies in Alzheimer's disease: correlation with cognitive severity. *Ann Neurol* 27:457-464.

Haass C and Selkoe DJ (1993): Cellular processing of ß-amyloid precursor protein and the genesis of amyloid ß-peptide. *Cell* 75:1039-1042.

Huber G et al. (1993): Involvement of amyloid precursor protein in memory formation in rat: an indirect antibody approach. *Brain Res* 603:348-352.

Kang J et al. (1987): The precusor of Alzheimer's disease amyloid A4 protein resembles a cell-surface receptor. *Nature* 325:733-736.

Löffler J and Huber G (1992): ß-Amyloid precursor protein isoforms in various rat brain regions and during brain development. *J Neurochem* 59:1316-1324.

Moya KL et al. (1994): In vivo neuronal synthesis and axonal transport of Kunitz Protease Inhibitor-containing forms of the amyloid precursor protein. *J Neurochem* 63: 1971-1974.

Schubert D et al. (1989): The regulation of amyloid ß protein precursor secretion and its modulatory role in cell adhesion. *Neuron* 3:689-694.

Schubert W et al. (1991): Localization of Alzheimer ßA4 amyloid precursor protein at central and peripheral synaptic sites. *Brain Res* 563:184-194.

Shivers BD et al. (1988): Alzheimer's disease amyloidogenic glycoprotein: expression pattern in rat brain suggests a role in cell contact. *EMBO J* 7:1365-1370.

Weidemann A et al. (1989): Identification, biogenesis and localization of precursor of Alzheimer's disease A4 amyloid protein. *Cell* 57:115-126.

Alzheimer Disease: From Molecular Biology to Therapy
edited by R. Becker and E. Giacobini

PEPTIDES INHIBITOR OF AMYLOIDOGENESIS IN ALZHEIMER'S DISEASE

Claudio Soto[1,2], Frances Prelli[2], and Blas Frangione[2]
[1]Departments of Neurology and [2]Pathology'
New York University Medical Center, New York, NY

Mark S. Kindy[3]Frederick de Beer[4]
[3]Departments of Biochemistry and [4]Medicine,
University of Kentucky, Lexington, KY

INTRODUCTION

Alzheimer's disease (AD) is the most common form of dementia in adults. It determines a progressive loss of intellectual function and appears both sporadically and in an autosomal dominant (familial) form. Presently there is no treatment for AD.

An early manifestation of the disease is the deposition of amyloid as insoluble fibrous masses in extracellular senile plaques. The main component of amyloid is a 4.3 kDa hydrophobic peptide, amyloid beta-peptide (Aβ), that is encoded within the larger amyloid precursor protein (βPP) gene, localized on chromosome 21 (Sisodia and Price, 1995). In addition to amyloid deposits, Aβ is also found in a soluble form (sAβ) in cerebrospinal fluid and plasma from normal individuals and AD patients (Wisniewski et al., 1994b). The mechanism by which amyloid deposits are associated with the dementia in AD is unknown, although controversial reports have suggested that when Aβ aggregates into amyloid fibrils, it becomes neurotoxic leading to cell death (Lorenzo and Yankner, 1994; Pike et al., 1993).

Peptides containing the sequence 1-40 or 1-42 of Aβ and shorter derivatives can form amyloid-like fibrils *in vitro*. It has been suggested that Aβ fibrillogenesis is modulated not only by concentration, pH and clearance rate, but also by the interaction with several Aβ-binding proteins, including: apolipoproteins (apo) E and J, α1-antichymotrypsin, serum amyloid P, and heparan sulfate proteoglycans (Wisniewski et al., 1994b). Immunohistochemical analysis indicated that these proteins are co-deposited with amyloid in senile plaques and *in vitro* studies showed that Aβ-binding proteins affect the fibrillogenesis rate. Whether they are inert bystanders or participating actively in the amyloidogenesis process is unknown. Several lines of evidence suggest that at least the two apolipoproteins interact *in vivo* with amyloid or sAβ. The high-affinity interaction (Kd= 2 nM) between Aβ and apoJ, the identification of the complex apoJ-sAβ in CSF, the co-localization of sAβ with apoJ-containing high density plasma lipoproteins, the participation of apoJ in the transport of Aβ through the blood-brain barrier and the demonstration that the interaction of Aβ with apoJ prevents peptide aggregation and degradation *in vitro* (Matsubara et al., 1996), indicate that this protein plays an important role as a physiological carrier for sAβ. On the other hand, apoE also binds with high affinity to Aβ (Kd = 10-20 nM), but this interaction appears to induce a change in Aβ conformation and increases the ability of the peptide to form amyloid fibrils *in vitro*, under certain conditions (Wisniewski et al., 1994a; Ma et al., 1994; Soto et al., 1996). Genetic studies have shown a close association between the allele ε4 of apoE and both sporadic and familial late-onset AD (Corder et al., 1993). In addition, it has been shown that apoE and C-terminal fragments derived from it are tightly complexed with Aβ fibrils in AD amyloid deposits (Naslund et al., 1995; Wisniewski et al., 1995). The wide association of apoE with clinically and biochemically distinct types of amyloid deposits had led to the suggestion that apoE may act as pathological chaperone in amyloidogenesis, binding to the amyloid precursors and modulating fibril formation, possibly by inducing or stabilizing β-pleated sheet conformation (Wisniewski and Frangione,1992).

Several studies have shown that amyloid formation is driven by hydrophobic interactions and is dependent on conformational changes of amyloid precursors. The influence of hydrophobicity and β-sheet secondary structure on amyloid formation is also evident by comparison of the sequence of other amyloidogenic proteins. It is likely that hydrophobicity facilitates monomeric interaction and that β-sheet content drives this interaction to β-sheet oligomers and amyloid fibrils.

In this study, we designed inhibitors of fibrillogenesis based on the hypothesis that amyloid formation could be inhibited by peptides homologous to Aβ, with a similar degree of hydrophobicity but with a very low propensity to adopt a β-sheet conformation. The aim was to design a short Aβ homologous peptide with sufficient similarity to interact with Aβ and with the ability to block the formation of intermolecular β-pleated sheets and, therefore, the formation of amyloid fibrils.

Design of inhibitor peptides

We focused on the central hydrophobic region within the N-terminal domain of Aβ, amino acids 17-21 (LVFFA), as a model for our inhibitor peptide (Fig. 1A). Proline residues were introduced in the inhibitor peptide in order to block β-sheet structure and charged residues were added at the ends of the peptide to increase solubility. Proline was chosen to block β-sheet structure since it rarely forms part of this conformation and does not occur in the interior of antiparallel β-sheets (Wouters and Curmi, 1994), due to the extraordinary characteristics of this amino acid. Moreover, recent data showed that the introduction of proline residues into short peptides homologous to Aβ resulted in non-amyloidogenic analogues (Wood et al., 1995).

Based on these criteria, we designed an 11 amino acid peptide, called inhibitor of Aβ fibrillogenesis peptide (iAβ), which has low probability to adopt a β-sheet conformation due to the presence of proline residues (Fig 1A). The circular dichroism spectrum of iAβ in aqueous solution was typical of unordered structures; indeed, iAβ did not aggregate even at high concentrations (4 mg/ml) or after long periods of incubation (more than 30 days). (Data not shown).

The interaction between Aβ and iAβ was studied by monitoring the quenching of Tyr^{10} fluorescence of Aβ (Fig. 1B). Aβ excited at 280nm showed a fluorescence spectrum with a maximum at 309nm (Fig. 1B, inset), which is typical of tyrosine emission. The presence of iAβ induced a saturable quenching of the fluorescence, reaching a maximum of 12.6% of the total fluorescence at approximately 4 μM of iAβ (Fig. 1B). Non-linear regression analysis of the binding data to a rectangular hyperbola allowed calculation of a relative dissociation constant of 75.9±6.5 nM.

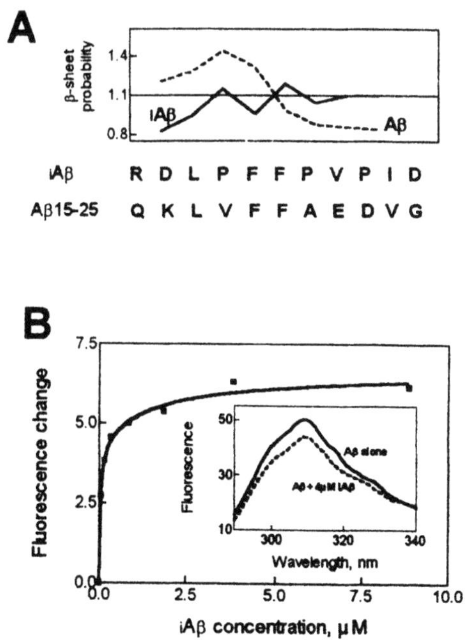

FIGURE 1. Sequence of iAβ and interaction Aβ- iAβ.
A: amino acid sequence and β-sheet probability for iAβ and for the region Aβ used as a template for iAβ.

B. iAβ binding to Aβ as characterized by fluorescence spectroscopy. The inset shows the fluorescence spectra of Aβ incubated alone or in the presence of 4μm of iAβ.

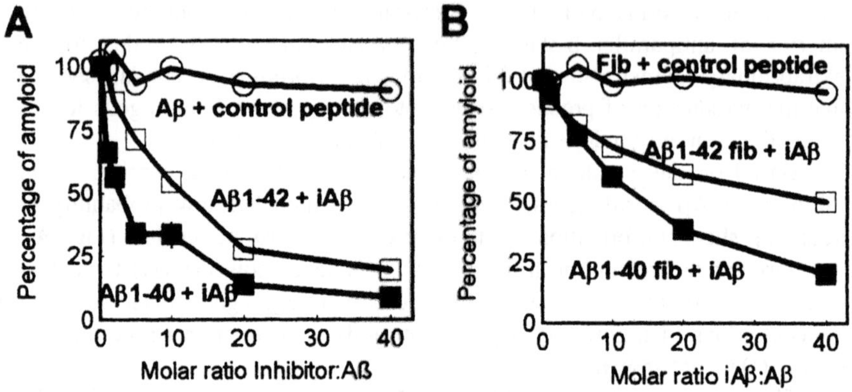

FIGURE 2. Inhibition of Aβ fibrillogenesis and dissolution of preformed fibrils by iAβ *in vitro*. A: Dose-dependent inhibition of Aβ1-40 and Aβ1-42 fibrillogenesis by iAβ. The Aβ concentration was 1 mg/ml and the incubation time 24 hr. B: Fibril dissolution induced by different molar ratios of iAβ after 24 hr of incubation.

Inhibition of Aβ amyloid formation and dissolution of preformed fibrils in vitro

The quantitative evaluation of the effect of iAβ on *in vitro* Aβ fibrillogenesis was based on a fluorometric assay that measures thioflavine T (ThT) fluorescence emission (Soto et al., 1995). Figure 2A shows the influence of different concentrations of iAβ on fibrillogenesis of the two major variants of Aβ (Aβ1-40 and Aβ1-42). iAβ inhibited in a dose-dependent manner *in vitro* amyloid formation by both Aβ variants. After 24 hr of incubation in the presence of a 5-fold or 20-fold molar excess of iAβ, Aβ1-40 formed only 33.9% and 13.7%, respectively, of the amyloid detectable in the absence of inhibitor (Fig. 2A). Although the inhibitor is less efficient with Aβ1-42, a 5-, 20-, and 40-fold molar excess of iAβ over Aβ1-42 resulted in a 28.7%, 72.3% and 80.6% of inhibition respectively (Fig. 2A). Several non-related peptides had no effect on fibrillogenesis or slightly increased Aβ amyloid formation, probably by incorporation into the fibrils.

In order to evaluate the ability of iAβ to disassemble preformed Aβ fibrils, Aβ1-40 or Aβ1-42 (1 mg/ml) were preincubated for 5 days at 37°C before the addition of inhibitor peptide. Fig. 2B shows the dissolution of Aβ1-40 or Aβ1-42 fibrils after 24hr incubation with different iAβ concentrations. The inhibitor efficiently affected disaggregation of Aβ1-40 fibrils, achieving almost complete dissolution when used in a 40-fold molar excess. Conversely, only ~51% of Aβ1-42 fibril reduction was obtained with the same molar excess of iAβ (Fig 2B). The maximum level of fibril dissolution was obtained after 2 days of incubation with iAβ and remained unaltered thereafter. The inhibition of fibril formation and the dissolution of preformed fibrils by iAβ was also analyzed by negative-staining electron microscopy, giving similar results (Data not shown).

Inhibition of in vivo fibrillogenesis using a mouse model of amyloidosis related to amyloid-A

We used a well-characterized mouse model for systemic amyloid-A (AA) deposition. This model has been used to test the role of amyloid-associated components such as proteoglycans and apolipoprotein E (Snow et al., 1991; Kindy et al., 1995) and to test inhibitors of amyloid deposition *in vivo* (Merlini et al., 1995; Kisilevsky et al., 1995). Secondary or reactive amyloidosis is an inflammation-associated disorder in which AA protein is deposited in several organs. The AA protein is a 76 residues N-terminal fragment derived from proteolysis of a precursor called serum amyloid A (SAA) protein. Experimental amyloidosis in mice was induced by injection of amyloid enhancing factor (AEF) and silver nitrate. Under these conditions, the animals developed amyloid deposits in the spleen after 36-48 hr of the injection. We examined the effect of iAβ on AA amyloid formation after 5 days. When 5 mg of iAβ were injected together with AEF, after 24 hr of preincubation, the area occupied by amyloid in the spleen was decreased in approximately 86.4% in comparison with the animals treated without the inhibitor (Table 1).

Animal	AEF/SN	Control* untreated	AEF/SN** +iAβ
1	29.4	0.05	4.35
2	31.65	0.1	3.68
3	32.97	0.21	5.32
4	30.77	0.04	3.67
Average ± SE	31.2± 0.75	0.10± 0.04	4.26± 0.39

TABLE I. Effect of iAβ on *in vivo* amyloid deposition, using animal model of amyloidosis related to amyloid A. Amyloid was induced by injections of Amyloid Enhancing Factor (AEF) and silver nitrate (SN). BALB/c mice were sacrificed after 5 days and amyloid quantitated by immunohistochemistry and image analysis. *Represent the group of animals non-treated with AEF and SN. **AEF was preincubated with 5 mg of iAβ for 24 hr and both injected together into the mice plus SN.

DISCUSSION

The results of this study support the concept that the formation of a β-sheet secondary structure is important for fibrillogenesis. A short peptide partially homologous to the central hydrophobic region of Aβ (residues 17-21), and containing amino acids which block the adoption of a β-sheet structure, binds Aβ, inhibits amyloid formation *in vitro* and partially dissolves preformed Aβ fibrils. Furthermore, the inhibitor is able to block the *in vivo* deposition of AA in the mouse model of amyloidosis.

We postulate that iAβ inhibits amyloid formation by binding to monomeric Aβ peptides, thereby blocking the formation of the oligomeric β-sheet conformation precursor of the fibrils. The dissolution of preformed fibrils induced by iAβ may indicate that the monomeric peptide is in equilibrium with the fibrils, as previously suggested (Maggio et al., 1992). A molar excess of the inhibitor may bind to monomeric peptide thus displacing the equilibrium, and leading to fibril disaggregation. However, we can not rule out direct iAβ binding to fibrils leading to amyloid destabilization.

Although short peptides have been used extensively as drugs in medicine, they commonly exhibit two undesirable properties; namely, proteolytic degradation and generation of immune response. These problems can be minimized by: designing peptide derivatives (such as cyclic peptides or peptides containing D-amino acids) or by generating a peptide mimetic molecule. Preliminary results indicate that a peptide with the same sequence of iAβ but containing D-amino acids has similar ability as iAβ to inhibit Aβ amyloid formation.

We propose that our inhibitor peptide or its derivatives, if proven non-toxic, non-degradable and capable of crossing the blood-brain barrier, may be used as therapeutic agents to prevent and/or retard amyloidogenesis in AD. This approach is an alternative to the most common and popular approach, namely the inhibition of βPP processing pathway II given rise to heterogeneous Aβ peptides.

ACKNOWLEDGMENTS

This research was supported by NIH grants AG05891, AG08721, AG10953 (LEAD), AG09690, AG12981 and NS31220.

REFERENCES

Corder EH, Saunders AM, Strittmatter WJ, Schmechel DE, Gaskell PC et al. (1993): Gene dose of apolipoprotein E type 4 allele and the risk of Alzheimer's disease in late onset families. *Science* 261:921-923.

Kindy MS, King AR, Perry G, De Beer MC and De Beer FC (1995): Association of apolipoprotein E with murine amyloid A protein amyloid. *Lab Invest* 73:469-475.

Kisilevsky R, Lemieux LJ, Fraser PE, Kong X, Hultin PG et al. (1995): Arresting amyloidosis *in vivo* using small-molecule anionic sulphonates or sulphates: implications for Alzheimer's disease. *Nature Med* 1:143-148.

Lorenzo A and Yankner BA (1994): Beta-amyloid neurotoxicity requires fibril formation and is inhibited by congo red. *Proc Natl Acad Sci USA* 91:12243-12247.

Ma J, Yee A, Brewer HB, Jr., Das S and Potter H (1994): Amyloid-associated proteins alpha 1-antichymotrypsin and apolipoprotein E promote assembly of Alzheimer beta-protein into filaments. *Nature* 372:92-94.

Maggio JE, Stimson ER, Ghilardi J, Allen CJ, Dahl CE et al. (1992): Reversible *in vitro* growth of Alzheimer's disease β-amyloid plaques by deposition of labeled amyloid peptide. *Proc Natl Acad Sci USA* 89:5462-5466.

Matsubara E, Soto C, Governale S, Frangione B and Ghiso J (1996): Apolipoprotein J and Alzheimer amyloid solubility. *Biochem J* (In Press)

Merlini G, Ascari E, Amboldi N, Belloti V, Arbustini E et al. (1995): Interaction of the anthracycline 4'-iodo-4'-deoxydoxorubicin with amyloid fibrils: Inhibition of fibrillogenesis. *Proc Natl Acad Sci USA* 92:2959-2964.

Naslund J, Thyberg J, Tjernberg LO, Wernstedt C, Karlstrom AR et al. (1995): Characterization of stable complexes involving apolipoprotein E and the amyloid beta peptide in Alzheimer's disease brain. *Neuron* 15:219-228.

Pike CJ, Burdick D, Walencewicz AJ, Glabe CG and Cotman CW (1993): Neurodegeneration induced by beta-amyloid peptides *in vitro*: the role of peptide assembly state. *J Neurosci* 13:1676-1687.

Sisodia SS and Price DL (1995): Role of the β-amyloid protein in Alzheimer's disease. *FASEB J* 9:366-370.

Snow AD, Bramson R, Mar H, Wight TN and Kisilevsky R (1991): A temporal and ultrastructural relationship between heparan sulphate proteoglycans and AA amyloid in experimental amyloidosis. *J Histochem Cytochem* 39:1321-1330.

Soto C, Castaño EM, Frangione B and Inestrosa NC (1995): The alpha-helical to beta-strand transition in the amino-terminal fragment of the amyloid beta-peptide modulates amyloid formation. *J Biol Chem* 270:3063-3067.

Soto C, Golabek AA, Wisniewski T and Castaño EM (1996): Amyloid β-peptide is conformationally modified by apolipoprotein E *in vitro*. *Neuroreport* (In Press)

Wisniewski T, Castaño EM, Golabek A, Vogel T and Frangione B (1994a): Acceleration of Alzheimer's fibril formation by apolipoprotein E *in vitro*. *Am J Pathol* 145:1030-1035.

Wisniewski T, Ghiso J and Frangione B (1994b): Alzheimer's disease and soluble A. *Neurobiol Aging* 15:143-152.

Wisniewski T, Lalowski M, Golabek A, Vogel T and Frangione B (1995): Is Alzheimer's disease an apolipoprotein E amyloidosis? *Lancet* 345:956-958.

Wisniewski T and Frangione B (1992): Apolipoprotein E: a pathological chaperone protein in patients with cerebral and systemic amyloid. *Neurosci Lett* 135:235-238.

Wood JD, Wetzel R, Martin JD and Hurle MR (1995): Prolines and amyloidogenicity in fragments of the Alzheimer's peptide β/A4. *Biochem* 34:724-730.

Wouters MA and Curmi PMG (1994): An analysis of side chain interactions and pair correlations within antiparallel beta-sheets: the differences between backbone hydrogen-bonded and non-hydrogen bonded residue pairs. *Protein Sci* 3:43S.

Alzheimer Disease: From Molecular Biology to Therapy
edited by R. Becker and E. Giacobini

Apoptosis and Alzheimer's Disease: A Self-Assembly and the Initiation of Apoptosis by Plasma Membrane Receptor Cross-Linking

C.W. Cotman, D. H. Cribbs, and J. H. Su
Institute of Brain Aging and Dementia,
University of California, Irvine

The discovery that β-amyloid (Aβ) has a powerful biological activity in cultured neurons and other cells has greatly stimulated many new lines of investigation. Only a few years ago it was generally thought that Aβ itself was metabolically inert. It is now clear that Aβ places neurons at risk for injury, damages neuronal processes, and is one of the stimuli that initiates cell death by apoptosis. Aβ can also affect signal transduction processes in non-neuronal cells and thereby reprogram the microenvironment of the brain in the course of aging and neurodegenerative diseases.

In this chapter, we will briefly discuss the hypothesis that Aβ has an active role in neuronal injury and cell death. Specifically, we suggest that Aβ acts on neurons in a manner analogous to a superantigen, and causes a form of programmed cell death, termed activation-induced cell death, through the cross-linking of membrane receptors. We further critique the evidence that Aβ is one of the mechanisms which places neurons at risk and initiates their degeneration during aging. This hypothesis is consistent with our recent evidence showing that the accumulation of Aβ in the brain correlates with the decline of function with age as measured by global rating scales (Cummings and Cotman, 1995).

THE TRANSFORMATION OF Aβ INTO POLYMERIZED AND AGGREGATED STRUCTURES ALTERS ITS BIOLOGICAL ACTIVITY

Several years ago we suggested that senile plaques stimulated an abberent sprouting of neurites preceding their degeneration (Geddes et al., 1985). This suggested that plaques contain a substance, possibly Aβ, that can stimulate growth and contribute to process and cellular degeneration. Initial investigations into the action of Aβ *in vitro* revealed that Aβ peptides stimulate process outgrowth and enhance survival over short time intervals in cultured hippocampal neurons (Whitson et al., 1989; Yankner et al., 1990). Concurrent studies paradoxically showed that these peptides can also enhance neuronal death in response to excitotoxins (Koh et al., 1990) and induce neurodegeneration in culture (Yankner et al., 1989, 1990; Pike et al., 1991a, b; Behl et al., 1992), suggesting that Aβ peptides are capable of exerting multiple bioactivities, i.e. enhancing growth or inducing toxicity. Further studies have clarified this issue, demonstrating that the *in vitro* activity of Aβ peptides is dependent on the assembly state of these peptides (Pike et al., 1991a, b; 1993; Busciglio et al., 1992; Mattson et al., 1993).

As Aβ ages, it polymerizes into higher order structures, similar to those observed in the Alzheimer brain, and in this form, stimulates neuronal cell death. After incubation (aging) of synthetic $A\beta_{1-42}$ peptides for several days *in vitro,* sheet-like structures are visible at the light microscopic level and an altered electrophoresis profile is evident on reducing gels, indicating that insoluble aggregates have formed (Pike et al., 1991a; Burdick et al., 1992). Aβ peptides that exhibit aggregation as assessed by electrophoresis or sedimentation assays demonstrate toxicity in cultured neurons, whereas Aβ peptides that do not exhibit an aggregated state by these measures do not exhibit toxicity (Pike et al., 1991b; 1992).

There are two mechanisms by which Aβ may initiate degeneration: a necrotic or apoptotic pathway. Several lines of morphological and biochemical data provide strong evidence for the hypothesis that Aβ induces degeneration in primary neuronal cultures via an apoptotic or programmed cell death (PCD) pathway (Loo et al., 1993; Watt et al., 1994), a finding confirmed by (Forloni et al., 1993).

Aβ AND ACTIVATION-INDUCED PCD

A common feature of extracellular activators of PCD is that they involve interactions with membrane proteins that either produce a novel signal that leads to PCD (Dellabona et al., 1990; Marrack and Kappler, 1990; Nagata and Golstein, 1995) or disrupt a constitutive cellular survival signal required to suppress an intrinsic cell suicide program (Raff et al., 1993). The loss of a constitutively required signal can occur via a reduction in the level of soluble trophic factors or a disruption of cell-cell or cell-extracellular matrix interactions (Raff et al., 1993; Ruoslahti and Reed, 1994). The active generation of a death signal is dependent on the cross-linking of specific membrane receptors by multivalent ligands, which initiates a signal transduction event that triggers PCD, and is thus called activation-induced cell death (Shi et al., 1989; Lenardo, 1991; Radvanyi et al., 1993). A number of well-characterized extracellular agents that bind to the cell surface have been reported to initiate activation-induced cell death, and in several cases, the receptor involved has been identified, e.g. bacterial superantigens (Dellabona et al., 1990; Marrack and Kappler, 1990), the Fas ligand (Nagata and Golstein, 1995), and certain viruses (Banda et al., 1992).

Aβ-initiated PCD may be another example of activation-induced cell death. This hypothesis is based on several characteristics of Aβ-initiated apoptosis. Aβ induces cell death only when in a polymerized or fibrillar β-sheet assembly state (Loo et al., 1993; Cotman and Anderson, 1995) and when bound to neuronal surfaces, suggesting that Aβ may initiate activation-induced cell death in a manner analogous to superantigens and Fas receptor cross-linking. To begin to address this hypothesis, we have identified a reagent that can cross-link neuronal membrane receptors and have used this reagent to examine the cellular response.

CONCANAVALIN A CAUSES THE CROSS LINKING OF MEMBRANE RECEPTORS AND INITIATES ACTIVATION-INDUCED CELL DEATH

Although specific activated cell death-linked receptors have not yet been identified in neurons, lectins are capable of binding and cross-linking many different glycosylated membrane receptors on cells, activating the receptors and causing a wide variety of cellular responses, including the induction of cell death (Stanley, 1981; Janssen et al., 1993; Kim et al., 1993). We have examined the response of neurons to the lectin concanavalin A (Con A). Because neurons contain a high density of receptors for this lectin, it causes receptor cross-linking (Cotman and Taylor, 1974) and it is possible to examine binding with and without cross-linking by chemically modifying the Con A.

We have found that Con A is a powerful stimulus of cell death in neurons. It produces measurable cell death and morphological changes at concentrations greater than 10 nM, and at 100 nM, and induces massive cell death in neuronal cultures. Con A-induced neuronal death exhibits many of the hallmarks associated with PCD, such as membrane blebbing, nuclear condensation and margination, and internucleosomal DNA cleavage (Cribbs et al., 1996).

Succinylated Con A is not as effective at producing cell death as native Con A. Succinylated Con A is a stable dimer and is much less effective at forming membrane aggregates of Con A receptors than the native Con A. Native Con A is a tetramer and binds with other tetramers to induce receptor aggregration. Native Con A caused a clustering of Con A receptors in agreement with previous findings (Cotman and Taylor, 1974). In contrast, succinylated Con A does not produce receptor clusters or significant cell death even at micromolar concentrations. Thus, the subsequent downstream intracellular signaling that initiates PCD is blocked when the cross-linking of membrane receptors to form receptor clusters is blocked. These results suggest that receptor cross-linking appears to be a necessary part of the signal that initiates Con A-induced PCD.

The Con A model shares certain features with Aβ-induced PCD. The time courses for both the Con A- and Aβ-induced morphological changes and the subsequent PCD are comparable (Pike et al., 1993). Although most protocols for Aβ-induced cell death now employ pre-assembled peptide, Aβ in the soluble form will also cause cell death over a much extended time course. Shortly prior to cell death, the surface of the neurons exhibits clusters of the peptide (Busciglio et al., 1992; Pike et al., 1992), not unlike that observed with Con A.

Taken together, these findings raise the possibility that both Con A and Aβ act through a cross-linking of cell surface molecules to initiate activation-induced cell death. Neurons may be particularly sensitive to disturbances on the surface and to cross-linking of specific or even non-specific components.

DOES APOPTOSIS OCCUR IN THE ALZHEIMER BRAIN

Studies by us and others using cell culture models predict the existence of neuronal apoptosis in the Alzheimer brain. Our strategy has been to employ multiple markers to evaluate this hypothesis, because it is unlikely that one marker by itself is sufficient, and since multiple indices are necessary to dissect the relevent specific pathways involved.

It is clear that the Alzheimer brain shows extensive DNA damage in some neurons, nuclear morphology consistent with apoptosis (Su et al., 1994), and the presence and co-expression in some cells of c-jun (Anderson et al., 1996) as predicted from culture experiments (Anderson et al., 1995). Importantly, while many of the cells labelled by terminal deoxynucleotidyl transferase (TdT) in the Alzheimer brain exhibit clear morphological characteristics of apoptosis, studies in Alzheimer tissue reveal over half of the neurons have TdT-labelled nuclei in the entorhinal cortex/hippocampal formation. This is unexpected since once damage is initiated it would be expected that the cells would rapidly proceed into PCD and be removed in a few days. Therefore, the extensive labelling is either an artifact or indicates the existence of an unusual process.

If the large increase in TdT labelling in Alzheimer's disease (AD) is related to disease pathology and not a non-specific artifact, TdT labelling would be predicted to be found in brain areas typically exhibiting AD pathology and neuronal cell loss such as the entorhinal cortex, but not in areas in which such pathology is generally absent such as cerebellum. In general, TdT labelling is absent or present at very low levels in the cerebellum as compared to the entorhinal cortex of the Alzheimer cases examined (Anderson et al., 1996). These findings are consistent with previous results regarding ultrastructural neuronal injury and the formation of AD pathology (Yamazaki et al., 1992; Li et al., 1994), supporting the hypothesis that TdT labelling reflects AD-related pathological processes and neuronal loss. Furthermore, comparison of conventional post-mortem control and Alzheimer tissue with rapid autopsy tissue did not reveal differences in TdT labelling, again supporting the consistency and validity of TdT labelling of tissue with the post-mortem delays in these studies (Anderson et al., 1996). Thus, the extensive labelling does not appear to be an artifact.

Alternatively, the accumulation of DNA damage may be the result of deficient DNA repair, an unexpectedly slow course of PCD such as delayed clearing, or the induction of a set of compensatory mechanisms to arrest the program. In view of the large numbers of neurons showing DNA damage, it is unlikely that the clearing response is the only factor involved. The initiation of compensatory mechanisms in response to such events may be predicted to contribute to an active and prolonged process of DNA-damage/DNA-repair and cell death.

We have previously shown that overexpression of the cell death inhibitory protein bcl-2 blocks apoptosis induced by Aβ in cultured neurons (Cribbs et al, submitted). To test the hypothesis that compensatory mechanisms inhibiting cell death may be initiated in the Alzheimer brain, immunoreactivity for bcl-2 was examined in human post-mortem tissue and found to be elevated in Alzheimer brain (Satou et al., 1995). Furthermore, there is a strong, nearly 1 to 1 colocalization between bcl-2 and TdT labelling in AD brain (Su et al., 1996), suggesting that bcl-2 is induced in neurons that exhibit DNA damage.

It may be that bcl-2 and related mechanisms arrest the program and allow cells to remain viable. In any event, the presence of DNA damage may be one of the earliest markers of neuronal abnormality in the AD brain, and correspondingly, indicates an increased vulnerability of neuronal cells to a variety of insults.

CONCLUSION

Recent research has provided very clear evidence that a transition in Aβ peptides to higher order conformations leads to the accumulation of the product and the emergence of a new biological activity. In the case of Aβ, this conformation has self-assembly characteristics that, once initiated, appear to generate further Aβ accumulation (Cribbs et al., 1995) and a cascade that places neurons at risk and, over time, contributes to their degeneration. The critical factor is the generation of higher order protein assemblies which have biological activity and can contribute to tissue dysfunction and degeneration. In this respect, the Aβ model is not unique to the Aβ-peptide, but has parallels to prion disease and other forms of amyloidosis, such as pancreatic islet disease and amyloidosis in the cardiac or vascular system (Taubes, 1996). Taken together, Aβ accumulation appears to be a member of a family of protein conformation-dependent disorders that the organism must face as it ages. These disorders appear to be both genetically- and risk factor-based making their etiology difficult to characterize, but at the same time, increasing the possibility of points of therapeutic intervention.

REFERENCES

Anderson AJ, Pike CJ and Cotman CW (1995): Differential induction of immediate early gene proteins in cultured neurons by beta-amyloid (Aβ): association of c-jun with αβ-induced apoptosis. *J Neurochem* 65:1487-1498.

Anderson AJ, Su JH and Cotman CW (1996): DNA damage and apoptosis in Alzheimer's disease: colocalization with c-Jun immunoreactivity, relationship to brain area, and effect of postmortem delay. *J Neurosci* 16:1710-1719.

Banda NK, Bernier J, Kurahara DK, Kurrle R, Haigwood N, Sekaly R-P and Finkel TH (1992): Crosslinking CD4 by human immunodeficiency virus gp120 primes T cells for activation-induced apoptosis. *J Exp Med* 176:1099-1106.

Behl C, Davis J, Cole GM and Schubert D (1992): Vitamin E protects nerve cells from amyloid beta protein toxicity. *Biochem Biophys Res Commun* 186:944-950.

Burdick D, Soreghan B, Kwon M, Kosmoski J, Knauer M, Henschen A, Yates J, Cotman C and Glabe C (1992): Assembly and aggregation properties of synthetic Alzheimer's A4/β amyloid peptide analogs. *J Biol Chem* 267:546-554.

Busciglio J, Lorenzo A and Yankner BA (1992): Methodological variables in the assessment of beta amyloid neurotoxicity. *Neurobiol Aging* 13:609-612.

Cotman CW and Anderson AJ (1995): A potential role for apoptosis in neurodegeneration and Alzheimer's disease. *Mol Neurobiol* 10:19-45.

Cotman CW and Taylor D (1974): Localization and characterization of concanavalin A receptors in the synaptic cleft. *J Cell Biol* 62:236-242.

Cribbs DH, Davis-Salinas J, Cotman CW and Van Nostrand WE (1995): Aβ induces increased expression and processing of amyloid precursor protein in cortical neurons. *Alzheimer's Research (*in press).

Cribbs DH, Kreng VM, Anderson AJ and Cotman CW (1996): Crosslinking of membrane glycoproteins by Concanavalin A induces apoptosis in cortical neurons. *Neuroscience (*in press).

Cummings BJ and Cotman CW (1995): Image analysis of beta-amyloid load in Alzheimer's disease and relation to dementia severity. *Lancet* 346:1524-1528.

Dellabona P, Peccoud J, Kappler J, Marrack P, Benoist C and Mathis D (1990): Superantigens interact with MHC class II molecules outside of the antigen groove. *Cell* 62:1115-1121.

Forloni G, Chiesa R, Smiroldo S, Verga L, Salmona M, Tagliavini F and Angeretti N (1993): Apoptosis mediated neurotoxicity induced by chronic application of beta amyloid fragment 25-35. *Neuroreport* 4:523-526.

Geddes J, Monaghan D, Cotman C, Lott I, Kim R and Chui H (1985): Plasticity of hippocampal circuitry in Alzheimer's disease. *Science* 230:1179-1181.

Janssen O, Scheffler A and Kabelitz D (1993): In vitro effects of mistletoe extracts and mistletoe lectins. Cytotoxicity towards tumor cells due to the induction of programmed cell death (apoptosis). *Arzneimittelforschung* 43:1221-1227.

Kim M, Rao MV, Tweardy DJ, Prakash M, Galili U and Gorelik E (1993): Lectin-induced apoptosis of tumour cells. *Glycobiology* 3:447-453.

Koh JY, Yang LL and Cotman CW (1990): β-amyloid protein increases the vulnerability of cultured cortical neurons to excitotoxic damage. *Brain Res.* 533:315-320.

Lenardo MJ (1991): Interleukin-2 programs mouse alpha beta T lymphocytes for apoptosis. *Nature* 353:858-861.

Li YT, Woodruff PD and Trojanowski JQ (1994): Amyloid plaques in cerebellar cortex and the integrity of Purkinje cell dendrites. *Neurobiol Aging* 15:1-9.

Loo DT, Copani AG, Pike CJ, Whittemore ER, Walencewicz AJ and Cotman CW (1993): Apoptosis is induced by beta-amyloid in cultured central nervous system neurons. *Proc Natl Acad Sci USA* 90:7951-7955.

Marrack P and Kappler J (1990): The staphylococcal enterotoxins and their relatives [published erratum appears in Science 1990 Jun 1;248(4959)]:1066 [see comments]. *Science,* 248: 705-711.

Mattson MP, Tomaselli KJ and Rydel RE (1993): Calcium-destabilizing and neurodegenerative effects of aggregated β-amyloid peptide are attenuated by basic FGF. *Brain Res* 621:35-49.

Nagata S and Golstein P (1995): The Fas death factor. *Science* 267:1449-1456.

Pike CJ, Walencewicz AJ, Glabe CG and Cotman CW (1991a): Aggregation-related toxicity of synthetic β-amyloid protein in hippocampal cultures. *Euro J Pharm* 207:367-368.

Pike CJ, Walencewicz AJ, Glabe CG and Cotman CW (1991b): In vitro aging of β-amyloid protein causes peptide aggregation and neurotoxicity. *Brain Res* 563:311-314.

Pike CJ, Cummings BJ and Cotman CW (1992): β-amyloid induces neuritic dystrophy in vitro: similarities with Alzheimer pathology. *Neuroreport* 3:769-772.

Pike CJ, Burdick D, Walencewicz A, Glabe CG and Cotman CW (1993): Neurodegeneration induced by β-amyloid peptides in vitro: the role of peptide assembly state. *J Neurosci* 13:1676-1687.

Radvanyi LG, Mills GB and Miller RG (1993): Religation of the T cell receptor after primary activation of mature T cells inhibits proliferation and induces apoptotic cell death. *J Immunol* 150:5704-5715.

Raff MC, Barres BA, Burne JF, Coles HS, Ishizaki Y and Jacobson MD (1993): Programmed cell death and the control of cell survival: lessons from the nervous system. *Science* 262:695-700.

Ruoslahti E and Reed JC (1994): Anchorage dependence, integrins, and apoptosis. *Cell* 77:477-478.

Satou T, Cummings BJ and Cotman CW (1995): Immunoreactivity for BCL-2 protein within neurons in the Alzheimer's disease brain increases with disease severity. *Brain Res* 697:35-43.

Shi YF, Sahai BM and Green DR (1989): Cyclosporin A inhibits activation-induced cell death in T-cell hybridomas and thymocytes. *Nature* 339:625-626.

Stanley P (1981): Surface Carbohydrate Alterations of Mutant Mammalian Cells Selected for Resistance to Plant Lectins. In: *The Biochemistry of Glycoproteins and Proteoglycans.* Lennarz, WJ New York: Plenum Press 161-190.

Su JH, Anderson AJ, Cummings BJ and Cotman CW (1994): Immunohistochemical evidence for DNA fragmentation in neurons in the AD brain. *Neuroreport* 5:2529-2533.

Su JH, Satou T, Anderson AJ and Cotman CW (1996): Up-regulation of BCL-2 is associated with neuronal DNA damage in Alzheimer's disease. *Neuroreport* 7:437-440.

Watt J, Pike CJ, Walencewicz AJ and Cotman CW (1994): Ultrastructural analysis of β-amyloid-induced apoptosis in cultured hippocampal neurons. *Brain Res* 661:147-156.

Whitson JS, Selkoe DJ and Cotman CW (1989): Amyloid beta protein enhances the survival of hippocampal neurons in vitro. *Science* 243:1488-1490.

Yamazaki T, Yamaguchi H, Nakazato Y, Ishiguro K, Kawarabayashi T and Hirai S (1992): Ultrastructural characterization of cerebellar diffuse plaques in Alzheimer's disease. *J Neuropathol Exp Neurol* 51:281-286.

Yankner BA, Dawes LR, Fisher S, Villa KL, Oster GML and Neve RL (1989): Neurotoxicity of a fragment of the amyloid precursor associated with Alzheimer's disease. *Science* 245:417-420.

Yankner BA, Duffy LK and Kirschner DA (1990): Neurotrophic and neurotoxic effects of amyloid β-protein: reversal by tachykinin neuropeptides. *Science* 250:279-282.

Alzheimer Disease: From Molecular Biology to Therapy
edited by R. Becker and E. Giacobini

NMDA RECEPTOR DYSFUNCTION IN ALZHEIMER'S DISEASE

John Olney, David Wozniak, Masahiko Ishimaru, Nuri Farber
Department of Psychiatry, Washington University School of Medicine
St. Louis, MO

INTRODUCTION

Excessive activation of NMDA receptors (NR) by endogenous glutamate (Glu) causes excitotoxic neuronal degeneration in acute CNS injury syndromes such as stroke and trauma. It has been suggested that excessive NR activation may also play a role in Alzheimer's disease (AD). We propose that if excessive NR activation plays a role in AD, it is an indirect role in which excessive NR activation destroys NR, thereby rendering the NR system hypofunctional. According to our hypothesis, when the NR system becomes sufficiently hypofunctional it unleashes a complex excitotoxic process (fueled both by acetylcholine and glutamate) which is a major contributing factor to the widespread pattern of neuronal degeneration in AD. This hypothesis is derived in part from evidence that the NR system may, in fact, be hypofunctional in AD, and that experimental induction of NR hypofunction (NRH) in rat brain (by administration of NR antagonist drugs) triggers a pattern of neuronal degeneration resembling that seen in AD.

NRH NEURODEGENERATION: ITS NATURE AND PATTERN

NR antagonist neurotoxicity was first described (Olney et al., 1989) as a reversible reaction, but it was soon learned that it can be irreversible depending on the length of time NR are maintained in a hypofunctional state. Treating an adult rat with a low dose of an NR antagonist induces brief NRH which causes specific neurons in the posterior cingulate and retrosplenial (PC/RS) cortex to develop intracytoplasmic vacuoles that are conspicuous at 4 hrs but disappear by 24 hrs after treatment. However, administering an NR antagonist in high dosage or by continuous infusion for several days induces a prolonged NRH state which causes neurons in many cerebrocortical and limbic brain regions to degenerate unto death (Corso et al., 1994; Ellison

1994, 1995). Pyramidal and multipolar neurons are preferentially affected and the full pattern of damage includes the posterior cingulate, retrosplenial, frontal, temporal, entorhinal, perirhinal, piriform and prefrontal cortices, the amygdala and hippocampus. At 4 hrs, the reaction in PC/RS cortex consists of intracytoplasmic vacuole formation, but in other brain regions a spongiform reaction featuring edematous swelling of spines on proximal dendrites is the most prominent cytopathological change (Olney et al., unpublished). At 24 to 48 hrs, the affected neurons become argyrophilic (de Olmos cupric silver method) and immunopositive for heat shock protein kDa 72 and they begin to display cytoskeletal abnormalities, including a conspicuous corkscrew deformity of their apical dendrites. In the 72 to 96 hr interval the dying neurons undergo fragmentation and elicit an inconspicuous glial response.

NRH NEURODEGENERATION: MECHANISM AND PROTECTION

In a series of recent studies (reviewed in Olney & Farber, 1995) we have found that several classes of drugs effectively block the PC/RS neurotoxic

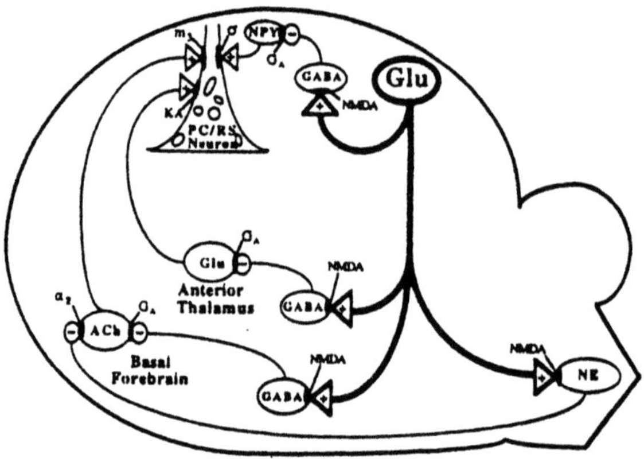

FIGURE 1. To explain NRH-induced degeneration of PC/RS neurons, we propose that Glu acting through NR on GABAergic and noradrenergic neurons maintains tonic inhibitory control over multiple excitatory pathways that convergently innervate PC/RS neurons. Systemic administration of an NR antagonist (or NRH produced by any other mechanism) would simultaneously abolish inhibitory control over multiple excitatory inputs to PC/RS neurons. This would create chaotic disruption among multiple intracellular second messenger systems, thereby causing derangement of cognitive functions subserved by the afflicted neurons, as well as eventual degeneration of these neurons. This circuit diagram focuses exclusively on PC/RS neurons. We hypothesize that a similar disinhibition mechanism and similar but not necessarily identical neural circuits and receptor mechanisms mediate damage induced in other corticolimbic brain regions by sustained NRH.

action of NR antagonists, including muscarinic receptor antagonists, $GABA_A$ receptor agonists, sigma receptor antagonists, non-NMDA glutamate receptor antagonists and α_2-adrenergic receptor agonists. These findings have provided new insight into the receptor mechanisms and neural circuitry (Fig. 1) involved in NRH neurodegeneration. Of greater potential import, if our hypothesis proves correct that an NRH mechanism triggers neuronal degeneration in AD, these findings may lead to new therapeutic or prophylactic approaches to AD.

To fully appreciate how a deficit in NR activity can trigger a neurodegenerative syndrome, it is necessary to begin thinking of Glu in a new light, as an agent that performs major inhibitory functions. By tonically activating NR on GABAergic neurons (Fig. 1), Glu regulates inhibitory tone and ordinarily protects the brain against its own self destructive potential; removing this inhibitory mechanism from certain networks unleashes excitotoxic forces that proceed to wreak self destruction within the network. The excitotoxic forces are complex and include glutamatergic, cholinergic and other less well defined components, but the key to understanding this neurotoxic process is to recognize that Glu, the master of all paradoxes, is not only an excitotoxic contributor to the pathological outcome, it is the driver of the inhibitory mechanism that normally holds the excitotoxic forces in check.

NRH: A MISSING LINK TO EXPLAIN AD?

If a deficit in NR activity can trigger widespread degeneration of corticolimbic neurons, it becomes important to inquire whether there are any factors operative in AD that might promote a deficiency in NR activity. The first major factor to consider is age. Several laboratories (Wenk et al., 1991; Tamaru et al., 1991; Gonzales et al., 1991; Magnusson and Cotman, 1993) have reported that the NR transmitter system in both rodents and primates becomes increasingly hypofunctional with advancing age. Thus, in the aging brain the stage is already set for widespread corticolimbic neurodegeneration to occur, and all that is required to explain why it occurs more ravagingly in the AD brain than in the "normal" aging brain is to identify one or more adjunctive conditions peculiar to the AD brain that may serve as catalysts.

Adjunctive conditions deserving priority attention include genetic factors that might promote amyloidopathy (unfavorable apoE genotype, Familial AD-linked mutations) and disturbances in energy metabolism. Both amyloid-opathy and defective energy metabolism are conducive to pathological activation of NR and excitotoxic degeneration of neurons bearing these receptors. Since this would result in loss of NR, the net effect would be decreased function of the NR system, i.e., NRH. Adding the NRH generated by these adjunctive conditions to the NRH of aging would cause the NRH index to exceed a critical value required for triggering widespread corticolimbic neurodegeneration. A logical priority for future research would be to explore specific mechanisms by which increased amyloidopathy (genetically driven) might act to convert the NRH condition present in the "normal" aging brain from a low grade simmering process to the more fulminating and widespread corticolimbic neurodegenerative syndrome that characterizes AD.

To what extent is the pattern of neurodegeneration unleashed by NRH relevant to the pattern seen in AD? The pathological changes are distributed diffusely throughout many cerebrocortical and limbic regions of the forebrain in both AD and NRH neurodegeneration. More specifically, the brain region most vulnerable to NRH degeneration (Fix et al., 1995), the PC/RS cortex, was recently described by Insausti et al. (1993) and Minoshima et al. (1994) as a region that is selectively affected early and sustains very severe damage in AD; interestingly, the anterior cingulate is relatively spared in both cases. Other brain regions preferentially affected by NRH (temporal, perirhinal, entorhinal and insular cortices) are among those most severely involved in AD. The neurons primarily involved in either case are specific pyramidal or multipolar neurons often distributed in a bilaminar pattern.

How does the cytopathology of AD compare with cytopathological changes induced by NRH? Key cytopathological processes that characterize AD as a distinctive disease entity are neurofibrillary tangles (NFT), amyloid plaques and loss of neurons and synaptic complexes. The cytopathological process described herein involving edematous excitotoxic degeneration of dendritic spines is a mechanism by which many synaptic complexes can be deleted from corticolimbic regions of the brain, and this can occur independent of neuronal loss or as a premonitory injurious antecedent to neuronal loss. Hyper-phosphorylation of tau microtubule-associated protein is thought to be the mechanism underlying NFT formation. The mechanism by which NRH injures neurons involves excessive activation of muscarinic (m1 or m3) receptors on the neuron that sustains injury. Since excessive stimulation of m1/m3 receptors excessively activates a protein kinase C second messenger system coupled to these receptors, and a major function of protein kinase C is to phosphorylate protein, it follows that sustained excessive m1/m3 receptor activity might lead to hyperphosphorylation of tau protein and thus to NFT formation. Whether the peculiar corkscrew deformity that occurs early in the degenerative process triggered by NRH reflects a destabilizing influence on microtubule proteins and, therefore, represents a cytoskeletal disturbance relevant to NFT mechanisms warrants further study. It is thought that HSP-72 functions as a chaperone protein that intercedes to correct abnormal protein folding. Excessive expression of HSP-72, which occurs in degenerating neurons both in the NRH model and in the AD brain (Hamos et al., 1991), might be interpreted in either case as a response to abnormal processing of microtubule proteins. We have seen no evidence of amyloid plaque formation in the NRH animal model, which is consistent with the assumption that genetic factors are primarily responsible for the amyloidopathy in AD. Our findings are consistent with the conclusion that NRH by itself can produce many of the neuropathological features of AD, and amyloidopathy (genetically driven) plus NRH can explain the entire AD syndrome.

SUMMARY

Here we have described a novel excitotoxic process potentially relevant to AD in which hypofunctional NMDA receptors cease driving GABAergic neurons which cease inhibiting the two major excitatory transmitters in the brain (glutamate and acetylcholine), and these disinhibited excitatory transmitters then act in concert to slowly hyperstimulate neurons in corticolimbic brain regions to death. Accepting this hypothesis featuring a multisystem network disturbance that can explain how a corticolimbic pattern of neurodegenerative events is triggered in AD does not require abandoning other candidate mechanisms. For example, currently popular genetic hypotheses that focus primarily on amyloid processing or deposition are compatible with our hypothesis and would gain in explanatory power if combined with our hypothesis.

ACKNOWLEDGMENTS

Supported in part by NIMH Research Scientist Award MH 33894 (JWO), Scientist Development Award for Clinicians DA 00290 (NBF) and a grant from the National Institute of Aging, AG 11355 (JWO, DFW).

REFERENCES

Corso TD, Wozniak DF, Sesma MA and Olney JW (1994): Neuron necrotizing properties of phencyclidine. *Soc Neurosci Abstr* 20:1531.

Ellison G (1994): Competitive and non-competitive NMDA antagonists induce similar limbic degeneration. *Neuroreport* 5:2688-2692.

Ellison G (1995): The N-methyl-D-aspartate antagonists phencyclidine, ketamine and dizocilpine as both behavioral and anatomical models of the dementias. *Brain Res Rev* 20:250-267.

Fix AS, Wozniak DF, Truex LL, McEwen M, Miller JP and Olney JW (1995): Quantitative analysis of factors influencing neuronal necrosis induced by MK-801 in the rat posterior cingulate/retrosplenial cortex. *J Brain Res* 696:194-204.

Gonzales RA, Brown LM, Jones TW, Trent RD, Westbrook SL and Leslie SW (1991): N-Methyl-D-Aspartate mediated responses decrease with age in Fischer 344 rat brain. *Neurobiol Aging* 12:219-225.

Hamos JE, Oblas B, Pulaski-Salo D, Welch WJ, Bole DG and Drachman DA (1991): Expression of heat shock proteins in Alzheimer's disease. *Neurol* 41:345-350.

Insausti R, Morena M, Insausti A, Ochoa L and Gonzalo LM (1993): The posterior cingulate cortex in man. Normal structure and changes in Alzheimer's disease. *Soc. Neurosci Abstr* 19:357.

Magnusson KR and Cotman CW (1993): Age-related changes in excitatory amino acid receptors in two mouse strains. *Neurobiol Aging* 14:197-206.

Minoshima S, Foster NL and Kuhl DE (1994): Posterior cingulate cortex in Alzheimer's disease. *The Lancet* 344:895.

Olney JW (1995): NMDA receptor hypofunction, excitotoxicity and Alzheimer's Disease. *Neurobiol Aging* 16:459-461.

Olney JW and Farber NB (1995): Glutamate receptor dysfunction and Schizophrenia. *Arch Gen Psych* 52:998-1009.

Olney JW, Labruyere J and Price MT (1989): Pathological changes induced in cerebrocortical neurons by phencyclidine and related drugs. *Science* 244:1360-1362.

Olney JW, Labruyere J, Wang G, Sesma MA, Wozniak DF and Price MT (1991): NMDA antagonist neurotoxicity: Mechanism and Protection. *Science* 254:1515-1518.

Tamaru M, Yoneda Y, Ogita K, Shimizu J and Nagata Y (1991): Age-related decreases of the N-methyl-D-aspartate receptor complex in the rat cerebral cortex and hippocampus. *Brain Res* 542: 83-90.

Wenk GL, Walker LC, Price DL and Cork LC (1991): Loss of NMDA, but not GABA-A, binding in the brains of aged rats and monkeys. *Neurobiol Aging* 12:93-98.

Alzheimer Disease: From Molecular Biology to Therapy
edited by R. Becker and E. Giacobini

ROLE OF NEURON-GLIA INTERACTIONS IN BRAIN ENERGY METABOLISM: IMPLICATIONS FOR NEURODEGENERATIVE DISORDERS

Pierre J. Magistretti, Philippe Bittar and Luc Pellerin
Laboratoire de Recherche Neurologique, Institut de Physiologie et Service de Neurologie du CHUV, Faculté de Médecine, Université de Lausanne, CH-1005 Lausanne, Switzerland

INTRODUCTION

There is increasing evidence indicating that impairments of energy metabolism are associated with neurodegeneration; indeed, the possibility that metabolic dysfunctions may be one of the causes of neuronal degeneration or at least could enhance neurodegenerative processes activated by other mechanisms such as for example glutamate-mediated excitotoxicity has received renewed attention (Beal, 1995; Novelli et al., 1988). Decreases in ATP or NADPH concentrations are caused by a variety of neurotoxins which produce neurodegeneration by inhibiting enzymes of energy metabolism (Coyle and Puttfarcken, 1993; Beal, 1995). Mitochondrial dysfunction and oxidative damage resulting from the metabolic formation of free radicals has been implicated in a variety of neurodegenerative processes (Beal, 1995). In particular, the activity of three mitochondrial enzymes is decreased in the most vulnerable cortical regions of patients with AD (Blass and Gibson, 1991). These enzymes are : pyruvate dehydrogenase, a-ketoglutarate dehydrogenase (KGDH) and cytochrome oxidase (Beal, 1995). The decreased activity of these enzymes is the consequence of a selective decrease in the expression of only certain of their subunits, thus ruling out the possibility that the impairment of their activity is due to a global cell loss (Simonian and Hyman, 1994). Among the various enzymes for which a dysfunction has been associated with AD, the most compelling evidence has been gathered for KGDH (Blass and Gibson, 1991).

IN VIVO IMAGING OF BRAIN ENERGY METABOLISM

A decrease in energy metabolism has been demonstrated in selected brain areas of patients presenting clinical symptoms corresponding to the diagnosis of probable AD (Herholz, 1995). In particular, decreases in glucose utilization (CMRglu) visualized with 2-fluorodeoxyglucose (FDG) Positron Emission Tomography (PET) are maximal in the temporal and parietal cortices of patients with probable AD (Jagust, 1996). This hypometabolism has generally been interpreted as reflecting the degree of cortical atrophy which is manifested by the clinical symptoms of dementia and is supported by the structural imaging acquisitions (CAT and MRI). However, a few studies have indicated that cortical hypometabolism is present at the very early stages of the disease, when cortical atrophy and neuropsychological deficits are absent (Haxby et al., 1990). The longitudinal nature of this analysis provided the initial evidence that cortical hypometabolism could precede cognitive impairment and not merely be the result of cortical atrophy. More compelling evidences have recently been provided through studies performed in asymptomatic at risk individuals (Small et al., 1995; Kennedy et al., 1995; Reiman et al., 1996). Indeed, two conditions put an individual at risk for developing AD: (i) being part of a pedigree with familial AD (FAD) (Bird, 1994) and (ii) the presence in the individual's genotype of the apolipoprotein (APO) E type 4 allele (Roses, 1995). A study has examined cerebral glucose metabolism in individuals members of a FAD pedigree (Kennedy et al., 1995). Using 2DG-PET, the study indicated a significant decrease in the global CMRglu of the asymptomatic at risk group when compared with age-matched controls; in addition, focal decreases in the parieto-frontal cortex were observed in the at risk group with a pattern that was similar, although less pronounced, to symptomatic members of FAD pedigrees (Kennedy et al., 1995).

Another recent study addressed the same issue; however the FAD at risk group was further subdivided into two sub-groups, one carrying the APOE 4 allele, the other not (Small et al., 1995). A third group, i.e. patients with probable AD was also examined (Small et al., 1995). PET analysis of cerebral glucose metabolism has indicated that among the subjects at risk, because of their belonging to a FAD pedigree, those with an APOE 4 allele had a significant decrease in parietal glucose metabolism (Small et al., 1995). The asymmetry in left-right parietal glucose metabolism was also higher in the at risk individuals with the APOE 4 genotype (Small et al., 1995). Patients with probable AD had even lower parietal metabolism and more pronounced left-right asymmetries.

Very recently a 2DG-PET study performed in middle-aged, cognitively normal individuals homozygous for the APOE 4 allele, has shown reductions of glucose metabolism in restricted brain areas which correspond to those areas in which glucose metabolism is decreased in patients with the clinical symptoms of probable AD (Reiman et al., 1996).

The results of these studies are quite instructive since they indicate, first, that cortical hypometabolism is present in asymptomatic individuals at risk for AD, and second that a decrease in glucose metabolism in selected brain areas is associated with the APOE 4 genotype.

METABOLIC INTERACTIONS BETWEEN GLIA AND NEURONS

The functional brain imaging techniques used in these *in vivo* studies are based on the fact that neuronal activity is coupled to energy metabolism (Raichle, 1993). Thus, when neuronal pathways are activated in a given area of the brain, blood flow and glucose uptake increase locally, in register with neuronal activity. Despite the fascinating observations made at the functional anatomical level, virtually nothing is known of the cellular localization of the signal detected with 2DG-PET. The assumption has generally been that glucose is taken up directly into neurons when their activity increases. A recent set of observations performed in purified preparations of neural cells, namely primary cultures of neurons or astrocytes, as well as analysis of the cytological relationships between intraparenchymal blood vessels, astrocytes and neurons has provided strong support for the notion that glucose is predominantly taken up by astrocytes which then release lactate as a metabolic substrate for neurons (Pellerin and Magistretti, 1994; Tsacopoulos and Magistretti, 1996).

Thus, when activation is mimicked *in vitro* by applying glutamate to primary cultures of neurons or astrocytes, 2DG uptake increases only in astrocytes (Pellerin and Magistretti, 1994). This action of glutamate is mediated by the transporters which mediate the cotransport of glutamate with sodium, not by glutamate receptors (Pellerin and Magistretti, 1994; Takahashi et al., 1995). The increase in intracellular sodium concentration which accompanies glutamate uptake into astrocytes activates the Na^+/K^+-ATPase which in several cellular systems, including astrocytes, is functionally-coupled to glucose uptake and lactate production, i.e. to glycolysis (Magistretti et al., 1995; Raffin et al., 1992). Indeed, ouabain inhibits the glutamate-evoked glucose uptake and lactate production by astrocytes. These data indicate therefore that glutamate stimulates glycolysis in astrocytes and suggest that during activation, when glutamate is released from activated pathways, glucose would enter primarily into astrocytes which would then release lactate as a metabolic substrate for neurons (Magistretti et al., 1995). Indeed, simple consideration of the fact that astrocytic end-feet surround brain capillaries, suggests that astrocytes constitute the first cellular barrier for glucose entering the brain. In addition, evidence accumulated over the years has clearly indicated that lactate is an adequate metabolic substrate for neurons (Schurr et al., 1988). In order to favor such a metabolic flux of lactate, a selective distribution of the enzyme lactate dehydrogenase (LDH) would be expected. Indeed, an isoform of LDH (LDH 1) is enriched in tissues like the heart which can utilize lactate as a substrate by converting it to pyruvate, which can then enter the tricarboxylic acid cycle; another isoform (LDH 5) is enriched in lactate-producing tissues like skeletal muscle. We have recently observed that neurons are stained exclusively by anti-LDH 1 antibodies (which recognize the "heart form" of LDH) and LDH 5 immunoreactivity is exclusively observed in astrocytes (Bittar et al., 1996).

What is the link between these *in vitro* data and functional brain imaging? The link lies in the evidence obtained from PET as well as proton magnetic resonance spectroscopy (MRS) which indicates that during activation of modality-specific pathways, glycolysis is transiently activated (Fox et al., 1988; Prichard et al., 1991). Thus it appears that glutamate-mediated glycolysis in astrocytes may provide the cellular mechanism for the activation-induced glycolysis demonstrated *in vivo* by PET and MRS. These considerations provide a cellular resolution to these functional brain imaging techniques and indicate that the signal that is visualized with 2DG-PET during activation, originates in astrocytes (Pellerin and Magistretti, 1994). This view does not challenge the validity of the 2DG-based techniques to monitor neuronal activation, since the uptake of the tracer into astrocytes is triggered by a neuronal signal, i.e. the activity-dependent release of glutamate.

HYPOTHESIS

Evidence is being provided indicating a decrease in glucose metabolism in selected brain areas of cognitively intact individuals at risk for AD. The functional brain imaging procedure used in these studies is based on the visualization of 2DG uptake into the brain parenchyma. Recent experimental evidence has been provided indicating that the 2DG signal originates in astrocytes and that it is linked to glutamate uptake into these cells (Pellerin and Magistretti, 1994; Takahashi et al., 1995). Thus the decrease in glucose metabolism detected in unaffected individuals at-risk for developing AD, may be linked to an impaired functionality of the various molecular steps that link glutamate clearance from the synaptic cleft to glucose uptake into and lactate release from astrocytes. These astrocyte-specific steps include glutamate reuptake and metabolism, the activity of the Na/K-ATPase as well as glucose uptake and metabolism. These considerations should stimulate further studies on neuron-glia interaction in neurodegenerative diseases.

ACKNOWLEDGMENTS

Research in PJM's laboratory is supported by FNRS grant # 31-40565.94, and by a grant from the Roche Foundation.

REFERENCES

Beal FM (1995): *Mitochondrial Dysfunction and Oxidative Damage in Neurodegenerative Diseases*. Heidelberg: Springer-Verlag.

Bird TD (1994): Familial Alzheimer's disease. *Ann Neurol* 36:335-336.

Bittar PG, Charnay Y, Pellerin L, Bouras C and Magistretti PJ (1996): Selective distribution of lactate dehydrogenase isoenzymes in neurons and astrocytes of the human brain. *J Cereb Blood Flow Metab*, in press.

Blass JP and Gibson GE (1991): The role of oxidative abnormalities in the pathophysiology of Alzheimer's disease. *Rev Neurol* 147:513-525.

Coyle JT and Puttfarcken P (1993): Oxidative stress, glutamate, and neurodegenerative disorders. *Science* 262:689-695.

Fox PT, Raichle ME, Mintun MA and Dence C (1988): Nonoxidative glucose consumption during focal physiologic neural activity. *Science* 241:462-464.

Haxby JV, Grady CL, Koss E, Horwitz B, Heston L, Schapiro M, Friedland RP and Rapoport S (1990): Longitudinal study of cerebral metabolic asymmetries and associated neuropsychological patterns in early dementia of the Alzheimer type. *Arch Neurol* 47:753-760.

Herholz K (1995): FDG PET and differential diagnosis of dementia. *Alzheimer Dis Assoc Disord* 9:6-16.

Jagust WJ (1996): Functional imaging patterns in Alzheimer's disease. Relationships to neurobiology. In: *The Neurobiology of Alzheimer's Disease*, Wurtman RJ, Corkin S, Growdon JH and Nitsch RM, eds. New York: The New York Academy of Sciences, pp. 30-36.

Kennedy AM, Frackowiak RSJ, Newman SK, Bloomfield PM, Seaward J, Roques P, Lewington G, Cunningham VJ and Rossor MN (1995): Deficits in cerebral glucose metabolism demonstrated by positron emission tomography in individuals at risk of familial Alzheimer's disease. *Neurosci Lett* 18:17-20.

Magistretti PJ, Pellerin L and Martin JL (1995): Brain energy metabolism. An integrated cellular perspective. In: Psychopharmacology : *The Fourth Generation of Progress*, Bloom FE and Kupfer DJ, eds. New York: *Raven Press*, pp. 657-670.

Novelli A, Reilly JA, Lysko PG and Henneberry RC (1988): Glutamate becomes neurotoxic via the N-methyl-D-aspartate receptor when intracellular energy levels are reduced. *Brain Res* 451:205-212.

Pellerin L and Magistretti PJ (1994): Glutamate uptake into astrocytes stimulates aerobic glycolysis : a mechanism coupling neuronal activity to glucose utilization. *Proc Natl Acad Sci USA* 91:10625-10629.

Prichard J, Rothman D, Novotny E, Petroff O, Kuwabara T, Avison M, Howseman A, Hanstock C and Shulman R (1991): Lactate rise detected by ^{1}H NMR in human visual cortex during physiologic stimulation. *Med Sci* 88:5829-5831.

Raffin CH, Rosenthal M, Busto R and Sick TJ (1992): Glycolysis, oxidative metabolism and brain potassium ion clearance. *J Cereb Blood Flow Metab* 12:34-42.

Raichle ME (1993): Visualizing the mind. *Sci Amer* 270:36-42.

Reiman EM, Caselli RJ, Yun LS, Chen K, Bandy D, Minoshima S, Thibodeau SN and Osborne D (1996): Preclinical evidence of Alzheimer's disease in persons homozygous for the e4 allele for apolipoprotein E. *N Engl J Med* 334:752-758.

Roses AD (1995): Apolipoprotein E genotyping in the differential diagnosis, not prediction, of Alzheimer's disease. *Ann Neurol* 38:6-14.

Schurr A, West CA and Rigor BM (1988): Lactate-supported synaptic function in the rat hippocampal slice preparation. *Science* 240:1326-1328.

Simonian NA and Hyman BT (1994): Functional alterations in Alzheimer's disease : selective loss of mitochondrial-encoded cytochrome oxidase mRNA in the hippocampal formation. *J Neuropathol Exp Neurol* 53:508-512.

Small GW, Mazziotta JC, Collins MT, Baxter LR, Phelps ME, Mandelkern MA, Kaplan A, La Rue A, Adamson CF, Chang L, Guze BH, Corder EH, Saunders AM, Haines JL, Pericak-Vance MA and Roses AD (1995): Apolipoprotein E Type 4 Allele and cerebral glucose metabolism in relatives at risk for familial Alzheimer disease. *JAMA* 273:942-947.

Takahashi S, Driscoll BF, Law DM and Sokoloff L (1995): Role of sodium and potassium ions in regulation of glucose metabolism in cultured astroglia. *Proc Natl Acad Sci USA* 92:4616-4620.

Tsacopoulos M and Magistretti PJ (1996): Metabolic coupling between glia and neurons. *J Neurosci* 16:877-885.

Alzheimer Disease: From Molecular Biology to Therapy
edited by R. Becker and E. Giacobini

NEUROTROPHINS, GROWTH FACTORS AND MIMETIC AGENTS AS NEUROPROTECTORS IN THE TREATMENT OF ALZHEIMER'S DISEASE

Alvin J. Glasky and Ronald F. Ritzmann
Education and Research Inst., Olive View
UCLA Medical Center, Sylmar, CA

Michel P. Rathbone, Pamela J. Middlemiss and Candice Crocker
Department of Biomedical Sciences
McMaster University, Hamilton, ONT, Canada

INTRODUCTION

Recent advances in understanding the action of growth factors on neurons have provided the opportunity for development of new strategies treatment of Alzheimer's disease (AD) (Hefti, 1994; Tuszynski and Gage, 1994). Since the original description of nerve growth factor (NGF) by Levi-Montalcini et al. (1995), the immutability of the mature central nervous system has been successfully challenged. The number of growth factors affecting neuronal cells has expanded from NGF to include brain-derived neurotrophic factor (BDNF), neurotrophin-3 (NT-3), neurotrophin-4/5 (NT-4/5), fibroblast growth factor (FGF), glial cell line-derived neurotrophic factor (GDNF) and insulin-like growth factors (IGF).

Trophic factors have been demonstrated to reverse cognitive deficits produced in animals by aging (Markowska et al., 1994) and by lesions (Cuello, 1993) and to induce improved cholinergic neuronal function (Tuszynski and Gage, 1995). Identification and cloning of the receptors for many of the growth factors has provided tools for elucidating the molecular basis for their action (Bothwell, 1995; Lindsay, 1994). The clinical development of neurotrophic agents has been difficult since these macromolecular factors are large proteins and do not cross the blood-brain barrier (see Tonnaer and Dekker, 1994). While positive clinical effects have been reported after intracranial infusion of NGF to Alzheimer's patients (Seiger et al., 1993), this method of neurotrophin administration is not practical. An alternative approach is to stimulate the production of neurotrophin *in situ* by an orally active inducing agent.

The hypoxanthine derivative AIT-082, 4-[[3-(1,6-dihydro-6-oxo-9-purin-9-yl)-1-oxopropyl]amino] benzoic acid, potassium salt, has been shown to enhance cognitive function (Ritzmann et al., 1993) and to mimic the effects of NGF to induce neuritogenesis in PC12 cells *in vitro* (Middlemiss et al., 1995). The ability of AIT-082 to increase production of mRNA's for several neurotrophic factors *in vitro* and *in vivo* will be illustrated here and the mechanism by which these effects are produced will be proposed.

RESULTS

The win-shift paradigm, a positive reinforcement delayed alternation model of working memory, has been shown to reflect deficits associated with aging, medial septal lesions, four vessel occlusion and scopolamine treatment in rats (Ordy et al., 1988) and to involve the septo-hippocampal cholinergic system. We have demonstrated the existence of age-induced deficits in mice with the same paradigm (Ritzmann et al., 1993) and have shown that AIT-082 reverses these age-induced deficits (Glasky et al., 1994).

One year old mice (male C57BL/6) were trained in the win-shift paradigm and tested once a month (10 trials) for the duration of their memory trace. Young mice and aged non-impaired mice have a memory trace ≥ 90 sec in this task. Administration of AIT-082 (30 mg/kg/day orally in the drinking water) was initiated at 13 months of age and continued for the duration of the experiment. A comparable control group received water without any drug.

Figure 1a shows that the mean duration of the memory trace declined more slowly in the treated group and that it tended to plateau after 20 months. Figure 1b shows that in the control group not a single subject had measurable memory at 20 months; the mean of the treated group did not reach zero through 21 months. At 22 months, 50% of the treated subjects exhibited no memory deficit. It appears that there are two subgroups in the treated mice, those with a short-term response to AIT-082 (a delay of 2-3 months in

a. Mean duration of memory trace b. Percent with no memory deficit

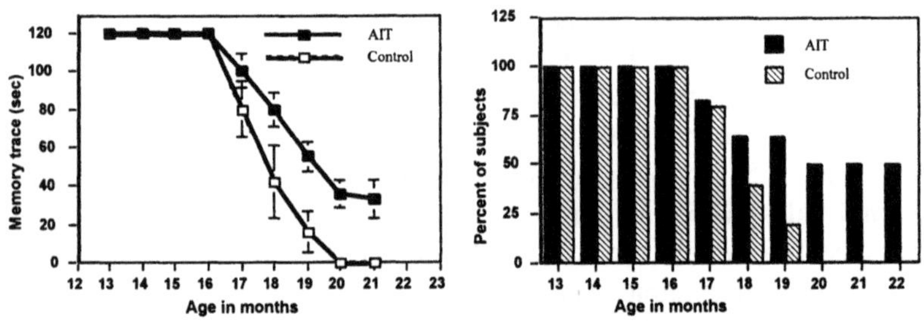

FIGURE 1. Effect of prophylactic oral treatment on memory in mice

memory decline) and those with a longer term response (no deficit at 22 months). Mice from these two groups were sacrificed, brain tissue collected from the frontal cortex, hippocampus and cerebellum, rapidly frozen and analyzed for mRNA.

The ability of AIT-082 to reverse the amnestic effect of various specific inhibitors in normal adult Swiss Webster mice was measured in the shuttle box passive avoidance paradigm. Subjects were placed in the lighted section of the shuttle box and allowed to explore the dark chamber for 60 seconds in the first or pre-test trial. Twenty-four hours later, on the second (acquisition) trial, all subjects received a five second foot shock one second after entering the dark chamber. Retention is measured 24 hours later by determining the latency to enter the dark chamber. AIT-082 (30 mg/kg, i.p.) or vehicle was administered one hour before the acquisition trial; inhibitor or vehicle was administered 20 minutes after the acquisition trial.

Data in Table 1 is presented as the percent of subjects retaining memory of the foot shock during the test trial. Since AIT-082 reverses the amnesia produced by scopolamine, L-nitroarginine methyl ester (L-NAME), NBQX, or MK-801, the site of action of AIT-082 must be "down-stream" from these receptors. AIT-082 appears to only partially reverse the effects of NPC-15437, a protein kinase C inhibitor.

AIT-082 has been shown to mimic the effects of NGF to promote neuritogenesis in PC12 cells (Middlemiss et al., 1995) and to increase the production of cGMP. The neuritogenic effect was blocked by methemoglobin (which captures and removes both carbon monoxide and nitric oxide), methylene blue (an inhibitor of soluble guanylyl cyclase, which produces cGMP) and zinc protoporphyrin (an inhibitor of heme oxygenase, an inducible enzyme which produces carbon monoxide and biliverdin) but was not affected by L-NAME, the inhibitor of nitric oxide production.

In astrocytes cultures we measured the production of mRNA by southern blot and found that AIT-082 increased the production of mRNA's for NGF, NT-3 and bFGF but not BDNF. This indicates that AIT-082 produces a broad but selective enhancement of neurotrophin mRNA production.

			% Correct Responses	
		Dose	Inhibitor	
Inhibitor	Site of Action	(mg/kg)	Alone	+ AIT-082
Control			100	
Scopolamine	Muscarinic antagonist	0.5	20	73
L-NAME	NO synthetase inhibitor	5.0	43	83
NPC 15437	Protein kinase C inhibitor	0.5	40	60
NBQX	AMPA antagonist	10.0	14	75
MK-801	NMDA antagonist	0.2	20	71

TABLE I. Effect of AIT-082 on amnesia induced by specific inhibitors in passive avoidance paradigm

Brain Region	mRNA NGF	mRNA NT-3
Cortex	2.21	5.26
Hippocampus	1.65	1.73
Cerebellum	0.84	0.91

TABLE II. Effect of AIT-082 on brain neurotrophin mRNA (ratio of mRNA in responders/mRNA in non-responders)

Using RT-PCR technology, mRNA was analyzed from the brain tissues obtained from the mice treated prophylactically with AIT-082 as described above. The preliminary results shown in Table II are expressed as the ratio of mRNA from long term responders/mRNA from short term responders, which had no memory at the time of sacrifice. The long-term responders had higher levels of mRNA for NGF and NT-3 in the frontal cortex and hippocampus but there was no difference in the cerebellum.

DISCUSSION AND CONCLUSION

It is proposed that AIT-082 acts at the site of heme oxygenase to generate carbon monoxide and by activation of guanylyl cyclase induces a cascade of biochemical reactions through the second messenger system leading to the production of mRNA for neurotrophins.

AIT-082 can delay or prevent age-induced working memory deficits when treatment begins at 13 months of age, before any impairments are detectable. In the subgroup in which no deficit is observed at 22 months of age, the level of mRNA for NGF and NT-3 is increased in the cortex and hippocampus but not the cerebellum.

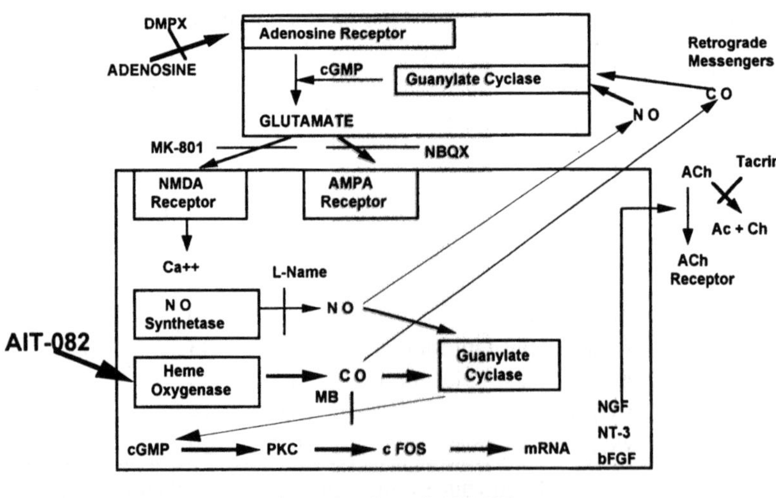

FIGURE 2. Proposed mechanism of action of AIT-082

In astrocyte cultures *in vitro*, we have observed that mRNA for NGF, NT-3 and NT-4/5 but not BDNF are elevated after treatment with AIT-082. Fischer et al. (1994) have reported that age-induced impairments in spatial learning and memory in rats was reversed by i.c.v administration of the neurotrophins NGF, NT-3 or NT-4/5 but not by BDNF. They also reported that NGF and NT-3 but not NT-4/5 or BDNF produced a significant reduction in cholinergic neuron atrophy in the septum, nucleus basalis and the striatum. Zhou et al. (1995) reported that late-phase long-term potentiation (LTP) in hippocampal slices from aged mice (18-30 months) was significantly decreased when compared to adult (10-12 weeks) or middle-aged (6-8 months) mice and NGF, NT-3 and NT-4/5 but not BDNF compensated for the loss of late-phase LTP. The consistency of the observations of the three laboratories suggests that there is a functional significance to the related effects of the NGF, NT-3 but not BDNF on memory. Since AIT-082 is orally active, penetrates the blood brain barrier rapidly, and induces the production of multiple neurotrophins, AIT-082 represents an interesting candidate for human clinical evaluation in the treatment of Alzheimer's disease by means of a unique mechanism of action.

ACKNOWLEDGMENTS

The research from our laboratories was supported by grants from the National Institute on Aging (AG 09911), the Ontario Mental Health Foundation, the ALS Society of Canada and the Canadian Spinal Research Organization.

REFERENCES

Bothwell M (1995): Functional interactions of neurotrophins and neurotrophin receptors. *Annu Rev Neurosci* 18:223-253.

Cuello AC (1993): Trophic responses of forebrain cholinergic neurons: a discussion. *Prog Brain Res* 98:265-277.

Fischer W et al. (1994): Reversal of spatial memory impairments in aged rats by nerve growth factor and neurotrophins 3 and 4/5 but not by brain-derived neurotrophic factor. *Proc Natl Acad Sci USA* 91:8607-9611.

Glasky AJ et al. (1994): Effect of AIT-082, a purine analog, on working memory in normal and aged mice. *Pharm Biochem Behav* 47:325-329.

Hefti F (1994): Neurotrophic factor therapy for nervous system degenerative diseases. *J Neurobio* 25:1418-1435.

Levi-Montalcini R et al. (1995): Update of the NGF Saga. *J Neurol Sci* 130:119-127.

Lindsay RM (1994): Neurotrophins and receptors. *Prog Brain Res* 103:3-14.

Markowska AL et al. (1994): Human nerve growth factor improves spatial memory in aged but not in young rats. J *Neurosci* 14:4815-4824.

Middlemiss PJ et al. (1995): AIT-082, a unique purine derivative, enhances nerve growth factor mediated neurite outgrowth from PC12 cells. *Neurosci Lttr* 199:1-4.

Ordy JM et al. (1988): An animal model of human-type memory loss based on aging, lesion, forebrain ischemia, and drug studies with the rat. *Neurobiol Aging* 9:667-683.
Ritzmann RF et al. (1993): Effect of age and strain on working memory in mice as measured by win-shift paradigm. *Pharm Biochem Behav* 44:805-807.
Seiger AA et al. (1993): Intracranial infusion of purified nerve growth factor to an Alzheimer patient: The first attempt of a possible future treatment strategy. *Behav Brain Res* 57:255-261.
Tonnaer JA and Dekker WC (1994): Nerve growth factor, neurotrophic agents and dementia. In: *Anti-Dementia Agents*. Academic Press Ltd. 139-165.
Tuszynski MH and Gage FH (1994): Neurotrophic factors and disease of the nervous system. *Ann Neurol* 35:S9-S12.
Tuszynski MH and Gage FH (1995): Bridging grafts and transient nerve growth factor infusions promote long-term central nervous system neuronal rescue and partial functional recovery. *Proc Natl Acad Sci USA* 92:4621-4625.
Zhuo M et al. (1995): Effects of dopamine D1 agonists and neurotrophins on age-related changes in the hippocampus. *Abstr 25th Ann Mtg Soc Neurosci* 111.3.

Alzheimer Disease: From Molecular Biology to Therapy
edited by R. Becker and E. Giacobini

CORTICAL SYNAPTOGENESIS AND BEHAVIOURAL CONSEQUENCES IN CNS LESIONED ANIMALS RECEIVING NEUROTROPHIC FACTOR THERAPY

A. Claudio Cuello
Dept. of Pharmacology & Therapeutics, McGill University,
Montreal, QC, Canada.

INTRODUCTION

The ultimate goal in neurodegenerative disease therapy is the preservation of neuronal somata and their synaptic connections. In this regard, there is overwhelming experimental evidence that these objectives can be met with the application of neurotrophic factors in diverse lesion models leading to anterograde or retrograde neuronal degeneration. The issue of whether neurotrophic factor administration can re-establish or regenerate new synaptic contacts has been recently reviewed in the context of future neurotrophic factor based therapies (Cuello and Thoenen, 1995). However, much has yet to be learned on the limits and possible undesirable effects of this approach. Thus, in recent attempts to apply purified mouse nerve growth factor (NGF) in Alzheimer's disease (AD) the treatment had to be interrupted due to weight loss and cefalalgias (Sieger et al., 1993). Indeed, excessive NGF offering provokes ectopic synaptic formation of peptide-containing terminals within the white matter of transgenic mice overexpressing chick NGF, produced by oligodendrocytes during early post-natal stages (Ma et al., 1995). Will trophic therapy, therefore, be undesirable in AD? I would like to propose that neurotrophic factors (NTFs) have a defined opportunity in AD therapy provided that some basic conditions are met. These should be, firstly, early treatment, at a stage when there is sufficient regenerative capacity left in the diseased brain to respond to NTFs, an approach which is currently hampered by the lack of presymptomatic biological diagnosis. Secondly, much has yet to be learned about the specificity, interactions and dosage of NTFs capable of provoking synaptogenesis in the cerebral cortex accompanied by desirable behavioural effects. Thirdly, the field anxiously awaits a suitable animal model for AD where these (and other) propositions can be properly and exhaustively investigated.

DE NOVO CORTICAL SYNAPTOGENESIS IN ANIMAL MODELS

The strong evidence that the loss of cortical presynaptic elements is best correlated with deterioration of mental functions in AD (DeKosky and Scheff, 1990; Masliah et al., 1991; Terry et al., 1991) is an important finding which implicitly makes synaptogenesis a therapeutic target. Such an idea is reinforced by the recent observation of Games et al., (1995) linking the formation of diffuse A-ß plaques in the cerebral cortex and hippocampus with loss of presynaptic elements (synaptophysin-immunoreactive) in transgenic mice carrying a minigen coding for the valine 717 mutation of the amyloid precursor protein (APP). If indeed this is the evolution of AD, it is reasonable to assume that synaptic attrition is an early component of its neuropathology, something which could be redressed by the timely application of NTFs in the absence of therapeutics capable of arresting plaque and tangle formation.

NTFs are unquestionably capable of inducing synaptogenesis during development (Lewin and Barde, 1996) and -- more importantly in the context of AD -- in the mature and fully differentiated CNS, as we have demonstrated in the cerebral cortex ipsilateral to areas suffering strokes when given NGF treatment (Garofalo et al., 1992, 1993).

In these experiments, we have focused on the NGF modulation of the cortical cholinergic nerve terminal network. The nucleus basalis neurons projecting to the cortex are extremely sensitive to the offering of this neurotrophin as they are well endowed with both the low, $p75^{LNGFR}$, (e.g. Pioro and Cuello, 1990) and high, $p140^{trkA}$, (e.g. Holtzman et al., 1992) affinity NGF receptors. They retrogradely retract their cortical processes and their cell somata becomes notably reduced in size after a partial cortical devascularizing (stroke) lesion (Cuello, 1994). NGF reverses these cortical changes, inducing an increase in the size of cholinergic boutons and the number of actual synapses (Garofalo et al., 1992). The ganglioside GM1 can also protect these cortical terminals to a lesser extent at higher doses (Garofalo et al., 1993). Interactions between these two agents have also been observed in the cerebral cortex in relation to the expression of neurochemical cholinergic presynaptic markers, such as choline acetyltransferase enzymatic activity and high affinity choline uptake sites (Cuello et al., 1989; Garofalo and Cuello, 1995). Such an interaction between a *bona fide* NTF and a glycosphyngolipid can presently be explained as a molecular interaction at the cell membrane level facilitating the NGF message by increased phosphorylation of the $p140^{trkA}$ receptors (Ferrari et al.,1995).

The observed synaptic changes in the cerebral cortex induced by NGF are consistent with the finding that this neurotrophin facilitates the invasion of cholinergic fibers and the formation of proximal cholinesterase positive synapses in the hippocampus with the axotomized fornix (Kawaja et al., 1992); and by the biochemical indication of up-regulation of presynaptic cholinergic markers in the partially deafferentated hippocampus (Lapchak and Hefti, 1991). These experimental therapies have been shown to facilitate cortical acetylcholine (ACh) output after lesions of the nucleus basalis (Scali et al., 1994) and the release of endogenous ACh *in vivo* in the cortex adjacent to stroke lesions (Maysinger et al., 1992.)

The question of whether a given trophic factor will affect a given type of synapse or whether it will have broader effects is a matter of current interest. Thus, BDNF or NT3, but not NGF, disrupts the synaptic patterning in the visual cortex of neonatal cats (Cabelli et al., 1995), while the application of anti-NGF antibodies can similarly interfere with the development of ocular dominance in the visual cortex (Berardi et al., 1994). The application of NGF has also been shown to produce a broad increase of synaptophysin (presynaptic sites indistinctively from the transmitter nature) in aged rats (Chen et al., 1995). This possibility has also been raised in primates (*Cercopithecus aethiops*) in a lesion, trophic factor therapeutic model in which human recombinant NGF (hr-NGF) was administered in the form of a gel for approximately a week rendering long term (5 months) biochemical and morphological preservation of the cholinergic neurons of the nucleus basalis of Meynert and their cortical projections (Liberini et al., 1994). More recently we have observed, in collaboration with Burgos and Masliah, that in the primate the synaptic changes induced by hr-NGF in the cerebral cortex also occur adjacent to the stroke lesion and most likely involve several neurotransmitter systems (Burgos et al., 1995).

Another issue of considerable relevance to AD is whether NTFs can modulate the organization of postsynaptic sites in the cerebral cortex. It is well documented that AD pyramidal neurons, which ultimately are the output route of cortical synaptic processing, suffer drastic aberrations of the dendritic pattern with loss of branches and spines (Buell and Coleman, 1981). NGF has been shown to maintain dendritic architecture in pyramidal neurons of aged rats (Mervis et al., 1991). In our lesion model of partial and unilateral cortical devascularization we have observed, in collaboration with Kolb and Gorny, that the partial dendritic atrophy induced by lesions in pyramidal neurons can be fully reverted by NGF (Kolb et al., 1996a) and further that this treatment stimulates the branching and number of spines in non-lesioned young adult rats (Kolb et al., 1996b). Thus, it is possible that NGF, and perhaps other NTFs, can provoke preferential phenotypic changes in those neurons possessing the specific receptors to these factors. However, some widespread trophic effects might be the consequence of enhanced neuronal activity resulting from NGF-induced synaptogenesis. NTFs *per se* are also capable of producing modifications of neuronal activity that can be best defined as improving synaptic efficiency (e.g. Schuman and Madison, 1994; Knipper et al., 1994). Another appealing explanation for these broad NTF effects, beyond the neuronal systems for which they are expected to act selectively, is based on the observation that a given NTF is capable of stimulating the synthesis and release of other NTFs. For example, *in vitro* basic fibroblast growth factor (b-FGF) stimulates the production and secretion of NGF by astrocytes (Yoshida and Gage, 1992). This phenomenon is not restricted to embryonic or neonatal neural cells *in vitro* but also occurs in the adult CNS. Thus, we have recently demonstrated with Otten and collaborators, that in the remaining cortex of rats bearing unilateral stroke lesions the application of acidic-FGF up-regulates the steady state levels of NGF mRNA and the expression of the corresponding peptide (up to eightfold) (Figueiredo et al., 1995).

BEHAVIOURAL CONSEQUENCES TO NEUROTROPHIC FACTOR THERAPY

We should obviously be cautious about extrapolating to Alzheimer's the effects of experimental NTF therapy in rodents in the absence of a universally accepted animal model of the disease. Undoubtedly, the intense research activity on this front will be of great value to assess the opportunities ahead once such models become available. At this stage it is important to point out that the administration of NGF has been shown to compensate behavioural deficits observed in aged, impaired rats (Fischer et al., 1987) a situation that correlates well with the correction of losses of cholinergic markers (Fischer et al., 1987, 1991; Williams et al., 1993). In the NGF-treated, cortically-lesioned rat model discussed above, we have observed that the deficits normally induced by the lesions on the retention of learned behaviours in both passive avoidance and Morris water maze tests are not present, and that the behaviours are fully retained (Garofalo and Cuello, 1994). Similarly, when untreated, cortically lesioned animals are exposed to new tasks they display poor performance, taking them nearly 10 days to reach the hidden platform in the Morris water maze compared to non-lesioned controls. Lesioned animals receiving NGF treatment behaved in an intermediate fashion, but with performance levels closer to controls than to the untreated, cortically lesioned cohort (Kolb et al., 1996a).

CONCLUDING REMARKS

NTF therapy in AD remains an attractive possibility, the extent of which depends on the development of new diagnostic tools, the availability of early biological diagnosis and the profound understanding of the spectrum of activities of NTFs in the CNS. The unquestionable evidence that these agents are capable of inducing profound pre- and post-synaptic remodeling in the cerebral cortex, including the generation of new synapses (synaptogenesis), opens up interesting possibilities for corrective therapies, not only in AD but also in other neurodegenerative conditions and CNS trauma.

ACKNOWLEDGMENTS

The research summarized above has been funded by grants from the Medical Research Council of Canada, The National Institute on Aging (NIH Grant # AG11903-01A1), and the Centres of Excellence, Canada. I would like to acknowledge the work of my collaborators in these efforts, B.C. Figueiredo, L. Garofalo, B. Kolb, U. Otten, A. Ribeiro-da-Silva, and W. Tetzlaff. I would like to also thank Sid Parkinson for editorial assistance in preparing this manuscript.

REFERENCES

Berardi N, Cellerino A, Domenici L, Fagiolini M, Pizzorusso T, Cattaneo A and Maffei L (1994): Monoclonal antibodies to nerve growth factor affect the postnatal development of the visual system. *Proc Natl Acad Sci USA* 91:684-688.

Buell SJ and Coleman P (1981): Quantitative evidence for selective dendritic growth in normal human aging but not in senile dementia. *Brain Res* 21:423-41.

Burgos I, Cuello AC, Liberini P, Pioro EP and Masliah E (1995): NGF-mediated synaptic sprouting in the cerebral cortex of lesioned primate brain. *Brain Res* 692:154-160.

Cabelli RJ, Hohn A and Shatz CJ (1995): Inhibition of ocular dominance column formation by infusion of NT-4/5 or BDNF. *Science* 267:1662-1666.

Chen KS, Masliah E, Mallory M and Gage FH (1995): Synaptic loss in cognitively impaired aged rats is ameliorated by chronic human nerve growth factor infusion. *Neuroscience* 68:19-27.

Cuello AC (1994): Trophic factor therapy in the adult CNS: remodelling of injured basalo-cortical neurons. *Progr Brain Res* 100:213-221.

Cuello AC, Garofalo L, Kenigsberg RL and Maysinger D (1989): Gangliosides potentiate *in vivo* and *in vitro* effects of nerve growth factor on central cholinergic neurons. *Proc Natl Acad Sci USA* 86:2056-2060.

Cuello AC and Thoenen H (1995): The pharmacology of neurotrophic factors. In: *Pharmacological sciences: perspectives for research and training in the late 1990s,* Cuello AC and Collier B, eds. Basel: Birkhauser, pp. 241-245.

DeKosky ST and Scheff SW (1990): Synapse loss in frontal cortex biopsies in Alzheimer's disease: correlation with cognitive severity. *Ann Neurol* 27:457-464.

Ferrari G, Anderson BL, Stephens RM, Kaplan DR and Greene LA (1995): Prevention of apoptotic neuronal death by GM1 ganglioside. Involvement of Trk neurotrophin receptors. *J Biol Chem* 270:3074-3080.

Figueiredo BC, Pluss K, Skup M, Otten U and Cuello AC (1995): Acidic FGF induces and its mRNA in the injured neocortex of adult animals. *Mol Brain Res* 33:1-6.

Fischer W, Björklund A, Chen K and Gage FH (1991): NGF improves spatial memory in aged rodents as a function of age. *J Neurosci* 11:1889-1906.

Fischer W, Wictorin K, Björklund A, Williams L, Varon S and Gage FH (1987): Amelioration of cholinergic neuron atrophy and spatial memory impairment in aged rats by nerve growth factor. *Nature* 329: 65-68.

Games D, Adams D, Alessandrini R, Barbour R, Berthelette P, Blackwell C, Carr T, Clemens J, Donaldson T, Gillespie F, Guido T, Hagoplan S, Johnson-Wood K,

Khan K, Lee M, Leibowitz P, Lieberburg I, Little S, Masliah E, McConlogue L, Montoya-Zavala M, Mucke L, Paganini L, Penniman E, Power M, Schenk D, Seubert P, Snyder B, Soriano F, Tan H, Vitale J, Wadsworth S, Wolozin B, Zhao J (1995): Alzheimer-type neuropathology in transgenic mice overexpressing V717F beta-amyloid precursor protein. *Nature* 37:3523-527.

Garofalo L and Cuello AC (1994): Nerve growth factor and the monosialoganglioside GM1: Analogous and different *in vivo* effects on biochemical, morphological, and behavioral parameters of adult cortically lesioned rats. *Exp Neurol* 125:195-217.

Garofalo L and Cuello AC (1995): Pharmacological characterization of nerve growth factor and/or monosialoganglioside GM1 effects on cholinergic markers in the adult lesioned brain. *J Pharmacol Exp Ther* 272:527-545.

Garofalo L, Ribeiro-da-Silva A and Cuello AC (1992): Nerve growth factor-induced synaptogenesis and hypertrophy of cortical cholinergic terminals. *Proc Natl Acad Sci USA* 89:2639-2643.

Garofalo L, Ribeiro-da-Silva A and Cuello AC (1993): Potentiation of nerve growth factor-induced alterations in cholinergic fibre length and presynaptic terminal size in cortex of lesioned rats by the monosialoganglioside GM1. *Neuroscience* 57:21-40.

Holtzman DM, Li Y, Parada LF, Kinsman S, Chen C-K, Valletta JS, Zhou J, Long JB and Mobley WC (1992): $p140^{trk}$ mRNA marks NGF-responsive forebrain neurons: Evidence that trk gene expression is induced by NGF. *Neuron* 9:465-478.

Kawaja MD, Rosenberg MB, Yoshida K and Gage FH (1992): Somatic gene transfer of nerve growth factor promotes the survival of axotomized septal neurons and the regeneration of their axons in adult rats. *J Neurosci* 12:2849-2864.

Knipper M, Leung LS, Zhao D and Rylett RJ (1994): Short-term modulation of glutamatergic synapses in adult rat hippocampus by NGF. *Neuroreport* 5:2433-2436.

Kolb B, Côté S, Ribeiro-da-Silva A and Cuello AC (1996a): NGF treatment prevents dendritic atrophy and promotes recovery of function after cortical injury. *Neuroscience* submitted.

Kolb B, Gorny G, Côté S, Ribeiro-da-Silva A and Cuello AC (1996b): Nerve growth factor stimulates growth of cortical pyramidal neurons in young adult rats. *Brain Res* Submitted.

Lapchak PA and Hefti F (1991): Effect of recombinant human nerve growth factor on presynaptic cholinergic function in rat hippocampal slices following partial septohippocampal lesions: Measures of [^{3}H]acetylcholine synthesis, [^{3}H]acetylcholine release and choline acetyltransferase activity. *Neuroscience* 42:639-649.

Lewin GR and Barde Y-A (1996): Physiology of the Neurotrophins. In: *Annual Review of Neuroscience*, Cowan WM, Shooter EM, Stevens CF and Thompson RF, eds. Palo Alto, CA: Annual Reviews Inc., pp. 289-317.

Liberini P, Pioro EP, Maysinger D and Cuello AC (1994): Neocortical infarction in subhuman primates leads to restricted morphological damage of the cholinergic neurons in the nucleus basalis of Meynert. *Brain Res* 648:1-8.

Ma W, Ribeiro-da-Silva A, Noel G, Julien J-P and Cuello AC (1995): Ectopic substance P and calcitonin gene-related peptide immunoreactive fibers in the spinal cord of transgenic mice over-expressing nerve growth factor. *Eur J Neurosci* 7:2021-2035.

Masliah E, Terry RD, Alford M, DeTeresa R and Hansen LA (1991): Cortical and subcortical patterns of synaptophysin-like immunoreactivity in Alzheimer's disease. *Am J Pathol* 138:235-246.

Maysinger D, Herrera-Marschitz M, Goiny M, Ungerstedt U and Cuello AC (1992): Effects of nerve growth factor on cortical and striatal acetylcholine and dopamine release in rats with cortical devascularizing lesions. *Brain Res* 577:300-305.

Mervis RF, Pope D, Lewis R, Dvorak RM and Williams LR (1991): Exogenous nerve growth factor reverses age-related structural changes in neocortical neurons in the aging rat. A quantitative Golgi study. *Ann NY Acad Sci* 640:95-101.

Pioro EP and Cuello AC(1990): Distribution of nerve growth factor receptor-like immunoreactivity in the adult rat central nervous system: Effect of colchicine and correlation with the cholinergic system-I. Forebrain. *Neuroscience* 24:57-87.

Scali C, Casamenti F, Pazzagli M, Bartolini L and Pepeu G (1994): Nerve growth factor increases extracellular acetylcholine levels in the parietal cortex and hippocampus of aged rats and restores object recognition. *Neurosci Lett* 170:117-120.

Schuman EM and Madison DV (1994): Locally distributed synaptic potentiation in the hippocampus. *Science* 263:532-536.

Seiger Å, Nordberg A, Von Holst H, Bäckman L, Ebendal T, Alafuzoff I, Amberla K, Hartvig P, Herlitz A, Lilja A, Lundqvist H, Långström B, Meyerson B, Persson A, Viitanen M, Winblad B and Olson L (1993): Intracranial infusion of purified nerve growth factor to an Alzheimer patient: The first attempt of a possible future treatment strategy. *Behav Brain Res* 57 255-261.

Terry RD, Masliah E, Salmon DP, Butters N, DeTeresa R, Hill R, Hansen LA and Katzman R (1991): Physical basis of cognitive alterations in Alzheimer's disease: synapse loss is the major correlate of cognitive impairment. *Ann Neurol* 30:572-580.

Williams LR, Rylett RJ, Ingram DK, Joseph JA, Moises HC, Tang AH and Mervis RF (1993): NGF affects the cholinergic biochemistry and behaviour of aged rats. *Prog Brain Res* 98:241-250.

Yoshida K and Gage FH (1992): Cooperative regulation of nerve growth factor synthesis and secretion in fibroblasts and astrocytes by fibroblast growth factor and other cytokines. *Brain Res* 569:14-25.

Alzheimer Disease: From Molecular Biology to Therapy
edited by R. Becker and E. Giacobini

APP KNOCKOUT AND APP OVER-EXPRESSION IN TRANSGENIC MICE

Hui Zheng, Gurparkash Singh, Minghao Jiang,
Myrna Trumbauer, Howard Chen and Lex Van der Ploeg
Department of Genetics & Molecular Biology,
Merck Research Laboratories, P.O. Box 2000, Rahway, NJ 07065, USA

David Smith, Dalip Sirinathsinghji , Gerard Dawson and Susan Boyce
Merck Sharp & Dohme Research Laboratories, Neuroscience
Research Center, Eastwick Road, Essex, CM20 2QR, UK

Connie Von Koch and Sangram Sisodia
Neuropathology Laboratory, The Johns Hopkins University
School of Medicine, Baltimore, MA 21205

INTRODUCTION

Theβ-amyloid peptide (Aβ), the major component of the neuritic plaques characterizing Alzheimer's disease (AD), is a 39 to 43 amino acid peptide derived from proteolytic cleavage of a larger β-amyloid precursor protein (APP). Mutations in the APP gene have been identified which cause familial, early onset AD, suggesting that APP metabolism is a central event in AD progression (Mullan & Crawford, 1993). APP is an abundant protein in the brain and Aβ is produced not only in AD patients, but also in cerebrospinal fluid (CSF) of normal individuals (Selkoe, 1994). APP has been implicated in numerous activities based on *in vitro* studies, such as regulation of cell adhesion, neurite outgrowth and intraneuronal calcium (Selkoe, 1994). However, the *in vivo* function of APP remains unclear.

To understand the *in vivo* function of APP and its processing, we have generated mice with null alleles of APP (Zheng et al., 1995). Homozygous APP deficient mice were viable and fertile. However, various abnormalities have been observed, including a decreased body weight and locomotor activity, a spontaneous seizure activity and an impaired behavioral performance. In addition, compared to the control mice, there seemed to be premature death in the APP null colony. Immunohistochemical analysis of the mice prior to their death showed reactive gliosis and a decrease in synaptophysin and synapsin staining. This result suggests that the premature death is the result of an impaired synaptic function. This finding highlights an important role of APP in synaptic function of CNS neurons.

The amyloid precursor proteins are a family of three members consisting of APP, APLP1 (APP like protein 1) and APLP2 (Slunt et al., 1994, Wasco et al., 1992). The expression pattern of APLP2 is very similar to that of APP. To determine the functional redundancy between APP and APLP2, we have produced mice deficient in either APLP2 alone or both APP and APLP2.

Unlike the APP null mice, the APLP2 knockout mice appeared normal with respect to body weight, histological analysis, and neurological evaluations. However, the APP and APLP2 double knockout mice exhibited a more severe phenotype than either of the single knockout mouse lines individually. Approximately 80% of the mice died within the first week of age. Those that survived beyond weaning were approximately 40% underweight, had difficulty in balancing and righting reflexes, and were prone to seizure. Detailed histochemical analysis are being performed to identify factors contributing to the early lethality of the mice.

In an attempt to produce a murine model for AD, we have generated transgenic mice overexpressing the human APP751cDNA with the familial AD mutation Val (V) to Iso (I) at amino acid codon 717 (hAPP751 FAD717V-I) under the transcriptional control of neuronal-specific human Thy-1 promoter. Western-blot analysis showed that the transgene is expressed at a similar level to that of the endogenous mouse APP. To avoid any potential inhibitory effect of mouse APP and Aβ, which has three amino acid differences compared with human Aβ, we have cross-bred these transgenic mice with the APP knockout mice and produced mice that express the human APP751 on the mouse APP-null background. The mice are being aged for immunohistochemical analysis.

RESULTS AND EXPERIMENTAL PROCEDURES

The APP and APLP2 knockout mice were produced as described (Zheng et al., 1995). Since Drosophila containing a deletion of its APP homolog (Appl) exhibits behavioral deficits, which can be partially rescued by the human APP695 (Luo et al., 1992), we performed both a general neurological examination and complex learning and memory evaluations on the mice. At 3 months of age, APP null mice have a normal body temperature and pain reactivity, and show the same swimming and balancing ability. However, APP knockout mice have lower body weights, reduced forelimb grip strength and locomotor activity than wild-type controls and approximately 30% of the mice show significant seizure activities.

Two tasks were performed for cognitive examinations: conditional avoidance and performance in a Morris water maze. In the conditioned avoidance paradigm, a light was presented in one half of the testing chamber signaling that a mild electric shock was about to be delivered. A period of 10 sec elapsed before the grid floor was electrified, during which time the mouse could successfully 'avoid' this shock by crossing through a doorway into the other half of the chamber. Mice failing to cross within the 10 sec received a shock. Repeated measures analysis of variance (ANOVA), conducted over the 16 days of training revealed that the APP knockout mice at 4 months of age performed significantly more 'avoids' than their wild-type controls and thus seemed to reflect a better ability to acquire the task.

The Morris water maze experiment was conducted to examine hippocampal-dependent spatial learning and memory ability. In this test, a white circular pool was filled with an opaque white solution and maintained at 26-28C. At a particular position a platform was submerged just below the surface of the solution onto which the mouse could climb. The mouse was released at one of four locations on the edge of the pool and allowed to swim freely. The trial ended when the mouse reached and climbed onto the platform. If it had not climbed onto the platform after 60 sec, the mouse was placed on the platform. The mouse remained on the platform for 30 sec after which it was removed and the next trial began. The mice received four trials each day. A video camera located directly above the pool was linked to an image analyzer and computer which recorded the movements of the mouse and provided a measure of the latency to reach the platform and swimming speed for each trial. The results showed that, in contrast to the conditioned avoidance test, at 8 months of age, the APP knockout mice learned the location of the platform more slowly and during probe trials had a poorer memory for its location than wildtype controls.

Transgenic mice have been produced that express various forms of the human APP gene (Price & Sisodia, 1994). We have generated mice overexpressing the hAPP751 isoform with a V-I familial AD mutation at amino acid 717 (hAPP751 FAD717V-I) driven by the neuronal-specific human Thy-1 promoter. A unique distribution of the hAPP751 FAD717V-I mRNA was observed in the central nervous system with high levels of expression in the cortex, hippocampus, amygdala, the superficial layers of the superior colliculus and the central grey, using in situ hybridization. Western blot analysis showed that the hAPP751 FAD717V-I protein is expressed at equivalent levels to that of the combined endogenous mouse APP isoforms in the line 14-2 derived homozygous animals. No plaque deposition has been observed in these mice at up to 18 month of age.

Successful AD-like neuropathology was reported in transgenic mice overexpressing the hAPP with a V-F mutation (Games et al., 1995). The requirement for a high level of hAPP expression in transgenic mice to produce plaque deposition may be explained by protective effects of the endogenous mouse APP or Aβ. To test this hypothesis, the APP-deficient mice were bred with the hAPP751 FAD 717V-I transgenic mice and mice were produced that express the human APP 751 cDNA with the 717V-I mutation on the mAPP null background. It is our hope that such an approach will help in generating a mouse model of AD that mimics human pathological conditions.

REFERENCES

Games D et al. (1995): Alzheimer-type neuropathology in transgenic mice overexpressing V717F b-amyloid precursor protein. *Nature* 373:523.

Luo L., Tully T and White K (1992): Human amyloid precursor protein ameliorates behavioral deficit of flies deleted for Appl gene. *Neuron* 9:595.

Mullan M and Crawford F (1993): Genetic and molecular advances in Alzheimer's disease. *Trends Neurosci* 16:398.

Price DL and Sisodia SS (1994): Cellular and molecular biology of Alzheimer's disease and animal models. *Ann Rev Med* 45:435.

Selkoe DJ (1994): Cell biology of the amyloid b-protein precursor and the mechanism of Alzheimer's disease. *Annu Rev Cell Biol* 10:373.

Slunt HH, Thinakaran G, Van Koch C, Lo ACY, Tanzi RE and Sisodia SS (1994): Expression of a ubiquitous, cross-reactive homologue of the mouse b-amyloid precursor protein (APP). *J Biol Chem* 269:2637.

Wasco W, Bupp K, Magendantz M, Gusella J, Tanzi R and Solomon F (1992): Identification of a mouse brain cDNA that encodes a protein related to the Alzheimer disease-associated amyloid b protein precursor. *Proc Natl Acad Sci USA* 89:10758.

Zheng H et al. (1995): β-Amyloid precursor protein-deficient mice show reactive gliosis and decreased locomotor activity. *Cell* 81:525.

Alzheimer Disease: From Molecular Biology to Therapy
edited by R. Becker and E. Giacobini

ALPHA2c-ADRENOCEPTOR OVEREXPRESSING MICE AS A MODEL TO STUDY COGNITIVE FUNCTIONS OF ALPHA2-ADRENOCEPTORS

Markus Björklund, Minna Riekkinen and Paavo Riekkinen Jr.
Department of Neurology, University of Kuopio,
Kuopio, Fin 70211, Finland

Jukka Sallinen and Mika Scheinin
Department of Pharmacology and Clinical Pharmacology,
University of Turku, Turku, Finland

Jouni Sirviö
Department of Neuroscience, A.I. Virtanen Institute,
University of Kuopio, Kuopio, Finland

Antti Haapalinna
Orion Corporation, Orion-Farmos, Turku, Finland

Richard E. Link and Brian Kobilka
Stanford University Medical Center, Stanford, CA, USA

INTRODUCTION

Locus coeruleus (LC) noradrenaline (NA) neurons innervate forebrain structures, such as hippocampus, thalamus and cortex, that are important for the regulation of cognitive processes (Foote, 1987). Dysfunction of LC NA cells has been associated with the cognitive decline observed in Parkinson's disease (PD) and in some patients with Alzheimer disease (AD). Indeed, the number of LC cells may be severely depleted in the brains of AD and PD patients.

Alpha2-adrenoceptors (A2-AR) were divided into three subtypes, termed alpha2a, b and c-adrenoceptors, based on both radioligand binding and results from gene cloning (Bylund, 1992). The A2-AR subtypes (2a, 2b, 2c) have different anatomical distribution in brain areas involved in separate functional systems. A2a-ARs are located in the LC, brain stem, cerebral cortex and amygdala. This subtype is located postsynaptically, but is also the most numerous presynaptic receptor and autoreceptor. A2b-ARs gene expression has been localized nearly exclusively at the thalamus, but the density of A2b-AR in brain appears to be low or undetectable. The A2c-ARs are found at the hippocampus, cortex and striatum. Importantly, in the caudate and accumbens nuclei A2c-ARs are the predominant A2-AR subtype. The unique distribution of A2c-ARs suggests that this receptor subtype may modulate the executive function of fronto-striatal/accumbens loops on learned motor responses and hippocampal processing of spatial information.

We have investigated the role of A2c-ARs in the regulation of cognitive functions by studying the behavior of genetically engineered mice that overexpress A2c-ARs in a tissue specific manner. We predicted that if A2c-AR overexpression (OE) is responsible for a performance alteration, the effects of atipamezole and dexmedetomidine should be increased in OE mice. We evaluated non-spatial and spatial reference memory in a water maze (WM) paradigm employing both the visible and hidden platforms. The effects of genetic manipulation on the baseline cortical electrical arousal and on the response of cortical EEG activity to dexmedetomidine were analyzed to study whether the observed WM defect is due to impaired arousal in the OE mice. Further, possible changes in exploratory activity and aversively motivated inhibitory avoidance were measured also to further characterize the selectivity of behavioral defects in OE mice.

MATERIALS AND METHODS

Production of KO and OE mice: The generation of mice with tissue specific overexpression of the A2c-adrenergic receptor (A2c-AR) gene

Mice with overexpression of the A2c-AR were generated by pronuclear injection using standard techniques (Palmiter & Brinster, 1986; Hanahan, 1989). In brief, 1-cell fertilized eggs were harvested from 5-week old super-ovulated FVB/N female mice. The microinjected constructs contained 4.5 kb of 5' flanking sequence and 5 kb of 3' flanking sequence surrounding the *Adra2c* open reading frame, which encodes the murine A2c-AR (Link et al., 1992). In addition, a tyrosinase minigene construct was co-injected for visual identification of the transgenic progeny by different coat color. This construct contained 2.2 kb of 5', 0.7 kb of 3' and 1.3 kb of coding sequence from pBS-Tyrosinase, which has been previously described (Overbeek et al., 1991). Eggs which survived microinjection, as judged by morphology, were transferred to the oviduct of pseudopregnant foster mothers. DNA was collected from tail biopsies of weanling animals from these transfers and analyzed by Southern blotting or PCR for the presence of the adrenergic transgene. The adrenergic transgene and dark coat color were found to cosegregate during breeding. In addition, the intensity of dark coat color was correlated with the number of copies of both the tyrosinase minigene and the adrenergic transgene as detected by Southern analysis. The transgenic mouse line (281#17) which had greatest amounts of gene copies was chosen for further breeding and experiments.

The expression of the *adra2C* gene was determined by receptor autoradiography and by radioligand binding experiments in striatal membranes. The results showed approximately 3-fold tissue-specific A2C-binding in the brains of the overexpressing mice, whereas the A2c-AR binding in the brains of the mice lacking the A2c-AR -gene were greatly reduced in corresponding areas (Link et al., 1995; Sallinen et al., unpublished results). The mice overexpressing the A2c-AR as well as the mice with targeted disruption of the A2c-AR are viable and fertile and appear grossly normal.

Drugs

Dexmedetomidine (Orion-Farmos, Ltd; 0.5-300 ug/kg) and atipamezole (Orion-Farmos Ltd; 30-1000 ug/kg) were dissolved in NaCl 0.9% and injected s.c. in a volume of 5 ml/kg. Drugs were injected 30 min before or immediately after behavioral training.

WM

The mice were trained during 5 consecutive training days to find a clearly visible platform. During all of the training days, five trials in a day with 30 s inter-trial interval were assessed trials. Next, 2 additional visible platform training days were assessed and no drugs were administered during these days. The location of the visible platform was altered every day. The hidden platform training followed the visible platform stage. The platform was located in a fixed place for the next 5 days. The platform was removed from the pool and the time the mice spent on the previous platform location was used as an index of retention (the longer time, the better performance). Drugs were injected every day either before (ati: 30 and 300 ug/kg; dex: 0.5 - 5 ug/kg) or after the daily training. One group of mice was tested with different protocol that started with a free swim trial (no platform) followed by hidden platform reference memory training. In this study, ati was administered to OE and WT mice for 10 days during training (post-training: ati 30 and 300 ug/kg; pretraining: ati 1000 ug/kg), and on the eleventh day treatment conditions were altered: previously post-training and pretraining treated groups received ati treatment before and after training trials, respectively.

PA

Ati (30 - 1000 ug/kg) and dex (2 - 20 ug/kg) were injected before the training trial. PA was evaluated with a step through inhibitory avoidance test. During the single training trial, mice received shock treatment in the dark compartment, and retention was tested 72 hr later.

Cortical EEG measurements
Cortical EEG activity measurements were made from freely moving OE and WT mice and the spectral analysis was calculated from 1 - 60 Hz band.

RESULTS

Effects of A2c-AR overexpression on WM
No change in the distribution of swimming during the free swim trial was observed between any of the OE and WT groups. The OE mice were impaired compared with the WT mice in developing a normal search strategy during the visible and hidden platform stages: OE mice swum more in the peripheral annulus than WT mice. The escape distance of OE mice was not longer than that of the controls during the visible and hidden platform training stages. However, the transfer test showed that OE mice did not cross over the previous platform location as often and swam more in the incorrect annulus that did not contain the escape platform during spatial training.

A2c-AR OE modulates the response of alpha2-drugs on spatial behavior.
In WT mice before training, injected dex 0.5 ug/kg non-significantly decreased escape distances to the visible platform, but the higher doses (2 and 5 ug/kg) had no effect. Dex had no effect on the distribution of swimming in WT mice. Contrary, dex 5 ug/kg further increased searching from the incorrect peripheral annulus only in OE mice during the visible platform stage. Smaller doses (0.5 and 2 ug/kg) had no effect.

During the hidden platform stage, dex 5 ug/kg slightly increased distance values of both WT and OE mice, but increased searching from the incorrect annulus only in the OE mice (5 ug/kg). Dex 0.5 and 2 ug/kg had no effect on escape distance or the distance swam in the incorrect peripheral sector of the pool.

Pretraining ati (30 and 300 ug/kg) treatment decreased dose dependent escape distance values to the visible platform in OE and WT mice, and the decrease was greater in OE mice. In OE mice ati at 30 and 300 ug/kg decreased the distance swum in the incorrect peripheral annulus during the visible platform stage.

During the hidden platform stage, ati (30 and 300 ug/kg) again decreased distance values in OE and WT mice. The excessive swimming in the peripheral zone of the pool by OE mice was normalized by ati treatment.

Post-training treatment with ati (30 and 300 ug/kg) had no effect on escape performance to a hidden platform in WT or OE mice.

Pretraining treatment with ati 1000 ug/kg decreased markedly escape distance values and the distance swum in the incorrect peripheral zone in WT and OE mice that were initially trained to find a hidden platform. In the 12th training day, pretraining ati treatment was discontinued and the distance swum increased robustly in OE mice and to smaller extent in WT mice. Importantly, those WT and OE mice that had been treated with post-training ati (30 and 300) were as accurate as their controls during the first eleven training days. However, on the 12th day, ati was injected before training at 300 and 1000 ug/kg for those WT and OE mice that had been previously treated only after daily training during days 1-10. Ati administered on the day 12 of testing fully normalized abnormal searching for the platform from the incorrect outermost zone in OE mice and decreased escape distance values.

Effects of A2c-AR overexpression on PA

Both WT and OE mice acquired the PA task rapidly and performed well during the testing trial. Dex 2 - 20 ug/kg did not disrupt PA behavior of WT or OE mice.

Cortical EEG activity

No differences were observed between the baseline EEG values of OE and WT mice. Dex (3 - 300 ug/kg) suppressed dose dependent desynchronized EEG activity equally effectively in OE and WT mice.

DISCUSSION

The present results indicate that overexpression of A2c-ARs in mice impair the development of cue and spatial navigation in WM paradigm, and that ati and dex modulate the distribution of exploration in OE mice more effectively than in WT mice. Therefore, it is likely that the increased searching of the platform from the incorrect peripheral zone by the OE mice is caused by dysfunctioning of A2c-ARs.

Our results further suggest that alteration of arousal or anxiety does not cause the WM defect observed in OE mice. **First**, baseline cortical EEG activity and the EEG response to active drugs was similar in OE and WT mice. The present result that the cortical electrical arousal was not adversely affected by A2c-AR overexpression is important, since several earlier studies have described that treatment with subtype non-selective alpha2-agonist, such as dex and clonidine, decreases vigilance (Riekkinen Jr, 1993). Indeed, we have previously observed that in rats dex, over a dose range that sedates rats, impairs EEG arousal. Therefore, the present result showing that overexpression of A2c-ARs has no effect on cortical EEG activity and does not modulate the EEG response to dex in mice, suggests that this receptor subtype is not important for the sedating effect of classical alpha2-agonists. **Second**, the identical behavior and dose responses of ati and dex treatments in the OE and WT mice during the PA training/testing trial and WM free swim trial suggest that the OE mice were not more anxious than WT mice. Therefore, the decrease induced by ati treatment and increase induced by dex treatment of searching from the peripheral zone of the WM pool during the escape training trials were probably not due to modulation of anxiety.

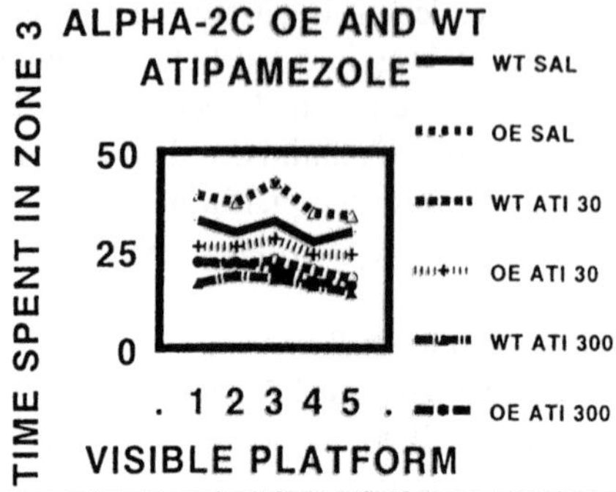

FIGURE 1. Alpha2c-overexpressing mice swim more in the incorrect peripheral annulus during visible and hidden platform stages of water maze training. Atipamezole, an alpha2-antagonist, decreased time spent in the incorrect zone and the effect of atipamezole was more pronounced in overexpressing than wild type mice. Y-axis: training days. X-axis: time spent in the incorrect zone. WT = wild type mice, OE = alpha2c-overexpressing, ATI 30, 300 = atipamezole 30 and 300 ug/kg sc. 15 min. before daily trainings, Sal = NaCl 0.9 %.

Our results suggest that overactivation of A2c-ARs may disrupt execution of navigation strategies dependent on the use of proximal intramaze (cue navigation) and distal extramaze (spatial navigation) cues. Therefore, it is relevant to note that the functioning of certain brain areas containing A2c-ARs, such as dorsal and ventral striatum and hippocampus, is important for the processing of specific components of cue and/or place navigation behavior. The involvement of the hippocampal formation in spatial navigation is supported by several studies. For example, lesion of the dorsal hippocampus disrupts cue navigation in WM pool. Importantly, frontal cortical and dorsal hippocampal inputs may interact at the level of ventral striatum (accumbens nucleus) to modulate motor responses important for spatial behavior (Pennartz, 1994). Contrary, the dorsal striatum is a component of memory system that mediates cue navigation performance in WM pool. Navigation to a clearly visible platform was disrupted by lesion of dorsal striatum, but escape performance to a hidden platform was unaffected. Actually, striatal lesioned rats that accurately learned a spatial navigation strategy and found the hidden escape platform, avoided escape to a clearly visible platform. Further, striatal lesioned rats continued to swim in the previous location of the hidden platform, suggesting that the rats could use only spatial navigation mediated by the hippocampal system. Interestingly, fimbria fornix (ff) damage induced a qualitatively opposite effect, as ff-lesioned rats did not learn the location of hidden platform and rapidly acquired escape to a visible platform.

Further studies should explore the involvement of dorsal and ventral striatal and hippocampal A2c-ARs in the modulation of cue and spatial navigation behavior to elucidate the site of action of A2c-AR active drugs to modulate cognitive processing. Furthermore, these studies will also further characterize the qualitative nature of the mnemonic processes that are under A2c-AR modulation. The results of these studies may reveal novel therapeutic potentials for A2c-AR active drugs in the treatment of cognitive dysfunctions associated with PD and AD.

REFERENCES

Bylund DB (1992) Subtypes of α1- and α2-adrenergic receptors, *FASEB J* 6:832-839.

Foote SL and Morrison JH (1987) Extrathalamic modulation of cortical functioning. *Annu Rev Neurosci* 10:67-95.

Hanahan D (1989): Transgenic mice as probes into complex systems. *Science* 246(4935):1265-75.

Link R, Daunt D, Barsh G, Cruiscinski A, Kobilka B (1992): Cloning of two genes encoding alpha2-andrenergic receptor subtypes and identification of a single amino acid in the mouse alpha 2-C10 homolog responsible for an interspecies variation in antagonist binding. *Mol Pharmacol* 42(1):16-27.

Link RE, Stevens MS, Kulatunga M, Scheinin M, Barsh GS, and Kobilka BK (1995) Targeted Inactivation of the Gene Encoding the Mouse alpha2C-Adrenoceptor Homolog. *Mol Pharmacol* 48:48-55.

Overbeek PA, Aguilar-Cordova E, Hanten G, Schafner DL, Patel P, Lebovitz RM, Lieberman MW (1991): Coinjection strategy for visual identification of transgenic mice. *Transgenic Res* 1:31-37.

Palmiter RD, Brinster RL (1986): Germ-line transformation of mice. *Ann Rev Genet* 20:465-499.

Pennartz CMA, Groenewegen HJ, and Lopes-daSilva FH (1994) The Nucleus accumbens as a complex of functionally distinct neuronal ensembles: an integration of behavioural, electrophysiological and anatomical data, *Progress in Neurobiology* 42: 719-761.

Reikkinen P Jr, Lammintausta R, Ekonsalo T, Sirvio J (1993): The effects of alpha 2-andrenoceptor stimulation on neocortical EEG activity in control and 6-hydroxydopamine dorsal noradrenergic bundle-lesioned rats. *Eur J Pharmacol* 238(2-3):263-272.

Alzheimer Disease: From Molecular Biology to Therapy
edited by R. Becker and E. Giacobini

β-APP-751 TRANSGENIC MICE: DEFICITS IN LEARNING AND MEMORY

Paula M. Moran
Marion Merrell Dow Research Institute, 16 rue D'Ankara, 67080 Strasbourg, CEDEX, France

Paul C. Moser
Synthelabo Recherche, Reuil-Malmaison, France

Linda S. Higgins and Barbara Cordell
Scios Nova Inc., Mountain View, CA 94043, USA

INTRODUCTION

While the etiology of Alzheimer's disease (AD) is unknown, one of the primary histopathological features of the disease is the accumulation in brain of extracellular deposits of the peptide ß-amyloid (ß-A4). This peptide is produced by the proteolytic processing of the larger precursor protein ß-amyloid precursor protein (ß-APP).

Producing an animal model with construct validity has been especially problematic in the AD research field, as very few animals naturally develop the histopathological changes that are seen in AD. Rodents do not develop amyloid deposits and the animals that do, namely aged primates and polar bears, are not practical experimental animals. Because of this, animal models relating to AD have largely centered on lesioning areas of the brain known to be especially vulnerable to synaptic loss in AD such as the basal forebrain or the septo-hippocampal system. Other approaches have been to induce memory deficits pharmacologically, most commonly using cholinergic antagonists, or to study behavioural and neurochemical changes that occur with normal aging.

One recent experimental approach to modeling disease states is the use of genetic engineering techniques in animals. This technology has allowed us to introduce a gene from one species into the germline of another. In the case of animal models of AD such techniques have now made it possible to express human ß-APP in mice in an attempt to produce histopathological features similar to those seen in AD.

We have produced transgenic mice which express the 751-amino acid isoform of human ß-APP. There are at least six different mRNA transcripts produced by splice variants of the ß-APP gene. Four of these isoforms (695, 714, 751 and 770 amino acids in length) contain the ß-A4 segment. Expression of the human ß-APP-751 cDNA from a neuron specific promoter in these mice leads to diffuse deposits of ß-A4 and increased tau protein immunoreactivity, particularly in the hippocampus, cortex and amygdala (Quon et al., 1991, Higgins et al., 1994). It was, therefore, of interest to determine whether these transgenic mice displayed any behavioural abnormalities that might be consistent with expression of human ß-APP-751.

Production of these mice has been described in detail elsewhere (Quon et al., 1991). Female JU mice homozygous for the transgene of human ß-APP-751 cDNA under the control of a neuron-specific enolase promoter were used. Two age groups were tested, one of 5-6 months (n=12 wild type(wt); n=11 transgenic) and one group of 9-12 months old (n=12 per group). Behavioural testing of these mice was divided into three categories of test: 1) Neurological tests; string test, rotarod test, circadian variation in body temperature; 2) General behavioural tests; 24 hr activity, spontaneous alternation, plus maze anxiety, behavioural despair; and 3) Learning test; water maze.

MATERIALS AND METHODS

Methods for neurological and behavioural tests and detailed description of water-maze experimental methods have been described previously (Moran et al., 1995).

Water Maze

Briefly, following a 60 sec habituation trial to verify their ability to swim, the mice were given eighteen training trials during which their latency to find the hidden platform was measured up to a maximum of 60 sec. A probe trial consisted of a 60 sec trial during which there was no platform present in the maze and during which its search pattern was analyzed. The percentage of time spent searching the quadrant of the maze that formerly contained the platform was calculated and compared with the percentage of time spent searching the other quadrants.

In the visible platform version of the task a white platform was raised above the surface of the water and a flag placed in the middle to make it more visible. Latency to locate the platform was recorded up to a maximum of 60sec.

RESULTS

There were no significant differences between ß-APP-751-transgenic and wild-type control mice in either 6 or 12 month groups in the string test of muscle strength, the rotarod test of motor coordination, in circadian variation in body temperature, the plus maze test of anxiety nor in the behavioural despair test. In 12 month mice the pattern of locomotor activity over 24 hr was similar in both groups, but the overall level of activity of the transgenic group was significantly lower than controls during the dark phase in both 6 and 12 month old mice ($P<0.01$). There were no significant differences between transgenic and wild-type groups in spontaneous alternation scores in six month old mice. In the 12 month old groups, transgenic mice alternated significantly less than wild-type control mice ($P < 0.05$ Mann-Whitney *U*-test).

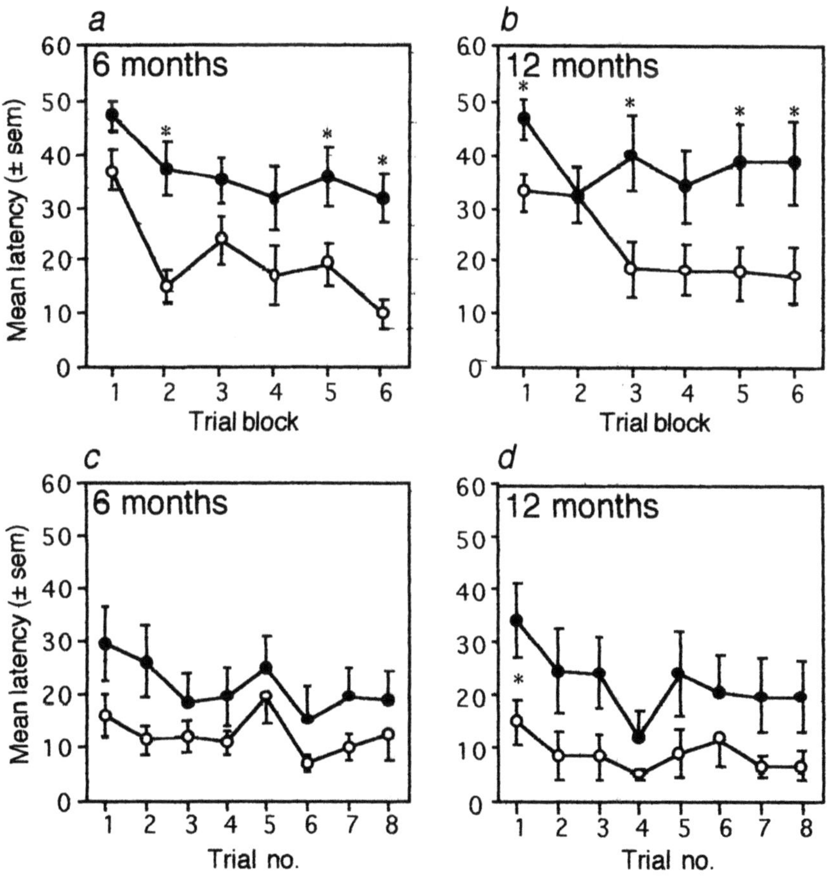

FIGURE 1 (*a & b*): Mean latencies to locate <u>hidden</u> platform in 6 and 12 month old mice. Filled circles are ß APP 751 mice, unfilled circles are wild-type mice. (*c & d*): Mean latencies to locate <u>visible</u> platform in 6 and 12 month old mice.

In the water-maze, latencies to locate the hidden platform by both the six month ß-APP-751 mice ($F_{1,22} = 12.31$, $P = 0.002$) and the 12 month APP-751 mice ($F_{1,20} = 6.19$, $P = 0.02$) were higher than in wild type controls (Fig.1 *a-b*). However, the latency of wild-type and APP-751 groups of both ages decreased similarly with training.

In the visible platform task there was no overall significant difference between APP-751 and wild-type mice (Fig.1 *c-d*).

Differences in time spent in each quadrant were analyzed for each probe trial (Figure 2). In 6 month old mice, by probe trials 2 ($F_{3,33}=3.9$, $P < 0.05$) and 3 ($F_{3,33}=22.56$, $P < 0.0001$), wt mice selectively searched the training quadrant. In the transgenic group this selective searching failed to reach significance on probe trial 2, but was again evident by probe trial 3 ($F_{3,33}=5.17$, $P < 0.005$).

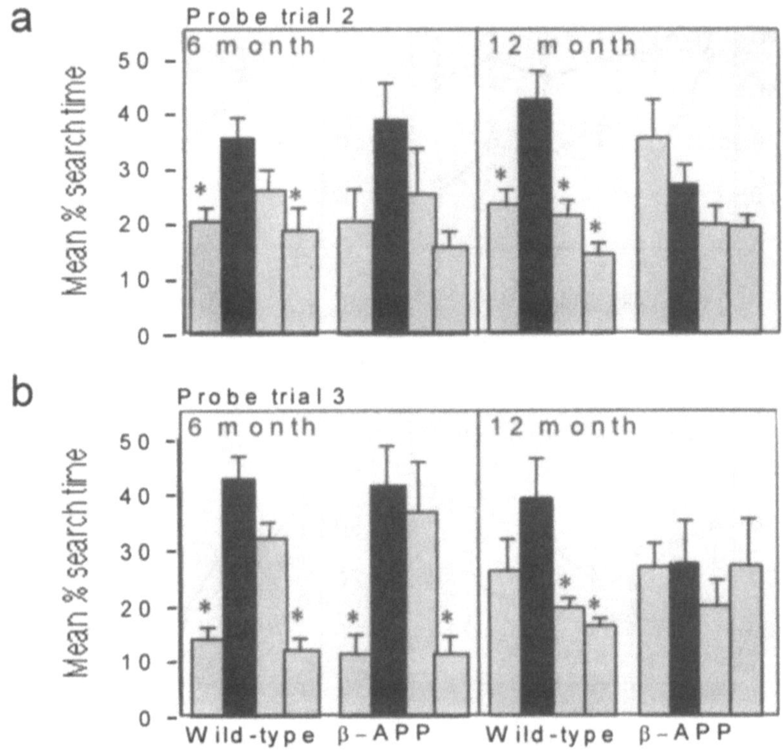

FIGURE 2. Mean % search times per maze quadrant on 2nd (*a*) and 3rd (*b*) probe trials in water maze. The columns represent the maze quadrants from left to right, Adjacent left, Training (filled column), Adjacent right and Opposite. * significant difference ($P < 0.05$) from training quadrant.

In 12-month old mice, during the second and third probe trials, the wt mice searched the training quadrant significantly more than other quadrants (Probe trial 2: $F_{3,33} = 8.8$, $P < 0.001$; Probe trial 3: $F_{3,33} = 5.14$, $P < 0.01$). No significant differences were apparent in this measure for the transgenic mice (Probe trial 1: $F_{3,27} = 1.32$, ns; Probe trial 2: $F_{3,27} = 1.91$, ns; Probe trial 3: $F_{3,27} = 0.28$, ns). Thus, 12 month old transgenic mice did not show any evidence of learning the spatial location of the platform.

DISCUSSION

Our results demonstrate that mice expressing human ß-APP-751 have learning deficits in the absence of marked neurological deficits or behavioural abnormalities. They also show that these deficits were present only in the 12 month old mice suggesting that they are age-related.

Previous behavioural studies on mice carrying a transgene containing the 695 isoform of human APP, have reported abnormalities in activity and sensorimotor function, but no marked spatial learning abnormalities (Yamagucci et al., 1991; Perry et al., 1995). As we have previously reported that APP-695 transgenic mice display no histopathology (Higgins et al., 1993), this indirectly supports a link between histopathology and the learning impairments that we have observed. However, direct confirmation of such a link awaits future studies examining the precise correlation between the histopathology and learning deficit.

It will be of interest in future studies to assess explicitly whether these mice also have difficulties with other types of information processing such as working memory or attention as the deficit in spontaneous alternation could be considered to have a rudimentary working memory component.

A reproducible behavioural deficit that can be linked to the primary histopathological features of AD will represent a major step forward in producing an animal model in which potential new therapies can be evaluated. To date, it has been practically impossible to test novel compounds targeting amyloid or tau protein in behavioural models of relevance to AD. While much characterization still remains to be done, models such as the one described here will make behavioural testing of potential new therapies of this kind possible in the future.

REFERENCES

Higgins LS, Catalano R, Quon D and Cordell B (1993): Transgenic mice expressing human ß-APP-751 but not mice expressing ß-APP 695, display early Alzheimer's disease-like pathology. *Ann NY Acad Sci* 695: 224-227.

Higgins LS, Holtzman DM, Rabin J, Mobley WC and Cordell B (1994): Transgenic mouse brain histopathology resembles early histopathology. *Ann of Neurol* 35: 598-607.

Moran PM, Higgins LS, Cordell B and Moser PC (1995): Age-related learning deficits in transgenic mice expressing the 751-amino acid isoform of human ß-amyloid precursor protein. *Proc Natl Acad Sci* 92:5341-5345.

Perry TA, Torres E, Czech C, Beyreuther K, Richards SJ and Dunnett SB (1995): Cognitive and motor function in transgenic mice carrying excess copies of the 695 and 751 amino acid isoforms of the amyloid precursor protein gene. *Alz Res* 1:5-14.

Quon D, Wang Y, Catalano R, Marian Scardina J, Murakami K and Cordell B (1991): Formation of ß-amyloid protein deposits in brains of transgenic mice. *Nature* 352:239-241.

Yamaguchi F, Richards SJ, Beyreuther K, Salbaum M, Carlson SA and Dunnett SB (1991): Transgenic mice for the amyloid precursor protein 695 isoform have impaired spatial memory. *Neuroreport* 2:781-784.

Part II

THE CHOLINERGIC SYSTEM IN BRAIN

Treatment of Alzheimer Disease
Cholinesterase Inhibitors in Alzheimer Disease Treatment

Alzheimer Disease: From Molecular Biology to Therapy
edited by R. Becker and E. Giacobini

CENTRAL AND PERIPHERAL CONSEQUENCES OF CHOLINERGIC IMBALANCE IN ALZHEIMER'S DISEASE

Daniela Kaufer-Nachum, Alon Friedman, Meira Sternfeld, Shlomo Seidman, Rachel Beeri, Christian Andres and Hermona Soreq
Department of Biological Chemistry, The Life Sciences Institute,
The Hebrew University of Jerusalem, Israel 91904

INTRODUCTION

Cholinergic neurotransmission in the central and the peripheral nervous system is a primary feature common to all vertebrates (Soreq and Zakut, 1993). Rapid, fine tuning sustains well balanced ratios between key cholinergic constituents during cognitive processes such as learning and memory. To assure such higher brain functions, the levels of acetylcholine (ACh) should be tightly regulated. This is achieved by maintenance of a dynamic equilibrium between the levels of the synthesizing enzyme choline acetyl transferase (ChAT), the hydrolyzing enzyme acetylcholinesterase (AChE; Schwarz et al., 1995) and the high affinity choline transporter responsible for the re-uptake of choline (Slotkin et al., 1994). Changes in this equilibrium occur in physiopathological syndromes such as Alzheimer's disease (AD) (Beeri et al., 1995, Andres et al., 1996). Cholinergic imbalance is associated with characteristic modifications in the electrophysiological properties of certain regional circuits and consequent disruption of central transmission systems. Therefore, perturbed cholinergic communication may cause long-term central responses affecting cognitive, autonomous and neuromotor performance. To delineate the molecular mechanisms leading from cholinergic imbalance to dysfunction, we combined transgenic and gene expression tests with electrophysiological and molecular biology analyses in in vivo and in vitro (brain slice) systems from various species.

RESULTS AND DISCUSSION

Transgenic overexpression of human AChE should enhance the rate of ACh hydrolysis, creating a cholinergic imbalance. We have used this experimental approach in two evolutionarily distinct organisms: developing tadpoles of the frog Xenopus laevis (Ben Aziz-Aloya et al., 1993; Shapira et al., 1994; Seidman et al., 1995) and the laboratory mouse (Beeri et al., 1995; Andres et al., 1996). The Xenopus system is versatile and economical (Seidman and Soreq, 1996). Therefore, it provides the opportunity for rapid changes of the microinjected DNA and enables us to demonstrate that the synaptic accumulation of AChE depends on the 3'-exon No. 6 of the human AChE gene. However, the transgenic DNA is not integrated in Xenopus, limiting the duration of the induced imbalance. In contrast, the human transgene has been stably introduced into the mouse genome, ensuring genetic transmission of the imbalance situation from one generation to the next.

Certain peripheral consequences of AChE overexpression were common to the frog and the mouse systems. In both systems, electron microscopy analyses demonstrated structural abnormalities in neuromuscular junctions (NMJs), which were similarly enlarged in Xenopus tadpoles and in adult mice. In addition, secondary changes in the levels of muscle nicotinic ACh receptors in Xenopus NMJs were indicated by the increased α-bungarotoxin binding (Shapira et al., 1994). In mouse NMJs, abnormal electrophysiology and post-excitation fatigue in older animals reflected progressive deterioration of neuromotor function (Andres et al., 1996).

Cholinergic pathways in the central nervous system (CNS) were studied in the transgenic mice (Beeri et al., 1995, Andres et al., 1996). These analyses demonstrated acquired resistance to the hypothermic responses induced by AChE inhibitors and cholinergic agonists as well as progressive deterioration of learning and memory capacities. Table I summarizes these consequences of transgenic overexpression of human AChE in the Xenopus and mouse systems.

In humans, natural mutagenesis provides a readily available example for the consequences of cholinergic imbalance (Loewenstein-Lichtenstein et al., 1995). Homozygous carriers of the "atypical" mutation in butyrylcholinesterase (BuChE) have lower activities of serum BuChE and hence lower capacities for scavenging anti-AChE drugs and poisons. Such individuals are therefore vulnerable to both organophosphorous insecticides and muscle relaxants under conditions which would be non-toxic to normal individuals (Prody et al., 1989; Ehrlich et al., 1994; Schwarz et al., 1995). Recently, we found that the "atypical" serum BuChE, unlike its normal counterpart, also displays a reduced capacity for interacting with tacrine, physostigmine and pyridostigmine. Reduced scavenging in blood may lead to increased concentration of such drugs in the brain. Thus, the paucity of brain AChE activity that is caused by excessive drug interactions in individuals with deficient BuChE interferes with appropriate cholinergic functions in a manner reciprocal to that observed for the consequences of transgenic AChE overexpression.

Property	Xenopus laevis tadpoles	adult transgenic mice
1. Transgene - host DNA interaction	None reported	Stable integration
2. Duration of transgene expression	Transient, embryonic	Indefinite, embryonic and adult life
3. Synaptic accumulation	+ (with 3'-exon 6)	+ (with 3'-exon 6)
4. Effects on NMJ structure	- Increased post-synaptic length - Wider synaptic cleft	- Increased synapse area - Changes in post-synaptic folds
5. NMJ function	- Double α-bungarotoxin binding - Visual observation of abnormal swimming pattern (Kaufer-Nachum and Soreq, unpublished).	- Delayed potentials; - Enlarged motor units; - Post-excitation fatigue; - Progressive deterioration
6. CNS cholinergic synapses	- Synaptic accumulation	- Synaptic accumulation - Acquired resistance to hypothermia induced by anti-AChEs and cholinergic agonists - Progressive deterioration of learning and memory

TABLE I. Central and Peripheral consequences of transgenic overexpression of human AChE

An opposite cholinergic imbalance can be achieved by exposing the CNS to AChE inhibitors, thus creating an overload of acetylcholine in the synapses. Such exposure, for example - in occupational poisoning by organophosphorous insecticides, was reported to cause delayed deterioration in cognitive function resembling Alzheimer's type dementia (quoted in Soreq and Zakut, 1993). In order to test the molecular mechanism(s) underlying the interconnection between the cholinergic imbalance and nervous system functions, we developed a convenient in vitro system by extending the use of brain slices (Friedman and Gutnick, 1989) for molecular neurobiology analyses. RNA extractions followed by reverse-transcriptase PCR amplifications revealed that these slices maintain stable levels of intact messenger RNAs encoding for synaptic proteins (i.e. synaptophysin), ion channels (i.e. L-type Ca^{++} channel), cholinergic proteins (i.e. AChE, ChAT) and early immediate genes (i.e. the oncogene c-fos), for at least 11 hours. Brain slices incubated with AChE inhibitors (organophosphates, pyridostigmine) for 1 hr displayed a >100 fold increase in the levels of c-fos mRNA, an accepted marker of enhanced neuronal excitability, a >50 fold increase in AChE mRNA and a 100-fold decrease in ChAT mRNA. No increase was detected in synaptophysin or L-type Ca^{++} channel mRNAs, demonstrating the restriction of this short-term response to transcripts related with cholinergic transmission. Parallel electrophysiological recordings displayed increased neuronal excitability in the CA1 area of the hippocampus following pyridostigmine addition. This observation indicates that brain slices respond to the addition of AChE inhibitors by enhanced excitability and synchrony of the local circuit associated with transcriptional changes leading to subsequent suppression of the capacity for ACh synthesis and enhancement of its hydrolysis, essentially causing a shut-down of cholinergic neurotransmission.

While similar transcriptional modulations were also observed in brain tissue from mice injected i.p. with 2 mg/kg pyridostigmine, the slice system offers clear advantages for testing the molecular mechanisms underlying the detected changes. Thus, we were able to demonstrate that the electrophysiological as well as the transcriptional response to cholinesterase inhibitors could be modulated by pre-incubating the brain slices with different agents such as the intracellular Ca^{++} chelator Bapta-AM, the Na^{+} channel blocker tetrodotoxin or the M_1 antagonist scopolamine. Therefore, excessive activation of cholinergic pathways appears to induce a selective feedback response, dependent on the activation of muscarinic receptors, Na^{+} influx and release of intracellular Ca^{++} stores, all leading to prolonged, secondary cholinergic imbalance. Similar analyses, now in progress, are aimed at evaluating the electrophysiological and post-transcriptional responses of brain slices from AChE-overexpressing transgenic mice to AD drugs. These studies will offer a closer look at the functional consequences of cholinergic imbalance in the mammalian brain.

ACKNOWLEDGMENTS

This research has been supported by the U.S. Army Medical Research and Development Command (Grant DAMD 17-94-C-4031, to H.S.) and by the Smith Foundation for Psychobiology (to A.F.).

REFERENCES

Andres C, Beeri R, Friedman A, Lev-Lehman, E, Timberg R, Shani M and Soreq H (1996): Transgenic AChE induces neuromuscular deterioration in mice. (Submitted for publication)

Beeri R, Andres C, Lev-Lehman E, Timberg R, Huberman T, Shani M and Soreq H (1995): Transgenic expression of human acetylcholinesterase induces progressive cognitive deterioration in mice. Curr Biol 5:1063-1071.

Ben Aziz-Aloya R, Seidman S, Timberg R, Sternfeld M, Zakut H and Soreq H (1993): Expression of a human acetylcholinesterase promoter-reporter construct in developing neuromuscular junctions of Xenopus embryos. Proc Natl Acad Sci USA, 90: 2471-2475.

Ehrlich G, Ginzberg D, Loewenstein Y, Glick D, Kerem B, Ben-Ari S, Zakut H and Soreq H (1994): Population diversity and distinct haplotype frequencies associated with AChE and B ChE genes of Israeli Jews from Trans-Caucasian Georgia and from Europe. Genomics 22:288-295.

Friedman A and Gutnick MJ (1989): Intracellular calcium and control of burst generation in neurons of guinea-pig neocortex in vitro. Eur J Neurosci 1:374-381.

Loewenstein-Lichtenstein Y, Schwarz M, Glick D, Norgaard-Pedersen B, Zakut H and Soreq H (1995): Genetic predisposition to adverse consequences of anti-cholinesterases in "Atypical" BChE carriers. Nature/Medicine 1:1082-1085.

Prody AC, Dreyfus P, Zamir R, Zakut H and Soreq H (1989): De novo amplification within a "silent" human cholinesterase gene in a family subjected to prolonged exposure to organophosphorous insecticides. Proc Natl Acad Sci USA. 86:690-694.

Schwarz M, Glick D, Loewenstein Y and Soreq H (1995): Engineering of human cholinesterases explains and predicts diverse consequences of administration of various drugs and poisons. Pharmacol and Therap 67:283-322.

Seidman S, Sternfeld M, Ben Aziz-Aloya R, Timberg R, Kaufer D and Soreq H (1995): Synaptic and epidermal accumulation of human acetylcholinesterase is encoded by alternative 3'-terminal exons. Mol Cell Biol 15:2993-3002.

Seidman S and Soreq H (1996): Transgenic Xenopus Microinjection Methods and Developmental Neurobiology. Humana Press, Neuromethods Series. (in press)

Shapira M, Seidman S, Sternfeld M, Timberg R, Kaufer D, Patrick JW and Soreq H (1994): Transgenic engineering of neuromuscular junctions in Xenopus laevis embryos transiently overexpressing key cholinergic proteins. Proc Natl Acad Sci USA 91:9072-9076.

Slotkin TA, Nemeroff CB, Bissette G and Seidler FJ (1994): Overexpression of the high affinity choline transporter in cortical regions affected by Alzheimer's disease. J Clin Invest 94:696-702.

Soreq H and Zakut H (1993): Human Cholinesterases and Anticholinesterases, Academic Press, San Diego.

Ehrlich G, Ginzberg D, Loewenstein Y, Glick D, Kerem B, Ben-Ari S, Zakut H and Soreq H (1994): Population diversity and distinct haplotype frequencies associated with ACHE and BCHE genes of Israeli Jews from Trans-Caucasian Georgia and from Europe. *Genomics* 22: 288-295

[illegible]

Alzheimer Disease: From Molecular Biology to Therapy
edited by R. Becker and E. Giacobini

THE ANATOMY OF MONOAMINERGIC-CHOLINERGIC INTERACTIONS IN THE PRIMATE BASAL FOREBRAIN

John F. Smiley and M.-Marsel Mesulam
Center for Behavioral and Cognitive Neurology and the Alzheimer Program, Northwestern University Medical School, Chicago, IL

All areas of the cerebral cortex receive a dense cholinergic innervation which arises from the medial septal nucleus (Ch1), vertical and horizontal limb nuclei of the diagonal band (Ch2 and Ch3), and nucleus basalis of Meynert (nbM, Ch4). Neuroanatomical experiments in the monkey have shown that the hippocampal complex receives its cholinergic input predominantly from Ch1 and Ch2 whereas the cerebral neocortex and amygdala receive their cholinergic inputs predominantly from Ch4. The cholinergic innervation of the cerebral cortex constitutes a major component of the Ascending Reticular Activating System and plays an important role in a variety of behavioral realms, including memory, learning, arousal, and attention.

One of the most consistent features of Alzheimer's disease (AD) is an early and severe degeneration of cholinergic axons in the cerebral cortex and a corresponding loss of cholinergic cells in the nbM (Geula and Mesulam, 1994). The role of cholinergic depletion in the pathophysiology and associated dementia of AD remains unresolved. However, the enhancement of cortical cholinergic transmission in AD remains a major goal for many of the pharmacological agents that are currently in use or under development.

The basal forebrain receives heterogeneous inputs from the brain stem and forebrain. These inputs provide potential therapeutic targets in AD because they can influence the activity of cholinergic neurons and the cortical release of acetylcholine. This chapter summarizes some observations on the monoaminergic input into cholinergic basal forebrain nuclei. Since the cortical cholinergic innervation displays major interspecies variations, we will confine our comments to observations made in the human and monkey brains.

METHODS

Monoamine axons were immunolabeled with the avidin-biotin peroxidase method as described previously (Geula et al., 1993) using antibodies that recognize tyrosine hydroxylase (TH), dopamine ß-hydroxylase (DBH), and serotonin or trypthophan hydroxylase. The resulting immunoreactivities were used as markers for dopaminergic, noradrenergic, and serotonergic pathways, respectively. For double labeling, tissue sections were processed for the concurrent labeling of one monoaminergic marker with cholinergic markers such as choline acetyl transferase (ChAT) or nerve growth factor receptor (NGFr). In these preparations, immunolabeling with a brown diaminobenzidine reaction product was followed by immunolabeling with a purple "VIP-red" reaction product (Vector Laboratories, Burlingame, CA). The antibodies to trypthophan hydroxylase, ChAT, and NGFr were gifts of John Haycock, Louis Hersh, and Mark Bothwell, respectively. Axon quantification was done by digitizing images of 0.25x0.33mm microscopic fields and counting axon intersections with lines of an overlaid grid of 9 vertical and 9 horizontal lines. Axon densities were classified as *very sparse* (0-20 intersections per) *light* (21-100 intersections) *intermediate* (101 to 200 intersections) *dense* (201-300 intersections) and *very dense* (>300 intersections).

RESULTS

Dopamine axons

Dopamine axons were detected with tyrosine hydroxylase (TH) immunoreactivity. Although TH is also present in noradrenergic axons, it is thought to be a marker predominantly of dopamine axons. In both human and monkey brains, double labeling showed cholinergic cells embedded in a thick tangle of varicose TH axons of heterogeneous sizes. The density of TH axons was at intermediate to very dense levels in all areas of the basal forebrain except Ch1 and Ch2, where it was light. The cholinergic cells of the Ch1 and Ch2 sectors project to the hippocampus (Mesulam et al., 1983), and these results suggest a possible differential dopaminergic control of cortical versus hippocampal cholinergic neurotransmission. Previous studies in rats (Zaborszky et al., 1993) and humans (Gaspar et al., 1985) also reported a low density of dopamine axons in the medial septal area.

Electron microscopy in the nbM showed that TH axons formed numerous, modestly asymmetric synapses which were comparatively large (approximately 0.21 to 0.56 microns in diameter). Double labeling electron microscopy demonstrated that TH synapses onto ChAT dendrites were common. Preliminary observations indicate that most (8 of 12) TH synapses were onto ChAT dendrites, suggesting that cholinergic cells constitute a major target of dopamine synapses in the nbM. Tract-tracing studies have concluded that the dopamine projections to the basal forebrain arise from the midbrain ventral tegmental area and/or substantia nigra (Russchen et al., 1985; Semba et al., 1988; Jones and Cuello, 1989).

TH-immunoreactive cells were also present in the hypothalamus and, on rare occasion, in the basal forebrain. These cells do not appear to synthesize dopamine, but may provide an alternative source of TH for the basal forebrain (Wisniowski et al., 1992).

Noradrenergic axons

Noradrenergic axons were visualized with dopamine ß-hydroxylase (DBH) immunoreactivity. In monkey and human brains, light to intermediate densities of these axons were detected among cholinergic cell groups throughout the basal forebrain. In double labeled sections, DBH axons did not form dense baskets around cholinergic cells as was reported in rats (Zaborszky et al., 1993). However, it was common to see complex DBH immunoreactive baskets on unlabeled cells, suggesting the presence of a substantial noradrenergic input to non-cholinergic cells of the basal forebrain.

Double labeling at the light microscopic level of analysis showed many apparent contacts between DBH axons and cholinergic cells, suggesting that there is abundant opportunity for direct synaptic input. Electron microscopy in the nbM showed that DBH-immunoreactive terminals make axodendritic synapses. These terminals have numerous dense core vesicles in addition to clear synaptic vesicles. Unlike TH axons, DBH terminals formed subtle symmetric synapses which were comparatively small (approximately 0.07 to 0.28 microns in diameter). Tract tracing in rats and monkeys indicates that the noradrenergic input into the basal forebrain arises from the nucleus locus coeruleus (Jones and Yang, 1985; Russchen et al., 1985).

Serotonergic axons

Serotonergic axons were labeled with serotonin antibodies in the monkey and with tryptophan hydroxylase antibodies in the human. The concurrent labeling of NGFr and serotonin in the monkey brain showed that serotonin-immunoreactive, bouton-like structures made apparent contacts with cholinergic cells throughout the basal forebrain. Serotonergic neurites were present as a dense sheet of small to medium sized varicose axons in Ch1-Ch4. In the human brain, initial experiments with tryptophan hydroxylase immunoreactivity showed a similar density of labeled axons in the nbM. Electron microscopy in the nbM of monkeys showed that serotonin terminals made relatively large symmetric synapses. Tract tracing studies have shown that both the medial and dorsal raphe nuclei contribute to the serotonergic input into the basal forebrain (Russchen et al., 1985; Zaborszky et al., 1993).

In Alzheimer's disease (AD)
Peliminary observations in AD (3 AD brains and 3 controls) were made on TH- and DBH-immunoreactive axons in the nbM. The density of DBH axons ranged from 8 to 38 intercepts in AD, and from 23 to 57 in controls. The density of TH axons ranged from 141 to 289 in AD and 141 to 240 in controls. The wide range of these values could reflect individual variations or variations in agonal and post-mortem parameters. Nevertheless, these results suggest that dopamine and norepinephrine axons are largely preserved in the nbM of AD.

DISCUSSION

This chapter reviews preliminary studies related to the dopaminergic, serotonergic, and noradrenergic innervation of the nbM in the human and monkey brains. At least in the rat, these monoamines have been shown to influence the activity of cholinergic cells in the basal forebrain and the release of acetylcholine in the cerebral cortex (Pepeu et al., 1983; Casamenti et al., 1986; Napier and Potter, 1989; Maslowski and Napier, 1991; Khateb et al., 1993; Day and Fibiger, 1993).

Our results indicated that the nbM in both the monkey and human brain receives substantial input from all three of these monoamine systems. Density measurements suggested that this input is fairly homogeneous throughout the Ch1-Ch4 sectors of the cholinergic basal forebrain, the only exception being a comparatively low density of dopamine axons in the Ch1 and Ch2 sectors. Double immunolabeling showed that all three axon systems come in close contact with cholinergic cells of the basal forebrain. Electron microscopic observations demonstrated that dopaminergic, serotonergic, and noradrenergic terminals make synaptic contacts with dendrites of nbM neurons. At this time, the electron microscopic analysis of double immunolabeling has been done only with TH and ChAT. These studies have shown that TH-immunoreactive (presumably dopaminergic) terminals make synapses onto dendrites of cholinergic nbM cells.

Experiments in rats showed that DBH-immunoreactive axons contacted most cholinergic cells of the basal forebrain except for the cortically projecting neurons of the ventral pallidum (Zaborszky et al.., 1993). Our results are slightly different since they indicated that all cholinergic cell groups of the basal forebrain are invested with numerous DBH-immunoreactive axons. We also detected many non-cholinergic nbM cells enveloped by DBH-immunoreactive axonal baskets. This finding is consistent with studies in the rat which showed numerous DBH synapses on non-cholinergic cells of the basal forebrain (Zaborszky et al., 1993).

Our preliminary observations suggest that dopaminergic and noradrenergic axons are generally preserved in the nbM of patients with AD. This is consistent with biochemical measurements which have found a preservation of dopamine and norepinephrine in the basal forebrain in AD (Beal et al., 1990; Sparks et al., 1992).

The light and electron microscopic immunohistochemical experiments outlined in this chapter show that the cholinergic nuclei of the basal forebrain receive an intense and widespread monoaminergic innervation. The nbM and associated cholinergic nuclei of the basal forebrain therefore provide pivotal relays where the cholinergic innervation of the entire cerebral cortex can be modulated by monoaminergic synapses. The relative preservation of at least some of these monoamine inputs in AD introduces alternative possibilities for the pharmacological manipulation of cortical cholinergic innervation in this disease.

REFERENCES

Beal MF, MacGarvey U and Swartz KJ (1990): Galanin immunoreactivity is increased in the nucleus.basalis of meynert in Alzheimer's disease. *Ann Neurol* 28:157-161.

Casamenti F Deffenu Abbamonki AL and Pepeu G (1986): Changes in cortical acetylcholine output induced by modulation of the nucleus basalis. *Brain Res Bull* 16:689-695.

Day J and Fibiger HC (1993): Dopaminergic regulation of cortical acetylcholine release: effects of dopamine receptor agonists. *Neurosci* 54:643-648.

Gaspar P, Berger B, Alvarez C, Vigny A and Henry JP (1985): Catecholaminergic innervation of the septal area in man: Immunocytochemical study using TH and DBH antibodies. J Comp Neurol 241:12-33.

Geula C, Schatz CR and Mesulam M-M (1993): Differential localization of NADPH-diaphorase and calbinding-D28k within the cholinergic neurons of the basal forebrain, striatum and brainstem in the rat, monkey, baboon and human. Neurosci 461-467.

Geula, C and Mesulam M-M (1994): Cholinergic systems and related neuropathological predilection patterns in Alzheimer disease. In: Alzheimer Disease, Terry RD, Katzman R and Bick KL, eds. New York: Raven Press, pp. 263-291.

Jones BE and Cuello AC (1989): Afferents to the basal forebrain cholinergic cell area from pontomesencephalic-catecholamine, serotonin, and acetylcholine-neurons. Neurosci 31:37-61.

Jones BE and Yang T-Z (1985): The efferent projections from the reticular formation and the locus coeruleus studied by anterograde and retrograde axonal transport in the rat. J Comp Neurol 242:56-92.

Khateb A, Fort P, Alonso A, Jones BE and Muhlethaler M (1993): Pharmacological and immunohistochemical evidence for serotonergic modulation of cholinergic nucleus basalis neurons. Eur J Neurosci 5:541-547.

Maslowski RJ and Napier TC (1991): Dopamine D1 and D2 receptor agonists induce opposite changes in the firing rate of ventral pallidal neurons. Eur J Pharmacol 200:103-112.

Mesulam M-M, Mufson EJ, Levey AI and Wainer BH (1983): Cholinergic innervation of cortex by the basal forebrain: Cytochemistry and cortical connections of the septal area, diagonal band nuclei, nucleus basalis (substantia innominata), and hypothalamus in the Rhesus monkey. J Comp Neurol 214:170-197.

Napier C and Potter PE (1989): Dopamine in the rat ventral pallidum/substantia innominata: biochemical and electrophysiological studies. Neuropharmacol 28:757-760.

Pepeu G, Casamenti F, Mantovani P and Magnani M (1983): Neurotransmitters that act on cholinergic magnocellular forebrain nuclei influence cortical acetylcholine output. *Adv Behav Biol* 30:43-50.

Russchen FT, Amaral DG and Price JL (1985): The afferent connections of the substantia innominata in the monkey, Macaca fascicularis. *J Comp Neurol* 242:1-27.

Semba K, Reiner PB, McGeer EG and Fibiger HC (1988): Brainstem afferents to the magnocellular basal forebrain studied by axonal transport, immunohistochemistry, and electrophysiology in the rat. *J Comp Neurol* 267:433-453.

Sparks LD, Hunsaker JC, Slevin JT, KeKosky ST, Kryscio RJ and Mardesbery WR (1992): Monoaminergic and cholinergic synaptic markers in the numcleus basalis of Meynert (nbM): Normal age-related changes and the effect of heart disease and Alzheimer disease. *Ann Neurol* 31:611-619.

Wisniowski L, Ridely RM, Baker HF and Fine A(1992): Tyrosine hydroxylase-immunoreactive neurons in the nucleus basalis of the common marmoset. *J Comp Neurol* 325:379-387.

Zaborszky L, Cullinan WE and Luine VN (1993): Catecholaminergic-cholinergic interaction in the basal forebrain. In: *Progress in Brain Research,* Cuello AC, ed. New York: Elsevier, (vol. 98) pp. 31-49

Alzheimer Disease: From Molecular Biology to Therapy
edited by R. Becker and E. Giacobini

PHARMACOLOGICAL INDUCTION OF CHOLINERGIC HYPOFUNCTION AS A TOOL FOR EVALUATING CHOLINERGIC THERAPIES

Israel Hanin
Dept of Pharmacology and Experimental Therapeutics, Loyola University Chicago, Stritch School of Medicine, Maywood, IL

INTRODUCTION

The cholinergic deficiency hypothesis as a cause of dementia and cognitive deficits in Alzheimer's Disease (AD) is still widely accepted, even though other neurotransmitters besides acetylcholine are also affected in AD. In fact, major current focus on therapies has emphasized reversal or attenuation of the cholinergic deficiency (as is evident from the contents of this book). Nevertheless, the only FDA approved drug currently on the market in the United States is the cholinesterase inhibitor, tacrine, although it induces only a modest curative effect. Availability of an animal in which a long-term cholinergic hypofunction has been pharmacologically induced would therefore be extremely useful as a tool for evaluating new cholinergic therapies. Such an animal model would also allow one to study the consequences of long-term cholinergic hypofunction on the vulnerability of associated physiological parameters.

ANIMAL MODELS OF LONG-TERM CHOLINERGIC HYPOFUNCTION

Several animal models of cholinergic hypofunction have been developed over the past few years, and are presently employed. Admittedly, most of these models do not address the issue of the actual neuropathological causes of AD, i.e. beta amyloid deposition and formation of senile plaques, neurofibrillary tangles, and neuronal loss in some, but not all cortical and subcortical regions. Nevertheless, these models can serve a purpose in addressing the consequences of a central cholinergic hypofunction -namely, a deficit in cognitive function.

Substance Tested	Behavioral Paradigm	Reference(s)
Physostigmine (cholinesterase inhibitor)	PA	Brandeis et al., 1986 Ogura et al., 1987 Yamazaki et al., 1991
	MWM	Nakamura et al., 1992
SDZ ENA 713 (cholinesterase inhibitor)	AA,RAM	Endo et al., 1993
Arecoline (muscarinic agonist)	PA MWM	Yamazaki et al., 1991 Nakamura et al., 1992
Oxotremorine (muscarinic agonist)	PA	Yamazaki et al., 1991
AF102B (FKS-508) (M_1 receptor agonist)	PA T-Maze RAM	Nakahara et al., 1988 Nakahara et al., 1989 Fisher et al., 1989
AF150 (s) (M_1 receptor agonist)	PA MWM, RAM	Brandeis et al., 1995
DuP 996 (cognition enhancer)	T-Maze	Murai et al., 1994
Nefiracetam (cognition enhancer)	AA, locomotor activity	Abe et al., 1994
MKC-231 (choline uptake enhancer)	T-maze	Murai et al., 1994
Efflexor® (antidepressant)	RAM	Moyer et al., 1995
α-sialylcholesterol (neurotrophic factor?)	PA	Abe et al., 1993
Vitamin E (antioxidant)	MWM	Wortwein et al., 1994
Ganglioside AGF2 (trophic factor?)	RAM	Emerich and Walsh, 1990
SR-3 (essential fatty acid preparation)	MWM	Yehuda et al., 1995

TABLE I. Pharmacological reversal of AF64A-induced behavioral deficits. PA = passive avoidance; AA = active avoidance; RAM = radial arm maze; MWM = Morris water maze

Animals with a cholinergic hypofunction display memory and learning disorders, and it has been generally accepted that strategies and drugs which are developed to reverse these deficits, or to improve the animal's cognitive abilities, might also ultimately be applicable in the treatment of AD. Approaches which have been attempted to develop such a model have been varied and imaginative. Pharmacological long-term cholinergic hypofunction can be achieved by targeting a variety of sites at the cholinergic neuron, or by directly damaging the cholinergic neuron (see Hörtnagl and Hanin, 1992). Much information is available in the literature which describes long-term cholinergic hypofunction induced by, for example, excitotoxins, the cholinotoxin AF64A, and more recently, 192 IgG-saporin (Nilsson et al., 1992). The various available models of cholinergic hypofunction have been surveyed and reviewed (Karczmar, 1991; Hanin, 1994; Hörtnagl and Hanin 1992). This report focuses on the AF64A model, with which we have had extensive experience.

AF64A, administered intracerebroventricularly (icv), selectively causes a central cholinergic hypofunction in the hippocampus, and results in cognitive deficits in adult experimental animals (Hanin, 1990). Various and diverse pharmacological approaches have been employed successfully in the reversal of these AF64A-induced deficits. Recent examples of these are listed in Table 1.

Long-term cholinergic deficiency *in vivo* induced by other means besides AF64A administration has also been reported to exhibit, in experimental animals, cognitive and behavioral deficits. The literature abounds with reports showing that, similar to the AF64A-treated animal, these deficits can be reversed or attenuated by cholinergic agonists, cholinesterase inhibitors, and other experimental pharmacological agents.

DISCUSSION AND SUMMARY

Two key questions can be asked about the utility of animal models of cholinergic hypofunction, for the evaluation of cholinergic therapies:

1) Have such models resulted in new drugs for the treatment of AD?
The answer, at this stage, is NO. However, availability of such animal models allows us to rule out those treatment strategies which may not be appropriate. They also enable the conduction of comparative analyses of promising therapeutic candidates.

2) Does long-term cholinergic hypofunction result in the neuropathological changes seen in AD?
We do not as yet have the answer to that question. Animal models of cholinergic hypofunction, as used currently, do not exhibit the neuropathological markers seen in AD. These animals have been exposed to relatively short-term cholinergic deficiency (weeks to a few months). While it has been demonstrated that such short-term cholinergic lesions in the brains of rats will induce an increase in amyloid precursor protein biosynthesis in target

regions of the lesioned brain (Wallace et al., 1991; 1993), it has not yet been determined whether longer term cholinergic hypofunction would exhibit persistence in such changes, or might have an effect on, for example, formation of plaques and tangles, the accepted hallmarks of AD. Clearly, more extensive studies need to be conducted in this arena.

In considering the above questions, it is essential that one keeps in mind that we are dealing with a "model." As such, by definition, the model is a "representation of something" or a "pattern of something existing" (paraphrased from Webster's Collegiate Dictionary), rather than an exact replica - which would be the ideal situation. As long as a mechanism by which to retard or prevent the neurodegeneration which occurs in the brain of AD patients is not available, we need to refine ways to alleviate the cognitive deficits of these patients. Keeping the limitations of any "model" in mind, one can at least identify the neurochemical and physiological consequences of the long-term cholinergic hypofunction and study risk factors for such consequences. This, in turn, would allow the development of strategies and approaches to reverse the cholinergic deficiency, and to stabilize the changes that occur as a result of the development of the cholino-deficient state.

REFERENCES

Abe E, Murai S, Masuda Y, Saito H, and Itoh T (1993): α-Sialyl cholesterol reverses AF64A-induced deficit in passive avoidance response and depletion of hippocampal acetylcholine in mice. *Brit J Pharmacol* 108:387-392.

Abe E, Murai S, Saito H, Masuda Y, Takasu Y, Shiotani T, Tachizawa H, and Itoh T (1994): Effects of nefiracetam on deficits in active avoidance response and hippocampal cholinergic and monoaminergic dysfunctions induced by AF64A in mice. *J Neural Transm* 95:179-193.

Brandeis R, Pittel Z, Lachman C, Heldman E, Luz S, Dachir S, Levy A, Hanin I, and Fisher A (1986): AF64A-induced cholinotoxicity: Behavioral and biochemical correlates. In: *Alzheimer's and Parkinson's Diseases: Strategies for Research and Development*, Fisher A, Hanin I and Lachman C, eds. New York: Plenum Press Inc., pp. 469-477.

Brandeis R, Sapir M, Hafif N, Abraham S, Oz N, Stein E, and Fisher A (1995): AF150(S): A new functionally selective M1 agonist improves cognitive performance in rats. *Pharmacol Biochem Behav* 51:667-674.

Emerich DF and Walsh TJ (1990): Ganglioside AGF2 promotes task-specific recovery and attenuates the cholinergic hypofunction induced by AF64A. *Brain Res* 527:299-307.

Endo H, Tajima T, Goto T, Ikari H, Kuzuya F, and Iguchi A (1993): Effects of SDZ ENA 713 on AF64A-induced rats. *Soc Neurosci Abstr* 19:1625.

Fisher A, Brandeis R, Karton I, Pittel Z, Gurwitz D, Haring R, Sapir M, Levy A, and Heldman E (1989): (±)-cis-2-methyl-spiro (1,3-oxathiolane-5,3') quinuclidine (AF102B): A new M1 agonist, attenuates cognitive dysfunctions in AF64A-treated rats. *Neurosci Lett* 102:325-331.

Hanin I (1990): AF64A-induced cholinergic hypofunction. In: *Cholinergic Neurotransmission: Functional and Clinical Aspects: Progress in Brain Research, Vol. 84*, Aquilonius S-M, and Gillberg P-G, eds. Amsterdam: Elsevier Science Publishers, pp. 289-299.

Hanin I (1994): The centrally cholino-deficient animal as a model of Alzheimer's Disease (AD). In: *Alzheimer Disease: Therapeutic Strategies*, Giacobini E and Becker R, eds. Boston: Birkhauser, pp.367-374.

Hörtnagl H and Hanin I (1992): Toxins affecting the cholinergic system. In: *Handbook of Experimental Pharmacology: Selective Neurotoxicity*. Herken H and Hucho A, eds. Berlin, Heidelberg: Springer Verlag, pp. 63-70.

Karczmar AG (1991): SDAT models and their dynamics. In: *Cholinergic Basis for Alzheimer Therapy*, Becker R and Giacobini E, eds. Boston: Birkhauser, pp. 141-152.

Moyer JA, Morris H and Boast CA (1995): Efflexor®, a new antidepressant, attenuates scopolamine- and AF64A-impaired radial maze performance in rats.I 21:164.

Murai S, Saito H, Abe E, Masuda Y, Odashima J and Itoh T (1994): MKC-231, a choline uptake enhancer, ameliorates working memory deficits and decreased hippocampal acetylcholine induced by ethylcholine aziridinium ion in mice. *J Neural Transm* 98:1-13.

Nakahara N, Iga Y, Mizobe F and Kawanishi G (1988): Amelioration of experimental amnesia (passive avoidance failure) in rodents by the selective M1 agonist AF102B. *Japan J Pharmacol* 48:502-506.

Nakahara N, Iga Y, Saito Y, Mizobe F and Kawanishi G (1989): Beneficial effects of FKS-508 (AF102B), a selective M1 agonist, on the impaired working memory in AF64A-treated rats. *Japan J Pharmacol* 51:539-547.

Nakamura S, Tani Y, Maezano Y, Ishihara T and Ohno T (1992): Learning deficits after unilateral AF64A lesions in the rat basal forebrain: Role of cholinergic and noncholinergic systems. *Pharm Biochem Behav* 42:119-130.

Nilsson A, Leanza G, Rosenblad C, Lappi DA and Robertson D (1992): Spatial learning impairments in rats with selective immunolesion of the forebrain cholinergic system. *Neuroreport* 3:1005-1008.

Ogura H, Yamanishi Y and Yamatsu K (1987): Effects of physostigmine on AF64A-induced impairment of learning acquisition in rats. *Japan J Pharmacol* 44:498-501.

Wallace WC, Bragin V, Robakis NK, Sambamurti K, VanderPutten D, Merril CR, Davis KL, Santucci AC and Haroutunian V (1991): Increased biosynthesis of Alzheimer amyloid precursor protein in the cerebral cortex of rats with lesions of the nucleus basalis of Meynert. *Mol Brain Res* 10:173-178.

Wallace W, Ahlers ST, Gotlib J, Bragin V, Sugar J, Gluck R, Shea PA, Davis KL and Haroutunian V (1993): Amyloid precursor protein in the cerebral cortex is rapidly and persistently induced by loss of subcortical innervation. *Proc Natl Acad Sci USA*. 90:8712-8716.

Wortwein G, Stackman RW and Walsh TJ (1994): Vitamin E prevents the place learning deficit and the cholinergic hypofunction induced by AF64A. *Exp Neurol* 125:15-21.

Yamazaki N, Kato K, Kurihara E and Nagaoka A (1991): Cholinergic drugs reverse AF64A-induced impairment of passive avoidance learning in rats. *Psychopharmacology* 103:215-222.

Yehuda S, Carraso RL and Mostofsky DI (1995): Essential fatty acid preparation (SR-3) rehabilitates learning deficits induced by AF64A and 5,7-DHT. *NeuroReport* 6:511-515.

Alzheimer Disease: From Molecular Biology to Therapy
edited by R. Becker and E. Giacobini

RATIONAL DESIGN OF NEW ACETYLCHOLINESTERASE INHIBITORS

Mario Brufani and Luigi Filocamo
Department of Biochemical Science "A. Rossi Fanelli"
University "La Sapienza", Rome, Italy

INTRODUCTION

Six different classes of drugs, theoretically, could be useful for the treatment of the cholinergic deficit which characterizes Alzheimer's Disease (AD):

- Cholinesterase inhibitors (ChEI), which increase the synaptic levels of acetylcholine (ACh) by retarding its hydrolysis.
- ACh precursors, such as phosphatidylcholine, which might enhance the availability of choline.
- ACh releasers, which should facilitate the release of ACh from presynaptic end terminals, thereby activating the second messenger PIP_2 hydrolysis.
- M_1 and M_3 receptor agonists, which mimic ACh on the postsynaptic end terminal receptors.
- M_2 receptor antagonists (M_2 generally are presynaptics and play a role in controlling ACh release via negative feedback).
- Nicotinic agonists or substances having nicotinic like effects, which should also favor the release of ACh.

So far the ChEIs are the only agents which have been shown to produce statistically significant improvements in large multicenter, double-blind, placebo controlled trials on both psychometric measures of cognitive performance and quality of life indices in Alzheimer patients.

Tacrine, whose proprietary name is Cognex®, was the first acetylcholinesterase inhibitor (AChEI) to be registered and is presently used for palliative treatment of AD in the USA and in some European countries.

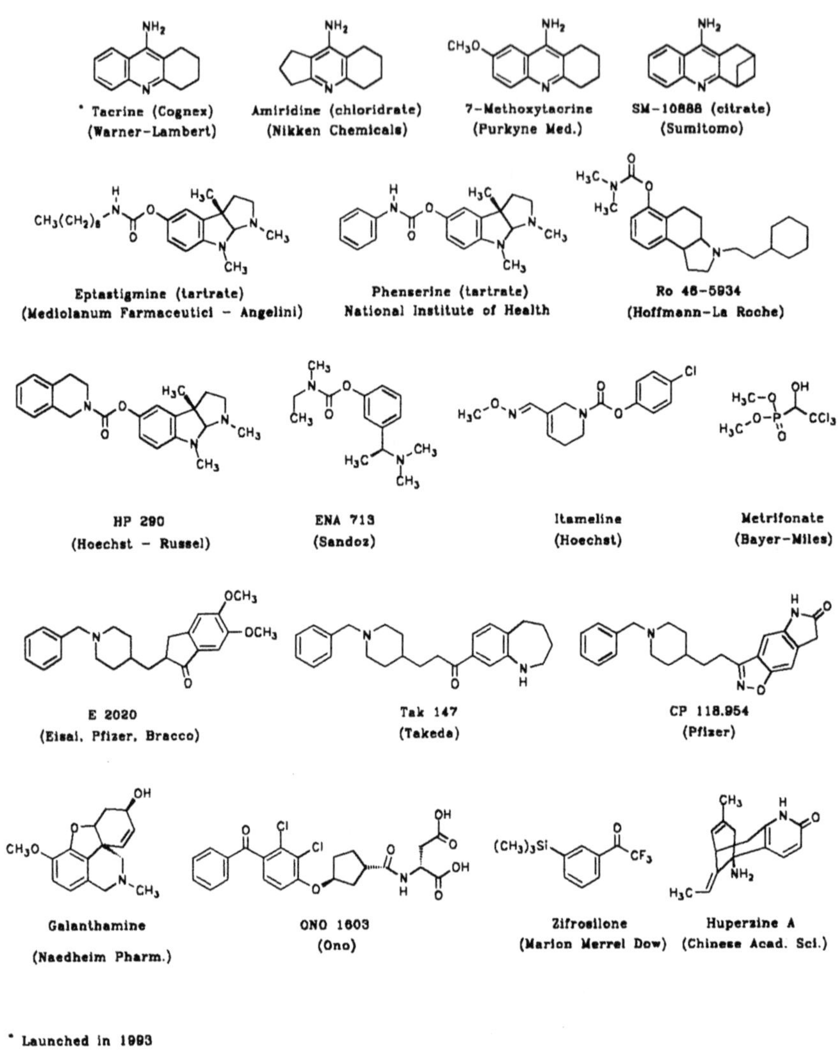

FIGURE 1. Acetylcholinesterase inhibitors in clinical and preclinical study.

SECOND AND THIRD GENERATION OF AChE INHIBITORS

The structural formulas of ChEI in clinical and preclinical study are shown In Figure 1. The references for these compounds are collected under "additional references."

Several tacrine analogues are under pre-clinical or clinical evaluation, including amiridine, SM-10888, and 7-methoxytacrine. These AChEIs act at a site close to the catalytic triad of the enzyme. The X-ray crystal structure of

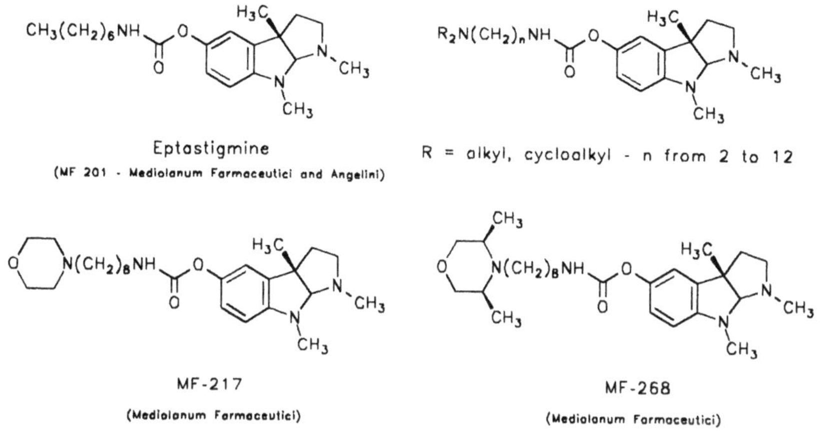

FIGURE 2. Eptastigmine and the aminoalkylcarbamoyl derivatives of eseroline.

the complex between tacrine and AChE was determined; therefore, the mechanism of action of these inhibitors is well known.

Six drugs: eptastigmine, phenserine, ENA-713, RO-46, 5943, HP/290 and itameline are either derivatives or structural analogues of the alkaloid physostigmine, or in some way related to this natural compound. They all inhibit the enzyme by pseudo-irreversibly carbamoylating the serine residue of the catalytic triad.

The AChE catalytic triad is composed of one glutamic, one histidine and one serine residue; via the histidine residue, the glutamic carboxylic group activates the serine hydroxy group, which in turn hydrolyzes ACh ester function by a nucleophilic attack. The intermediate serine acetic ester is then quickly hydrolyzed by an activated water molecule. When the substrate is physostigmine or one of its structural analogues, the intermediate carbamic

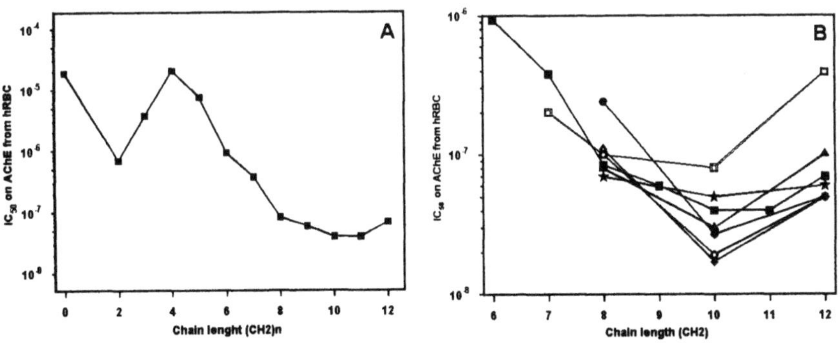

FIGURE 3. IC_{50} of morpholinoalkyl (A) and of aminoalkyl (B) derivatives vs. chain length. (■) Morpholinoalkyl, (□) o-methoxyphenylpiperazinylalkyl, (s) N,N-dipropylaminoalkyl, (O) piperidinylalkyl, (☆) N-methylpiperazinylalkyl, (●) N-acetylpiperazynilalkyl, (♦) N,N-diethylaminoalkyl derivatives.

ester is hydrolyzed very slowly, the nature of the carbamic substituent influencing the rate of the hydrolysis. Similarly, metrifonate, a fosfonic ester, irreversibly phosphorylates the enzyme.

On the contrary, E-2020 and other 4-substituted benzylpiperidines reversibly and specifically inhibit the AChE by forming a complex in which the N-benzylpiperidine group presumably interacts with the anionic site which recognizes the quaternary ammonium-group of ACh.

A series of other inhibitors, some of natural origin (galantamine and huperzine A), others of synthetic (ONO-1603, zifrosilone), inhibit the enzyme by presumably forming reversible complexes with it.

Since 1982, our group has been synthesizing and studying a large number of analogues and derivatives of physostigmine in which the methylcarbamic group of the alkaloid has been substituted with other alkylcarbamic and dialkylcarbamic groups having alkylic chains of different length and, as a consequence, different lipophilicity.

With respect to physostigmine our aims were to 1) reduce toxicity and side-effects; 2) increase oral absorption; 3) increase plasmatic half-life; 4) facilitate the penetration across the blood-brain barrier.

Among the compounds synthesized, eptastigmine (Fig. 2) was chosen since it showed a better pharmacological profile compared to physostigmine, and satisfied the four points indicated above (Brufani et al., 1986; Brufani et al., 1987). Indeed, in designing this compound we had expected to achieve improvements on the pharmacokinetic properties. On the contrary, eptastigmine mainly differs from physostigmine for its kinetic of enzymatic reactivation.

Fifty percent of AChE carbamoylated by physostigmine is reactivated in about 15 mins, whereas in the presence of eptastigmine the reactivation time is increased to over 3 hrs. The source of the enzyme is also important in determining these reactivation times. Reactivation is faster for erythrocyte derived enzyme and is very slow for cerebral enzyme (Moriearty and Becker, 1992).

When eptastigmine was submitted for the first clinical trial carried out in the USA, two cases of reversible neutropenia were observed following an oral dose of 40 mg, 3 times a day. The American study was interrupted and later restarted at much lower doses (about 10 mg, three times a day). At these doses, no other cases of neutropenia were reported.

With the aim of understanding the chemical and biochemical bases of ematotoxicity of eptastigmine, we substituted the heptyl side chain of the substance with more polar groups (Alisi et al., 1995). We have obtained interesting, even if somewhat unforeseen, results by inserting basic groups in the alkylic chain. The inhibitory potency on AChE depends on the length of the aliphatic chain which separates the amino group from the carbamic group. Activity is high with an ethylic bridge, decreases then to a minimum with a four terms chain, but then increases again to a maximum with chain of eight and ten carbon atoms (Fig. 3A). Inhibition is time-dependent and increases with the basicity of the amino group (Fig. 3B).

These data can be interpreted by assuming the existence of a gorge above the catalytic triad of AChE, which hinders the access of the inhibitors into the catalytic pocket of the enzyme, when in the proximity of the carbamic group there are large substituents. Furthermore, an "anionic site" must exist at a distance from the catalytic site equivalent to a chain of 8-10 carbon atoms. This peripheral "anionic site" can interact with the lateral amino group of the inhibitor.

This hypothesis is in agreement with the X-ray crystal structure of AChE from *Torpedo Californica* determined by Silman and colleagues (Sussman et al., 1991). It is also in agreement with mutagenic studies (Kronman et al., 1994), which supported the presence of a peripheral "anionic site", 15 Å away from the catalytic site in the enzyme.

It is likely that our derivatives bind AChE in a similar manner as do decametonium salts, at the same time interacting both with the main "anionic site" and with the peripheral "anionic site". On the basis of this model it is possible to design selective inhibitors both for AChE and for butyrylcholinesterase (BuChE).

Among the compounds synthesized in our group, we have chosen compounds MF-268 and MF-217 for a more accurate pharmacological evaluation. In MF-268 the basic group is dimethylmorpholin, in MF-217 the basic group is morpholin; in both compounds the alkyl chain is composed of 8 carbons. Both compounds are absorbed p.o. and cross the blood-brain barrier.

The choice of MF-268 and MF-217 was based upon pharmacokinetic considerations and was imposed from the need of reducing the basicity of the amino group of the lateral chain. Derivatives having strong basic amino-groups in this position, like alkylated piperidines, and piperazines, or dialkyl amines, are poorly absorbed following oral administration and do not cross the blood-brain barrier following a single bolus i.v. dose. A compromise between the conflicting requirements of a basic group for marked inhibitory potency and a balanced equilibrium, at physiological pH, between the protonated and not protonated forms was obtained by introducing into the chain an amine having a pKa of about 9.

MF-217 gives a strong and long-lasting inhibition of cerebral AChE and a high and durable increase of the levels of cortical ACh, as shown by Giacobini and co-workers using a microdyalysis technique, without the use of ChEI in the perfusion fluid (Zhu et al., 1995).

CONCLUSION

The knowledge of the structure and function of the components in the catalytic site and that of the two "anionic sites" allows the synthesis of inhibitors selective for both AChE or BuChE. Furthermore, some inhibitors presently undergoing clinical evaluation are highly selective for AChE. Nonetheless, most side effects, which limit the use of AChE inhibitors, do not seem to be associated to the inhibition of BuChE, but to the inhibition of whole peripheral tissue ChE. Moreover, the role played by BuChE in the body is at present unclear. Therefore, an ideal inhibitor should not only be selective for AChE, but also for cerebral enzyme with respect to peripheral enzyme.

In mammalian brain synapsis a hydrophobic-tailed tetrameric G4 form of AChE is dominant. It possesses a hydrophobic proteic subunit which anchors it to plasma membrane. This form may have lipophilic binding sites absent in other aggregation forms.

A difference in the reactivation time of human cerebral membrane bound AChE, with respect to that in erythrocytes upon inactivation with eptastigmine, could support this hypothesis. It is probably possible to design inhibitors which, by interacting with these lipophilic sites, are specific for the G4 form of the enzyme.

ACKNOWLEDGMENTS

The research from our laboratory reviewed here was supported by grants from the Mediolanum Farmaceutici S.p.A. (Milan, Italy), in which laboratories most of the pharmacological work was done.

REFERENCES

Alisi MA, Brufani M, Filocamo L, Gostoli G and Licandro E (1995): Synthesis and structure-activity relationships of new acetylcholinesterase inhibitors: morpholinoalkylcarbamoyleseroline derivatives. *Bioorg Med Chem Lett* 5:2077-2080.

Borroni E, Damsma G, Giovacchini C, Mutel V, Jakob-Rötne R and Da Prade M (1994): A novel acetylcholinesterase inhibitor, RO 46-5934, which interacts with muscarinic M_2 receptors. *Biochem Soc Trans* 22:755-758.

Brufani M, Marta M and Pomponi M (1986): Anticholinesterase activity of a new carbamate, heptylphysostigmine, in view of its use in patients with Alzheimer-type dementia. *Eur J Biochem* 157:115-120.

Brufani M, Castellano C, Marta M, Oliverio A, Pagella PG, Pavone F, Pomponi M and Rugarli PL (1987): A long-lasting cholinesterase inhibitor affecting neural and behavioral processes. *Pharmacol Biochem Behav* 26:625-629.

Craig NH, Pei XF, Soncrant TT, Ingram DK and Brossi A (1995): Phenserine and ring C hetero-analogues: drug candidates for the treatment of Alzheimer's disease. *Med Res Rev* 15:3-31.

E-2020 (1993): *Drug Future* 18:77.

HP-290 (1991): 201st ACS National Meeting, Division of Medicinal Chemistry. Atlanta, April 14-19, Abstract Book, n-74.

Huperzine A (1994): *Drugs Future* 19:595.

Kronman C, Ordentlich A, Barak D, Velan B and Shaffuman A (1994): The "back door" hypothesis for product clearance in acetylcholinesterase challenged by site-directed mutagenesis. *J Biol Chem* 269:27819-27822.

Lamy P (1994): The role of cholinesterase inhibitors in Alzheimer's disease. *CNS Drugs* 1:146-165. (Tacrine, velnacrine, galantamine, huperzine A, metrifonate, E713, SM-10888).

7-Methoxytacrine (1992): *Drug Future* 17:151.

Mihara M, Ohnishi A, Tomono Y, Hasegawa J, Shimamaura Y, Yamazaki K and Morishita N (1993): Pharmacokinetics of E2020, a new compound for Alzheimer disease, in healthy male volunteers. *Int J Clin Pharmacol, Ther Toxicol* 31:223-229.

Miyamoto M and Goto G (1994): Effects of Tak-147, a novel acetylcholinesterase inhibitor, on scopolamin-induced impairment of the delayed discrimination task in rats. *Japan J Pharmacol* 64 (Suppl 1):Abst 0-167.

Moriearty PL and Becker RE (1992): Inhibition of human RBC acetylcholinesterase by heptylphysostigmine. *Meth Find Exp Clin Pharmacol* 14:615-621.

Nik-247 (amiridine) (1994): *Drug Future* 19:343.

SDZ-ENA 713 (1994): *Drug Future* 19:656.

SM-10888 (1995): *Drug Future* 20:114.

Sussmann JL, Harel M, Frolow F, Oefner C, Goldman A, Toker L and Silman I (1991): Atomic structure of acetylcholinesterase from *Torpedo Californica*: a prototypic acetylcholine-binding protein. *Science* 253:872-874.

Tak-147 (1995): *Drug Future* 20:248.

Villabos A, Butler TN, Chapin DS, Chen YL, DeMattos CB, Ives JL, Jones SB, Liston DR, Nagel AA, Nason DM, Nielsen JN, Ramirez AD, Shalaby IA and White WF (1995): 5,7-Dihydro-3-[2-[(1-phenylmethyl)-4-piperidinyl]-ethyl]- 6H-pyrrolo[3,2-f]-1,2-benzisoxazol-6-one: a potent and centrally selective inhibitor of acetylcholinesterase with improved margin of safety. *J Med Chem* 38: 2802-2808. (CP 118.954).

Zhu X-D, Cuadra G, Brufani M, Maggi T, Pagella PG, Williams E and Giacobini E (1995): Effects of MF-268, a new cholinesterase inhibitor, an acetylcholine and biogenetic amines in rat cortex. J *Neurosc Res* 43:120-126.

Zifrosilone MDL-73754 (1994): *Drug Future* 19:854.

Froimowitz [illegible]
Lamy [illegible] (1994) [illegible]
[illegible]
Mysinger M and Gallo [illegible]
Myers [illegible]
[illegible]

Alzheimer Disease: From Molecular Biology to Therapy
edited by R. Becker and E. Giacobini

ADVANCES IN UNDERSTANDING CHOLINERGIC BRAIN NEURONS: IMPLICATIONS IN THE USE OF CITICOLINE (CDP–CHOLINE) TO TREAT STROKE

Richard J. Wurtman
Department of Brain & Cognitive Sciences, and Clinical Research Center, M.I.T., Cambridge, MA

Bobby W. Sandage, Jr.
Interneuron Pharmaceuticals, Inc., Lexington, MA

Steven Warach
Department of Neurology, Beth Israel Hospital, Boston, MA

INTRODUCTION

This article describes the use of oral Citicoline (CDP-Choline) to treat patients who have suffered a stroke, and explains the neurochemical mechanisms thought to underlie its therapeutic actions.

CDP-Choline is an endogenous compound that can also serve as a prodrug. It is formed in all cells, via the Kennedy Cycle, from pre-existing choline molecules during the biosynthesis of phosphatidylcholine (PC) (Kennedy and Weiss, 1956). When administered to people or rats, it is rapidly and completely broken down to the active compounds choline and cytidine; these appear in the circulation (Lopez G-Coviella et al., 1987), and are readily taken up into the brain (Blusztajn and Wurtman, 1983; Savci and Wurtman, 1995). The resulting increase in brain choline levels enhances the substrate saturation of the enzymes choline acetyltransferase and choline kinase thereby accelerating, respectively, the formations of acetylcholine (ACh) and phosphocholine (Blusztajn and Wurtman, 1983). Similarly, the increase in brain cytidine levels accelerates the formation of cytidine triphosphate (CTP) (Savci and Wurtman, 1995). The elevations in brain phosphocholine and CTP levels enhance, in turn, the substrate-saturation of the low-affinity enzyme choline-phosphate cytidylyltransferase (Weinhold and Feldman, 1992), thus promoting the biosynthesis of endogenous CDP-choline. This compound then quickly combines with available fatty acids, in the form of diacylglycerol (DAG), to form more PC.

Hence, exogenous CDP-Choline is able to improve brain function in three known ways: by making more of a neurotransmitter, ACh, which is critically important for cognitive function; by increasing brain levels of PC and, ultimately, of the amount of membrane per cell (which may accelerate the restoration of function in stroke-damaged neurons); and, by sopping up fatty acids released as a consequence of tissue damage (Trovarelli et al., 1981), which might otherwise be oxidized to neurotoxic compounds that would extend the stroke (Clemens et al., 1996). Some existing literature discussed below supported the view that exogenous CDP-Choline could diminish the severity of strokes, and/or improve the return of various functions during the recovery period. We now describe the drug's significant beneficial effects on stroke outcome as observed in a 259-patient multicenter Phase III study (Clark et al., 1996) recently completed in the United States.

NEUROCHEMICAL EFFECTS OF CDP-CHOLINE OR OF CHOLINE PLUS CYTIDINE

CDP-Choline affects the brain indirectly; its administration elevates plasma levels of its constituent moieties choline and cytidine which are then taken up into the brain and enhance the rates at which their products, phosphocholine and CTP, are formed (Lopez G-Coviella et al., 1987). Two hours after normal volunteers received CDP-Choline orally, plasma choline levels were increased by 48% and plasma cytidine by 136%. Intravenously-administered CDP-Choline was rapidly hydrolyzed in both the human and the rat. In subjects receiving the drug by a 30-minute intravenous infusion, plasma CDP-Choline fell to undetectable levels almost immediately after the end of the infusion period. (Plasma choline and cytidine peaked at that time, but their concentrations remained elevated for at least 6 hr). Five minutes after rats received a bolus injection of CDP-Choline, none of the unchanged compound could be detected in their plasma.

The uptake of cytidine by, and its fate in, cells was examined by exposing PC 12 cells to [3H]cytidine (Lopez G-Coviella and Wurtman, 1992). Uptake exhibited normal Michaelis-Menten kinetics (Km = μM) when concentrations in the medium were below 50 μM. (Normal plasma cytidine concentrations in human have been described as 1-10 μM; cf Lopez G-Coviella et al., 1987). Once inside the cell, the cytidine was converted mainly to cytidine triphosphate (CTP). Incubation of rat striatal slices with cytidine similarly elevated their CTP contents (Savci and Wurtman, 1995). The addition of cytidine to media caused a dose-related enhancement of the incorporation of isotopically-labeled choline into PC in PC-12 cells (Lopez G-Coviella and Wurtman, 1992), and in rat striatal slices (Savci and Wurtman, 1995).

Moreover, supplementing the media with both choline and cytidine elevated the absolute contents, per PC-12 cell, of all three major structural phospholipids (PC, phosphatidylethanolamine [PE], and phosphatidylserine [PS]) indicating that the cells now contained more membrane. Similar increases in phospholipid levels per mg protein were observed in rat striatal slices. The ability of CDP-Choline to enhance PC synthesis reflects the kinetic properties of choline-phosphate cytidylyltransferase. The enzyme's Michaelis-Menten constants for phosphocholine and CTP (0.234 and 0.22 mM, respectively) (Weinhold and Feldman, 1992) are so high that it very likely is unsaturated with either substrate under physiologic circumstances. Brain phosphocholine levels in rats and humans have been described as 0.2 - 0.5 mM (Millington and Wurtman, 1982; Blusztajn J., personal communication). CTP levels in rat brain were less than 0.1 mM (Hisanaga et al., 1986).

The chronic consumption of CDP-Choline via the diet increased brain PC levels (per mg protein) in 3 mo old mice (500 mg/kg/day of the drug for 27 months) or in 12 mo old animals treated for 3 mo (Lopez G-Coviella et al., 1992). Similar effects were seen, sometimes at lower doses and involving all three phospholipids, when young rats consumed the drug for 42 or 90 days (Lopez G-Coviella et al., 1995). This dose, given acutely, produces an increase in plasma choline levels (12 to 24 μM) about twice that observed when humans ingest 25-30 mg/kg of the drug (Lopez G-Coviella et al., 1987).

EFFECTS OF CDP-CHOLINE ON STROKE IN HUMANS

Five previously published placebo-controlled studies suggested that CDP-Choline would promote recovery from ischemic stroke. Boudouresques and Michel (1980), in a double-blind placebo-controlled study, compared responses to CDP-Choline (n=27; 250 mg tid intravenously given for 10 days) or placebo (n=25) among patients who had acutely suffered (i.e., within 48 hours) an ischemic stroke. Patients were classified by stroke severity, i.e., massive deficit, slight or partial deficit, or almost no deficit. Significantly more CDP-Choline-treated than placebo patients initially classified as having a massive deficit were found to have very good, good or fairly good results (78% vs. 41%). Patients with slight deficits showed similar rates of recovery. There were 2 cases of agitation in the CDP-Choline treated patients and 1 in the placebo treated patients; no other adverse findings were reported.

Goyas et al. (1980) also studied patients randomized to treatment with 48 hours of stroke onset. Patients assigned at random received either placebo or CDP-Choline (n=31) at two different doses, i.e., half received 750 mg/day intravenously for 10 days and then 250 mg/day for 10 days, and the other half received 750 mg/day for 3 days, 500 mg/day for 3 days, and then 250 mg/day for 4 days). Outcome measurements obtained at day 90 included clinical assessments of motor, muscular force, sensory, and higher functions; walking; and psychometric assessments. CDP-Choline treated patients demonstrated a

4.6 ± 4.8 point improvement on total scores compared with only a 1.6 ± 2.0 point improvement in those receiving placebo ($p<0.05$). Hypertonia and walking ability were also assessed. CDP-Choline treated patients had less hypertonia when compared with placebo-treated patients (58% vs. 20%, $p=0.03$), and significantly more ($p=0.02$) of them were walking normally or with some assistance (60% vs. 21%). A significant treatment effect was also observed with the psychometric tests. Mortality was not different between the treatment groups (28% CDP-Choline treated vs. 37% placebo–treated). Finally, global assessment of outcome revealed that 37% of CDP-Choline-treated experienced complete recovery versus 0.7% (1 patient) of the placebo group.

Corso et al. (1982) in a double–blind study, treated 33 acute stroke patients with either CDP-Choline (n=17; 1000 mg/day intravenously for 30 days) or placebo (n=16). Patients were followed for 30 days, and sensory, motor and speech deficits were assessed. A global assessment of recovery classified patients as follows: significantly improved, moderately improved, slightly improved, or exhibiting "normal progress". When patients were grouped as to improved (significant/moderate/slight) vs. normal progress, 76% of the CDP-Choline-treated patients demonstrated improvement but only 31% of those receiving placebo. No adverse events or changes in laboratory parameters were noted.

In the largest controlled study conducted outside the United States, Tazaki et al. (1988) studied the effect of CDP-Choline on acute stroke patients in a multicenter, double–blind, placebo-controlled study. Patients began treatment with 14 days of stroke onset; they were assessed using the Japanese Coma Scale (10 grades) and a global improvement rating scale (six categories: markedly improved, improved, slightly improved, unchanged, worse, and severely deteriorated). Fewer CDP-Choline–treated patients died than those receiving placebo. On day 14, 52% of the CDP-Choline-treated patients showed improvement (i.e., markedly improved, improved, or slightly improved) compared with 26% of the placebo-treated patients ($p<0.01$). Similarly, 51% of the CDP-Choline-treated patients showed improvement on the Japanese Coma Scale, compared with 33% of the placebo-treated patients ($p<0.01$). Four patients experienced adverse events in the CDP-Choline-–treated group (3 hepatic dysfunction, 1 renal dysfunction) compared with 11 of the placebo-treated patients (4 hepatic dysfunction, 3 skin rashes, 1 hot flash, 1 restlessness, 1 anemia, and 1 renal dysfunction).

Hazama et al. (1980) evaluated the effect of CDP-Choline in post-stroke recovery. One hundred sixty–two patients entered the study within one year of their stroke (almost half of the patients were within 3 months post- stroke); they were randomized to receive CDP-Choline 250 mg iv (n=55); CDP-Choline 1000 mg iv (n=54); or placebo (n=53), in a double-blind manner and treated for 8 weeks. Baseline demographics among the groups were similar with majority of patients having a diagnosis of ischemic stroke;

however, a third of the patients had experienced a hemorrhagic stroke. Recovery was assessed by hemiplegia function test (upper limb); basic activity test of lower limb; passive range of motion; neurological exam; and subjective symptoms. Improvements in one or more grades of function tests were 53% for the CDP-Choline (1000 mg) treated groups; 55% for the 250 mg treated group; and 32% for the placebo treated group (p=0.06). Similar findings were observed for lower limb activity, i.e., 43% vs. 36% vs. 29%, respectively, although these differences did not attain statistical significance. Four CDP-Choline-1000 mg patients, five CDP-Choline-250 mg treated patients; and two placebo treated patients experienced adverse effects (vertigo; heavy-headedness; insomnia; increased numbness).

A recently completed randomized, multicenter, placebo-controlled, double-blind dose-response trial evaluated three doses of CDP-Choline (500, 1000 and 2000 mg daily) versus placebo in 259 patients (Clark et al., 1996). Patients were treated for 6 weeks, and then followed for an additional 6 weeks. Patients with acute ischemic stroke involving middle cerebral artery evaluated; they began treatment within 24 hours of symptom onset. Efficacy variables included Barthel Index, Modified Rankin Scale, NIH Stroke Scale and a neuropsychological battery.

Baseline demographics were similar between the treatment groups, except that the 1000 mg dose group was significantly heavier than the other groups. The average time from symptom onset to initiation of treatment was 14-15 hrs. CDP-Choline (500 mg or 2000 mg) significantly improved functional recovery, as measured by Barthel Index. Significantly more CDP-Choline-treated patients (500 mg=53%, 2000 mg–45%) demonstrated a full recovery using this scale ($\geq$95) than those receiving placebo (33%). In addition, the rate at which full recovery occurred was significantly faster for the 500 mg and 2000 mg treatment groups than for placebo (on average 14 days faster to full recovery). Likewise, significantly more CDP-Choline treated patients (500 mg=34%) achieved full recovery as measured by NIH stroke scale ($\leq$1) when compared with those receiving placebo (16%). Global improvement, measured by the Rankin Scale, was found to be significantly (p=0.036) different among the four treatment groups in favor of CDP-Choline (500 mg=3.1, 1000 mg=2.48, 2000 mg=3.11, and placebo=2.63).

The Mini-Mental State Exam, an overall clinical assessment of the patients' cognitive states showed a favorable effect in CDP-Choline treated groups. Significantly higher percentages of patients in the 500 mg CDP-Choline treated groups (59%) and 2000 mg CDP-Choline treated group (50%) exhibited normal or almost normal MMSE ($\geq$25) scores than placebo treated patients (35%).

CDP-Choline was well tolerated by these patients. There were no differences in mortality rates among the treatment groups, and the only statistically significant, treated-related adverse events were dizziness and accidental injuries (which were described primarily as falls). Hence, CDP-Choline shows promise as a safe and effective treatment for stroke.

ACKNOWLEDGMENTS

These studies were supported in part from grants from the NIH (MO1–RR–000–88 and MH–28783) and the Center for Brain Sciences & Metabolism Charitable Trust.

REFERENCES

Blusztajn JK and Wurtman RJ (1983): Choline and cholinergic neurons. *Science* 221:614–621.

Boudouresques BA and Michel B (1980): Therapeutic conduct in light of a cerebral vascular accident and the use of CDP–Choline. *Intl Symp Brain Suffering and Precursors of Phospholipids* Paris, January 18:109–121.

Clark W, Warach S and Citicoline Study Group (1996): Randomized Dose Response Trial of Citicoline in Acute Ischemic Stroke Patients. *Neurology* 46(2):S62.004.

Clemens JA, Stephenson DT, Smalstig EB, Roberts EF, Johnstone EM, Sharp JD, Little SP, and Kramer RM (1996): Reactive glia express cytosolic phospholipase A2 after transuient global forebrain ischemia in the rat. *Stroke* 27:527– 535.

Corso EA, Arena M, Ventimiglia A, Bizzarro G, Camp G, and Rodolico F (1982): CDP-Choline for cerebrovascular disorders: Clinical evaluation and evaluation of electrophysiological symptomatology. *Cl Therap* 102:379–386.

Goyas JY, Bastard J, Missourm A (1980): Results after 90 days of stroke treatment with CDP–Choline concerning a double blind test. *Intl Symp Brain Suffering and Precursors of Phospholipids* Paris, January 18:123-128.

Hazama T, Hasegawa T, Ueda S, and Sakuma A (1980): Evaluation of the effect of CDP-Choline on post- stroke hemiplegia employing a double-blind controlled trial. Assessed by a new rating scale for recovery in hemiplegia. *Int J Neurosci* 11:211–225.

Hisanaga K, Onodera H, and Kogure K (1986): Changes in levels of purine and pyrimidine nucleotides during acute hypoxia and recovery in neonatal rat brain. *J Neurochem* 47:1344–1350.

Kennedy EP and Weiss SP (1956): The function of cytidine coenzymes in the biosynthesis of phospholipids. *J Biol Chem* 222:193–214.

Kramer RM (1966): Reactive glia express cytosolic phospholipase A2 after transuient global forebrain ischemia in the rat. *Stroke* 27:527–535.

Lopez G-Coviella I, Agut J, Von Borstal R, and Wurtman RJ (1987): Metabolism of cytidine (5')–diphosphocholine (CDP-choline) following oral and intravenous administration to the human and the rat. *Neurochem Intl* 11:293–297.

Lopez G-Coviella I, Agut J, Ortiz JA and Wurtman RJ (1992): Effects of orally administered cytidine 5'diphosphate choline on brain phospholipid content. *J Nutr Biochem* 3:313–315.

Lopez G-Coviella I, Agut J, Savci V, Ortiz JA and Wurtman, RJ (1995): Evidence that 5'cytidinediphosphocholine can affect brain phospholipid composition by increasing choline and cytidine levels. *J Neurochem* 65:889–894.

Lopez G-Coviella I and Wurtman RJ (1992): Enhancement by cytidine of membrane phospholipd synthesis. *J Neurochem* 59:338–343.

Millington WR and Wurtman RJ (1982): Choline administration elevates brain phosphorylcholine concentrations. *J Neurochem* 38:1748–1752.

Savci V and Wurtman RJ (1995): Effect of cytidine on membrane phospholipid synthesis in rat striatal slices. *J Neurochem* 64:378–384.

Tazaki Y, Sakai F, Otomo E, Kutsuzawa T, Kameyama M, Omae T, Fujishima M, and Sakuma A (1988): Treatment of acute cerebral infarction with a choline precursor in a multicenter double-blind placebo-controlled study. *Stroke* 9:211–216.

Trovarelli G, DeMediio GE, Dorman RV, Piccinin GL, Horrocks LA,and Porcellati G (1981): Effect of cytidine diphosphate choline (CDP– choline) on ischemia-induced alterations in brain lipid in the gerbil. *Neuron Res* 6:821–833.

Weinhold PA and Feldman DA (1992): Choline-phosphate cytidylytransferase. *Methods in Enzymology* 209:248–258.

Alzheimer Disease: From Molecular Biology to Therapy
edited by R. Becker and E. Giacobini

CHOLINESTERASE INHIBITORS DO MORE THAN INHIBIT CHOLINESTERASE

Ezio Giacobini
Department of Geriatrics, University Hospitals of Geneva,
CH-1226 Thonex, Geneva, Switzerland
Department of Pharmacology, Southern Illinois University
School of Medicine, Springfield, Illinois, USA

INTRODUCTION

The cholinergic system plays an important role in learning and memory processes. The crucial role of acetylcholine (ACh) is supported by three lines of evidence. The first line is the effects of pharmacological manipulation using agonists or antagonists at both nicotinic and muscarinic receptors (Decker and McGaugh, 1991; Murray and Fibiger, 1985; Mandel and Thal, 1988; Mandel et al., 1989; Drachman and Leavitt, 1974; Drachman, 1982; Flicker et al., 1992; Vanderwolf et al., 1990; Wesnes et al., 1990). Second is the fact that adverse effects of lesioning cholinergic nuclei are ameliorated by intracerebral transplantation of fetal cholinergic cells or genetically modified tissue (Alkon et al., 1991; Dekker et al., 1991; Page et al., 1991; Berger-Sweeney et al., 1994; Dunnett et al., 1985; Gage and Bjorklund, 1986; Nilsson et al., 1987). The third is the fact that deficits in cholinergic cortical innervation and decreases in nicotinic receptors are seen in humans during aging and Alzheimer's disease (AD) (Perry et al., 1978; Whitehouse et al., 1982; Bartus et al., 1982; Giacobini et al., 1989; Schröder et al., 1995). Results from these investigations are consistent with the concept that cholinergic function is required for learning and memory. A recent article of Winkler et al. (1995) demonstrates that cerebral ACh is not only necessary for cognitive behavior in the rat but its presence and function within the neocortex is also sufficient to improve learning deficits and restore memory in experimental animals following severe damage to the nucleus basalis of Meynert. By analogy, in Alzheimer patients, restoration of cholinergic neurotransmission should be sufficient to ameliorate impaired learning and memory. The formulation of this hypothesis has been followed by numerous clinical trials using various types of cholinergic drugs (Giacobini, 1994). This chapter will focus mainly on cholinesterase inhibitors and on effects of these drugs other than cholinesterase inhibition in CNS.

COMPOUND	COMPANY	CHARACTERISTICS	PHASE*
YM 796	Yamanouchi	M_3 weak agonist M_1 selective agonist	II Japan
RS 86	Sandoz	M_1 agonist	exper.
AF-102B FKS-508	Snowbrand Israel I.B.R. - Forest	M_1 (M_3)agonist	II-III Japan US
BIBN-99	K. Thomae GmbH. Boehringer	M_2 antagonist	exper.
Cl-979/RU35926 Milameline	Warner Lambert Roussel	Partial M_1 agonist (non selective)	II
LY287041	Eli Lilly	M_1 agonist	exper.
SR-46559	Sanofi	M_1 agonist. (not sel.) M_2 antagonist	exper
CI-1002	Parke Davis	AChE inhibitor M_1 antagonist	exper.
PD-151832	Parke Davis	M_1 agonist	exper.
PD-142505	Parke Davis	M_1>M_2 agonist	exper.
LY246708 Xanomeline	Lilly/Novo- Nordisk	M_1 agonist	II-III
SB202026	Smith Kline Beecham	M_1 agonist	I-II Eur
PDC 008.004	Pharm Disc Corp.	M_2 antagonist	exper.
RO 46-5934	Hoffman La Roche	AChE inhibitor M_2 antagonist	exper.

TABLE I. Muscarinic agonists and antagonists of clinical interest *Clinical Phase in USA; Eur = Clinical Phase in Europe

Cholinergic Therapy: Which way to go? Muscarinic agonists and antagonists

For as long as development of ChEI has been in progress, a parallel line of research has attempted to develop muscarinic drugs such as agonists to stimulate selectively postsynaptic M_1 receptors or antagonists to inhibit the effect of M_2 presynaptic receptors in order to improve ACh release in brain. Neither approach has produced a highly selective drug; therefore, many compounds have never reached clinical trials or are still at early stages (Table I). One major obstacle continues to be the presence of severe side effects, particularly of gastro-intestinal and cardiac nature. Because of the present limitations, future drugs need to demonstrate higher receptor selectivity. The number of muscarinic agonists presently in clinical trial is lower than that of ChEI (Table I). Data from clinical trials with muscarinic agonists are still scanty, particularly if compared to the rich literature about ChEI. Some differences are starting to emerge between the clinical potential of these two classes of drugs. Cholinesterase inhibitors seem to exert a predominantly cognitive effect (attention, memory, concentration) while muscarinic agonists seem to act mainly on behavioral aspects of the diseases (Table II). If confirmed, combination of properties of both agents may prove to be of benefit (Table VI) Also, there may be differences in side effects (Table II). For muscarinic agonists, the main obstacle is still to overcome autonomic side effects which seem to be substantial even in the lately developed products.

Nicotinic Agonists

Reduction in nicotinic acetylcholine receptor (nAChR) pharmacology and expression have been among the first reported neurochemical landmarks of AD (cf. DeSarno et al., 1982; Giacobini et al., 1989; Schröder et al., 1995). Therapeutic strategies based on the findings of impaired nicotinic cholinergic transmission are being developed aimed at stimulating decreased nAChR function (Schröder et al., 1995). Studies with microdialysis have shown that nicotine as well as several analogues investigated in our laboratory may stimulate the release of norepinephrine, dopamine and serotonin together with ACh (Summers et al., 1994; Summers and Giacobini, 1995). The different effects of various nicotinic agonists on cortical neurotransmitters suggest a

CLASS	CLINICAL EFFECTS	SIDE EFFECTS
Cholinesterase inhibitors	Predominantly cognitive	very low incidence with new drugs
Muscarinic agonists	Predominantly behavioral	significant cholinergic side effects

TABLE II. Differences in clinical potential and side effects between cholinesterase inhibitors and muscarinic agonists.

Compound	Country	Company	Clinical Phase****	Side Effects Comments
Physostigmine slow release	USA	Forest	III	N.A.
ENA 713	USA/Europe	Sandoz	III	Low side effects
Eptastigmine	USA/Italy	Mediolanum	III	Low side effects
E-2020	USA/Japan	Eisai	III	Low side effects
MDL 73,745	USA/Europe	Marion Merrell Dow	II	Low side effects
Metrifonate	USA/Germany	Bayer/Miles	III	Low side effects
Tacrine (THA) *	USA/Europe	Warner-Lambert	IV	Hepatotoxicity
Velnacrine (HP029)**	USA/Europe	Hoechst-Roussel	II	Hematology ***
Suronacrine (HP128)**				Hepatotoxicity
Galanthamine	Germany	Shire Pharm.	II	Low side effects
	USA	Ciba-Geigy		
Huperzine A	China	Chinese Acad. Sci.	III	N.A.
NX-066	England/USA	Astra Arcus	II	N.A.
CP-118,954	USA	Pfizer	II	N.A.
KA-672	Germany	Schwabe	I	N.A.
NIK 247	Japan	Nikken	III	Low side effects
TAK 147	Japan	Takeda	III	Low side effects

TABLE III. Cholinesterase inhibitors: AD clinical trials (1996). * other indications: HIV, tardive dyskinesia; ** withdrawn; ***neutropenia or *** agranulocytosis. ****Clinical Phase in USA. N.A. = data not available differential action on subtypes of receptors and specific pre- and post-synaptic interactions (Summers et al., 1995). A second approach to nicotinic cholinergic therapy of AD has been the development of cholinergic channel activators such as ABT-418 (Arnerić et al., 1994). This compound is undergoing clinical studies.

Cholinesterase Inhibitors

ChEI are, so far, the only drugs demonstrating clinical efficacy in the treatment of AD (c.f. Giacobini, 1994, 1995). The principle used behind indirect cholinomimetic therapy with ChEI is to reduce ACh hydrolysis in central nervous system (CNS) nerve terminals by means of ChEI (Becker et al., 1991). The resulting increase in extracellular ACh concentration should restore central cholinergic hypofunction and improve memory and cognition (Becker and Giacobini, 1988). The use of a ChEI (THA, tacrine, tetrahydroaminoacridine) has resulted in a dose-dependent clinical efficacy in 20-30% of AD patients (Knapp et al., 1994). Since 1988, the number of ChEI in development for AD treatment has increased from 6 to 13 in 1996 (Table III). In spite of this fact, up to 1995 tacrine has been the only drug approved for the indication of AD, both in the USA and Europe (France, Sweden, Italy, Finland, Switzerland).

A 1996 list of ChEI (Table III) in clinical trials includes at least 13 drugs, most of which have already advanced to clinical phase III. The next two-year period (1996-1998) should be the most crucial in this process of selection. The use of tacrine in several thousand patients in Europe and USA has taught a precious lesson (Table IV). Drug companies and research laboratories have profited both pre-clinically and clinically from this experience. The new generation ChEI to replace tacrine in the market will have to fulfill certain requirements which are listed in Table V. It remains to be demonstrated whether or not such a drug(s) is(are) already present among the dozen in clinical trials (Table III).

The major focus in developing a successor to tacrine is obviously avoiding toxicity including liver, bone marrow and CNS effects. In order to [illegible] gain, the therapeutical effect should be extended to at least half of the patients and should be maintained for a period of at least 2-3 years. It is also important that the improvement seen in cognitive performance translates into a significant enhancement of activities of daily living and in a demonstrable delay in institutionalization. The goal of slowing down deterioration is clearly in the mind of researchers. To test this effect, we are still missing crucial experimental models and selective clinical markers. In addition, clinical assessment tools are not perfected enough to measure a neuroprotective effect. Consequently, a drug with genuine neuroprotective effects may well not be recognized as such in clinical trials. Another problem with ChEI is the identification of those patients most likely to benefit from therapy. Choosing the stage of disease at which to start medication may also be crucial for the success of the future ChEI.

Combinations of ChEI with muscarinic agonists or antagonists to obtain potentiating effects

Using microdialysis, we observed that in rat cortex the extracellular concentration of ACh following AChE inhibition is regulated through muscarinic receptors (Messamore et al., 1993). These data suggest that a combination of an AChE inhibitor and a presynaptically acting selective muscarinic antagonist could represent a useful strategy to: 1) enhance the release of ACh, and 2) simultaneously elevate its extracellular concentration. Based on this principle, various combinations of ChEI and muscarinic acting drugs or a new drug combining both actions can be suggested (Table VI). Some of these approaches are being attempted (Table I) using either agonists or antagonists. Also interesting would be to explore the combination of an M2 selective antagonist with a ChEI to augment ACh release (Tables I and VI). Other responses may also become attenuated. Tolerance, or a kind of "wearing off" phenomenon, to the clinical effect could develop as a result of receptor desensitization or down-regulation following a prolonged ChEI treatment. Modulation of ACh release, up-regulation of nicotinic and down-regulation of muscarinic receptors have been reported in the CNS of rats following prolonged administration of physostigmine (PHY) (De Sarno and Giacobini, 1989). Therefore, it seems useful to test, experimentally and clinically, various combinations which may prevent or reduce tolerance to drug effect. It is probable that high doses of ChEI would cause tachyphylaxis and enhanced side effects. This will make it necessary to individualize doses for each patient and use an appropriate mode of administration.

---

1. Low toxicity levels of a ChEI can be tolerated provided there is no fatal outcome (side effects/benefit ratio).
2. A modest improvement in the patient may be seen as a significant advantage from the caregiver's point of view.
3. The patient can be taken off tacrine and then be put back again without totally losing efficacy of the drug.
4. Tacrine should not be discontinued abruptly (high risk for withdrawal with psychotic symptoms).
5. There is a high individual variability in size of effective dose and in occurrence of side effects.
6. Estrogens may have a synergistic effect with tacrine.
7. The cost of the drug should be such to make it accessible to a vast number of patients, including those who are not insured (pharmaco-economic question).

TABLE IV. What did we learn from the tacrine experience?

Compared to Tacrine	General Prerequisites
Be less toxic	Slow down deterioration
Show stronger clinical efficacy	Improve performance (ADL)
Benefit more than 25% of patients	Delay institutionalization Be sold at a moderate price for long-term treatment (5-10 yrs)

TABLE V. Prerequisites for a new cholinesterase inhibitor to replace tacrine.

Methodological advances in the study of ChEI pharmacology

The study of the cholinergic system has been made possible through the development of sensitive micromethods during the last forty years. Lately, a sensitivity for ACh determinations in the low femtomole (fmole) range was reached using highly sensitive electrochemical detectors (ECD) (Table VII).

Based on observations in animals we postulated that ChE inhibition in plasma, erythrocytes or in brain could not be considered as an accurate predictor of changes in cortical ACh (Messamore et al., 1993). Therefore, it became important, following administration of a ChEI, to measure directly CNS ACh levels. This measurement allows one to evaluate the potential of the drug to elevate the neurotransmitter to therapeutically relevant concentrations. Microdialysis *in vivo* is the only method which allows one to carry on such measurements in the awake animal without interference of anesthesia. In particular, when studying the effect of a ChEI it is important to avoid the interaction with a second ChEI. We developed a microdialysis technique which allows fmol range measurement of ACh without introduction of a second ChEI in the probe to artificially magnify ACh levels (Messamore et al., 1993; Cuadra et al., 1994). This technique has been used extensively in our laboratory to examine the effect of several ChEI being tested in clinical trials or to develop novel compounds.

Cholinesterase inhibitors effect on extracellular concentrations of cortical neurotransmitters

Clinical and experimental evidence indicates involvement and interactions between the cholinergic system and the biogenic amine systems in the

COMPOUND TO BE COMBINED WITH A ChEI	PHARMACOLOGICAL EFFECTS TO BE EXPECTED
M_1 and M_3 partial agonist	Potentiate postsynaptic effects Decrease tolerance & desensitization
M_1 antagonist	Counteract cholinergic toxicity and desensitization
M_2 antagonist	Enhance ACh levels and increase its release

TABLE VI. Combinations of ChEI with muscarinic agonists and antagonists and their effects.

	Method	Sample Size	Sensitivity (moles)	Reference
AChE Activity	microdiver	one cell	10^{-12}	Giacobini and Zajicek, 1956
	gasometric			Giacobini, 1957
	radiometric	one cell	10^{-12}	Koslow and Giacobini, 1969
CAT Activity	radiometric	one cell	10^{-12}	Buckley et al., 1967
				McCaman and Hunt, 1965
				McCaman and Dewhurst, 1970
		homog.	10^{-12}	Goldberg and McCaman, 1973
ACh Level	HPLC-ECD	10 µl of dialysate	10^{-14}	Cuadra et al., 1994
			10^{-15}	Giacobini (1996)

TABLE VII: Cholinergic system: forty years of development of micromethod (1956-1996). AChE: acetylcholinesterase; CAT: cholineacetyltransferase

cognitive impairments observed in AD (Hardy et al., 1985; Decker and McGaugh, 1991). A brain region of particular interest is the frontal cortex because in both humans and rodents it represents the major cholinergic projection of the nucleus basalis magnocellularis (NBM) of the basal forebrain (Mesulam and Geula, 1988). Of the NBM neurons that project to the cerebral cortex, 80-90% are cholinergic in the rat (Rye et al., 1984). Similarly, the major, if not sole, noradrenergic projection to the cortex is the locus coeruleus (LC) (Parnavelas, 1990). Pharmacological alleviation of combined cholinergic NBM/noradrenergic LC lesion-induced memory deficits in rats has been reported (Santucci et al., 1991).

Table VIII compares the effects on ACh, norepinephrine (NE) and dopamine (DA) levels as well as AChE inhibition after systemic administration of six ChEI studied in our laboratory. With the exception of MF-268, not yet tested, they have all shown clinical efficacy. The difference in chemical structure among these compounds is a striking characteristic of new ChEI. Our results show a significant increase in cortex for all three neurotransmitters and for all six ChEIs investigated.

The results reported in Table VIII also suggest that extracellular ACh levels in cortex are not directly related to ChE inhibition, supporting results of previous microdialysis studies showing comparable elevations of ACh levels in spite of different magnitudes of ChE inhibition (Messamore et al., 1993). As a consequence, CNS ChE inhibition can not be considered as a reliable predictor of its effect on concentrations of extracellular ACh in cerebral cortex.

A new aspect of ChEI pharmacology is the effect on neurotransmitters other than ACh (Cuadra et al., 1994). This effect depends not only on dose but also on the type of compound and could be of therapeutic significance.

Compound	Dose mg/kg	ChE max.% inhib.	ACh	NE	DA
Physostigmine	0.3	60	4000	75	120
Heptyl-physost.	2	75	2500	25	75
E 2020	2	35	2100	100	80
MF-268	2	40	2500	100	60
MDL 73,745	2	65	1020	120	370
Metrifonate	80	70	1700	60	75

TABLE VIII. ChEI effects on ACh, NE, DA levels and ChE activity in rat brain cortex after s.c. administration. E 2020 = (R,S)-1benzyl-4-(5,6 dimethoxy-1-idanon)-2-yl-methylpiperidine (Giacobini et al., 1996); MF-268 = 2, 6-dimethylmorfolin-octyl-carbamoyl eseroline (Zhu et al., 1996); Metrifonate =0,0-dimethyl-(1-hydroxy-2,2,2 trichloroethyl-phosphate) (Mori et al., 1994, 1995b); MDL 73,745=2,2,2-trifluoro-1-(3-trimethylsilylphenyl)ethanone (Zhu et al., 1995)

Co-administration of ChEI with adrenergic agonists and antagonists demonstrates the interaction between cholinergic and adrenergic systems

Several studies have indicated close interactions between cholinergic and noradrenergic systems (Decker and McGaugh, 1991). NE decreases the release of ACh from cholinergic terminals in cortex (Vizi, 1980; Moroni et al., 1983). This effect is mediated both directly via alpha-adrenergic receptors on cholinergic terminals and indirectly via NE modulation of gamma aminobutyric acid (GABA) release (Beani et al., 1986). There is also evidence that NE and ACh interact with each other, influencing learning and memory (Santucci et al., 1991). The interaction between ACh and NE appears to be reciprocal as ACh is also able to modulate NE function (Roth et al., 1982; Egan and North, 1985, 1986; Hörtnagl et al., 1987). In a previous study (Cuadra et al., 1994; Giacobini and Cuadra, 1994), we have shown that systemic administration of low doses of PHY and HEP elicit a significant and simultaneous increase in ACh and NE levels. It is possible that the NE elevation seen in our studies could down-regulate ACh levels and decrease the therapeutic effect of these drugs.

Effect of adrenergic antagonist co-administration

To investigate this putative cholinergic-adrenergic interaction, we studied the effect of PHY and its analog heptylphysostigmine (HEP) in animals pretreated with idazoxan (IDA), a selective $\alpha 2$-antagonist, on the extracellular levels of ACh, NE, DA and 5-hydroxytryptamine (5-HT (serotonin) in cerebral cortex using microdialysis (Cuadra and Giacobini, 1995a).

In this study, we found that IDA administered either systemically or locally into the brain has no effect on extracellular levels of ACh. This suggests NE may not be involved in tonic regulation of cortical cholinergic activity. The increase of cortical NE release seen after local or systemic IDA administration agrees with the results of L'Heureux et al. (1986) and Dennis et al. (1987). This suggests the effects of IDA on NE release are mediated primarily by α2-adrenoceptors located presynaptically on noradrenergic nerve terminals.

The possibility of further prolonging the effect of ChEI with selective α2-antagonist co-administration and additive DA-ACh interaction may be of therapeutic interest. Specifically, our data suggest that a combination of cholinergic and adrenergic drugs may improve the pharmacological effects of ChEI on several cortical neurotransmitter functions which may represent a significant advantage in AD treatment because of the multiple transmitter deficits seen in the disease.

Effect of adrenergic agonist co-administration

In order to obtain further information on cortical neurotransmitter interaction, we evaluated the effect of PHY and its analogue HEP on the extracellular levels of ACh, NE, DA and 5-HT in animals pre-treated with clonidine (CLO), a selective α2-agonist (Cuadra and Giacobini, 1995b).

In agreement with our previous observations (Cuadra et al., 1994; Cuadra and Giacobini, 1995a), which suggested that NE may not be involved in the tonic regulation of cortical cholinergic activity, we detected no effect on extracellular levels of ACh after either systemic or local administration of CLO but NE, DA and 5-HT levels were all decreased. CLO co-administration reduced the effect of PHY on ACh levels, however, HEP administered to animals pre-treated with CLO produced a stronger effect than HEP alone.

The reduction in cortical NE release observed after local or systemic CLO (54% and 57%, respectively) is in agreement with results previously reported by L'Heureux et al. (1986) and Van Veldhuizen et al. (1993). The CLO data, together with our previous results (Cuadra and Giacobini, 1995a) obtained in rats pre-treated with IDA, suggest that ChEI effects on cortical NE release might be mainly mediated by α2-autoreceptors located on noradrenergic nerve terminals (Ong et al., 1991; Coull, 1994).

In analogy, both routes of CLO administration (s.c. and local through the probe) also decreased extracellular levels of DA. This effect of CLO on cortical release of DA might indicate an activation of α2-heteroreceptors localized presynaptically on terminals of dopaminergic neurons which have been demonstrated to modulate its release (Ueda et al., 1983; Dubocovich, 1984). It is well established that DA participates in the control of cognitive function (Brozoski et al., 1979) and plays a role in attention and reward mechanisms (Wise, 1978; Beninger, 1983).

In conclusion, our data suggest that co-administration of a selective α2-agonist such as CLO with ChEI does not represent a favorable pharmacological and therapeutical alternative. Furthermore, the decrease of extracellular DA may represent a negative effect in the treatment of cognitively impaired AD patients. Considering our previous results with IDA (Cuadra and Giacobini, 1995a), we suggest that a combination of an α2-antagonist with HEP may represent a more favorable approach to improve the clinical efficacy of ChEIs in AD treatment.

Cholinesterase inhibitors and APP secretion: a possible slowing effect of deterioration ?

The β-amyloid peptide (βA4), one of the major constituent proteins of neuritic plaques in the brain of AD patients, originates from a larger polypeptide denominated amyloid precursor protein (APP) (Kang et al., 1987). APP is widely distributed throughout the mammalian brain including rat brain with a prevalent neuronal localization (Beeson et al., 1994). APP can be processed by several alternative pathways, but the mechanisms responsible for this processing are not completely understood. A secretory pathway is believed to generate non-amyloidogenic soluble derivatives (APPs) following cleavage within the βA4 segment (Sisodia et al., 1990; Esch et al., 1990). Cholinergic agonists regulating processing and secretion of APPs by increasing, as demonstrated *in vitro*, protein kinase C (PKC) activity of target cells (Nitsch et al., 1992; Buxbaum et al., 1992; Nitsch and Growdon, 1994) could decrease potentially amyloidogenic derivatives. We suggested that long-term inhibition of ChE having the effect of increasing the level of synaptic ACh may result in the activation of normal APP processing in AD brain (Giacobini, 1994). This phenomenon could slow down the formation of amyloidogenic APP fragments.

To determine whether ChEI could alter the release of APP we used superfused brain cortical slices of the rat (Mori et al., 1995a) following the method described by Nitsch et al. (1993). Three short- and long-lasting ChEI were tested for their ability to enhance the release of non-amyloidogenic soluble derivatives (APPs) (Mori et al., 1995a). These included: PHY, HEP and DDVP (dichlorvos, a metabolite of metrifonate) at concentrations producing ChE inhibitions ranging from 5% to 95%. All three ChEI elevated

Drug	Conc. (μM)	Increase (% of basal)	ChE Act. (% Inhib.)	APP-KPI mRNA (% of basal)
Bethanechol	1	48	0	-
	100	53	0	-
Physostigmine	.1	48	25	-
Heptyl-physostygmine	.1	41	61	-35*
Dichlorvol	.02	33	95	-
Phorbol myristate	.1	-	-	+50

TABLE IX. Drug-stimulated changes of basal APPs release and APP-KPI mRNA from rat brain (Mori et al., 1995a); *from Giacobini et al., 1995 (5 mg/kg s.c. 48 hrs)

APPs release significantly above control levels (Table IX). Electrical field stimulation significantly increased the release of APPs within 50 min. Similar increase was observed after muscarinic receptor stimulation with bethanechol (BETHA). Tetrodotoxin (TTX) completely blocked the effect of electrical stimulation (Mori et al., 1995a).

The levels of total APP mRNAs in rat cortical slices did not change after incubation with BETHA, DDVP and PHY, but activation of PKC with phorbol 12-myristate-13-acetate (100 nM) increased the level of total APP mRNA by 50% (Table IX) (Giacobini et al., 1995). PHY and MTF administration (0.3 mg/kg and 80 mg/kg s.c., respectively) for 3-48 hrs did not significantly change the levels of APP 695 and APP-KPI (Kunitz-type) protease inhibitor mRNAs (Table IX). HEP administration (5 mg/kg s.c., 3-48 hrs) decreased by 35% the level of APP-KPI mRNA in rat cerebral cortex (Giacobini et al., 1995). AD pathology has been associated with an increase of the KPI-containing forms of APP and the propensity across species to develop neuritic plaques in the cortical regions (Anderson et al., 1989). Our findings suggest that administration of ChEI to AD patients by increasing secretion of APP and inhibiting formation of specific APP mRNAs may exert a neuroprotective effect by activating normal APP processing through a muscarinic mechanism and decreasing amyloid deposition in brain cells.

CONCLUSIONS

ChEIs, particularly second generation, post-PHY and post-tacrine compounds, affect cortical and presumably sub-cortical neurotransmitters other than ACh. Co-administration of ChEI with adrenergic agonists and antagonists clearly demonstrate a coupling between cholinergic and non-cholinergic systems. This effect depends not only on the dose but also on the type of compound. It might be of additional therapeutical value by activating pathways and circuits other than cholinergic ones which are also hypofunctional in AD. It also represents a possibility of prolonging the effect of ChEI by means of double function hybrid-compounds or co-administration of two drugs. A newly demonstrated *in vitro* feature of ChEI is their ability to enhance the release of non-amyloidogenic soluble derivatives of APP and possibly slow down the formation of β-amyloid deposition in brain. This might slow down cognitive deterioration of the patient treated with ChEI. Recent clinical trials of ChEI extending beyond 36 mo. duration should be able to demonstrate whether or not this pharmacological effect on APP metabolism is of clinical significance. Cholinomimetic alternatives other than ChEI exist and are also being explored pharmacologically and clinically. The most common are based on direct stimulation of muscarinic or nicotinic receptors. However, also with these compounds we suggest combinations of drugs to potentiate the cognitive effect and to decrease side effects.

ACKNOWLEDGMENTS

The author wishes to thank Diana L. Smith for typing the manuscript and Joyce M. Barton for editing the text. Supported by the National Institute on Aging #P 30 AG 08014.

REFERENCES

Alkon DL, Amaral DG, Bear MF, Black J, Carew TJ, Cohen NJ, Disterhoft JF, Eichenbaum H, Golski S, Gorman LK et al. (1991): Learning and memory. FESN Study Group Brain Res. *Brain Res Reviews* 16(2):193-220.

Anderson JP, Refolo LM, Wallace W, Mehta P, Krishnamurthi M, Gotlib J, Bierer L, Haroutunian V, Perl D and Robakis NK (1989): Differential brain expression of the Alzheimer's amyloid precursor protein. *EMBO J* 8:3627-3632.

Arneric SP, Sullivan JP, Decker MW, Brioni JD, Briggs CA, Donnelly-Roberts D, Marsh KC, Rodrigues AD, Garvey DS, Williams M and Buccafusco JJ (1994): ABT-418: A novel cholinergic channel activator (ChCA) for the potential treatment of Alzheimer disease. In: *Alzheimer Disease: Therapeutic Strategies.* Giacobini E and Becker R, eds. Boston: Birkhauser, pp. 196-205.

Bartus RT, Dean RL, Beer B and Lippa AS (1982): The cholinergic hypothesis of geriatric memory dysfunction. *Science* 217:408-417.

Beani L, Tanganelli S, Antonelli T and Bianchi C (1986): Noradrenergic modulation of cortical acetylcholine release is both direct and gamma-aminobutyric acid-mediated. *J Pharmacol Exp Ther* 236: 230-236

Becker RE and Giacobini E (1988): Mechanisms of cholinesterase inhibition in senile dementia of the Alzheimer type: Clinical, Pharmacological, and Therapeutic aspects. *Drug Dev Res* 12:163-195.

Becker RE, Moriearty P and Unni L (1991): The second generation of cholinesterase inhibitors: clinical and pharmacological effects. In: *Cholinergic Basis for Alzheimer Therapy,* Giacobini E and Becker R, eds. Boston: Birkhauser, pp. 263-296.

Beeson JG, Shelton ER, Chan HW and Gage FH (1994): Differential distribution of amyloid protein precursor immunoreactivity in the rat brain studied by using five different antibodies. *J Comp Neurol* 342:78-96.

Beninger RJ (1983): The role of dopamine in locomotor activity and learning. *Brain Res Rev* 6:173-196.

Berger-Sweeney J, Heckers S, Mesulam MM, Wiley RG, Lappi DA and Sharma M (1994): Differential effects on spatial navigation of immunotoxin-induced cholinergic lesions of the medial septal area and nucleus basalis magnocellularis. *J Neurosci* 14(7):4507-4519.

Brozoski TJ, Brown RM, Rosvold HE and Goldman PS (1979): Cognitive deficit caused by regional depletion of dopamine in the prefrontal cortex of rhesus monkey. *Science* 205:929-932.

Buckley G, Consolo S, Giacobini E and McCaman R (1967): A micromethod for the determination of choline acetylase in individual cells. *Acta Physiol Scand* 71:341-347.

Buxbaum JD, Oishi M, Chen Hl, Pinkas-Kramarski R, Jaffe EA, Gandy SE and Greengard P (1992): Cholinergic agonists and interleukin 1 regulate processing and secretion of the Alzheimer βA4 amyloid protein precursor. *Proc Natl Acad Sci USA* 89:10075-10078.

Coull JT (1994): Pharmacological manipulation of the alpha-2-noradrenergic system - effects on cognition. *Drugs and Aging* 5(2):116-121.

Cuadra G and Giacobini E (l995a): Coadministration of cholinesterase inhibitors and Idazoxan: effects of neurotransmitters in rat cortex in vivo. *J Pharm Exp Ther* 273 (1):230-240.

Cuadra G and Giacobini E (1995b): Effects of cholinesterase inhibitors and clonidine coadministration on rat cortex neurotransmitters in vivo. *J Pharm Exp Ther* 275(1):228-236.

Cuadra G, Summers K, and Giacobini E (1994): Cholinesterase inhibitor effects on neurotransmitters in rat cortex in vivo. *J Pharm Exp Ther* 270(1):277-284.

Decker MW and McGaugh JL (1991): The role of interactions between the cholinergic system and other neuromodulatory systems in learning and memory. *Synapse* 7(2):151-168.

Dekker AJ, Connor DJ and Thal LJ (1991): The role of cholinergic projections from the nucleus basalis in memory. *Neurosci & Biobehav Rev* 15(2):299-317.

Dennis T, L'Heureux R, Carter C and Scatton B (1987): Presynaptic alpha-2-adrenoceptors play a major role in the effects of idazoxan on cortical noradrenaline release. *J Pharmacol Exp Ther* 241:642-649.

De Sarno P and Giacobini E (1989): Modulation of acetylcholine release by nicotinic receptors in the rat brain. *J Neurosci Res* 22:l94-200.

De Sarno P, Giacobini E and Clark B (1982): Changes in nicotinic receptors in human and rat CNS. *Fed Proc* 2:364.

Drachman DA (1982): Aging and dementia: insights from the study of anticholinergic drugs. In: *Biological aspects of Alzheimer disease.* Katzmann R. ed. *Banbury Report* 15:363-369.

Drachman DA and Leavitt J (1974): Human memory and the cholinergic system: a relationship to aging? *Arch Neurol* 30:113-121.

Dubocovich ML (1984): Presynaptic alpha-adrenoceptors in the central nervous system. *Ann New York Acad Sci* 430:7-25.

Dunnett SB, Toniolo G, Fine A, Ryan CN, Bjorklund A and Iversen SD (1985): Transplantation of embryonic ventral forebrain neurons to the neocortex of rats with lesions of nucleus basalis magnocellularis II. Sensorimotor and learning impairments. *Neuroscience* 16(4):787-797.

Egan TM and North RA (1986): Actions of acetylcholine and nicotine on rat locus coeruleus neurons *in vitro*. *Neuroscience* 19:565-571.

Egan TM and North RA (1985): Acetylcholine acts on M_2-muscarinic receptors to excite rat locus coeruleus neurones. *Brit J Pharmacol* 85:733-735.

Esch FS, Keim PS, Beattie EC, Blacher RW, Culwell AR, Oltersdorf T, McClure D and Ward PJ (1990): Cleavage of amyloid beta peptide during constitutive processing of its precursor. *Science* 248:1122-1124.

Flicker C, Ferris SH and Serby M (1992): Hypersensitivity to scopolamine in the elderly. *Psychopharmacology* 107:437-441.

Gage FH and Bjorklund A (1986): Cholinergic septal grafts into the hippocampal formation improve spatial learning of memory in aged rats by an atropine-sensitive mechanism. *Neuroscience* 6(10):2837-2847.

Giacobini E (1957): Quantitative determination of cholinesterase in individual sympathetic cells. *J Neurochem* 1:234-244.

Giacobini E (1994): Cholinomimetic therapy of Alzheimer disease: does it slow down deterioration? In: *Recent Advances in the Treatment of Neurodegenerative Disorders and Cognitive Dysfunction.* Racagni G, Brunello N and Langer SZ, eds. *Int Acad Biomed Drug Res,* New York: Karger 7(23): pp. 51-57.

Giacobini E (1995): Cholinesterase inhibitors. From preclinical studies to clinical efficacy in Alzheimer disease. In: *Enzymes of the cholinesterase family.* Quinn D, Balasubramaniam AS, Doctor BP and Taylor P, eds. New York: Plenum Press, pp 463-469

Giacobini E (1996): Cholinesterase Inhibitors do more than Inhibit Cholinesterase. In: *Alzheimer Disease: From Molecular Biology to Therapy.* Becker R and Giacobini E, eds. Boston: Birkhauser.

Giacobini E and Cuadra G (1994): Second and third generation cholinesterase inhibitors: from preclinical studies to clinical efficacy. In: *Alzheimer Disease: Therapeutic Strategies.* Giacobini E and Becker R, eds. Boston: Birkhauser, pp. 155-171.

Giacobini E and Zajicek J (1956): Quantitative determination of acetylchlinesterase activity in individual nerve cells. *Nature* 177:185-186.

Giacobini E, DeSarno P, Clark B and McIlhany M (1989): The cholinergic receptor system of the human brain: neurochemical and pharmacological aspects in aging and Alzheimer. In: *Progress in Brain Research.* Nordberg A, Fuxe K, Holmstedt B and Sundwall A. eds. Amsterdam: Elsevier, Vol 79 pp. 335-343.

Giacobini E, Mori F, Buznikov A and Becker R (1995): Cholinesterase inhibitors alter APP secretion and APP mRNA in rat cerebral cortex. *Soc Neurosci Abstr.* 21:988.

Giacobini E, Zhu X-D, Williams E, and Sherman KA (1996): The effect of the selective reversible acetylcholinesterase inhibitor E2020 on extracellular acetylcholine and biogenic amines levels in rat cortex. *Neuropharm* 35(2):205-211.

Goldberg AM and McCaman RE (1973): The determination of picomole amounts of acetylcholine in mammalian brain. *J Neurochem* 20:1-8.

Hardy J, Adolfsson R, Alafuzoff I, Bucht G, Marcusson J, Nyberg P, Perdhal E, Wester P and Winblad B (1985): Transmitter deficits in Alzheimer's disease. *Neurochem Intl* 7:545-563.

Hörtnagl H, Potter PE and Hanin I (1987): Effect of cholinergic deficit induce ethylcholine aziridinium (AF64A) on noradrenergic and dopaminergic parameters in rat brain. *Brain Res* 421:75-84.

Kang J, Lemaire H-G, Unterbeck A, Salbaum JM, Master CL, Grzeschil K-H, Multaup G, Beyreuther K and Muller-Hill B (1987): The precursor of Alzheimer disease amyloid A4 protein resembles a cell-surface receptor. *Nature* 325: 733-736.

Knapp MJ, Knopman DS, Solomon PR, Pendlebury WW, Davis CS and Gracon SI (1994): A 30-week randomized controlled trial of high-dose Tacrine in patients with Alzheimer's disease. *J Amer Med Assoc* 271(13):985-991.

Koslow SH and Giacobini E (1969): An isotopic micromethod for the measurement of cholinesterase activity in individual cells. *J Neurochem* 16:1523-1528.

L'Heureux R, Dennis T, Curet O and Scatton B (1986): Measurement of endogenous noradrenaline release in the rat cerebral cortex in vivo by transcortical dialysis: Effects of drugs affecting noradrenergic transmission. J *Neurochem,* 46:1794-1801.

Mandel RJ and Thal LJ (1988): Physostigmine improves water maze performance following nucleus basalis magnocellularis lesions in rats. *Psychopharmacology* 96(3):421-425.

Mandel RJ, Gage FH and Thal LJ (1989): Enhanced detection of nucleus basalis magnocellularis lesion-induced spatial learning deficit in rats by modification of training regimen. *Behav Brain Res* 31(3):221-229.

McCaman RE and Dewhurst SA (1970): Choline acetyltransferase in individual neurons of Aplysia californica. *J Neurochem* 17:1421-1426.

McCaman RE and Hunt JM (1965): Microdetermination of choline acetylase in nervous tissue. *J Neurochem* 12:253-259.

Messamore E, Warpman U, Williams E and Giacobini E (1993): Muscarinic receptors mediate attenuation of extracellular acetylcholine levels in rat cerebral cortex after cholinesterase inhibition. *Neurosci Lett* 158:205-208.

Mesulam MM and Geula C (1988): Nucleus basalis (Ch4) and cortical cholinergic innervation in the human brain: observations based on the distribution of acetylcholinesterase and choline acetyltransferase. *J Comp Neurol* 275:216-240.

Mori F, Cuadra G, and Giacobini E (1995b): Metrifonate effects on acetylcholine and biogenic amines in rat cortex. *Neurochem Res* 20(9):1081-1088.

Mori F, Cuadra G, Williams E, Giacobini E and Becker R (1994): Effects of metrifonate on acetylcholine and monoamine levels in rat cortex. *Soc Neurosci Abst* 20:83 (No. 4016).

Mori F, Lai CC, Fusi F and Giacobini E (1995a): Cholinesterase inhibitors increase secretion of APPs in rat brain cortex. *Neuro Report* 6(4):633-636.

Moroni F, Tanganelli S, Antonelli T, Carlá V, Bianchi C and Beani L (1983): Modulation of cortical acetylcholine and gamma-aminobutyric acid release in freely moving guinea pigs: effects of clonidine and other adrenergic drugs. *J Pharmacol Exp Ther* 236:230-236.

Murray CL and Fibiger HC (1985): Learning and memory deficits after lesions of the nucleus basalis magnocellularis: reversal by physostigmine. *Neuroscience* 14(4):1025-1032.

Nilsson OG, Shapiro ML, Gage FH, Olton DS and Bjorklund A (1987): Spatial learning and memory following fimbria-fornix transection and grafting of fetal septal neurons to the hippocampus. *Experimental Brain Res*. 67(1):195-215.

Nitsch RM and Growdon JH (1994): Role of neurotransmission in the regulation of amyloid beta-protein precursor processing. *Biochem Pharmacol* 47(8):1275-1284.

Nitsch RM, Farber SA, Growdon JH and Wurtman RJ (1993): Release of amyloid beta-protein precursor derivatives by electrical depolarization of rat hippocampal slices. *Proc Natl Acad Sci USA*, 90:191-193.

Nitsch RM, Slack BE, Wurtman RJ and Growdon JH (1992): Release of Alzheimer precursor derivatives stimulated by activation of muscarinic acetylcholine receptors. *Science* 258:304-307.

Ong ML, Ball SG and Vaughn PFT (1991): Regulation of noradrenaline release from rat occipital cortex tissue chops by alpha-2-adrenergic agents. *J Neurochem* 56:1387-1393.

Page KJ, Everitt BJ, Robbins TW, Marston HM and Wilkinson LW (1991): Dissociable effects on spatial maze and passive avoidance acquisition and retention following AMPA- and ibotenic acid-induced excitotoxic lesions of the basal forebrain in rats: differential dependence on cholinergic neuronal loss. *Neuroscience* 43(2-3)::457-472.

Parnavelas JG (1990): Neurotransmitters in the cerebral cortex. In: *Progress in Brain Research.* Uylings HBM, Van Eden CG, De Bruin JPC, Corner MA and Feenstra MGP, eds. The Netherlands: Elsevier, Vol 85, pp. 13-29.

Perry EK, Tomlinson BE, Blessed G, Bergmann K, Gibson PH and Perry RH (1978): Correlation of cholinergic abnormalities with senile plaques and mental test scores in senile dementia. *Brit Med J* 2(6150):1457-1459.

Roth KA, McIntire SL and Barchas JD (1982): Nicotinic-catecholaminergic interactions in rat brain: evidence for cholinergic nicotinic and muscarinic interactions with hypothalamic epinephrine. *J Pharmacol Exp Ther* 221:416-420.

Rye DB, Wainer BM, Mesulam MM, Mufson EJ and Saper CB (1984): Cortical projections arising from the basal forebrain. *Neuroscience* 13:627-643.

Santucci AC, Haroutunian V and Davis KL (1991): Pharmacological alleviation of combined cholinergic - noradrenergic lesion-induced memory deficits in rats. *Clin Neuropharmacol* 14:1-8

Schröder H, Giacobini E, Weers A, Birtsch C and Schutz U (1995): Nicotinic receptors in Alzheimer's disease. In: *Brain Imaging of Nicotine and Tobacco Smoking.* Domino EF. ed. Ann Arbor, MI NPP Books, pp. 73-93.

Sisodia SS, Koo EH, Beyreuther K, Unterbeck A and Price DL (1990): Evidence that beta amyloid protein in Alzheimer disease is not derived by normal processing. *Science* 248:492-495.

Summers KL and Giacobini E (1995): Effects of local and repeated systemic administration of (-)nicotine on release of acetylcholine, norepinephrine, dopamine and serotonin in rat cortex. *J Neuroscience Res* 20(6):683-689.

Summers KL, Cuadra G, Naritoku D and Giacobini E (1994): Effects of nicotine on levels of acetylcholine and biogenic amines in rat cortex. *Drug Dev Res* 31:108-119.

Summers KL, Lippiello PM, Verhulst S and Giacobini E (1995): 5-Fluoronicotine, noranhydroecgonine and pyridyl-methylpyrrolidine release acetylcholine and biogenic amines in rat cortex *in vivo. Neurochem Res* 20 (9):1089-1094.

Ueda H, Goshima Y and Misu Y (1983): Presynaptic mediation by alpha$_2$-, beta$_1$- and β_2-adrenoceptor of endogenous dopamine release from slices of rat hypothalamus. *Life Sci* 33:371-376.

Van Veldhuizen MJ, Feenestra MG, Heinsbroek RP and Boer GJ (1993): In vivo microdialysis of noradrenaline overflow: effects of alpha-adrenoceptor agonists and antagonists measured by cumulative concentration-response curves. *Brit J Pharmacol* 109:655-660.

Vanderwolf CH, Dickson CT and Baker GB (1990): Effects of p-chloro-phenylalanine and scopolamine on retention of habits in rats. *Pharamcol Biochem Behav* 35:847-853.

Vizi ES (1980): Modulation of cortical release of acetylcholine by noradrenaline released from nerves arising from the rat locus coeruleus. *Neuroscience* 5:2139-2144.

Wesnes K, Anand R and Lorscheid T (1990): Potential of moclobomide to improve cerebral insufficiency identified using a scopolamine model of aging and dementia. *Acta Psychiatr Scand (Suppl)* 360:71-72.

Whitehouse PJ, Price DL, Struble RG, Clark AW, Coyle JT and Delon MR (1982): Alzheimer disease and senile dementia: loss of neurons in the basal forebrain. *Science* 215(4537):1237-1239.

Winkler J, Suhr ST, Gage FH, Thal LJ and Fisher LJ (1995): Essential role of neocortical acetylcholine in spatial memory. *Nature* 375:484-487.

Wise RA (1978): Catecholamine theories of reward: A critical review. *Brain Res* 152:215-247.

Zhu X-D, Cuadra G, Brufani M, Maggi T, Pagella PG, Williams E, and Giacobini E (1996): Effects of MF-268, a new cholinesterase inhibitor, on acetylcholine and biogenic amines in rat cortex. *J Neurosci* Res 43:120-126.

Zhu X-D, Giacobini E, and Hornsperger J-M (1995): Effect of MDL 73,745 on acetylcholine and biogenic amine levels in rat cortex. *Europ J Pharmacol* 276:93-99.

Alzheimer Disease: From Molecular Biology to Therapy
edited by R. Becker and E. Giacobini

LONG-TERM TACRINE TREATMENT: EFFECT ON NURSING HOME PLACEMENT AND MORTALITY

Stephen Gracon, Fraser Smith, and Toni Hoover
Parke-Davis Pharmaceutical Research,
Division of Warner-Lambert Company, Ann Arbor, MI

David Knopman
Department of Neurology, University of Minnesota, Minneapolis, MN

Lon Schneider
Department of Psychiatry, University of Southern California, Los Angeles, CA

Kenneth Davis
Department of Psychiatry, Mt. Sinai School of Medicine, New York, NY

Sheela Talwalker
G.D. Searle and Co., Skokie, IL

INTRODUCTION

The most important question arising from controlled studies of tacrine was whether the acute, symptomatic effects of tacrine treatment observed over 30 weeks (Knapp et al., 1994) translated into effects on long-term outcomes of the natural history of Alzheimer's disease (AD) such as nursing home placement and mortality. Figure 1 shows a schematic representation of the natural history of AD. MMSE score is used as a surrogate of disease severity, and disease milestones are indicated.

Almost 75% of patients in the United States will eventually enter a nursing home with an average stay of more than 3 years (Welch et al., 1992). Although nursing home placement may be a somewhat imprecise endpoint, it is a major decision point for the family. Duration of illness, declining cognitive function, lost activities of daily living, and severity of behavioral symptoms are important risk factors for nursing home placement and are good indicators for the severity of illness and, therefore, caregiver burden.

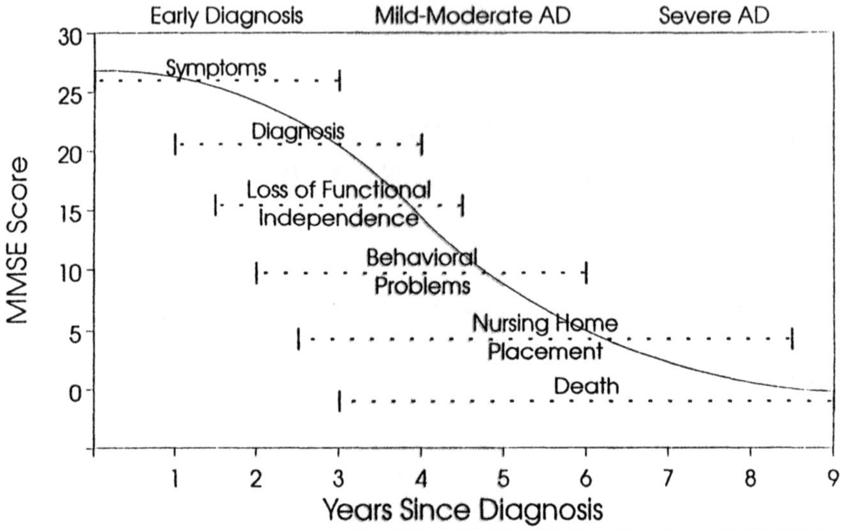

FIGURE 1. Natural history of Alzheimer's disease

Patients who completed the 30-week double-blind study and those who terminated early were eligible to receive long-term, open-label tacrine treatment at the discretion of the study physician in consultation with the family.

Approximately 2 years after the last patient completed the double-blind phase, the protocol was amended to allow collection of follow-up information on nursing home placement and mortality. Attempts were made, through the study centers, to contact the families of all 663 patients who originally entered the study. Data were collected to determine whether patients continued to take tacrine; at what daily dose; and whether they were living at home, in a nursing home, or had died (Knopman et al., 1996).

RESULTS

Follow-up data on nursing home placement were available for 595 (90%) of the 663 patients randomized to treatment in the 30-week study. Mortality data were available for all 663 patients who entered the study, 81 of whom had died.

Data were analyzed by logistic regression. Nursing home placement and mortality were analyzed first at the end of double-blind treatment (Week 30). Follow-up data were then analyzed based on patients' treatment status: all patients by last tacrine dose taken regardless of time off drug prior to follow-up ("all patients"); and patients who were on tacrine at follow-up or who had been off drug for 60 days or less prior to an event, by last dose taken ("on tacrine").

Comparison	Odds Ratio	95% CI	Difference Favors	p-Value
80 mg/day vs Placebo[a]	1.1	[0.3, 3.8]	80 mg/day	0.833
120 mg/day vs Placebo	1.6	[0.6, 4.1]	120 mg/day	0.338
160 mg/day vs Placebo	2.8	[1.0, 7.8]	160 mg/day	0.046*

TABLE I. 30-Week Study: logistic regression analysis of nursing home placement or mortality at week 30. [a] The sample size for the 80-mg/day treatment group was considerably smaller, approximately one-third that of the placebo group. * $p < 0.05$ Adapted from Knopman et al., 1996.

The "on tacrine" results are presented here because all patients continued to have adequate caregiver support, and the only obvious potential source of bias was patients' ability to tolerate gastrointestinal side-effects. The results of both analyses were, however, consistent.

Nursing Home Placement and Mortality at Week 30

At the end of the 30-week double-blind period, there was evidence of a trend toward a reduced probability of nursing home placement or death with increasing tacrine dose, which was statistically significant at 160 mg/day tacrine (Table I).

Nursing Home Placement and Mortality for Patients on Tacrine at Follow-up

The "on tacrine" analysis of nursing home placement included data for 320 patients, and the analysis of mortality included data for 308. Analyses of the follow-up data indicated that patients taking tacrine at doses >120 mg/day were more likely to remain at home compared with patients who were taking tacrine 80 mg/day or less (Table II and Figure 2). A significant result was also seen for patients taking >80 to ≤120 mg/day versus ≤80 mg/day.

Kaplan-Meier estimates of the 25th percentile for nursing home placement were 378 days for patients with a dose of 20 to ≤80 mg/day, 841 days for patients with a dose of >80 to ≤120 mg/day, and 772 days for patients with a dose of >120 to ≤160 mg/day. This represents a significant delay in time to entry into the nursing home of more than 400 days, as indicated by dashed lines in Figure 2.

Estimates and significance levels for analyses of mortality should be interpreted with caution, because fewer than 10% of tacrine-treated patients died and results may be highly influenced by one or two events.

Treatment assignment for these analyses was not random or blinded. Patients who were destined to decline more slowly could also have been the ones who could titrate to and tolerate higher tacrine doses. Therefore, these

Comparison	Odds Ratio	95% CI	Difference Favors	p-Value
NHP				
>80 to ≤120 vs 20 to ≤80	2.7	[1.4, 5.2]	>80 to ≤120	0.003*
>120 to ≤160 vs 20 to ≤80	2.8	[1.5, 5.2]	>120 to ≤160	0.001*
Mortality				
>80 to ≤120 vs 20 to ≤80	2.1	[0.7, 6.4]	>80 to ≤120	0.180
>120 to ≤160 vs 20 to ≤80	2.9	[1.0, 8.5]	>120 to ≤160	0.045*

TABLE II. 30-Week Study: logistic regression analyses of NHP and mortality at follow-up, by tacrine dose (mg/day) for patients who were on tacrine. *$p < 0.05$. Adapted from Knopman et al., 1996.

patients would have had a better outcome regardless of treatment compared with those who were more frail and could neither tolerate tacrine treatment nor be maintained at home. The present data make this an unlikely explanation given that the analysis of patients who continued to take tacrine demonstrated a significant delay in time to nursing home entry on the same order of

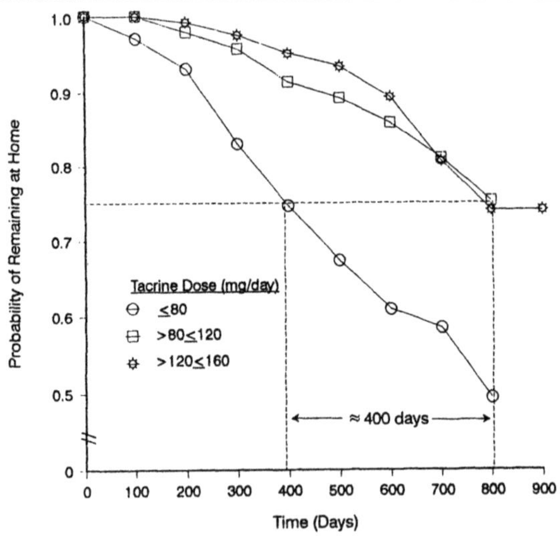

FIGURE 2. 30-Week Study: Probability of remaining at home, patients on tacrine by dose (from Cox proportional hazards regression). (Adapted from Knopman et al., 1996.)

magnitude as other analyses. Reanalysis of these data adjusted for apolipoprotein E genotype were also significant for delay in nursing home entry but not mortality.

CONCLUSION

Knopman et al. (1996) conclude that there is an association between high-dose tacrine treatment and a reduced risk of nursing home placement and mortality. The most likely explanation is a salutary effect of tacrine on the symptoms of AD rather than a selection bias. The authors conclude that to establish causality would require a long-term, multi-year, prospective, randomized trial, the ethics of which demand alternative approaches.

REFERENCES

Knapp MJ, Knopman DS, Solomon PR et al. (1994): A 30-week randomized controlled trial of high-dose tacrine in patients with Alzheimer's disease. JAMA 271: 985-991.

Knopman D, Schneider L, Davis K et al. (1996): Long-term tacrine (Cognex) treatment: effects on nursing home placement and mortality. Neurology, In Press.

Welch HG, Walsh JS, Larson EB (1992): The cost of institutional care in Alzheimer's disease: nursing home and hospital use in a prospective cohort. J Am Geriatr Soc 40:221-224.

Alzheimer Disease: From Molecular Biology to Therapy
edited by R. Becker and E. Giacobini

CHOLINESTERASE INHIBITORS: AN OVERVIEW OF THEIR MECHANISMS OF ACTION

Albert Enz and Philipp Floersheim
Preclinical Research, Sandoz Pharma Ltd., Basel, Switzerland

INTRODUCTION

Alzheimer's disease (AD) is a degenerative disorder of the CNS which will become a major health problem in the following decades. AD is associated with a decrease of cholinergic activity in the cortex and other brain regions. Of the different approaches to reversing the cholinergic deficit, the inhibition of acetylcholinesterase (AChE) seems to produce symptomatic improvements in clinical trials (Whitehouse, 1993).

AChE catalyzes the hydrolysis of acetylcholine (ACh) into choline and acetic acid. The active site of AChE comprises an esteratic subsite containing the catalytic machinery, and an anionic subsite which binds the quaternary group of ACh (Quinn, 1987). A second anionic subsite exists at a peripheral site so called because it is distant to the active site. The binding of ACh by the anionic site serves to bring the ester-group of ACh in apposition to the esteratic site of the enzyme. The catalytic action involves an active serine residue which brings about the hydrolysis of choline esters through proton transfer in a catalytic triad. The existence of additional peripheral anionic binding sites of AChE has been predicted. Cationic ligands bind to these sites and may induce conformational changes in the enzyme. These multiple binding sites for reversible ligands are the reason for the different inhibitory mechanisms seen with different classes of AChE inhibitors.

MECHANISMS OF INHIBITION

AChE has long been an attractive target for the rational design of mechanism-based inhibitors because of the pivotal role it plays in the nervous system. To reduce the metabolic breakdown of ACh and maintain cholinergic transmission in the brain of AD patients, several AChE inhibitors are currently undergoing clinical testing in this disease.

AChE inhibitors can be grouped into three broad classes, based on their structure and mode of inhibition. These are the tertiary and quaternary amines, the carbamates and the organophosphates. The inhibitors from the different groups have distinct mechanisms of AChE inhibition.

The organophosphates interact with the enzyme by forming a covalent bond with the serine of the catalytic site, which results in an extremely stable enzyme-inhibitor complex (irreversible inhibition). Metrifonate, a drug currently under clinical investigation in AD patients, belongs indirectly to this class of AChE inhibitors. This compound is transformed non-enzymatically *in vitro* and *in vivo* into the active organophosphate dichlorvos, which inhibits AChE irreversibly (Nordgren et al., 1978). The onset of inhibition is rapid (mins), and the duration of inhibition is more than 6 hrs (Hallak and Giacobini,1987).

Carbamate inhibitors like SDZ ENA 713 ((-)(S)-N-ethyl-3-[(1-dimethyl-amino)ethyl]-N-methylphenylcarbamate) mimic the substrate by forming a carbamoylated instead of acylated complex with the enzyme that is hydrolyzed considerably slower than the acylated form. Sequestration of AChE in its carbamoylated form thus precludes further enzyme-catalyzed hydrolysis of ACh for an extended period of time. *In vivo,* this time is several hrs (Enz et al., 1993).

The mechanism of AChE inhibition by the two tertiary amine compounds tacrine and E2020 is completely distinct from that already described. Enzyme kinetics experiments have assigned a so-called linear mixed type of inhibition to these drugs, comprising both competitive and noncompetitive inhibition (Fig.1a). The apparent inhibition constant (K_i value) of tacrine for AChE is in the nM range (Fig.1b), indicating high affinity for the enzyme. However, the doses used *in vivo* are much higher (up to 160 mg in humans) than are needed to produce such an inhibitor concentration. The reason for this discrepancy might be competitive interaction with AChE and butyrylcholinesterase (B ChE) and/or by rapid drug metabolism. The duration of the reversible inhibition by tacrine is, however, short and directly dependent on the drug concentration.

Fig. 1a: General Scheme for mixed type inhibition

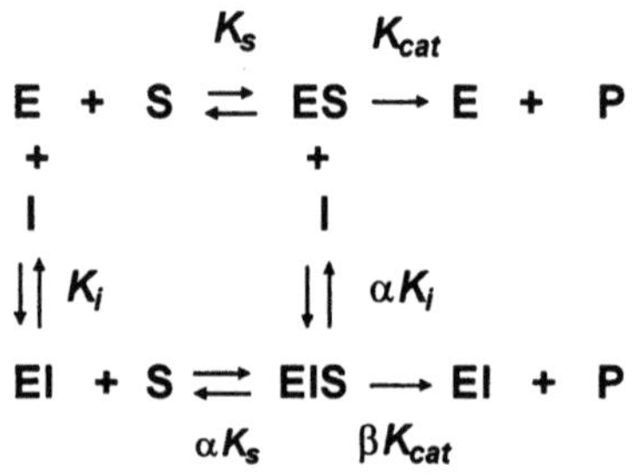

Fig 1.b.: Mixed linear AChE inhibition by Tacrine

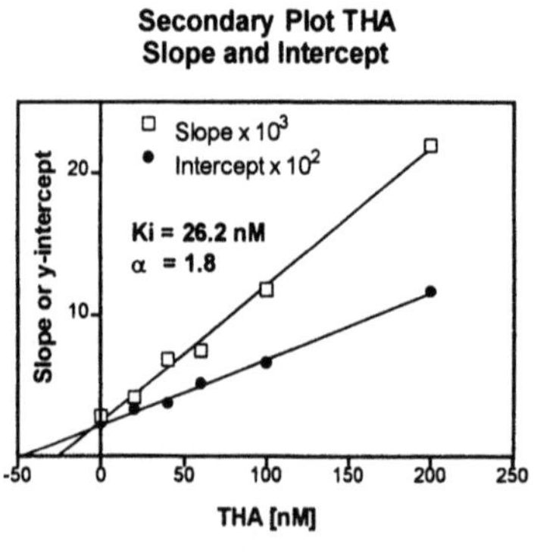

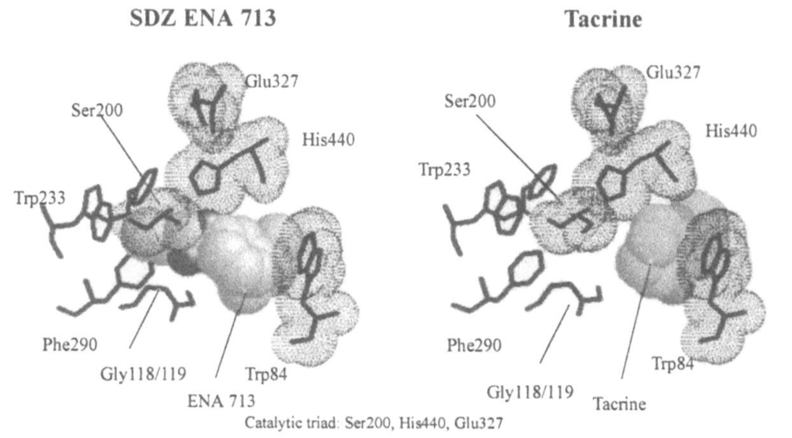

FIGURE 2.

Molecular model of SDZ ENA docked onto the AChE molecule

We built a molecular model of the adduct of SDZ ENA 713 to the X-ray structure of AChE isolated from T.californica (Sussman et al., 1991), as deposited in the Brookhaven Protein Data Bank (entry 1ACE), in order to understand its mode of inhibition in more detail. We oriented a model of SDZ ENA 713 in the active site of AChE with the hydroxyl oxygen of Ser 200 bound to the tetrahedral carbon atom of the carbamoyl group such that i) the anionic carbonyl oxygen points towards the NH groups Gly 118 and Gly 119; ii) the ester oxygen of SDZ ENA 713 is close to the imidazol ring of His 440; and iii) the ethyl group of the N-ethyl carbamate of SDZ ENA 713 is directed towards the aromatic moieties of Trp 233, Phe 288 and Phe 290.

Starting from this geometry, simulated annealing with a molecular mechanics force field, and with the residues of AChE kept fixed at their positions in the X-ray structure, leads to a final geometry of low energy in which the non-covalent interactions between SDZ ENA 713 and the aforementioned residues were retained. Furthermore, the ethyl group of dimethylamino ethyl substituent formed a favorable hydrophobic contact with the aromatic rings of Phe 330 and Trp 84. Also in this model the amino group, although modeled in the protonated state, does not form a salt bridge with any of the carboxylate groups of AChE. However, a conformational change in the side-chain of ASP 72 with its carboxylate group at 6.5Å would readily lead to a direct ionic or water-mediated contact with the ammonium group of ENA 713. Its benzene ring, in turn, does not show strong interactions with AChE. Its primary role is to rigidly align both its substituents favorably in the active site. This is in contrast to the experimental structure of the complex of AChE with tacrine (Harel et al., 1993), in which the aromatic moieties are strongly bound solely to the hydrophobic sites. The derived model shows one possible mode by which ENA 713 fits into the active site of AChE fulfilling chemical and structural requirements for a stereoselective active-site-directed inhibition.

PROPERTIES OF INDIVIDUAL AChE INHIBITORS

Adverse effects associated with AChE inhibitors are related to their mechanisms of action, metabolic properties and preferential sites of action:

i) the desired action appears to be determined by the mechanism of inhibition both of duration and selectivity;
ii) the potential for organ toxicity is due to the presence of active metabolites and is determined by the metabolism of the drug;
iii) peripheral cholinergic effects can be minimized by the use of brain selective inhibitors.

In the case of reversible AChE inhibitors like tacrine and E2020, the metabolism of the drug is not the result of an interaction with the target enzyme. The extensive metabolism of tacrine results in highly active metabolites that contribute to the observed elevation of liver enzymes (Spaldin et al., 1994). Carbamates, however, are decomposed by their action at the target enzyme, AChE. For SDZ ENA 713, this is almost the only metabolic pathway by which the drug is degraded, the product being a phenolic derivative. This phenolic cleavage product has been found to be the major metabolite in experiments in humans and animals. This enzymatic cleavage product is rapidly excreted via the kidney following sulfate conjugation. This is probably the reason why SDZ ENA 713 is so free of organ toxicity. The advantageous pharmacological profile of this brain selective AChE inhibitor, can be summarized as follows: SDZ ENA 713 selectively inhibits brain AChE with inhibition being more pronounced in the cortex and hippocampus relative to other brain regions such as pons/medulla and striatum (Enz et al., 1993). As a consequence, SDZ ENA 713 has no effects on cardiovascular parameters at doses at which clear central effects can be demonstrated.

Study in humans to assess the relationship between central and peripheral effects of SDZ ENA 713

Employing a fixed sequence, placebo-controlled, single-blind design, eight healthy male volunteers underwent pharmacokinetic blood sampling in conjunction with a 49 hr period of continuous lumbar CSF sampling. A maximal inhibition of 40% in CSF AChE activity occurred 2.4 hrs after a single oral administration of 3 mg SDZ ENA 713 and activity had recovered to baseline levels 8.8 hrs after the dose. Only a very small, statistically insignificant reduction in B ChE activity was measured in the CSF and plasma. These data supports the proposition that SDZ ENA 713 is a brain selective inhibitor of AChE also in man.

SUMMARY AND CONCLUSIONS

Clinical experience with cholinergic drugs in the treatment of AD has not yet shown clear cut relevant symptomatic improvements. The main reason for this might be that peripheral cholinergic effects and liver toxicity of some of these drugs limit their use and prevent confirmation of the cholinergic hypothesis. Drugs that cause selective and long-lasting inhibition of AChE in the brain by a mechanism of action linked to an uncomplicated drug metabolism have a real chance to improve symptoms in AD. SDZ ENA 713 fulfills these criteria and ongoing clinical trials worldwide should demonstrate the usefulness of the drug in treatment of AD.

REFERENCES

Becker RE, Colliver J, Elble R, Feldman E, Giacobini E, Kumar V, Markwell S, Moriearty P, Parks R, Shillcutt SD, Unni L, Vicary S, Womack C and Zec RF (1990): Effect of Metrifonate, a long-acting cholinesterase inhibitor in Alzheimer's disease. *Drug Dev Res* 19:425-434.

Enz A, Amstutz R, Boddeke H, Gmelin G and Malanowsky J (1993): Brain selective inhibiton of acetylcholinesterase: a novel approach to therapy for Alzheimer's disease. In: *Cholinergic Function and Dysfunction, Progress in Brain Research* Vol 98. Cuello AC, ed. Amsterdam, Elsevier Science Publisher, pp 431-438.

Hallak M and Giacobini E (1987): A comparison of the effects of two inhibitors on brain cholinersterase. *Neuropharmacol* 26:521-530.

Harel M, Schalk I, Ehrat-Sabatier L, Bouet F, Goeldner M, Hirth C, Axelsen PH, Silman I and Sussman JL (1993): Quaternary ligand binding to aromatic residues in the active-site gorge of acetylcholinesterase. *Proc Nat Acad Sci* 90:9031-9035.

Nordgren I, Bergstroem M, Holmstedt B and Sandoz M (1978): Transformation and action of metrifonate. *Arch Toxicol* 41:31-41.

Quinn DM (1987): Acetylcholinesterase: Enzyme structure, reaction dynamics, and virtual transition states. *Chem Rev* 87:955-979.

Spaldin V, Maddan S, Pool WF, Woolf TF and Park K (1994): The effect of enzyme inhibition on the metabolism and activation of tacrine by human liver microsomes. *Br J Pharmac* 38:15-22.

Sussman JL, Harel M, Frolow F, Oefner C, Toker L and Silman I (1991): Atomic structure of acetylcholinesterase from Torpedo californica: a prototypic acetylcholine-binding protein. *Science* 253:872-879.

Whitehouse PJ (1993): Cholinergic therapy in dementia. *Acta Neurol Scand* (Supp.) 149:42-45.

Alzheimer Disease: From Molecular Biology to Therapy
edited by R. Becker and E. Giacobini

PRECLINICAL PHARMACOLOGY OF METRIFONATE: A PROMISE FOR ALZHEIMER THERAPY

Bernard H. Schmidt, Volker C. Hinz, Arjan Blokland and Franz-Josef van der Staay
Institute for Neurobiology, Troponwerke GmbH & Co. KG, Köln, FRG

Richard J. Fanelli
Institute for Dementia Research, Bayer Corp., West Haven, CT

INTRODUCTION

Among the multiple transmitter deficits which have been described in Alzheimer's Disease (AD), the degeneration of brain cholinergic cell bodies is the most sensitive, specific and severe, as indicated by the good correlation between the cholinergic pathology and dementia. Therefore, current drug development strategies in AD therapy focus on the enhancement of cholinergic neurotransmission. The most advanced class of compounds in this respect are cholinesterase (ChE) inhibitors which aim to restore the concentration of acetylcholine (ACh) in the synaptic cleft.

Early clinical testings with first generation ChE inhibitors such as tacrine or physostigmine revealed major limitations to the use of these compounds because of low oral bioavailablility, a relatively low percentage of clinical responders, gastrointestinal side effects, and, with tacrine, additional hepatic transaminitis (Weinstock, 1995). At first glance, it seems that cholinergic side effects cannot be avoided if critical levels of ChE inhibition are achieved. However, as shown by Becker and Giacobini (1988), there is no consistent relationship among the degree of ChE inhibition, changes in brain ACh concentration, and adverse effects following administration of various ChE inhibitors to animals. The appearance of toxic side effects is determined rather by drug-specific pharmacological and pharmacokinetic properties, which regulate the fluctuations in enzyme activity. Hence, in case of rapidly absorbed, direct ChE inhibitors such as tacrine or physostigmine, cholinergic neurons may not have enough time to accommodate to and benefit from enhanced ACh concentrations, especially if the ChE inhibition is only of short duration.

In line with these considerations, the cholinergic AD therapeutic of choice should induce a long-lasting ChE inhibition with a slow onset, allowing for a safe and well-tolerated titration of ChE inhibition over long periods of time. As outlined in the previous work of Becker and Giacobini (1988) and this chapter, metrifonate (O,O-dimethyl-(1-hydroxy-2,2,2-trichloroethyl) phosphonate) fulfills these criteria.

The prodrug concept of slow-release plus slow-onset ChE inhibition

Metrifonate is unique among ChE inhibitors in AD treatment in that it is not a ChE inhibitor by itself (see Holmstedt et al., 1978). In aqueous solutions it is spontaneously transformed into the active metabolite, dichlorvos (2,2-dichlorovinyl dimethyl phosphate). This reaction does not require an enzyme. In human blood *in vitro*, the half-life time of metrifonate is about 60 min compared to only 10 min for dichlorvos (Villén et al., 1990). Therefore, it is not possible to accumulate dichlorvos after *in vivo* administration of the parent drug, metrifonate. The plasma levels of the active transformation product observed in mammals including man are consistently in the range of 1-2% of the respective concentration of metrifonate (Villén et al., 1990). The transformation of metrifonate into dichlorvos can be precipitated at alkaline pH values and can be blocked by acidification. A similar pH dependency was found for metrifonate-induced ChE inhibition using ChE from various tissues *in vitro* (Hinz et al., 1996a). ChE inhibition by dichlorvos does not depend on the pH. It is mediated by a competitive drug interaction at the catalytic site of ChE followed by dimethyl-phosphorylation of a serin residue located in the active site of the enzyme (Main, 1979). The result is a stable drug-enzyme complex. However, reactivation is possible either spontaneously (Reiner and Plestina, 1979) or, if required, by oxime therapy (Moriearty and Becker, 1992). It should be noted that dimethylphosphorylated ChE is not sensitive to the dealkylation step observed with organophosphates bearing longer ester chains. The latter reaction is also known as the "aging phenomenon" and leads to truly irreversible ChE inhibition (Main, 1979).

As outlined above, the major determinant for the tolerability of drug-induced ChE inhibition in mammals is the speed of the drop in ChE activity. When added to rat brain membranes or human erythrocytes *in vitro*, metrifonate requires up to 60 min incubation to equilibrate ChE inhibition (Hinz et al., 1996a; Moriearty and Becker, 1992). In contrast, tacrine leads to an apparently immediate drop in ChE activity. However, the difference in time-dependency between metrifonate and tacrine-induced ChE inhibition is not solely due to the time required for the transformation of metrifonate into dichlorvos. Dichlorvos-mediated ChE inhibition itself is also time-dependent, and a period of at least 30 min is required to achieve equilibrium ChE inhibition *in vitro* (Hinz et al., 1996a).

As a consequence of the strong time-dependency of ChE inhibition mediated by metrifonate/dichlorvos, it may be assumed that low levels of the active metabolite, slowly released from the parent compound, will lead to equivalent ChE inhibition levels as high, but short-lasting, as levels of directly administered dichlorvos. This coherence between time, drug concentration, and inhibitory efficacy is supported by the finding that the concentration of dichlorvos required to achieve a 50% ChE inhibition in rat brain homogenates *in vitro* decreased from 6 μM to 0.32 μM when total incubation time was increased from 6 to 36 min. A similar increase in potency was observed with metrifonate. However, dichlorvos was about 100 times more potent than metrifonate at any incubation time tested (Hinz et al., 1996a).

The loading concept for long-lasting ChE inhibition in vivo

When administered systemically to animals or humans, metrifonate mediates ChE inhibition in blood and brain in a dose- and time-dependent manner. After oral administration to rats, peak inhibition levels occur within 20-45 min in blood and within 45-60 min in brain (Hinz et al., 1996b). Most of the inhibition achieved recovers spontaneously within 10 hours after drug administration, confirming previous studies using different routes of application (Reiner and Plestina, 1979; Soininen et al., 1990). Dichlorvos-induced ChE inhibition follows similar kinetics, but with a slightly earlier onset of action (Hinz et al., 1996b; Reiner and Plestina, 1979).

Metrifonate-induced ChE inhibition in rat brain is closely followed by an increase in the bioavailability of ACh in the extracellular fluid of the cerebral cortex. This effect is specific, because there are no comparable changes in the release of other neurotransmitters, such as norepinephrine, dopamine, or serotonin (Mori et al., 1995a).

Although the major component of ChE inhibition recovers within 24 hrs, a distinct amount of ChE inhibition is maintained for longer periods. For rat brain, this amount has been estimated by Reiner and Plestina (1979) to about 20% of peak inhibition. This component does not recover spontaneously, but can be reactivated at least in part by oxime therapy (Moriearty and Becker, 1992). In the absence of precipitated enzyme reactivation, the recovery kinetics follow enzyme resynthesis (Reiner and Plestina, 1979; Becker et al., 1994).

Due to the long-lasting component of ChE inhibition, metrifonate offers the opportunity to load up significant amounts of stable enzyme inhibition, which will outlast the presence of the active drug in the body. This therapeutically interesting option has been previously observed in rats (Soininen et al., 1990), and has been successfully applied to titrate ChE inhibition levels with metrifonate in clinical trials (Becker et al., 1994). In rats and rabbits, there is an excellent correlation between erythrocyte and brain ChE inhibition levels, with slightly higher levels of inhibition being achieved in the blood upon oral administration of metrifonate (Kronforst et al., 1995;

Hinz and Schmidt, unpublished observations). On the other hand, plasma is not suitable for estimating brain trough ChE inhibition because of a much faster recovery time.

Safety-efficacy considerations

After acute administration of metrifonate to young adult rats, cholinergic symptoms are dose-dependently observed at oral doses exceeding 30 mg/kg. Aged rats are more sensitive. Their threshold dose is 10 mg/kg (Blokland et al., 1995). The effects are characterized by salivation and tremor, with less frequent observations of diarrhea, ptosis and limb abduction. All symptoms are transient. They peak within 30 min after dosing and disappear within 90 min post-administration, i.e., well before the reversal of ChE inhibition. This side effect profile is consistent with a selective stimulation of the cholinergic system by metrifonate.

Interestingly, symptoms of overdosing of metrifonate are greatly reduced upon repeated administration (Blokland et al., 1995). Only five oral doses of metrifonate, 100 mg/kg administered at 24 hrs intervals, reduced the incidence of side effects to less than 50% of that obtained after a single administration.

On the other hand, the efficacy of metrifonate to improve cognitive performance in various animal models (Blokland et al., 1995; van der Staay et al., 1996; Kronforst et al., 1995) was not diminished after subchronic administration. Metrifonate improved Morris Water Escape performance in rats in the same dose range, independent of an acute or subchronic administration schedule (unpublished data). Similarly, eye-blink conditioning in rabbits was still improved after 6 weeks of continuous administration (Kronforst et al., 1995). Hence, the therapeutic window, which is already rather broad compared to directly acting ChE inhibitors (Blokland et al., 1995), is even broader upon long-term treatment.

CONCLUSION

Metrifonate is unique among ChE inhibitors in that it acts as a prodrug. It induces a slow-onset ChE inhibition which can be loaded up to high levels of well-tolerated ChE inhibition. In animals, the compound exerts cognition enhancing properties which are maintained along with a reduction in side effects upon long-term treatment. These features, together with the recent evidence regarding potential additional beneficial effects of the active metabolite, dichlorvos, on the secretion of neuroprotective amyloid precursor protein species (Mori et al., 1995b) render metrifonate a promising candidate for AD therapy.

REFERENCES

Becker RE and Giacobini E (1988): Mechanisms of cholinesterase inhibition in senile dementia of the Alzheimer type: clinical, pharmacological, and therapeutic aspects. *Drug Dev Res* 12:163-195.

Becker RE, Moriearty P, Surbeck R, Unni L, Varney A and Vicari SK (1994): Second and third generation cholinesterase inhibitors: clinical aspects. In: *Alzheimer Disease: Therapeutic Strategies*, Giacobini E and Becker R, eds. Boston: Birkhäuser, pp. 172-178.

Blokland A, Hinz V and Schmidt BH (1995): Effects of metrifonate and tacrine in the spatial Morris task and modified Irwin test: Evaluation of the efficacy/safety profile in rats. *Drug Dev Res* 36:166-179.

Hinz V, Grewig S and Schmidt BH (1996a): Metrifonate induces cholinesterase inhibition exclusively via slow release of dichlorvos. *Neurochem Res*, in press.

Hinz V, Grewig S and Schmidt BH (1996b): Metrifonate and dichlorvos: Effects of a single oral administration on cholinesterase activity in rat brain and blood. *Neurochem Res*, in press.

Holmstedt B, Nordgren I, Sandoz M and Sundwall A (1978): Metrifonate: Summary of toxicological and pharmacological information available. *Arch Toxicol* 41:3-29.

Kronforst MA, Moriearty PL, Ralphs M, Becker RE, Schmidt B, Thompson LT and Disterhoft JF (1995): Facilitation of learning in the aging rabbit after chronic metrifonate treatment. *Soc Neurosci Abstr* 21:196.

Main AR (1979): Mode of action of anticholinesterases. *Pharmacol Ther* 6:579-628.

Mori F, Cuadra G and Giacobini E (1995a): Metrifonate effects on acetylcholine and biogenic amines in rat cortex. *Neurochem Res* 20:1081-1088.

Mori F, Lai CC, Fusi F and Giacobini E (1995b): Cholinesterase inhibitors increase secretion of APPs in rat brain cortex. *Neuro Report* 6:633-636.

Moriearty PL and Becker RE (1992): Inhibition of human brain and RBC acetylcholinesterase (AChE) by heptylphysostigmine (HPTL). *Meth Find Clin Pharmacol* 14:615-621.

Reiner E and Plestina R (1979): Regeneration of cholinesterase activities in humans and rats after inhibition of O,O-dimethyl-2,2-dichlorovinyl phosphate. *Toxicol Appl Pharmacol* 49:451-454.

Soininen HS, Unni L and Shillcutt S (1990): Effect of acute and chronic cholinesterase inhibition on biogenic amines in rat brain. *Neurochem Res* 15:1185-1190.

van der Staay FJ, Hinz VC and Schmidt BH (1996): Effects of metrifonate on escape and avoidance learning in young and aged rats. *Behav Pharmacol* 7:56-64.

Villén T, Aden-Abdi Y, Ericsson Ö, Gustafsson LL and Sjöqvist F (1990): Determination of metrifonate and dichlorvos in whole blood using gas chromatography and gas chromatography-mass spectrometry. *J Chromatogr* 529:309-317.

Weinstock M (1995): The pharmacotherapy of Alzheimer's disease based on the cholinergic hypothesis: an update. *Neurodegeneration* 4:349-345.

REFERENCES

[illegible]

Alzheimer Disease: From Molecular Biology to Therapy
edited by R. Becker and E. Giacobini

EPTASTIGMINE: A CHOLINERGIC APPROACH TO THE TREATMENT OF ALZHEIMER'S DISEASE

Bruno P. Imbimbo
Medical Department, Mediolanum Farmaceutici, Milan, Italy

INTRODUCTION

Cholinesterase inhibitors have a well established role in the symptomatic treatment of patients with Alzheimer's disease. However, clinical results obtained over the past 20 years have been contradictory. This is perhaps due to the poor systemic tolerability, short duration of action, unknown target cholinesterase inhibition and tolerance following chronic use of current drugs.

Eptastigmine (heptylphysostigmine, MF 201, L-693,487) is a new cholinesterase inhibitor (Brufani et al., 1986). Chemically, it is a derivative of physostigmine in which the carbamoyl-methyl group in position 5 of the indole ring has been substituted with a carbamoyl-heptyl group.

In vitro, eptastigmine produces a marked concentration-dependent inhibition of acetylcholinesterase (AChE) (Brufani et al., 1987). The IC_{50} for human red blood cell and brain AChE is $1.2 \cdot 10^{-8}$ M and $7.6 \cdot 10^{-9}$ M, respectively. The AChE inhibition induced in human brain does not spontaneously reverse as in human red blood cells (Moriearty et al., 1992).

In vivo, AChE inhibition is dose-dependent, long-lasting and particularly marked in the brain (De Sarno et al., 1989). Compared to physostigmine and tacrine, eptastigmine has a much longer duration of action. The drug improves performance in behavioral cognitive tests in both rodents (Dawson et al., 1991) and primates (Rupniak et al., 1992).

The objective of this paper is to review Phase I, Phase II, and Clinical Pharmacology studies. Only double-blind, placebo-controlled studies are considered.

PHASE I CLINICAL STUDIES

Phase I studies have demonstrated that eptastigmine is well tolerated in young (Goldberg et al., 1991; Imbimbo et al., 1995) and elderly (Swift et al., 1991) healthy volunteers up to doses of 30 mg following single administration.

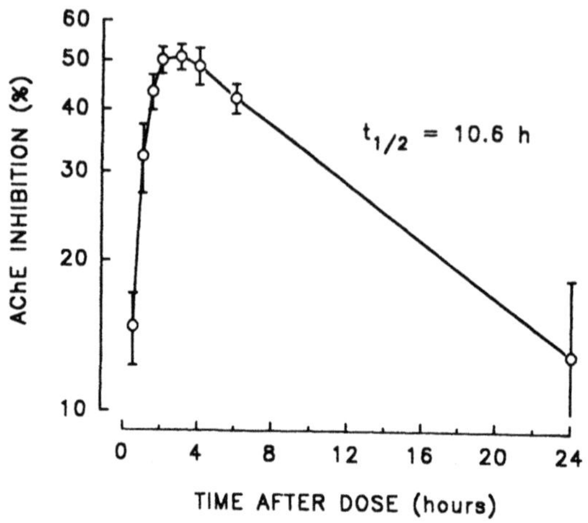

FIGURE 1. Time course of red blood cell AChE inhibition after a 30 mg oral dose of eptastigmine in healthy volunteers.

Following multiple administration, a maximum tolerated dose was not identified (highest dose tested was 32 mg t.i.d.). Side-effects are related to the level of cholinesterase inhibition (generally >70% inhibition). The drug inhibits red blood cell AChE in a dose-dependent fashion. The mean AChE recovery half-life was about 10 hrs with a mean residual inhibition of 13% 24 hrs after a 30 mg dose (Figure 1).

CLINICAL PHARMACOLOGY STUDIES

The activity of eptastigmine on the CNS was assessed in two studies.

In the first study, the effects of eptastigmine on rapid eye movement (REM) sleep in 12 healthy volunteers were evaluated. Latency to REM sleep decreased dose-dependently after eptastigmine compared to placebo (Figure 2).

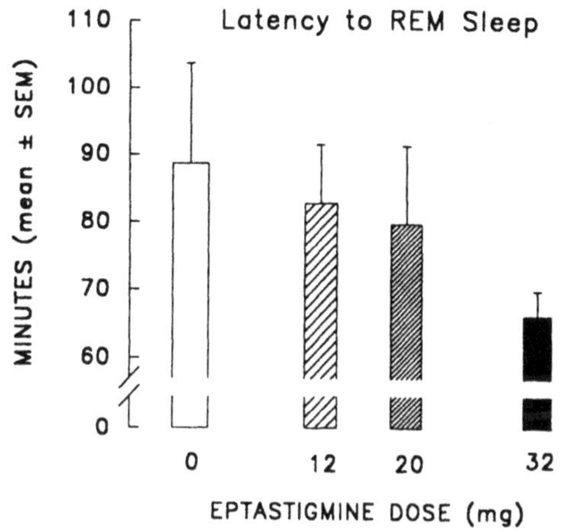

FIGURE 2. Effects of eptastigmine on REM sleep in healthy volunteers.

In the second study, we evaluated the ability of eptastigmine to reverse the cognitive deficits induced by intramuscular injection of scopolamine (0.4 mg). Twenty-four healthy subjects received single oral doses of 20 and 32 mg eptastigmine and placebo according to a crossover design (Lines et al., 1993). Eptastigmine attenuated memory impairment induced by scopolamine with an inverted U shaped dose-response relationship (Figure 3).

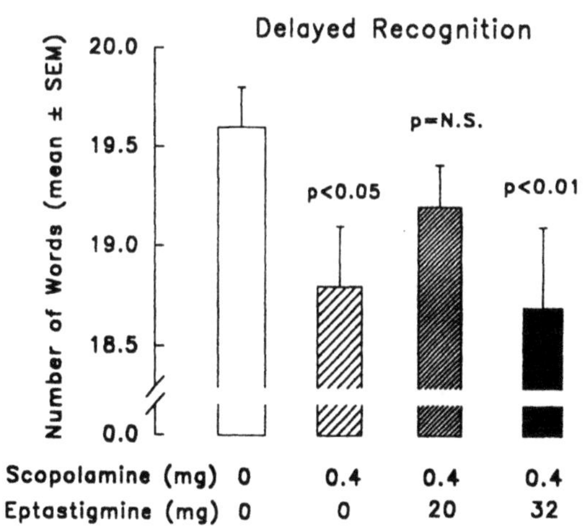

FIGURE 3. Effects of eptastigmine on scopolamine-induced cognitive deficits in healthy volunteers.

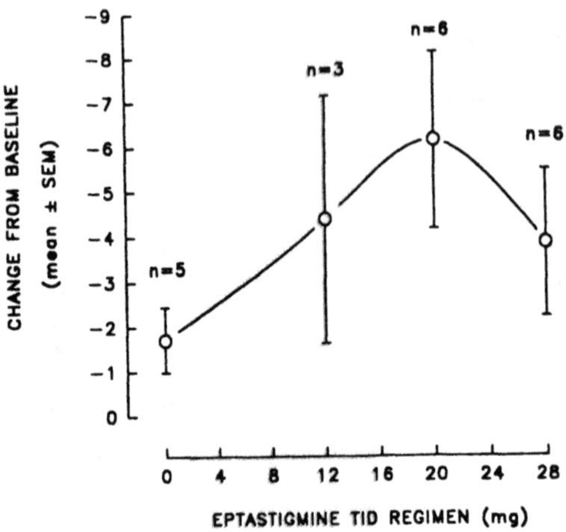

FIGURE 4. ADAS-Cog subscores at 2 weeks after different dose regimens of eptastigmine in Alzheimer patients.

PHASE II CLINICAL STUDIES

Three Phase II clinical studies in Alzheimer patients were carried out.

The first involved 25 patients treated for up to 4 weeks with doses up to 52 mg t.i.d. (Sramek et al., 1995). Following a step-wise dose escalation, the maximum tolerated dose was found to be 40 mg t.i.d. At two weeks, the effect of eptastigmine on ADAS-Cog followed an inverted U-shaped dose relationship, the 20 mg t.i.d. producing the greatest advantage versus placebo (Figure 4).

In a second study, 240 patients initially were to be enrolled in four groups (60 patients each). Patients had to undergo a rapid titration over a three week period to fixed dosage regimens of placebo t.i.d. or eptastigmine 20 mg t.i.d., 30 mg t.i.d. and 40 mg t.i.d. Following the enrollment of 59 patients, the study was discontinued because two patients (one on 30 mg t.i.d. and the other on 40 mg t.i.d.) developed reversible neutropenia. Data on ADAS-Cog are available at 3 weeks for 37 patients. All eptastgmine-treated groups showed a statistically significant positive effect compared to placebo. Once again, the drug's effect on ADAS-Cog followed an inverted U-shaped dose relationship, 20 mg t.i.d. producing the greatest advantage versus placebo (Figure 5).

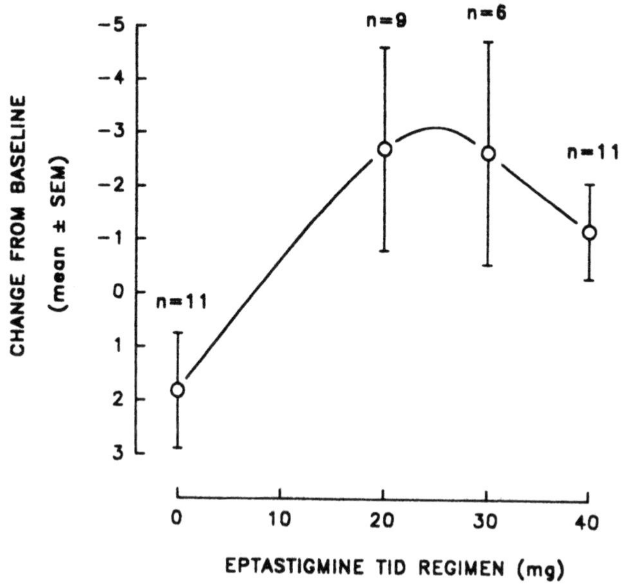

FIGURE 5. ADAS-Cog subscores at 3 weeks after different dose regimens of eptastigmine in 37 Alzheimer patients.

The third study involved 103 patients (Canal et al., 1996). Patients received from 30 mg to 60 mg per day for 4 weeks. Nine patients, all from the eptastigmine group, did not complete treatment. Thirty-one percent of patients on eptastigmine reported adverse events mainly of the cholinergic type. Cholinergic side effects were generally associated with peak red blood cell AChE inhibition exceeding 50% after the first dose or 70% at steady-state. At steady-state, average daily AChE inhibition ranged from 13% to 54%. Overall, 31% of patients on eptastigmine versus 0% on placebo ($p=0.006$) improved on the Physician CGIC. This percentage increased to 46% in the subgroup of patients with average daily AChE inhibition ranging from 30% to 35% (Figure 6). Patient performance on IADL also improved significantly compared to the placebo group ($p=0.019$). In the eptastigmine group, performances on all tests and scales improved with an inverted U-shaped relation to average daily AChE inhibition.

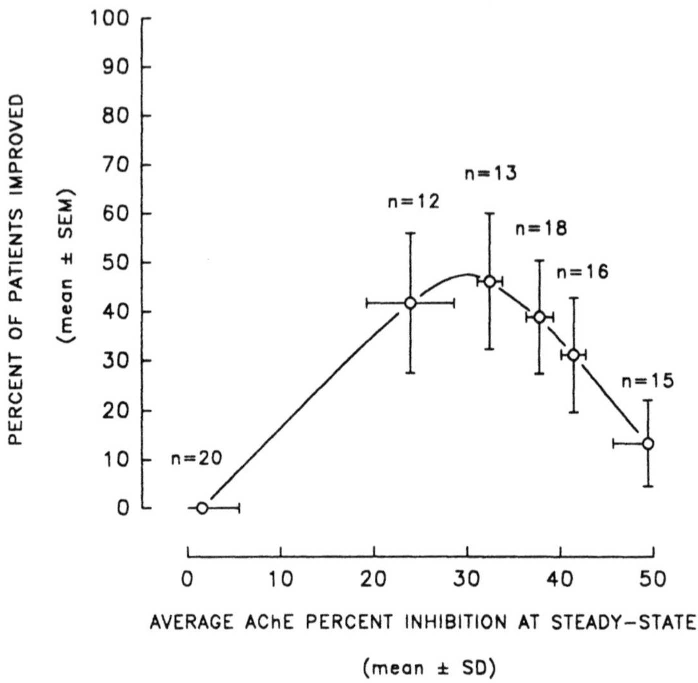

FIGURE 6. Relationship between average AChE inhibition at steady-state and Physician CGIC in 94 Alzheimer patients treated with eptastigmine (30-60 mg/day) or placebo for 4 weeks.

CONCLUSIONS

Studies in healthy volunteers show that eptastigmine has good tolerability and a long duration of action.

Clinical pharmacology studies suggest that after oral administration, eptastigmine penetrates the CNS. The effect on central cholinergic activity appears to be linearly related to the dose, while that on memory appears to be non-linear.

Short-term studies in Alzheimer patients indicate that doses of 40-60 mg per day of eptastigmine are relatively safe and well tolerated and that moderate AChE inhibition is associated with maximal cognitive efficacy.

Three multicenter studies in more than 1,000 patients are presently underway to evaluate the long-term efficacy of eptastigmine.

REFERENCES

Brufani M, Marta M, Pomponi M (1986): Anticholinesterase activity of a new carbamate, heptylphysostigmine, in view of its use in patients with Alzheimer-type dementia. *Eur J Biochem* 157:115-120.

Brufani M, Castellano C, Marta M, Oliverio A, Pagella PG, Pavone F, Pomponi M, Rugarli PL (1987): A long-lasting cholinesterase inhibitor affecting neural and behavioral processes. *Pharmacol Biochem Behav* 26: 625-629.

Canal N, Imbimbo BP for the Eptastigmine Study Group (1996): Relationship between pharmacodynamic activity and cognitive effects of eptastigmine in Alzheimer patients. *Clin Pharmacol Ther* (in press).

Dawson GR, Bentley G, Draper F, Rycroft W, Iversen SD, Pagella PG (1991): The behavioural effects of heptyl-physostigmine, a new cholinesterase inhibitor, in tests of long-term and working memory in rodents. *Pharmacol Biochem Behav* 39:865-871.

De Sarno P, Pomponi M, Giacobini E, Tang XC, Williams E (1989): The effect of heptyl-physostigmine, a new cholinesterase inhibitor, on the central cholinergic system of the rat. *Neurochem Res* 14:971-977.

Goldberg MR, Barchowsky A, McCrea J, Ben-Maimon C, Capra N, Fitzpatrick V, Bjornsson TD (1991): Heptylphysostigmine (L-693,487): safety and cholinesterase inhibition in a placebo-controlled rising-dose healthy volunteer study. *2nd Int Springfield Symp Alzheimer Ther* Abstract P22.

Imbimbo BP, Licini M, Schettino M, Mosca A, Zecca L, Radice D, Giustina A (1995): Relationship between pharmacokinetics and pharmacodynamics of eptastigmine in young healthy volunteers. *J Clin Pharmacol* 35:285.

Lines CR, Ambrose JH, Heald A, Traub M (1993): A double-blind, placebo-controlled study of the effects of eptastigmine on scopolamine-induced cognitive deficits in healthy male subject. *Psychopharmacology* 8:271-278.

Moriearty PL, Becker RE (1992): Inhibition of human brain and RBC acetylcholinesterase (AChE) by heptylphysostigmine (HPTL). *Meth Find Exp Clin Pharmacol* 14: 615-621.

Rupniak NMJ, Tye SJ, Brazell C, Heald A, Iversen SD, Pagella PG (1992): Reversal of cognitive impairment by heptylphisostigmine, a long lasting cholinesterase inhibitor, in primates. *J Neurol Sci* 107:246-249.

Sramek JJ, Block GA, Reines SA, Sawin SF, Barchowsky A, Cutler NR (1995): A multiple-dose safety trial of eptastigmine in Alzheimer's disease, with pharmacodynamic observations of red blood cell cholinesterase. *Life Sci* 56: 319-326.

Swift RH, Mant TG, Heald A, Sciberras DG, Imbimbo B, Morrison PJ, Goldberg MR (1991): A study to investigate the safety, tolerance, & pharmacodynamics of single rising doses of L-693,487 (heptylphysostigmine tartrate) in healthy elderly volunteers. *British Pharmacological Society. Joint Meeting with the Nordic Society. University of Southampton.*

Alzheimer Disease: From Molecular Biology to Therapy
edited by R. Becker and E. Giacobini

PHENSERINE: A SELECTIVE, LONG-ACTING AND BRAIN-DIRECTED ACETYLCHOLINESTERASE INHIBITOR AFFECTING COGNITION AND β-APP PROCESSING

Nigel H. Greig, Donald K. Ingram, William C.Wallace, Tadanobu Utsuki, Qian-Sheng Yu, Harold W. Holloway and Xue-Feng Pei
Natl Inst on Aging, Gerontology Research Center, NIH, Baltimore, MD

Vahram Haroutunian
Department Psychiatry, Mt. Sinai School Medicine, New York, NY

Debomoy K. Lahiri
Department Psychiatry/Neurology, Indiana University, School of Medicine, Indianapolis, IN

Arnold Brossi
Department Chemistry, University North Carolina, Chapel Hill, NC

Timothy T. Soncrant
Cure Pharmaceuticals Inc., Silver Spring, MD

Alzheimer's disease (AD) is characterized by (i) the presence of β-amyloid (Aβ) plaques and angiopathy, neurofibrillary tangles (NFTs), neuronal loss and brain atrophy; (ii) a time-course of degenerative changes; and (iii) a correlation of some measures of pathological changes with clinical course (Wisniewski and Weigel, 1994). Although the neuropathological quantification of plaques and NFTs in the Alzheimer's brain is the basis of confirming diagnosis after death (Khachaturian, 1985), patients initially present themselves to physicians as a consequence of a time-dependent and progressive memory loss. Indeed, it is counts of neocortical synapses rather than of plaques and NFTs that correlate best to psychometric indices of cognitive performance in AD (Terry et al., 1991). The loss of cholinergic neuronal markers, specifically the enzymes choline acetyltransferase (ChAT) and acetylcholinesterase (AChE), in selected brain regions, the forebrain and its projection sites in the cortex and hippocampus, remains one of the earliest known schedule of events leading to AD. These are markers of synaptic loss of the most affected neurotransmitter system involved in AD and in memory processing. Whereas multiple neurotransmitter deficiencies also occur in AD, these do so always in addition to, and not instead of, the cholinergic system deficiency.

These findings led to the hypothesis that cholinergic augmentation might improve cognition in AD. The most widely explored approach to achieve this has involved prolonging the life of endogenously released acetylcholine (ACh) by inhibiting its degrading enzyme, AChE. This approach is theoretically preferable to receptor agonist therapy as it may amplify the natural spatial and temporal pattern of ACh release, rather than globally and tonically stimulating either muscarinic or nicotinic receptors. Clinical studies utilizing relatively unsophisticated first and second generation cholinesterase (ChE) inhibitors have demonstrated encouraging, transient improvements in cognitive measures in AD patients (Becker et al., 1994). In general, patients have better responses with higher drug doses, concomitant with higher levels of brain AChE inhibition, but the appearance of side effects, which are dose-limiting and primarily peripherally mediated, precludes the use of most available ChE inhibitors in adequate doses to dramatically improve cognition. We hypothesized that it was possible to dissociate the beneficial actions of antiChE therapy in brain from its adverse brain and peripheral ones by designing a new class of enzyme subtype-selective and brain-directed inhibitors, phenserines. We incorporated into these agents characteristics to provide optimal pharmacokinetic and pharmacodynamic profiles for use in the elderly to maximize their action and assess the true utility of cholinergic augmentation in the therapy of AD.

CHEMISTRY

Extensive chemistry was undertaken to modify the structure of the naturally occurring alkaloid and unselective ChE inhibitor, (-)-physostigmine (Fig. 1), to develop phenserine and related agents. The catalysis of ACh by AChE and butyrylcholinesterase (BChE) involves its binding to an esteratic enzyme subsite. This involves its interaction within a catalytic triad of amino acids resulting in formation of an acylated enzyme intermediate and the release of choline. The former becomes rapidly hydrolyzed, rejuvenating active enzyme, by liberating acetic acid. The same sites interact with (-)-physostigmine to create a carbamylated enzyme, which is hydrolyzed at a far slower rate and results in transient enzyme inhibition (Fig. 2A). Improvements in the duration of inhibition were achieved, following extensive structure-activity and X-ray crystallography studies, by modulating the carbamate group to provide more stable, yet reversible, enzyme carbamylation for a longer action. Selectivity for AChE likewise was achieved by modifying the carbamate moiety.

Compound	IC50 [nM]		AChE
	AChE	BChE	Selectivity
Physostigmine	27.9	16.0	0.5
Phenserine	22.2	1552	70
Tolserine	10.3	1948	189
Cymserine 758	51.3	0.06	
Tacrine	190	47	0.25
Eptastigmine	21.7	5	0.23
Galanthamine	800	7313	9

TABLE 1. *In vitro* human ChE inhibition, IC50= concentration for 50% enzyme inhibition.

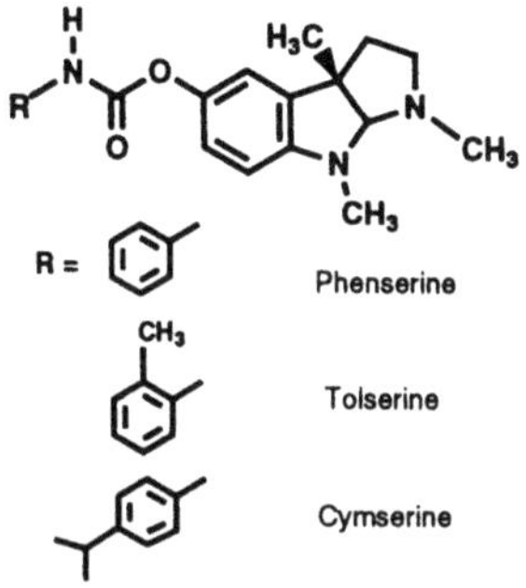

FIGURE 1. Structures of phenserines

Although AChE and BChE are products of different genes, they share a remarkable degree of amino acid homology. Recent studies have reported a critical difference between their amino acids that form a 3-dimensional structural pocket around their esteratic site (Soreq and Zakut, 1993). We capitalized on this to incorporate into our agents AChE or BChE selective inhibition. For example, unsubstituted (phenserine) and ortho-substituted (tolserine) phenylcarbamates are highly selective for AChE inhibition (Fig. 1 and Table 1), whereas para-alkylphenylcarbamates (cymserine) are the reverse. This latter substitution elongates the molecule to allow its bulkier phenylcarbamate to readily fit into the longer pocket of BChE, achieving its selective inhibition.

Extensive chemical modifications of the methylamino groups of physostigmine have resulted in development of novel hetero congeners with different and potentially useful characteristics. For example, replacement of the (N1)-methylamino group by a sulfur provides thia-ester analogs with characteristics applicable to transdermal delivery. Extensive reviews of all these novel compounds and of their high-yield, commercially-viable total syntheses have recently been published (Greig et al., 1995; Brossi et al., 1996).

PHARMACOLOGY

The unsubstituted phenylcarbamate of (-)-physostigmine, phenserine, as its water-soluble L-tartrate salt is being developed towards clinical studies.

Potency and AChE selectivity

Phenserine is a potent inhibitor of AChE and, on an equimolar basis, is more active than tacrine, galanthamine and E2020. Unlike the majority of ChE inhibitors, it is highly selective for AChE vs BChE (70-fold). BChE, present in brain, plasma and liver, is not involved in cholinergic neurotransmission. Recently, some adverse effects of unselective ChE inhibitors have been linked to co-BChE inhibition. Phenserine also has demonstrated a lack of activity in *in vitro* screens encompassing all common receptors.

Duration of action
As illustrated (Fig. 2A), phenserine administration to rats achieves high and steady-state, selective inhibition of AChE for approximately 8 hr. Higher doses of physostigmine and tacrine produce a lower and unselective inhibition of shorter duration, which match their pharmacokinetic half-lives. In contrast to phenserine's long pharmacodynamic half-life of 8 hr, it is rapidly cleared from plasma with a short pharmacokinetic half-life of 12.6 min (Fig. 2B). This unusual independence between pharmacodynamics and pharmacokinetics is not found for other reversible AChE inhibitors and allows optimally long duration of enzyme inhibition, requiring reduced dosing frequency, with rapid drug disappearance, which minimizes total body drug exposure. Once drug selectively binds and long-term reversibly carbamylates the esteratic site of AChE, continued presence of free drug in the body for drug action is no longer required. This is important in the elderly, which represents the fraction of the population afflicted with AD. The high variability and slowing of drug metabolism, commonly associated with age, often results in a gradual overdosing and toxicity in the elderly as one dose is often administered before a prior one is fully cleared (Schmacker, 1985). The dissociation between pharmacokinetics and pharmacodynamics minimizes this, as drug clearance (measured in minutes) can change dramatically without impacting on drug action (measured in hours).

Brain Selectivity
Phenserine rapidly enters brain after its administration and maintains 10-fold higher levels in brain vs plasma (Fig. 2B). Most other ChE inhibitors achieve only equal partition. For phenserine, high brain uptake with rapid brain disappearance, half-life 8.5 min, was achieved by providing it a balanced lipophilicity (octanol/water partition coefficient 7.2, physostigmine 1.2 and

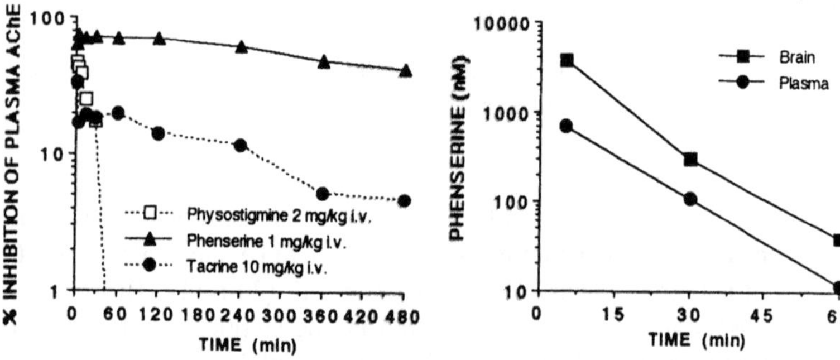

FIGURES 2A & B. Duration of AChE inhibition (*left*) and pharmacokinetics (*right*) after i.v. administration of phenserine(1 mg/kg).

eptastigmine 1150). This allows preferential brain uptake without the long-term sequestration associated with highly lipophilic compounds. Phenserine is highly potent (IC50 36 and 2500 nM AChE and BChE, respectively) and selective (70-fold AChE) against human cortex-derived enzyme. Its administration to rats results in a dose-dependent, long-term increase in brain extracellular levels of ACh, as determined by *in vivo* microdialysis.

Behavioral effects
Phenserine has demonstrated dramatic effects on cognition in rats. Utilizing the 14-unit T-maze, it attenuates scopolamine-induced memory impairments in adult rats over a broad dose range, not previously seen with other agents (Iijima et al., 1993). In addition, phenserine improves memory acquisition in elderly rats, which show an age-related decline in working memory (Fig. 3A)(Ikari et al., 1995).

Toxicology
As a consequence of its AChE and brain selectivity, phenserine has a highly favorable toxicological profile. Its maximum tolerated dose (15 and 40 mg/kg i.p. and p.o., vs 0.5 and 4 mg/kg, respectively, for physostigmine) is dramatically higher than its maximally effective one. The compound lacked toxicity in 28 day rat studies at doses up to a third of its single maximun tolerated level, and is presently undergoing FDA-approved 90 day studies in rats and dogs to allow initiation of clinical studies.

β-APP PROCESSING

The duration of AD likely extends some 20 years and only results in clinically observable detriments in cognitive performance once the normal redundancy and plasticity of brain is fully used. Interventive therapy at this point in the disease process is clearly hampered and, likely would prove more effective if initiated at an early time, prior to significant neurodegeneration. Recent *in vitro* studies suggest that cholinergic stimulation may enhance cell survival. Cholinergic stimulation of cholinoceptive cells alters their processing of β-APP (Nitsch et al., 1992; Lahiri et al., 1994) shifting its metabolism to favor a non-amyloidogenic pathway. Recent studies by Lahiri have demonstrated that phenserine and in particular cymserine, like tacrine but not physostgmine, reduce both secreted and cellular levels of β-APP in neuronal cell lines *in vitro*. Other *in vitro* studies indicate that cholinergic stimulation by the antiChEs promotes the survival of cerebellar cells (Mount et al., 1994), possibly by blocking apoptosis. Furthermore, chronic *in vivo* administration of antiChEs protects against age-related declines in cholinergic markers in rat cortex.

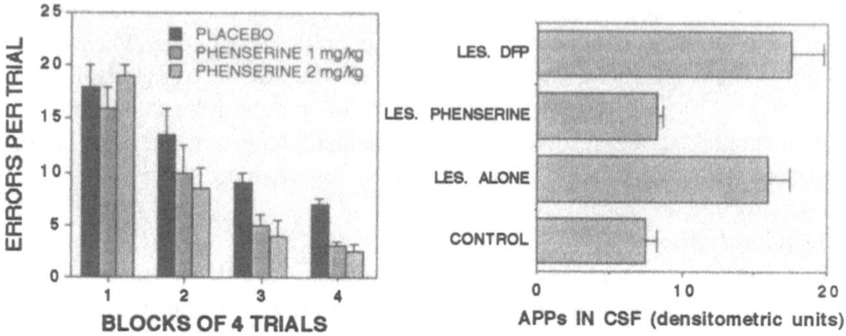

FIGURES 3A & B. Phenserine improves performance of elderly rats in Stone maze learning (*left*), and protects against APP rises after cholinergic lesions.

In rodents, disruption of cortical cholinergic inputs by destruction of the nucleus basalis magnocellularis results in reductions in cortical levels of ChAT and AChE (a partial model of AD) and a sustained increase in cortical production of β-APP, with its increased seceretion into CSF (Fig. 3B) (Wallace et al., 1995). Despite this elevation, very little ß-amyloid is detected in rat CSF as, unlike man, proteolytic cleavage of Aβ from β-APP does not occur in rat. Instead, secreted β-APP contains a significant portion of Aß's sequence. Recent studies with phenserine have demonstrated that it can completely protect against cholinergic lesion-induced increases in β-APP (Fig. 3B) (Haroutunian et al., 1996), whereas classical antiChEs (i.e., DFP) have no action. Taken together, these experimental findings suggest that optimal augmentation of cholinergic stimulation with phenserine and analogs, especially if undertaken early in the course of AD, may not only directly improve cognition, but also retard pathological processes responsible for its progression.

CONCLUSION

Anticholinesterases clearly possess the potential to intervene in AD at more than one level. They can immediately impact to directly improve cognitive performance and, if provided sufficiently early, likely can slow the disease process by reducing Aβ production through its precursor protein. Phenserine and analogs represent a new class of selective long-acting, brain-directed AChE inhibitors with a wide therapeutic window and low toxicity. They were specifically designed as therapeutics for use in the elderly and will hopefully allow us to assess the full utility of antiChE therapy in AD.

REFERENCES

Becker RE, Moriearty P, Surbeck R et al. (1994): Second and Third generation cholinesterase inhibitors: clinical aspects. In: *Alzheimer's Disease: Therapeutic Strategies*, Giacobini E and Becker R, eds. Boston: Birkhauser, pp. 172-178.

Brossi A, Pei XF and Greig NH (1996): Phenserine, a novel anticholinesterase related to physostigmine: total synthesis and biological properties. *Aust J Chem* 49:171-181.

Greig NH, Pei XF, Soncrant TT and Brossi A (1995): Phenserine and ring C hetero-analogues: drug candidates for the treatment of Alzheimer's disease. *Med Res Rev* 15:3-31.

Haroutunian V, Greig NH, Davis KL and Wallace WC (1996): Pharmacological modulation of Alzheimer's β-amyloid precursor protein levels in CSF of rats with forebrain cholinergic system lesions. *Mol Brain Res* Submitted

IIjima S, Greig NH, Garofalo P et al (1993): Phenserine: a physostigmine derivative that is a long-acting inhibitor of AChE and demonstrates a wide dose-range for attenuating a scopolamine-induced learning impairment of rats in a 14-unit T-maze. *Psychpharmacol* 112:415-420.

Ikari H, Spangler EL, Greig NH et al. (1995): Performance of aged rats in a 14-unit T-maze is improved following chronic treatment with phenserine, a novel long-acting anticholinesterase. *NeuroReport* 6:481-484.

Khachaturian ZS (1985): Diagnosis of Alzheimer's disease. *Arch Neurol* 42:1097-1105.

Lahiri DK, Lewis S and Farlow MR (1994): tacrine alters the secretion of the beta-amyloid precursor protein in cell lines. *J Neurosci Res* 37:777-787.

Mount HTJ, Dreyfus CF and Black IB (1994): Muscarinic stimulation promotes cultured purkinje cell survival: a role for acetylcholine in cerebellar development. *J Neurochem* 63:2065-2073.

Nitsch RM, Slack BE, Wurtman RJ and Growdon JH (1992): Release of Alzheimer amyloid precursor derivatives is stimulated by activation of muscarinic acetylcholine receptors. *Science* 258:304-307.

Schmacker DK (1985): Aging and drug disposition:an update. *Pharmacol Rev* 37:133-148.

Soreq H and Zakut H (1993): Human cholinesterases and anticholinesterases. San Diego: Academic Press.

Terry RD, Masliah E, Salmon DP et al. (1991): Physical basis of cognitive alterations in Alzheimer's disease. Synapse loss is the major correlate of cognitive impairment. *Ann Neurol* 30:572-580.

Wallace WC, Lieberburg I; Schenk D et al. (1995): Chronic elevation of secreted amyloid precursor protein in subcortically lesioned rats, and its exacerbation in aged rats. *J Neurosci* 15:4896-4905.

Wisniewski and Weigel (1994): Neuropathological basis of Alzheimer's disease, implications for treatment. In: *Alzheimer's Disease: Therapeutic Strategies*, Giacobini E and Becker R, eds. Boston: Birkhauser, pp. 17-22.

Alzheimer Disease: From Molecular Biology to Therapy
edited by R. Becker and E. Giacobini

AN OVERVIEW OF THE DEVELOPMENT OF SDZ ENA 713, A BRAIN SELECTIVE CHOLINESTERASE INHIBITOR

Ravi Anand, Richard D. Hartman and Peggy E. Hayes
CNS Department, Sandoz Pharmaceuticals Corporation
East Hanover, NJ 07936-1080, USA

Marguirguis Gharabawi
CNS Department, Sandoz Pharma LTD
4002 Basel, Switzerland

INTRODUCTION

Of the many approaches evaluated to treat the symptoms of Alzheimer's disease (AD), only cholinesterase inhibition has produced positive results. The clinical usefulness of the acetylcholinesterase inhibitors (AChEIs) has been limited by either an extremely short or excessively long half-life, hepatotoxicity, excessive peripheral cholinergic side effects, and the potential for drug-drug interactions. To obtain greater (maximum) therapeutic benefit, newer AChEIs that circumvent these problems must be developed. SDZ ENA 713 (ENA 713), a carbamate derivative, is a novel compound that appears to meet the criteria for an ideal AChEI. It is currently in late stages of clinical development worldwide.

NON-CLINICAL PHARMACOLOGY

As a carbamate, ENA 713 belongs to a series of miotine derivatives all having AChE inhibitory activity *in vitro* and *in vivo.* ENA 713 is more potent in inhibiting AChE in rat brain cortex and hippocampus than in other brain regions, such as the striatum and pons/medulla (See Table I). Whereas the brain enzyme was inhibited to about 80%, the AChE activity in peripheral organs, such as liver, lung, and heart, was only slightly affected (<20%).

As with AChE inhibition, ENA 713 induced a greater ACh accumulation in cortical tissue than in other brain regions (Enz et al., 1993). The duration of

Region	IC_{50} (µmol/kg p.o.)
Cortex	3
Hippocampus	4
Stratum	7
Pons, Medulla	7
Heart	40

TABLE I. Brain regional AChE inhibition by SDZ ENA 713

the central effect after a single oral administration of ENA 713 (0.75 mg/kg p.o.) was greater than 6 hours (Enz et al., 1990).

In vitro, ENA 713 inhibits purified rat brain AChE with an apparent $K_i$value of 1.5 µM. This inhibition is 100-1000 times weaker than that of physostigmine and tacrine (Enz et al., 1993). AChE exists in several distinct molecular forms in human brain. The most abundant form is the tetrameric G4, while the monomeric GG1 is present in smaller amounts. However, during aging and more dramatically in AD, the G4 is decreased in neocortex and hippocampus, while no change or a smaller decrease occurs in G1. ENA 713 has greater selectivity for inhibiting the G1 isozyme of AChE in human brain with 4 to 6 times greater potency than for inhibiting the G4 isozyme. The G1 and G4 forms are not differentially inhibited by tacrine.

CLINICAL PHARMACOLOGY

ENA 713 is rapidly (t_{max} 0.5-2 hr) and completely absorbed. It is weakly bound to plasma proteins (ca. 43%), and cleared its interaction with AChE with no relevant involvement of the major cyctochrome P450 isoenzmes. ENA 713 is also rapidly eliminated ($t_{½}$ 0.6-2 hr). Therefore, ENA 713 should not interact with concurrent medications taken by the elderly. In man, an oral 3.0 mg dose decreases AChE activity in CSF by approximately 40% and enzyme activity returns to baseline levels in approximately 9 hrs. The same dose is associated with a minimal (~10%) inhibition of the peripheral enzyme, butyrylcholinesterase. These results support the preclinical profile of ENA 713 in man, i.e. preferential AChE inhibition in the CNS compartment.

EARLY PHASE STUDIES

Early Phase 1 and 2 studies explored a dose range of 2-12 mg/day given b.i.d. or t.i.d. In an early 13-week dose-efficacy study, conducted at fixed daily doses of 2 or 3 mg b.i.d. in 402 mild to moderately demented patients with AD, a "valid patient" analysis indicated a significant difference in the number of responders in the 3 mg b.i.d. dose group, compared to placebo. (See Table II). Both 2 and 3 mg doses b.i.d. were well tolerated, indicating a possibility to administer higher doses.

Test	ENA 713 3 mg b.i.d.	ENA 713 2 mg b.i.d.	Placebo
CGI-C (% Success)[a]	42.2*	31.5	29.9
Digit symbol substitution[b] (number correct)	2.3 ± 7.8*	1.6 ± 5.6	0.4 ± 6.6
Fuld OME[b]			
total storage	1.0 ± 6.0*	0.5 ± 6.3*	-0.7 ± 5.6
total retrievals	1.3 ± 4.3*	1.0 ± 4.6	0.1 ± 4.3

TABLE II. Study B103: Efficacy results at Week 13. a) % Success = percent of patients with CGI-C scores of 1 (markedly improved) or 2 (moderately improved); b) Mean change in score from baseline ± S.D.; *) $p < 0.05$ vs. placebo

This study was followed by another double-blind placebo-controlled study in 114 mild to moderately demented patients treated for 18 weeks. Ninety patients were treated with ENA 713 in the dose range of 2-12 mg/day (45 b.i.d.; 45 t.i.d.). A "valid patient" analysis indicated that a significantly larger number of patients responded when treated with ENA 713 b.i.d., compared to placebo, as determined by the CIBIC-Plus (See Table III). Supportive results were also seen for the ADAS-Cog and NOSGER Memory subscale.

In a 10-week maximum-tolerated dose (MTD) study performed in 50 patients with mild to moderate dementia, which involved titration to a maximum dose of 12 mg/day, over 90% of the ENA 713-treated patients attained the MTD of 12 mg/day given b.i.d. or t.i.d. (Sramek et al., 1996).

An overall analysis of all safety data from placebo-controlled early phase trials indicated that the dropout rate due to adverse events was very low (< 15%). The most frequently reported adverse events included nausea, vomiting, dizziness and diarrhea. No signs of treatment-induced change in blood pressure, heart rate, or hepatic, renal, hematological or urinary indices was detected.

Results from these early studies indicated a signal for efficacy for ENA 713 in the dose range of 6-12 mg/day, given b.i.d., despite shortcomings

Measure	ENA 713 b.i.d.	ENA 713 t.i.d.	PLACEBO
CIBIC - Plus (% Responders)	57*	36	16
ADAS - COG (Mean change)	-2.74+	0.17	2.12
NOSGER - memory (Mean change)	-0.7**	-0.3**	1.4

TABLE III. Study B104: Efficacy results at Week 18 +) $p = 0.054$; *) $p<0.05$ vs. placebo; **) $p<0.01$ vs. placebo in the study designs and conduct, the limited number of patients and the short duration of treatment.

PHASE 2/3 PROGRAM

The design of the Phase 2/3 program, currently nearing completion, was finalized following meetings with the health authorities in Canada, France, Germany, United Kingdom and the United States. The program consists of four placebo-controlled potentially pivotal studies, followed by their open-label, long-term extensions, and two studies evaluating a faster rate of dose titration (See Table IV).

The trial program, known as ADENA (Alzheimer's Dementia with ENA 713), has included more than 2,800 patients enrolled in 116 clinical centers and nine countries (Australia, Austria, UK, US, Canada, Switzerland, Germany, France, Italy, and South Africa). More than 65% of those patients who have completed the initial 26-week treatment protocols have chosen to participate in open-label studies in which all patients will receive ENA 713 for up to one and one-half years. Patients included in the trials had a clinical diagnosis of probable AD according to NINCDS-ADRDA, and DSM-IV criteria, and were 50 years of age or older. Patients had mild to moderately severe dementia (Mini-Mental State Examination score of 10-26, inclusive). Only patients with severe and unstable physical diseases were excluded, and no restrictions were made on the use of concomitant medication, other than the prohibition of psychotropics and certain cytotoxic agents.

In the potentially pivotal studies, efficacy was measured at Weeks 12, 18 and 26 using the ADAS-Cog, CIBIC-Plus and Progressive Deterioration Scale (PDS). The CIBIC-Plus Scale used in the these trials was developed in conjunction with academia and is a semi-structured instrument incorporating a comprehensive review of the domains of cognition, behavior and functioning. Changes in each domain over the course of treatment are incorporated in rating the overall global change in accordance with standardized "change guidelines."

Study Type	Study No.	Location	Dose Range (mg/day)
Potentially Pivotal	B351	U.S.	0, 3, 6, 9
	B352, B303	U.S., Global	0, 1 - 4, 6 - 12
	B304	Global	0, 2 - 12 (bid, tid)
Extension	B353, B305	U.S., Global	2 - 12
Titration	B354, B355	U.S., Global	3 - 12 (3, 6, 9, 12)

TABLE IV. SDZ ENA 713 Phase 2/3 Program

Global patient enrollment in the potentially pivotal ENA 713 Phase 2/3 studies completed 3-6 months ahead of schedule, with over 2,800 patients being enrolled in 15 months. This may reflect the clinician and caregiver interest in new remedies believed to be medical advances.

An International Safety Monitoring Board was established to oversee and review all safety findings from the ENA 713 Phase 2/3 program on an ongoing basis. To minimize risk to patients, the Board is empowered to unblind data and to modify or terminate the studies if necessary. To date, all unblinded analyses performed by the Board have not detected any increased risk to patients receiving ENA 713.

Of the patients enrolled in the Phase 2/3 pivotal studies to date, 84% have either completed or are currently in treatment, 8.5% have dropped out due to adverse events, and 7.6% discontinued for other reasons. The most common adverse events which led to discontinuation were nausea, vomiting and dizziness.

Blinded data collected to date supports good tolerability for ENA 713, as well as a smooth dose titration rate. Further evidence supporting good tolerability and perception of efficacy is the high percentage of patients from the double-blind studies enrolling in the long-term extension study. Treatment with ENA 713 in these patients having multiple medical illnesses, and in patients receiving concurrent treatment with other drugs, has provided further support for the drug's tolerability, as well as its lack of drug-drug interactions.

REFERENCES

Enz A, Amstutz R, Hofmann A, Gmelin G and Kelly PH (1990): ENA 713, an acetylcholinesterase-inhibitor with preference for central activity: animal studies. In: *Basic, Clinical and Therapeutic Aspects of Alzheimer's and Parkinson's diseases.* Nagaatsu T, Fischer A and Yoshida M, New York: Plenum Press, pp.425-428.

Enz A, Amstutz R, Boddeke H, Gmelin G and Malanowski J (1993): Brain selective inhibition of acetylcholinesterase: a novel approach to therapy for Alzheimer's desease. *Progr Brain Res.* 98:431-438.

Sramek JJ, Anand R, Wardle TS, Irwin P, Hartman RD and Cutler NR (1996): Safety/tolerability trial of SDZ ENA 713 in patients with probable Alzheimer's disease. *Life Sciences* 58:1201-1207.

SDZ ENA 713 Investigator's Brochure (1994): On file at Sandoz Pharmaceuticals Corp.

Alzheimer Disease: From Molecular Biology to Therapy
edited by R. Becker and E. Giacobini

PRECLINICAL AND CLINICAL PROGRESS WITH HUPERZINE A: A NOVEL ACETYLCHOLINESTERASE INHIBITOR

Yifan Han
Department of Biochemistry, Hong Kong University of Science and Technology, Hong Kong

Xican Tang
Department of Pharmacology, Shanghai Institute of Materia Medica, Chinese Academy of Sciences, Shanghai 200031, China

INTRODUCTION

Alzheimer's disease (AD) is a progressive neurodegenerative disease that attacks the brain and results in impaired memory, thinking and behaviour in the elderly. This disease is pathologically characterized by the degeneration of the basal forebrain cholinergic system and the deposit of amyloid plaques in the brain (Whitehouse et al., 1982; Selkoe, 1991). It has become the fourth leading cause of death in developed nations. In the USA, the costs related to AD exceed those of related to cancer and heart disease. Yet, at present, there is no effective treatment available for this disorder.

Currently, only the cholinesterase (ChE) inhibition approach, which enhances the function of central cholinergic neurons by permitting acetylcholine to remain in the synaptic cleft longer through inhibiting ChE with inhibitors, has produced the most encouraging clinical results, and holds promise for the treatment of AD (Giacobini and Becker, 1994). This strategy is used with the goal of alleviating the central cholinergic deficiency which has been shown to occur in patients with AD. Thus far only two ChEIs, physostigmine (Phys) and tacrine, have been evaluated on a large scale in patients with AD. However, the short half-life and peripheral cholinergic effects of physostigmine limit its therapeutic value; clinical trials of tacrine, have confirmed its dose-dependent hepatotoxicity. Consequently, there has been a major rush among giant pharmaceutical firms, as well as top research institutions world-wide to search for new and better therapeutic agents of AD.

Huperzine A (Hup-A), originally discovered from the Chinese herbal medicine *Huperzia serrata* (also known as *Qian Chen Ta*), represents one highly promising compound (Liu et al., 1986). Hup-A, a novel *Lycopodium* alkaloid chemically unique from other agents under study for AD, is a reversible and mixed competitive ChEI, its potency and duration of acetylcholinesterase (AChE) inhibition rival those of Phys, galanthamine (Gal) (Tang et al., 1994) and tacrine (DeSarno et al.,1989). In addition, Hup-A has been found to be an effective memory enhancer in a number of different animal models (Tang et al., 1986; Hanin et al., 1991; Tang et al., 1994). Preliminary clinical trials have demonstrated that intramuscular injection and oral administration of Hup-A induced significant improvement of memory deficiencies in the aged and demantia patients being devoid of any remarkable side effects (Hanin et al., 1991; Tang et al., 1994). To further clarify the therapeutic potential of Hup-A for treatment of AD, we have conducted a comprehensive characterization of memory-enhancing effects of Hup-A and tacrine in various animal models with central cholinergic deficiencies. In this chapter, we first report some of our findings comparing effects of Hup-A with tacrine on spatial memory in rats with central cholinodeficiencies induced by AF64A and NBM lesions. This is followed by a review of reported clinical efficacy of Hup-A on memory, cognition, and behavior in patients with AD conducted at multicenters in China.

EFFECTS OF HUP-A ON RADIAL MAZE PERFORMANCE IN CENTRALLY CHOLINODEFICIENT RATS

The i.c.v. injection of AF64A (ethylcholine mustard aziridinium ion, a specific cholinergic neurotoxin) in rodents has been suggested as a valuable animal model for examining the potential effectiveness of cholinergic drugs which might be used to ameliorate the symptoms of AD (Hanin, 1990). Using a delayed non-match-to-sample radial maze task, the effects of Hup-A on spatial memory were assessed in rats with a lesion induced by AF64A (Xiong et al., 1995). Male Sprague-Dawley rats were trained to achieve the criterion level (less than one error in each trial for 3 consecutive trials) within about 90 days. Following the completion of training, rats were given AF64A (2nmol per side, i.c.v.), and exhibited substantial impairment in the performance compared with saline-treated rats. Analysis of CC indicated a significant treated effect at 5, 15, 30 and 60 min. delay intervals compared to the saline treated group. Similarly marked increases in the number of PDE in AF64A-treated group at the above delay intervals were observed. No significant difference in the latency per choice of AF64A-treated rats to perform the RAM task was observed. Neurochemical analysis demonstrated that these behavioral impairments were associated with about 50% decrease in hippocampal choline acetyltransferase (ChAT) activity, suggesting an important role of

Region	Lesion	Sham-operated	% of sham
Hippocampus	19.0±.5**	41.2±1.2	46
Striatum	88.6±.4	92.0±4.6	96
Cortex	32.0±1.1	34.3±0.9	93
Hypothalamus	15.9±0.5	16.1±0.7	99

TABLE I. Cholinergic acetyltransferase activity in the different brain regions after intraventricular administration of AF64A. Values represent Means±S.E.M. expressed as nmol ACh formed mg^{-1} protein h^{-1} . $^{**}P<0.01$ vs sham-operated group, independent *t*-test.

hippocampal cholinergic system in maintenance of spatial working memory. In contrast, there was no marked decrease in ChAT activity in other brain regions (Table 1).

The effects of Hup-A on the performance of AF64A-treated rats are presented in Figure 1. ANOVA on the CC and PDE indicated significant treatment effects with Hup-A at 15 min delay compared with AF64A-treated alone. *Post hoc* comparisons indicated significant improvement in the number of CC in the 0.4 mg kg^{-1} group as well as in the 0.5 mg kg^{-1} group. Similarly, the number of PDE in the 0.4 mg kg^{-1} group and 0.5 mg kg^{-1} group were substantially decreased.

The efficacy of Hup-A and tacrine on nucleus basalis of Meynert (NBM) lesion-induced spatial memory deficit was evaluated in rats (Yeung and Han, 1996). Male SD rats were trained to achieve the criterion level of

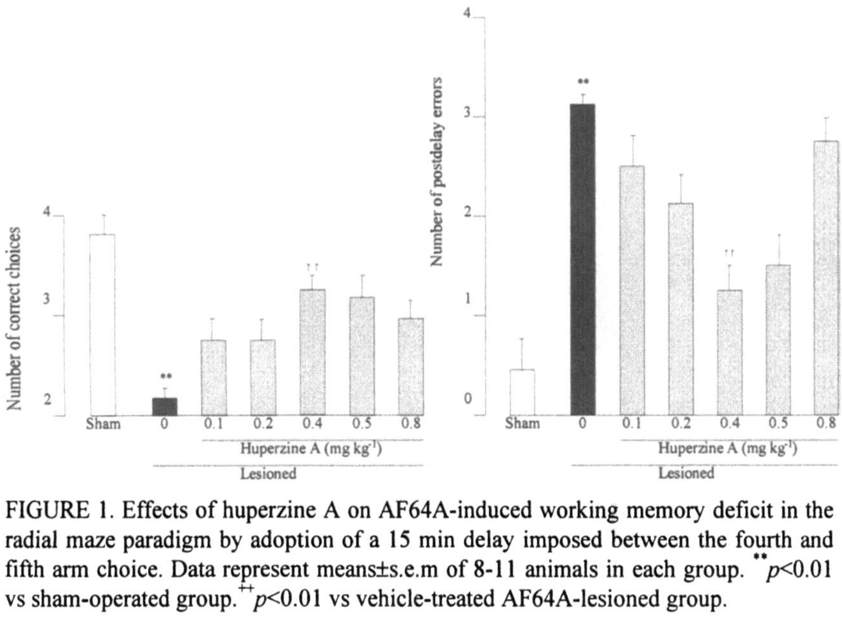

FIGURE 1. Effects of huperzine A on AF64A-induced working memory deficit in the radial maze paradigm by adoption of a 15 min delay imposed between the fourth and fifth arm choice. Data represent means±s.e.m of 8-11 animals in each group. $^{**}p<0.01$ vs sham-operated group. $^{++}p<0.01$ vs vehicle-treated AF64A-lesioned group.

performance in the radial maze in which eight arms were baited with reinforcement. Bilateral NBM lesion with ibotenic acid (6µg/µl, per side) caused significant impairments in rats' ability to perform this working memory task as evidenced by fewer correct choices before first error, more total errors to complete the task and therefore lower percent efficiency (number of reinforced arm entries/total number of arm entries x 100%). The behaviour impairment was correlated to about 40% decrease of ChAT activity in neocortex, but not in hippocampus, hypothalamus, and striatum. Figure 2 shows that Hup-A (0.3-0.6 mg/kg, IP, 30 minutes before the task) significantly reversed these memory deficits induced by NBM lesion. Tacrine (3-6 mg/kg) also ameliorated the memory impairment. These results indicate that the integrity of NBM is critical for spatial memory processing, and this working memory impairment induced by NBM lesion can be reduced by Hup-A,which is approximately 5 times more potent than tacrine.

EFFECTS OF HUP-A IN THE TREATMENT OF AD

Four separate clinical studies with Hup-A have been reported to date on favorable effects in treatments for patients suffering from age related memory dysfunctions or dementia. The treatment responses were evaluated according to either the method of Buschke and Fuld(1974) or Wechsler memory scale.

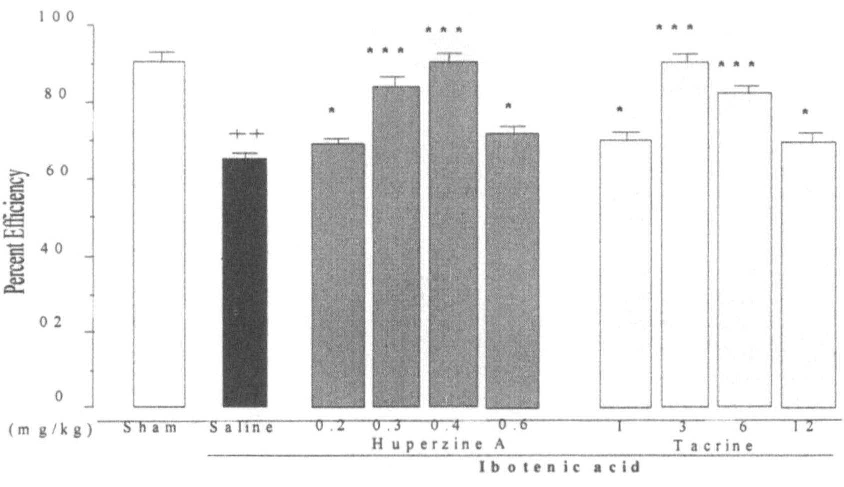

FIGURE 2. Effects of Huperzine A and Tacrine on reversing the radial maze performance deficits induced by bilateral NBM injection of ibotenic acid in rats. $^{++}p<0.01$ vs sham-operated. $^{*}p>0.05$ vs NBM lesioned group, $^{***}p<0.01$ vs NMB lesioned group. Means±SEM.

Comparison of all rating scores from Hup-A with placebo or hydergine showed significant therapeutic effects with minimal observed side effects (Hanin et al.,1991; Tang et al., 1994). To further confirm as well as expand these initial encouraging findings, a more comprehensive Phase II study has recently been conducted in 103 patients with AD at 7 mental hospitals nationwide in China (Xu et al., 1995). Fifty patients (age: 66±11; 28M: 22F) were administered orally 0.1 mg *bid* for 8 weeks, while another group of 53 patients (age: 67±11; 29M: 24F) were given placebo. All patients were assessed with Wechsler memory scale, mini-mental state examination scale (MMS), Hasegawa dementia scale and activity of daily living scale. The results of this trial demonstrated that 58% of patients treated with Hup-A showed marked improvements in their memory, cognitive and behavioral functions (all $p<0.01$, TabLE 2). What is more noteworthy is that better improvement in the MMS evaluation was observed from patients treated with Hup-A compared to those administered with tacrine (160 mg daily for 30 weeks) (Knapp et al., 1994). During the period of this study only mild side effects, i.e. dizziness, nausea and sweating occurred in a few patients, and no toxicity on liver and kidney was detected. Further clinical studies with longer term and especially direct comparison with tacrine are underway in order to establish the efficacy and safety of Hup-A.

Based on the above encouraging preclinical and clinical progress, it is reasonable to expect that Hup-A is a highly promising candidate for development as second generation of ChEI in the treatment of various memory disorders related to central cholinodeficiency.

Administration	(weeks)	Pla (n=53)	Hup (n=50)
MQ	0	47..9±21..5	55.8±21.1
	8	$51.6\pm25.6^{**}$	$64.4\pm26.2^{***\uparrow}$
MMS	0	14.4±4.7	16.0±5.0
	8	14.9±6.4	$18.9\pm6.2^{***\uparrow}$
MDS	0	15.6±5.3	16.1±5.6
	8	15.4±6.7	$19.7\pm6.5^{***\uparrow\uparrow}$
ADL	0	30.7±9.3	32.6±9.6
	8	31.9±0.7	29.1±9.3

TABLE II. Therapeutic effects of oral Hup-A on patients with AD. $^{**}P<0.05$, $^{***}P<0.01$ vs before treatment (0 week); $^{\uparrow}P<0.05$, $^{\uparrow\uparrow}P<0.01$ vs placebo. MQ=memory quotient; MMS=mini mental state examination scale; HDS=Hasegawa dementia scale; ADL= activity of daily living scale.

REFERENCES

Giacobini E and Becker R (1994): Development of drugs for Alzheimer therapy: a decade of progress. In: *Alzheimer Therapy: Therapeutic Strategies*. Giacobini E and Becker R eds. Boston: Birkhauser.

DeSarno P, Pomponi M, Giacobini E, Tang XC and Williams E (1989): The effect of heptyl-physostigmine, a new cholinesterase inhibitor, on central cholinergic system of the rat. *Neurochem Res* 14: 971-997.

Hanin I (1990): In: *Progress in Brain Research: Cholinergic Neurotransmission: Functional and Clinical Aspects,* Vol 84, Aquilonius S-M and Gilberg P-G, eds. Amsterdam: Elsevier, pp. 289-299.

Hanin I, Tang XC and Kozikowski AP (1991): Clinical and preclinical studies with huperzine A. In: *Cholinergic Basis for Alzheimer Therapy*, Giacobini E and Becker R, eds. Boston: Birkhauser, pp. 305-313.

Knapp MJ, Knopman DS, Solomon PR, Pendleberg WW and Davis CS et al. (1994): A 30-week randomized controlled trial of high-dose tacrine in patients with Alzheimer's disease. *J Am Med Assoc* 27: 985-989.

Liu JS, Zhu YL,Yu CM, Han YY and Wu FW et al. (1986): The structure of huperzine A and B, two new alkaloids exibiting marked anticholinesterase activity. *Can J Chem* 64: 837-839.

Selkoe DJ (1991): The molecular pathology of Alzheimer's disease. *Neuron* 6: 487-498.

Tang XC, Xiong ZQ, Qian BC, Zhou ZF and Zhang LL (1994): Cognition improvement by oral huperzine A: a novel acetylcholinesterase inhibitor. In: *Alzheimer Therapy: Theraputic Strategies*, Giacobini E and Becker R, eds. Boston: Birkhauser, pp. 113-119.

Tang XC, Han YF, Chen X and Zhu XD (1986): Effects of huperzine A on learning and retrieval process of discrimination performance in rats. *Acta Pharmacol Sin* 7(6): 507-512.

Whitehouse PJ, Pricc DL, Struble RG, Clark AW and Coylc JT et al. (1982): Alzheimer's disease and senile dementia: loss of neurons in the basal forebrain. *Science* 215: 1237-1239.

Xiong ZQ, Han YF and Tang XC. (1995): Huperzine A ameliorates the spatial working memory impairments induced by AF64A. *Neuroreport* 6: 2221- 2224

Xu SS, Gao ZZ, Weng Z, Du ZM and Xu WA et al.(1995): Efficacy of tablet huperzine-A on memory, cognition, and behavior in Alzheimer's disease. *Acta Pharmacologica Sinica* 16(5): 391-395.

Yeung OY and Han YF (1996): Reversal effects of Huperzine A and Tacrine on memory deficits in bilateral NBM lesioned rats. *Soc for Neurosci Abst.*

Alzheimer Disease: From Molecular Biology to Therapy
edited by R. Becker and E. Giacobini

P11467: AN ORALLY-ACTIVE ACETYLCHOLINESTERASE INHIBITOR AND α_2-ADRENOCEPTOR ANTAGONIST FOR ALZHEIMER'S DISEASE

Hugo M. Vargas, Craig P. Smith, Mary Li, Gina M. Bores, Andrew Giovanni, Lily Zhou, Dana Cunningham, Karen M. Brooks, Fernando Camacho, James T. Winslow, David E. Selk, Eva Marie DiLeo, Douglas J. Turk, Larry Davis, David M. Fink and Douglas K. Rush
Hoechst Marion Roussel, Inc; Neuroscience TA
Bridgewater, New Jersey, USA

Anne Dekeyne and Claude Oberlander
Roussel Uclaf, Centre de Recherches, Romainville, France

INTRODUCTION

Alzheimer's disease (AD) is a complex and multifaceted neurodegenerative disease characterized by cognitive and behavioral abnormalities. The primary cognitive deficit has been correlated with extensive cholinergic dysfunction and the efficacy of cholinergic therapies in this disease validates and supports the cholinergic hypothesis of AD (Weinstock, 1995). To date, tacrine has shown clinical efficacy in 20 to 30% of AD patients and remains the only acetylcholinesterase inhibitor (AChEI) to receive FDA approval for symptomatic treatment (Knapp et al., 1994). However, the limited efficacy of pure cholinergic therapies suggests that other neurochemical deficiencies are involved.

Non-cholinergic neurochemical abnormalities have been identified that may contribute to the cognitive and behavioral disorders associated with AD. Postmortem findings from AD brain show a loss of noradrenergic cell bodies in the locus ceruleus, the major source for hippocampal and cortical norepinephrine (NE) (Bondareff et al., 1982, 1987). The loss of noradrenergic function could lead to cognitive impairment since cholinergic-noradrenergic interactions are involved in memory processing (Decker and McGaugh, 1991; Camacho et al., 1996). In addition, noradrenergic dysfunction may underlie early-onset or aggressive forms of AD (Bondareff et al., 1987) and correlate with the occurrence of depression in AD patients (Zweig et al. 1988; Zubenko et al., 1990). However, monotherapy with adrenergic agents, like NE uptake inhibitors or MAO-B inhibitors, has proven ineffective in alleviating cognitive impairment in non-depressed AD patients (Cutler et al., 1985; Burke et al., 1993).

For these reasons, a mechanism-based therapeutic which activates both cholinergic and noradrenergic activity may be beneficial in the treatment of AD. Therefore, we discovered and developed a series of novel combined AChEIs with α_2-adrenoceptor antagonist properties. Of these, P11467 was chosen for further characterization. Through its ability to simultaneously enhance brain cholinergic (i.e., AChEI) and noradrenergic (i.e., pre-postsynaptic α_2-adrenoceptor blockade) neurotransmission, this agent represents a new multifaceted approach to AD therapy.

RESULTS

P11467 is a potent inhibitor of rat striatal AChE activity *in vitro* (IC_{50}: 0.12 ± 0.02 μM) and was 2-times more potent than tacrine (Table 1). When evaluated one hr after oral dosing, P11467 inhibited rat striatal (ED_{50}: 10 mg/kg) and mouse forebrain (ED_{50}: 7 mg/kg) AChE activity in a dose-dependent fashion. Inhibition of brain AChE was observed up to 6 hrs after dosing in both species, with AChE activity returning to baseline by 18 hrs.

P11467 also demonstrated the ability to completely displace [^{3}H]clonidine from rat cortical α_2-adrenoceptors *in vitro*, a property which differentiates P11467 from other AChEIs (Table 1). In comparison, P11467 had lower

Class	AChEI: IC_{50} (μM)	α_2-AR: K_I (μM)
AChEI		
Tacrine	0.23 ± 0.02	>10
Velnacrine	1.26 ± 0.15	>10
E-2020	0.027	>10
Galanthamine	0.92 ± 0.15	>10
Heptylphysostigmine	0.008 ± 0.001	>10
Physostigmine	0.004 ± 0.001	>10
Combined AChEI-α_2 AR Antagonist		
P11467	0.12 ± 0.02	0.95 ± 0.23
α_2–AR Antagonist		
Idazoxan	>100	0.005 ± 0.002
Yohimbine	>100	0.012
P7480	>100	0.018 ± 0.006

TABLE 1. Potency of P11467 and other preclinical and clinical reference AChE inhibitors and α_2-adrenoceptor (AR) antagonists *in vitro*.

affinity than the α_2-adrenoceptor antagonists idazoxan, yohimbine and P7480 (Camacho et al., 1996). Functional assays showed P11467 to be a competitive antagonist at both presynaptic and postsynaptic α_2-adrenoceptors. Antagonism of BHT 920 (selective α_2-agonist) induced contractions of the dog saphenous vein was used to measure postsynaptic α_2-adrenoceptor blockade *in vitro* (DeMey and Vanhoutte, 1981). In this assay, P11467 (1-10 μM) competitively antagonized vascular α_2-adrenoceptors and had a dissociation constant (K_B) of 1.8 ± 0.4 μM, a value close to its radioligand binding affinity. Enhancement of electrically-stimulated [^{3}H]NE release from rat cortical slices was an indicator of presynaptic α_2-adrenoceptor antagonism (Smith et al., 1994). P11467 (10 μM) significantly elevated [^{3}H]NE release in this preparation and had an equivalent effect to that of idazoxan (0.1 μM) and P7480 (1 μM). In contrast, the AChEI heptylphysostigmine (HEP; 1 μM) did not alter release. P11467 had affinity for α_1-adrenoceptors (K_I: 2.2 μM), but did not affect monoamine oxidase A or B, NE/DA/5-HT uptake nor bind muscarinic or dopaminergic receptors at 10 μM.

In pithed rats, pretreatment with P11467 (5 & 10 mg/kg, i.p.; 30 min) antagonized vascular postsynaptic α_2-adrenoceptors and caused a competitive shift of the i.v. BHT-920 pressor-response curve *in vivo*. In anesthetized and vagotomized dogs, cardiac nerve stimulation (1-10 Hz) causes a tachycardic response that is mediated by evoked NE release. The stimulated release of presynaptic NE (i.e., tachycardia) was significantly inhibited by the α_2-agonist clonidine (20 μg/kg, i.v.). P11467 (1 mg/kg, i.v.), but not HEP (1 & 3 mg/kg, i.v.), directly reversed the neuroinhibitory effect of clonidine and indicated that this agent antagonized presynaptic α_2-adrenoceptors *in vivo*.

Microdialysis studies of P11467 (1-5 mg/kg, i.p.) on hippocampal acetylcholine (ACh) and NE release were conducted in free-moving rats to confirm that its AChEI and α_2-adrenoceptor antagonist properties would elevate extracellular levels of both neurotransmitters. For these studies, an AChEI was not added to the microdialysis fluid. Systemic P11467 (5 mg/kg) significantly raised both ACh (peak: 400% at 1 hr) and NE (peak: 240% at 1 hr) over baseline levels. In contrast, HEP (5 mg/kg, i.p.) raised only ACh (peak: 700% at 1 hr), but not NE levels (peak: 125% at 1 hr). The maximal effect of P11467 on NE release *in vivo* was similar in magnitude to 5 mg/kg idazoxan (peak: 225% at 0.5 hr). These observations show that in the brain, the AChEI and α_2-adrenoceptor antagonist properties of P11467 are manifested as increased ACh and NE levels *in vivo*.

P11467 enhanced performance in specific memory tasks of cholinergic and noradrenergic function. For example, P11467 reversed scopolamine-induced deficits of passive avoidance in mice (0.03-0.3 mg/kg, s.c.) and enhanced step-down passive avoidance in rats (1.25 & 2.5 mg/kg, i.p.). This latter test is sensitive to central cholinergic, but not adrenergic, stimulation (Camacho et al., 1996). P11467 (0.63-2.5 mg/kg, i.p.) also enhanced olfactory investigation in a social recognition paradigm in mice (Winslow and

Camacho, 1995). In the linear maze task in rats, P11467 (1.25 mg/kg, i.p.) was found to be as active as idazoxan (2 mg/kg, i.p.). This latter memory task is known to be specifically facilitated by central noradrenergic stimulation (Sara and Devauges, 1989) and relatively insensitive to cholinergic stimulation (e.g., tacrine & muscarinic agonists are not active; C. Oberlander, unpublished). Therefore, the combined cholinergic and adrenergic activity of P11467 can be demonstrated in behavioral assays of memory.

The pharmacokinetic profile of P11467 was evaluated in rat and dog following oral and i.v. administration. Samples were collected over 24 hrs, extracted and analyzed by HPLC-UV. The results showed that P11467 was rapidly absorbed with estimated plasma half-lives for absorption on the order of 0.40 hrs in both species and terminal phase elimination of P11467 from the target brain tissue (rat) was 3.1 hrs. The parent compound was shown to be the major circulating form with maximal levels (C_{max}) in rat plasma of 482, 2728 and 5118 ng/mL and in rat brain of 170, 1100 and 1954 ng/g following 1, 5 and 10 mg/kg (p.o.), respectively. The C_{max} levels in dog plasma were 327, 526, 1083 and 3081 ng/mL following 0.5, 1, 5 and 10 mg/kg (p.o.), respectively. Comparison of area under the curve ($AUC_{0\text{-}24hrs}$) values in both rat and dog following oral and i.v. administration indicated an absolute bioavailability on the order of 100% in both species. There was a linear increase in both the C_{max} and the $AUC_{0\text{-}24hrs}$ as the dose increased for rat (plasma and brain) and dog (plasma). In addition, the rat brain to plasma ratio remained constant at 0.4 across the dose range. The pharmacokinetic evaluation of P11467 shows that this agent has excellent oral bioavailability and good brain penetration.

CONCLUSION

The pharmacological data provided above demonstrates that P11467 is a novel AChEI with combined activity as a pre- and postsynaptic α_2-adrenoceptor antagonist. In addition, this agent has a favorable pharmacokinetic profile, including complete oral bioavailability. In the central nervous system, this compound causes the activation of both cholinergic and noradrenergic neurotransmission. Recent neurochemical and behavioral studies have shown a synergistic effect of α_2-adrenoceptor blockade on cholinergic function (Cuadra and Giacobini, 1995; Camacho et al., 1996). Therefore, P11467 represents a novel therapeutic approach to treat cognitive deficits associated with AD.

REFERENCES

Bondareff W, Mountjoy CQ and Roth M (1982): Loss of neurons of origin of the adrenergic projection to cerebral cortex (nucleus locus ceruleus) in senile dementia. *Neurology* 32:164-168.

Bondareff W, Mountjoy CQ, Roth M, Rossor MN, Iversen LL, Reynolds GP and Hauser DL (1987): Neuronal degeneration in locus ceruleus and cortical correlates of Alzheimer Disease. *Alz Dis Assoc Disor* 1:256-62.

Burke WJ, Roccaforte WH, Wengel SP, Bayer BL, Ranno AE and Willcockson NK (1993): L-Deprenyl in the treatment of mild dementia of the Alzheimer type: results of a 15-month trial. *J Am Geriatr Soc* 41:1219-1225.

Camacho F, Smith CP, Vargas HM and Winslow JT (1996): α_2-Adrenoceptor antagonists potentiate acetylcholinesterase inhibitor effects on passive avoidance learning in the rat. *Psychopharmacol* 124 (in press).

Cuadra G and Giacobini E (1995): Coadministration of cholinesterase inhibitors and idazoxan: effects of neurotransmitters in rat cortex in vivo. *J Pharmacol Exp Ther* 273:230-240.

Cutler NR, Haxby J, Kay AD, Narang PK, Lesko LJ, Costa JL, Ninos M, Linnoila M, Potter WZ, Renfrew JW and Moore AM (1985): Evaluation of zimeldine in Alzheimer's Disease. *Arch Neurology* 42:744-748.

Decker MW and McGaugh JL (1991): The role of interactions between cholinergic systems and other neuromodulatory systems in learning and memory. *Synapse* 7:151-168.

DeMey J and Vanhoutte PM (1981): Uneven distribution of postjunctional $alpha_1$- and $alpha_2$-like adrenoceptors in canine arterial and venous smooth muscle. *Circ Res* 48:875-884.

Knapp MJ, Knopman DS, Solomon PR, Pendlebury WW, Davis CS, Gracon SI and the Tacrine Study Group (1994): A 30-week randomized controlled trial of high-dose tacrine in patients with Alzheimer's disease. *J Am Med Assoc* 271:985-991.

Sara SJ and Devauges V (1989): Idazoxan, an α_2-antagonist, facilitates memory retrieval in the rat. *Behav Neural Biol* 51:401-411.

Smith CP, Petko WW, Kongsamut S, Roehr JE, Effland RC, Klein JT and Huger FP (1994): Mechanisms for the increase in electrically stimulated [^{3}H]norepine-hrine release from rat cortical slices by N-(n-propyl)-N-(4-pyridinyl)-1H-indol-1-amine. *Drug Dev Res* 32:13-18.

Weinstock M (1995): The pharmacotherapy of Alzheimer's disease based on the cholinergic hypothesis: an update. *Neurodegeneration* 4:349-356.

Winslow JT and Camacho F (1995): Cholinergic modulation of a decrement in social investigation following repeated contacts between mice. *Psychopharmacol* 121:164-172.

Zubenko GS, Mossy J and Kopp U (1990): Neurochemical correlates of major depression in primary dementia. *Arch Neurology* 47:209-214.

Zweig RM, Ross CA, Hedreen JC, Steele C, Cardillo J, Whitehouse PJ, Folstein MF and Price DL (1988): The neuropathology of aminergic nuclei in Alzheimer's Disease. *Ann Neurology* 24:233-242.

Alzheimer Disease: From Molecular Biology to Therapy
edited by R. Becker and E. Giacobini

CHOLINESTERASE INHIBITORS AS THERAPY IN ALZHEIMER'S DISEASE: BENEFIT TO RISK CONSIDERATIONS IN CLINICAL APPLICATION

Robert E. Becker, Pamela Moriearty, Latha Unni, Sandra Vicari
Division of Research, Department of Psychiatry, Southern Illinois University School of Medicine, Springfield, IL

Currently available data about cholinesterase inhibitors in Alzheimer's disease (AD) indicate that knowledge of differences in pharmacology among the inhibitors is essential to physicians to maximize clinical benefits to patients. We express the concern that commercial promotion may not recognize the relevant issues. We caution physicians to consider carefully the rationale of their use of compounds from this class.

PREMISES AND STRATEGIES

Over a decade ago we decided to give a fair test to the potential for benefit in Alzheimer's disease (AD) from acetylcholine (ACh) supplementation through acetylcholinesterase (AChE) inhibition (Becker and Giacobini, 1988a). The available cholinesterase inhibitors, physostigmine and tacrine, did not allow the full potential benefit in AD to be explored because of the onset of distressing adverse effects, nausea, vomiting, diarrhea, at low levels of AChE inhibition. Our review of the pharmacology of the AChE inhibitors indicates that the appearance of the typical cholinergic syndrome--nausea, vomiting, diarrhea, progressing to bronchospasm, convulsions --and animal mortality--are not systematically related to either levels of AChE inhibition or tissue concentrations of ACh (Becker and Giacobini, 1988b,c). The possibility that toxic effects could be dissociated from therapeutic effects initiated a search for a compound without adverse characteristics which could be administered to humans to test the efficacy of cholinesterase inhibition in AD. We also sought other properties we think important: ability to achieve high levels of AChE inhibition both to test for maximum benefit and to overcome therapeutic resistance from tissue tolerance; long duration of action to maintain steady state of tissue effects and to reduce the need for dosing to at most once daily; a close relationship between clinical dose and therapeutic tissue concentrations to simplify dosing and increase safety; and a large clinical therapeutic index to provide safety in inadvertent overdose, a concern in the confused, memory-impaired AD patient (Becker, 1991). Our assumption is that the full potential efficacy of cholinesterase inhibitors in AD cannot be known until adverse effect-free inhibition of AChE can be maintained at steady state throughout the total range of physiological function of AChE.

PHARMACOKINETICS AND PHARMACODYNAMICS

Pharmacodynamic and pharmacokinetic studies in animals (Hallak and Giacobini, 1987; Hallak and Giacobini, 1989; Mori et al., 1995; Moriearty and Becker, 1992; Moriearty et al., 1993; Kronforst-Collins et al., 1996; Unni et al., 1994) documented the favorable properties of metrifonate, a prodrug for O,O-dimethyl O-(2,2-dichlorovinyl) phosphate (DDVP), an irreversible inhibitor of both butyrylcholinesterase (BuChE) and AChE. In an open study of 20 patients we documented low rates of occurrence of adverse effects with repeated doses of metrifonate (Becker et al., 1990). In this study, we also achieved up to 88% inhibition of red blood cell (RBC) AChE at steady state for one week with once-weekly dosing of metrifonate. We then documented levels of cerebrospinal fluid AChE inhibition in relation to RBC AChE inhibition in two patients (Becker et al., 1990) similar to those in earlier rat brain studies (Hallak and Giacobini, 1987; Hallak and Giacobini, 1989). We also reported the lack of hematological and genetic effects after repeated dosing (Moriearty et al., 1991; Bartels et al., 1994).

CLINICAL PHARMACOLOGY OF METRIFONATE IN ALZHEIMER'S DISEASE

Recently we reported results from two double blind studies in AD patients in which we compared metrifonate to placebo and from an open followup of patients after the first double blind study (Becker et al., 1996a,b). The initial double blind placebo-controlled study was intended to enter 100 AD patients to determine the effects of metrifonate on the cognitive deficit in AD patients. Many critics argue that statistical significance for treatment differences in a rating scale may be of little clinical significance. To assure our work focused on clinically significant drug effects we adopted, as our standard for self-evident clinical significance, a 25% improvement in the cognitive function presumed lost in our AD patients, a standard we used previously to indicate a clinically non-trivial statistical association between treatment and response. Thus a mean pretreatment Alzheimer Disease Assessment Scale (ADAS) (Rosen et al., 1984) score of 28 for the patients entered would require a 7 point reduction in score (improvement); a mean pretreatment Mini Mental State Examination (MMSE) (Folstein et al., 1975) score of 18 would require a 3 point increase in score (improvement) to meet our requirement of clinical significance. We did not intend that any statistically significant effects less than these clinical levels would be regarded as potentially lacking clinical utility, but only that lesser effects complicate clinical evaluation of drug effects in individual patients (Becker and Markwell, 1996). Mere statistical significance can be misleading if taken as evidence of magnitude of clinical effect when large groups are compared.

Our initial double blind study was the first long term administration of metrifonate to humans. Therefore, we conducted an interim analysis after 50 patients were completed to monitor safety and determine if significant effects could be achieved by the full study. In this analysis we found a .75 point improvement in ADAS-Cognitive (-C) scores in the metrifonate-treated patients and significant differences from the placebo group, 2.6 points, in ADAS-C scores. Statistically significant declines in the placebo but not in the metrifonate groups on the ADAS-C and MMSE led us to hypothesize a robust beneficial effect of metrifonate on cognitive decline rate in AD. We then carried out our second double blind placebo controlled study over a six month period. We focused on the decline in cognitive function of metrifonate treated patients. We treated all patients at baseline for one month with metrifonate and then randomly assigned all patients to an additional six months, double-blind treatment with metrifonate or placebo. Again we found a decline in the placebo-treated patients compared to metrifonate-treated patients in ADAS-C and MMSE scores. We found statistically significant decline in cognitive function of placebo-treated AD patients in ADAS-C and MMSE scores (each $p \leq 0.01$) and 0.00 and 0.24 point differences in metrifonate-treated patients' ADAS-C and MMSE scores respectively. We projected these three and six month study results to predict deterioration at one year for patients treated with metrifonate and placebo (Figures 1 and 2). In Figure 2, the metrifonate-treated three- and six- month projected one year rates of decline, and the rate

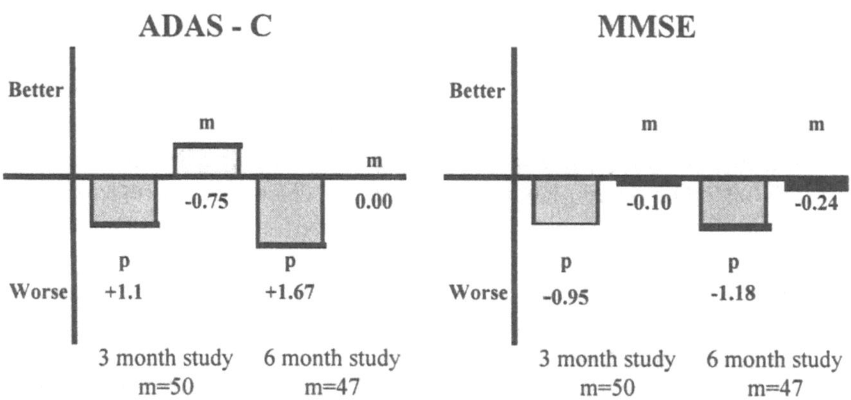

FIGURE 1. Mean changes in test scores with metrifonate (m) and placebo (p) treatments

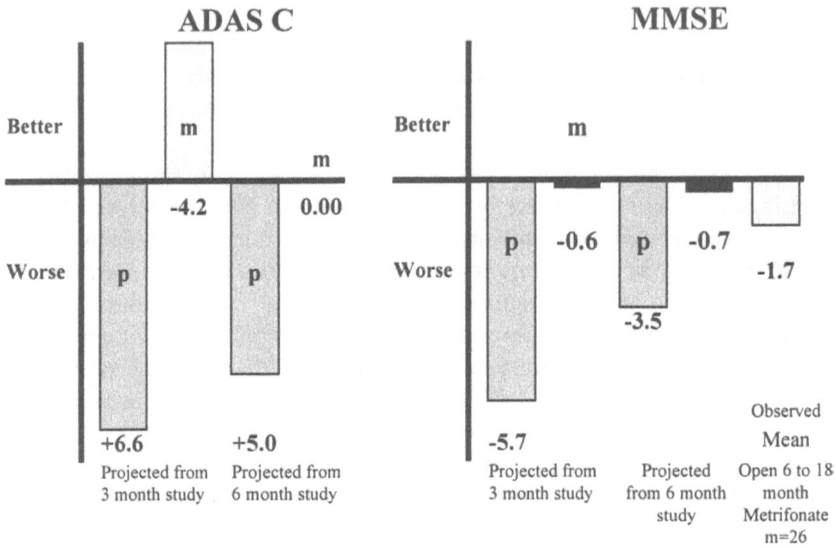

FIGURE 2. Projected and observed changes at one year from metrifonate (m) and placebo (p) treatments

from our six- to- eighteen month open metrifonate treatment of a small group of AD patients (Becker et al., 1996a), are less than the projected one year rates of decline in our placebo-treated patients. The placebo projected rates are consistent with rates of decline of untreated AD patients reported in the literature (Corey-Bloom and Thal, 1991). Each of these three studies suggests a robust, clinically significant, beneficial effect of metrifonate on cognitive decline in AD patients.

ADVERSE EFFECTS AND SAFETY

In the three and six month studies we found no difference in adverse events between the metrifonate- and placebo-treated patients, and no patients left the studies because of adverse events. We achieved targeted 40 to 60% RBC AChE inhibition levels, having used results from earlier pharmacokinetic-dynamic studies to determine the dosing regimen. In a 6- to 18-month open follow-up at 50 to 70% RBC AChE inhibition levels we found no interfering adverse effects associated with the use of metrifonate. In our studies of metrifonate in over 100 AD patients, we had only two patients who required adjustments in drug dose.

One patient developed adverse effects which resolved when the loading and maintenance dosing levels were reduced to 25% of the usual dose which achieves 50 to 70% RBC AChE inhibition in this patient. This patient's RBC AChE and plasma BuChE are more inhibited by DDVP *in vitro* than pooled enzyme and the patient's enzymes do not undergo the spontaneous reactivation seen *in vitro* prior to aging. The patient has normal levels of carboxyesterase activity in plasma compared to age matched controls. Carboxyesterase scavenges DDVP, reducing tissue concentrations. We have not yet excluded slowed DDVP metabolism as a cause of the differences in AChE inhibition in this patient. Our experience with this patient suggests a small number of patients may require lower than usual dosing with metrifonate. In these patients, effective levels of inhibition can be estimated with blood cholinesterase activity measurement.

The second patient was left to her own care when her husband was hospitalized, and she mistakenly ingested a three month supply of metrifonate over seven days without adverse effects. This wide margin of safety is consistent with the high rates of reported survival from massive overdose (WHO, 1981). If overdose with metrifonate results in toxic effects from cholinesterase inhibition, enzyme activity can be restored with 2-PAM (Moriearty and Becker, 1992). The ability to reverse inhibition and the rapid bodily clearance of DDVP means that toxic effects from overdose are partially reversible and can be expected to resolve more quickly than with drugs with long elimination half lives.

RESULT FROM CHOLINESTERASE INHIBITION

Our inability to demonstrate an immediate, robust, 25% reversal of cognitive loss at the high levels of AChE inhibition we achieved with metrifonate is consistent with the results reported with tacrine (Knapp et al., 1994) and with the cholinesterase inhibitors that have reached double blind evaluation. Reported immediate effects on cognitive loss do not significantly differ among compounds (Bieber et al., 1996; Imbimbo, 1996; Rogers and Friedhoff, 1996; Schmidt et al., 1996; Schwartz et al., 1996). We suggest that the disappointing news with AChE inhibition--that robust immediate effects on the cognitive loss in AD cannot be detected in group studies--is not a basis for discouragement. Our subjective impressions and the impressions reported to us by the patients who have participated in our studies are that immediate meaningful benefits accrue to patients and secondarily to their caretakers with metrifonate administration. Studies of drugs of this class reported in this volume document behavioral and cognitive benefits for patients with the variability among studies primarily a function of adverse events, although other pharmacokinetic and pharmacodynamic differences are also important. In our interpretation, the lack of robust, immediate clinical efficacy in this class directs the physician to avoid interfering adverse properties by selective use of drugs from this class.

RECOMMENDATIONS FOR THE USE OF CHOLINESTERASE INHIBITORS

We suggest thoughtful attention is required of the physician to achieve maximum clinical benefit for patients, due to the differences among drugs of this class. The cholinesterases currently under development differ significantly in metabolism; duration of action; rate, type and severity of adverse effects; dosing schedules; and ability to achieve and maintain steady state of enzyme inhibition. Each of these properties has implications for the treatment of patients. Dependence on hepatic metabolism increases the risks of drug interaction. Slow drug clearance complicates the management of adverse effects or overdose. Adverse events interfere with full efficacy. Adverse events range from high (30 to 60%) with tacrine (Cognex™) (Davis et al., 1992; Knapp et al., 1994) and physostigmine salicylate, medium (10 to 30%) (Schwartz et al., 1996) with ENA 713 (Anand et al., 1996) and heptylphysostigmine (Eptastigmine) (Imbimbo, 1996), to low or equivalent to rates reported by placebo-treated patients (0 to 10%) with E-2020 (Donepezil) (Rogers and Friedhoff, 1996) and metrifonate. The onset of adverse effects limits physicians to doses that achieve only 30 to 60% RBC AChE inhibition with most of these compounds. A sharp increase in the occurrence of adverse effects above the 50 to 80% levels of inhibition may preclude the use of drugs of this class in the higher range with the exception of metrifonate, for which maximum clinical effect was reported at 75% inhibition (Bieber et al., 1996) with no increase in adverse effects at higher levels of inhibition. Appearance of adverse effects with dosage increase may limit efficacy if higher doses are needed to compensate for tolerance. Short duration of action requiring more than once daily dosing complicates day to day care of AD patients who cannot remember dosing instructions. Inability to comply with dosing instructions can also lead to self administered accidental overdose of drug by memory-impaired AD patients. Therefore, a high therapeutic index, ability to reverse toxicity from overdose, and a short duration of toxicity from rapid drug clearance offer important additional safety advantages in the use of cholinesterase inhibitors in AD.

CLINICAL PROSPECTS FOR METRIFONATE

The efficacy and safety profile of metrifonate compares very favorably to other cholinesterase inhibitors. The documented advantages we find from our work with metrifonate are: the freedom from adverse effects including cholinergic symptoms; once daily dosing with the ability to administer forgotten doses without clinical dosage adjustment; the resistance to loss of clinical effect from missed doses due to its long duration of action; the greatly reduced risk of drug interactions due to metrifonate's hydrolysis to an active compound and lack of interaction with the P450 system; a wide margin of safety and the ability to compensate for loss of efficacy from tolerance by increasing dose because of the lack of adverse effects at high levels of enzyme inhibition.

THE GOOD NEWS

The good news with the cholinesterase inhibitors is the probability that a selection of inhibitors with improved properties will be available for prescription before the year 2000. The cholinesterase inhibitors uniformly show cognitive and behavioral benefits to patients in research studies and investigators report positive changes in patients' and caregivers' lives. To provide full advantage to patients, using drugs of this class, physicians need to be aware of the clinical importance of the differences in clinical properties among drugs of this class.

FUTURE PROSPECTS

Among the tasks remaining in research is the further study of this class for its possible effects on rate of decline in AD. This holds the potential for additional good news. Our 3, 6 and 12 month studies project a greater than 35% slowing in rate of cognitive decline in AD patients we have treated with metrifonate. We regard this as a robust, clinically significant effect that needs to be considered in therapeutic planning for AD patients if this effect can be supported in further long-term studies. If further evidence of substantial effects on decline is developed then study of possible effects in memory-impaired elderly and high risk persons free of clinical symptoms will be warranted.

An effect of metrifonate treatment on rate of decline in AD would also have research interest (Becker et al., 1996a). Both AChE and BuChE activities are related to processes that have been implicated in the pathology of AD (Koh et al., 1991; Nitsch et al., 1992; Buxbaum et al., 1992; Wright et al., 1993). Metrifonate may also inhibit serine proteases involved in amyloid or tau metabolism. If an effect of metrifonate on rate of decline in AD can be conclusively demonstrated, it may be informative to compare this to the effects from other cholinesterase inhibitors to determine if decline effects are related to selective BuChE or AChE inhibition or to metrifonate's activities elsewhere.

Hopefully, in this time of restricted funding, adequate support will be available to allow an orderly investigation-based determination of the full clinical effects of cholinesterase inhibition in AD. It appears relatively certain that a number of drugs from this class will become available by prescription. Rational use requires fuller understanding of their potential for beneficial effects on cognitive and behavioral decline and disease progression.

ACKNOWLEDGMENTS

This work was supported by National Institute of Mental Health Research Grants (RO1 MH1AG41821-01 and RO1 MH 41821) and by National Institute of Aging Alzheimer's Disease Center Grant (P30 AG08014), Robert E Becker, Principal Investigator.

Robert Becker holds a patent on metrifonate which is assigned to Southern Illinois University School of Medicine and has been licensed by them for commercial development.

REFERENCES

Anand R, Hartman R, Hayes P, and Gharabawi M (1996): An overview of the development of SDZ ENA 713, a brain selective cholinesterase inhibitor. In: *Alzheimer Disease: from Molecular Biology to Therapy.* Becker R and Giacobini E, eds. Boston:Birkhauser.

Bartels CF, Moriearty PL, Becker RE, Mountjoy CP and Lockridge O (1994): Acetylcholinesterase gene sequence and copy number are normal in Alzheimer's disease patients treated with the organophosphate metrifonate. *Proceedings*, Fifth International Symposium on Cholinesterases, Madras, India.

Becker R (1991): Therapy of the cognitive deficit in Alzheimer's disease: The cholinergic system. In: *Cholinergic Basis for Alzheimer Therapy.* Becker R and Giacobini E, eds. Boston, Birkhauser.

Becker R and Giacobini E (1988a): Advances in the therapy of Alzheimer's disease: Issues in the study and development of cholinomimetic therapies. In: *Current Research in Alzheimer Therapy.* Giacobini E and Becker R, eds. New York: Taylor and Francis.

Becker R and Giacobini E (1988b): Mechanisms of cholinesterase inhibition in senile dementia of the Alzheimer type. *Drug Dev Res* 12:163-195.

Becker R and Giacobini E (1988c): Pharmacokinetics and pharmacodynamics of acetylcholinesterase inhibition: Can acetylcholine levels in the brain be improved in Alzheimer's disease? *Drug Dev Res* 14:235-246.

Becker R and Markwell S (1996): A statistical view of the clinical significance of treatment in individual patients. In preparation.

Becker R, Colliver J, Elble R, Feldman E, Giacobini E, Kumar V, Markwell S, Moriearty P, Parks R, Schillcutt S, Unni L, Vicari S, Womack C and Zec R (1990): Effects of metrifonate, a long-acting cholinesterase inhibitor, in Alzheimer disease: Report of an open trial. *Drug Dev Res* 19:425-434.

Becker R, Colliver J, Markwell S, Moriearty P, Unni L and Vicari S (1996a): A double-blind, placebo-controlled study of metrifonate, an acetylcholinesterase inhibitor, for Alzheimer's disease. *Alzheimer Disease and Associated Disorders,* Vol 10, No. *3.*

Becker R, Colliver J, Markwell S, Moriearty P, Unni L and Vicari S (1996b): Effects of metrifonate on cognitive decline in Alzheimer's disease over six months. In preparation.

Becker R, Moriearty P, Surbeck R, Unni L, Varney A and Vicari S (1994): Second and third generation cholinesterase inhibitors: From preclinical studies to clinical efficacy. In: *Alzheimer Disease: Therapeutic Strategies*. Giacobini E and Becker R, eds. Boston, Birkhauser.

Bieber F, Mas J, Orazem J and Gulanski B (1996): Results of a dose finding study with metrifonate in patients with Alzheimer's disease. Abstract. 4th International Symposium on Advances in Alzheimer Therapy, Nice, France, April 10-14.

Buxbaum J, Oishi M, Chen H, et al. (1992): Cholinergic agonists and interleukin 1 regulate processing and secretion of the Alzheimer B/A4 amyloid protein precursor. *Proc Natl Acad Sci* 89:10075-10078.

Corey-Bloom J and Thal L (1991): Rate of cognitive decline in Alzheimer's disease. In: *Diagnostic and Therapeutic Assessments in Alzheimer's Disease*. Gottfries C, Levy R, Clinck G, and Tritsmen L, eds. United Kingdom: Wrightson Biomedical.

Davis KL, Thal L and Gamza ER, et al. (1992): A double-blind, placebo-controlled multicenter study of tacrine for Alzheimer's disease. *N Engl J Med* 327:1253-1259.

Folstein M, Folstein S and McHugh P (1975): Mini-mental state: a practical method for grading the cognitive state of patients for the clinician. *J Psychiat Res* 12:189-198.

Hallak M and Giacobini E (1987): A comparison of the effects of two inhibitors on brain cholinesterase. *Neuropharmacol* 26:521-530.

Hallak M and Giacobini E (1989): Physostigmine, tacrine and metrifonate - the effect of multiple doses on acetylcholine metabolism in rat brain. *Neuropharmacol* 28:199-206.

Imbimbo B (1996): Eptastigmine: A cholinergic approach to the treatment of Alzheimer's disease. In: *Alzheimer Disease: from Molecular Biology to Therapy*. Becker R and Giacobini E, eds. Boston:Birkhauser.

Knapp M, Knopman D and Solomon P et al. (1994): A 30-week randomized controlled trial of high-dose tacrine in patients with Alzheimer's disease. *JAMA* 271:985-991.

Koh J, Palmer E and Cotman C (1991): Activation of the metabotropic glutamate receptor attenuates N-methyl-D-aspartate neurotoxicity in cortical cultures. *Proc Natl Acad Sci* 88:9431-9435.

Kronforst-Collins MA, Moriearty PL, Ralph M, Becker RE, Schmidt B, Thompson LT and Disterhoft JF (1996): Metrifonate treatment enhances acquisition of eyeblink conditioning in aging rabbits. *Pharmacol Biochem Behav*, in press.

Mori F, Lai CC, Fusi F and Giacobini E (1995): Cholinesterase inhibitors increase secretion of APPs in rat brain cortex. *Neuroreport* 6:633-636.

Moriearty PL and Becker RE (1992): Inhibition of human brain and RBC acetylcholinesterase (AChE) by heptylphysostigmine (HLPL). *Meth Find Exp Clin Pharmacol* 14: 615-621.

Moriearty PL, Thornton SL and Becker RE (1993): Transdermal patch delivery of acetylcholinesterase inhibitors. *Meth Find Exp Clin Pharmacol* 15:407-412.

Moriearty PL, Womack CL, Dick BW, Colliver JA, Robbs RS and Becker RE (1991): Stability of peripheral hematological parameters after chronic acetylcholinesterase inhibition in man. *Am J Hematol* 37:280-282.

Nitsch R, Slack B, Wurtman R and Growdon J (1992): Release of Alzheimer amyloid precursor derivatives stimulated by activation of muscarinic acetylcholine receptors. *Science* 258:304-307.

Rogers S and Friedhoff L (1996): Donepezil (E2020) produces long-term clinical improvement in Alzheimer's disease. Abstract. 4th International Symposium on Advances in Alzheimer Therapy, Nice, France, April 10-14.

Rosen W, Mohs R and Davis K (1984): A new rating scale for Alzheimer disease. *Am J Psychiatry* 141:1356-1364.

Schmidt B, Hinz V, Fanelli R, Blokland A, and van der Staay F (1996): Preclinical pharmacology of metrifonate: A promise for Alzheimer therapy. In: *Alzheimer Disease: from Molecular Biology to Therapy.* Becker R and Giacobini E, eds. Boston: Birkhauser.

Schwartz G, Bodenheimer S and Thal L (1996): Extended-release physostigmine salicylate in the treatment of Alzheimer's disease. Abstract. 4th International Symposium on Advances in Alzheimer Therapy, Nice, France, April 10-14.

Unni L, Womack C, Hannant M and Becker R (1994): Pharmacokinetics and pharmacodynamics of metrifonate in humans. *Meth Find Exp Clin Pharmacol* 16:285-289.

WHO (1981): Dichlorvos, Environmental Health Criteria 79. Geneva, World Health Organization.

Wright C, Geula C and Mesulam M (1993): Neurological cholinesterases in the normal brain and in Alzheimer's disease: relationship to plaques, tangles, and pattern of selective vulnerability. *Ann Neurol* 34:373-384.

Part III

NICOTINIC AND MUSCARINIC CHOLINERGIC AGONISTS

Nicotinic Agonists
Muscarinic Agonists and Antagonists
The Potential of Antioxidant Therapy
The Potential for Anti-Inflammatory Therapy

Alzheimer Disease: From Molecular Biology to Therapy
edited by R. Becker and E. Giacobini

MOLECULAR HISTOCHEMISTRY OF NICOTINIC RECEPTORS IN HUMAN BRAIN

Hannsjörg Schröder, Andrea Wevers, Elke Happich, Ulrich Schütz and Natasha Moser
Department of Anatomy, University of Köln, D-50931 Köln, F.R.Germany

Robert A.I. de Vos, Gerard van Noort
Laboratorium Pathologie Oost Nederland, Enschede, NL-7512 AD Enschede, The Netherlands

Ernst N.H. Jansen
Medisch Spectrum Twente, Department of Neurology, NL-7500 KA Enschede, The Netherlands

Ezio Giacobini
Hôpitaux Universitaires de Genève, Dept. de Geriatrie
CH-1226 Thonex-Geneva, Switzerland

Alfred Maelicke
Institut für Physiologische Chemie und Pathobiochemie,
University of Mainz, D-55128 Mainz, F.R.Germany

INTRODUCTION

Only a decade ago the existence and functional significance of central nervous nicotinic acetylcholine receptors (nAChR) was still a subject of controversy. Today, the importance of this receptor class for signal transduction in human brain in normal and pathological conditions has become quite evident. nAChRs have turned out to be important pharmacological targets in disorders like Alzheimer's disease (AD) (Arneric et al., 1994). One prerequisite to understand nAChR function is a detailed study of the cellular distribution of nAChR subtypes. In recent years several human-specific data have been made available. This paper attempts to show actual developments in this field, summarizing the expression and localization of nAChRs in human development, age and neurodegenerative states.

Prenatal Development

Investigations on pre- and early postnatal nAChR expression in the human CNS are still in their beginnings. Recent data from animal *in vitro* studies point to an important role of nAChRs in the regulation of neuronal migration and growth cone direction (Zheng et al., 1994). It was, however, only very recently that the developmental *in vivo* expression of nAChRs has attracted much interest. Own studies on the rat CNS have shown the expression of nAChR isoform mRNAs in cells of the cortical, hippocampal and olfactory bulb neuroepithelium very early in prenatal development (E14) (Ostermann et al., 1995; Moser et al., 1996).

The human frontal cortex (A9) (Court et al., 1992), and hippocampus (CA1) (Court and Clementi, 1995), displays [^{3}H]nicotine binding sites in the 22nd week of gestation. Their concentration follows a distinct, region-specific time course during further fetal development, decreasing until birth by approximately 60% in the CA1 region, but remaining almost unchanged in A9. Own preliminary *in situ* hybridization data show α4 mRNA in the human cortex in the 19th week in many neuroepithelial and cortical plate cells (Fig. 1a,b). At birth, the distribution pattern resembles that of adult cerebral cortex (vide infra) (Fig. 1c,d).

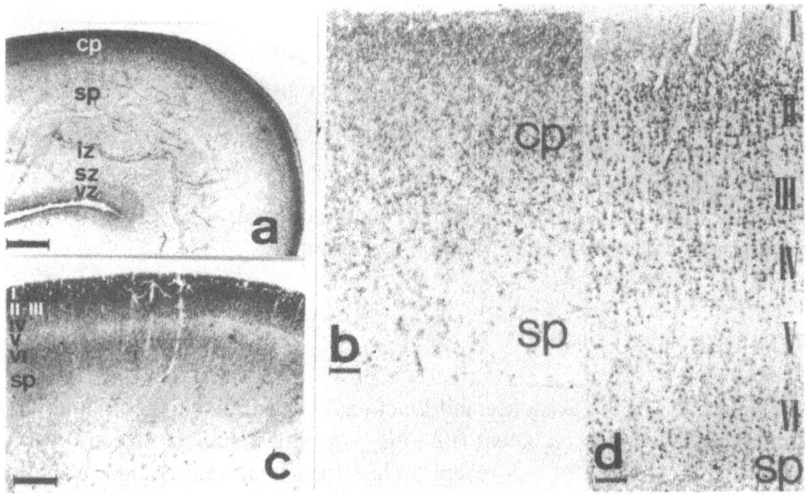

FIGURE 1. Non-isotopic *in situ*-hybridization for α4 nAChR transcripts in the developing human frontal neocortex using a digoxigenin-labeled riboprobe. 1a,b. 19th week. Note the transcript-expressing cells in the ventricular zone (vz) (survey 1a) and the cortical plate (cp). 1c,d. At birth, the distribution of mRNA-expressing neurons resembles very much that of adult age (cf. Table I). sz = subventricular zone iz = intermediate zone sp = subplate. Calibration bars: a: 600 µm b: 40 µm c: 540 µm d: 64 µm

Area	A10	A10	A10	A4	A4	A4
Layer	α3	α4	α7	α3	α4	α7
I	(+)	(+)	+	(+)	(+)	+
II	+-++	++-+++	++-+++	+-++	+-++	++-+++
III	+-++	++-+++	++-+++	+-++	+-++	++-+++
IV	(+)	(+)	(+)	(+)	(+)	(+)
V	+++	+++	++	+++	+++	++
VI	+++	+++	++	+++	+++	++

TABLE I. Cellular distribution of nAChR mRNA in the frontal cortex of adult man

It is noteworthy that changes in the expression of the presynaptic marker choline acetyltransferase, during both fetal and postnatal development, almost never parallel those of nicotinic binding sites (Court et al., 1992). This may be explained by the fact that cholinergic terminals on the postsynaptic site are facing not only nicotinic but also muscarinic receptors. A complete understanding of developmental changes at central cholinergic synapses requires a currently not available account of all receptor subunits and subtypes involved.

Postnatal Development and Aging

In postnatal life, decrements in binding sites are only small in the CA1 region, but amount to 46% from birth to age 50, and 48% between the 59th and 89th year of life. In the ninth decade, levels reach a minimum in both CA1 and A9 (Court et al., 1992; Court and Clementi, 1995)

The distribution of α3, α4, and α7 nAChR subunit mRNAs in different human cortical areas is shown in Table I. α4 transcripts are expressed in numerous neurons of all layers, α3 signals are present in a smaller number of neurons mainly in deeper layers. Transcripts for the α-bungarotoxin-sensitive subunit α7 are seen in numerous, mainly pyramidal-like neurons and rather many layer I neurons (Wevers et al., 1995). Only a few A4 giant pyramidal neurons express α7 mRNA.

The layer-specific distribution of nicotine binding sites has been shown to be mainly to the deeper cortical layers by means of receptor autoradiography.

On the cellular level using an immunohistological approach, mainly pyramidal neurons of layers II/III and V displayed α-nAChR immunoreactivity. Studies are underway in our laboratory to explore the cellular and subcellular distribution of nAChRs in more detail using subunit-specific antibodies.

In aging brains (55th to 75th year) a decrease by 50% in the density of nAChR-expressing neurons in human frontal cortex has been shown using antibodies to the nAChR α-subunit (Schröder, 1992). Immunolabel was found mainly in layer II/III and V pyramidal neurons which is in agreement with our previous results. This decrease is accompanied by a 22% loss of [3H]nicotine binding in the A9 region between the 50th and 70th year of life (Court et al., 1992).

Neurodegenerative diseases

The decrease of nAChR expression is dramatic in neurodegenerative diseases like Alzheimer's disease (AD) or Parkinson dementia (PDD). Decreases in nicotinic cerebrocortical binding up to 70% compared to age-matched controls have been described (cf. Schröder, 1992). Using [^{3}H]acetylcholine as ligand, Whitehouse et al. (1988) found reduced densities of binding sites in all cortical layers in AD as compared to controls in frontal, temporal and occipital cortices.

In AD patients, nAChR-immunoreactivity was found mainly in layer II/III and V pyramidal neurons. The cortices showed marked decreases of nAChR-protein expressing neurons as compared to age-matched controls. The loss of immunoreactive neurons was not confined to a particular layer but involved all layers. Neuron density did not show significant differences. Taking into account the pattern of muscarinic receptor (mAChR) protein expression in AD cortices (Schröder, 1992) the results suggest a selective impairment of nAChR protein expression, consistent with nicotine binding loss in AD (Whitehouse et al., 1988).

At the mRNA level, as assessed by *in situ* hybridization, the most common nAChR subunit, a4, does not appear to be decreased in PDD (Schröder et al., 1995) and AD (Own unpublished observations).

Conclusions and Perspectives

Nicotinic acetylcholine receptors are widespread in the human brain, closely related to cognitive functions reflected by their preferential localization in the human cerebral cortex. Histochemical studies have demonstrated differential distributions of several nAChR subunits (α3, α4, α7) at the mRNA and partly at the protein level. Mainly pyramidal neurons are equipped with these kind of receptors.

The expression of cortical nicotine binding sites is markedly decreased in neurodegenerative diseases, related to the loss of cognitive function. The deficit in nAChR expression may be due to disturbances at the level (1) of subunit transcription, (2) protein synthesis and (3) receptor transport and membrane insertion.

Currently, we are testing these different possibilities. While mRNA expression of the a4 subunit does not appear to be decreased in neurodegeneration (vide supra), accumulation of proteins like hyperphosphorylated tau and b-amyloid may hinder the transport of nAChRs. Investigations are underway in our laboratory to analyze the expression of other nAChR subunits and their relation to intraneuronal transport.

ACKNOWLEDGMENTS

This study was supported by the Deutsche Forschungsgemeinschaft, the Deutsch-Italienische Stiftung für Augen- und Hirnforschung / Co-Foundation of the Alexander von Humboldt-Stiftung and the Deutsche Parkinsonvereinigung. We are greatly indebted for excellent technical assistance to S. Nowacki, K. Pilz and H. Begthel and for the secretarial work to A. Kirchmayer.

REFERENCES

Arneric SP, Sullivan JP, Decker MW, Brioni JD, Briggs CA, Donnelly-Roberts D, Marsh KC, Rodrigues AD, Garvey DS, Williams M, Buccafusco JJ (1994): ABT-418: A novel cholinergic channel activator (ChCA) for the potential treatment of Alzheimer's disease. In: *Alzheimer Disease: Therapeutic Strategies*, Giacobini E, R Becker, eds., Boston: Birkhäuser, pp. 196-200.

Court J, Piggott MA, Perry EK, Barlow RB, Perry RH (1992): Age associated decline in high affinity nicotine binding in human brain frontal cortex does not correlate with the changes in choline acetyltransferase activity. *Neurosci Res Comm* 10:125-133.

Court J, Clementi F (1995): Distribution of nicotinic subtypes in human brain. *Alzheimer Dis Assoc Disorders* 9 (Suppl. 2):6-14.

Moser N, Wevers A, Lorke DE, Reinhardt S, Maelicke A, Schröder H (1996): α4-1 subunit mRNA of the nicotinic acetylcholine receptor in the rat olfactory bulb:cellular expression in adult, pre- and postnatal stages. *Cell Tissue Res* (in press)

Ostermann C-H, Grunwald J, Wevers A, Lorke DE, Reinhardt S, Maelicke A, Schröder H (1995): Cellular expression of α4 subunit mRNA of the nicotinic acetylcholine receptor in the developing rat telencephalon. *Neurosci Lett* 192:1-4.

Schröder H (1992): Immunohistochemistry of cholinergic receptors. *Anat Embryol* 186:407-429.

Schröder H, de Vos RAI, Jansen EH, Birtsch Ch, Wevers A, Lobron Ch, Nowacki S, Schröder R, Maelicke A (1995): Gene expression of the nicotinic acetylcholine receptor α4 subunit in the frontal cortex in Parkinson's disease patients. *Neurosci Lett* 187:173-176.

Wevers A, Arneric S, Giordano T, Birtsch Ch, Nowacki S, Schröder H (1995): Cellular distribution of the mRNA for the α7 subunit of the nicotinic acetylcholine receptor in the human cerebral cortex - A non-isotopic in situ hybridization study. *Drug Devel Res* 36:103-110.

Whitehouse PJ, Martino AM, Wagster MV, Price DL, Mayeux R, Atack JR, Kellar KJ (1988): Reductions in [^{3}H]nicotinic acetylcholine binding in Alzheimer's disease and Parkinson's disease: an autoradiographic study. *Neurology* 38:720-723.

Zheng JQ, Felder M, Connor JA, Poo M-M (1994): Turning of growth cones induced by neurotransmitters. *Nature* 368:140-144.

Alzheimer Disease: From Molecular Biology to Therapy
edited by R. Becker and E. Giacobini

THE NICOTINIC CHOLINERGIC SYSTEM AND ß-AMYLOIDOSIS

Jennifer A Court[1], Stephen Lloyd[1], Robert H. Perry[2], Martin Griffiths[1], Christopher Morris[1], Mary Johnson[1], Ian G. McKeith[3], Elaine K. Perry[1]
[1]MRC Neurochemical Pathology Unit
[2]Department of Neuropathology
[3]Department of Old Age Psychiatry
Newcastle General Hospital, Newcastle upon Tyne, UK

INTRODUCTION

Nicotinic cholinergic receptors in the CNS include a family of cationic channels consisting of a variety of α (α_2-α_9) and ß(ß$_2$-β_4) subunits. Although the pharmacology of individual combinations of subunits has not yet been fully elucidated, high affinity acetylcholine or nicotine binding is predominantly associated with the $\alpha_4\beta_2$ combination and high affinity α-bungarotoxin (αBT) binding with α_7 subunits (Nakayama et al., 1991; Flores et al., 1992; Clarke, 1992).

Increasing evidence linking the $\alpha_4\beta_2$ receptor subtype with the pathology of Alzheimer's disease (AD) suggests that down regulation of this receptor promotes ß-amyloidosis:

- Nicotine or epibatidine, but not αBT, binding is diminished in the cortex in normal aging and in AD (Warpman and Nordberg, 1995; Perry et al., 1995; Griffiths et al., in preparation).
- Nicotine binding in the human cerebral cortex is concentrated in the entorhinal cortex - an area particularly susceptible to Alzheimer-type pathology (Perry et al., 1992).
- Tobacco smoking, which like nicotine results in elevated nicotine binding in the brain, reduces the risk of developing AD in most studies (van Duijn and Hofman, 1991; Lee, 1994).
- In normal aging, nicotine binding is lower in individuals with, as opposed to without, ß amyloid plaques in the entorhinal cortex (Perry et al., 1996).

We have further explored the nicotinic cholinergic 'hypothesis' in AD by comparing neurochemical and neuropathological indices in a series of normal elderly individuals with established tobacco smoking (or non-smoking) histories.

METHODS

Three groups of normal elderly individuals with no neurological or psychiatric history included smokers, ex-smokers and non-smokers matched for mean age and postmortem delay. There were gender differences between the groups, non-smokers being predominantly female and smokers male although ex-smokers were equally represented. Smoking histories were established by retrospective case-note assessment or direct contact with relatives and only cases with clear-cut evidence on tobacco use were included. At autopsy, the right hemisphere was formalin fixed for neuropathological assessment, including quantitation of senile plaques stained with Von Braunmühl in 6 areas derived from the 4 cortical lobes (Perry et al., 1990). Coronal sections of the left hemisphere were flash frozen in liquid arcton, stored at -70°C and blocks of hippocampal and adjacent cortical tissue at or around the level of the lateral geniculate nucleus subdissected for cryostat sectioning and autoradiographic analysis of [^{3}H] nicotine and [^{125}I] αBT receptor binding (Clarke et al., 1985) and choline acetyltransferase.

RESULTS AND DISCUSSION

As previously reported (Benwell et al., 1988), nicotine receptor binding was elevated in smokers compared to non-smokers (Figure 1). There was no such elevation in ex-smokers, consistent with the relatively short term influence of nicotine in upregulating receptor binding. The elevation of binding in smokers was evident throughout the hippocampus, subiculum and entorhinal cortex being most pronounced in dentate granule layer. By contrast, there was no alteration of αBT binding in smokers compared with non-smokers (Figure 1). In addition to elevated nicotine binding, choline acetyltransferase tended to be higher in the hippocampus of smokers and ex-smokers compared with non-smokers (Figure 1).

Smokers were distinguished from non-smokers by the density of senile plaques (Figure 1). Plaques were on average 10 x higher in the non-smoking group, a difference that reached statistical significance in frontal and parietal cortex; and in parietal cortex ex-smokers were also significantly higher than non-smokers. The scatter was wide in each group, indicating heterogeneity in the relationship. This was not due to apoE genotype which bore no relation to plaque density or nicotine binding in any group.

The neuropathological data reported here confirm and extend previous findings on a smaller series (Perry and Perry, 1993) and are also consistent with a preliminary report that Braak staging (assessed using tangle density in hippocampal CA1 and plaque area) correlated inversely with the extent of smoking expressed in pack years in non-demented and demented individuals (Ulrich et al., 1994). Whilst our findings are suggestive of a protective effect of tobacco smoking on the evolution of ß-amyloidosis in the

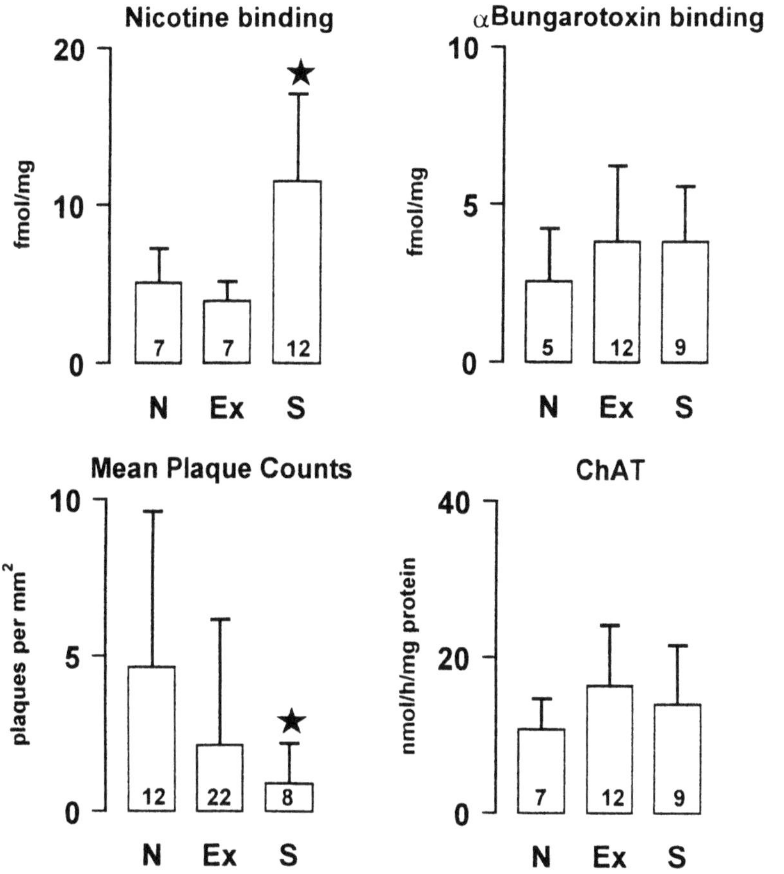

Figure 1. Comparison of nicotinic receptor binding, neocortical plaque density and choline acetyltransferase (ChAT) in normal individuals (over the age of 63), grouped as non-smokers (N), ex smokers (Ex) and smokers (S) matched for mean age. Receptor binding was measured in entorhinal cortex, ChAT in hippocampus and plaques as mean density obtained from 4 cortical areas. Significant differences between the groups were as follows: Nicotine binding in smokers versus non-smokers and ex-smokers ($p = 0.03$), plaques in smokers versus non-smokers ($p = 0.03$), choline acetyltransferase in smokers and ex smokers versus non-smokers ($p = 0.05$).

aging human brain, they need to be interpreted with caution in view of the gender differences between the groups. However, amongst ex-smokers there was no evidence of lower plaques in males compared with females.

An obvious mechanism of neuroprotection in smokers is through the action of nicotine, although possible actions of other components of tobacco smoke need to be considered. It has recently been reported that monoamine oxidase B is reduced in the brains of smokers (Fowler et al., 1996). Protective effects of monoamine oxidase B inhibition may involve reduced levels of hydrogen peroxide (a by-product of monoamine oxidation) and it has been proposed that formaldehyde and cyanide in smoke may reduce catalytic activity of the enzyme. To resolve the issue of whether nicotine is the active component in protection against AD, long term effects of the administration of nicotine or a related nicotinic agonist on the development and progress of the disease need to be investigated.

In pursuing the hypothesis that nicotinic receptor stimulation is protective against the pathological process in AD possible physiological mechanisms need to be considered. Nicotine enhances fast excitatory transmission in the CNS via presynaptic receptors, including the α_7 subunit (McGehee et al., 1995) and enhanced glutamate transmission (via NMDA) potentiates NGF-induced sprouting of septal cholinergic fibres (Heisenberg et al., 1994). Nerve growth factor also induces cleavage and release of extracellular portions of amyloid precursor protein (APP) in PC12 cells (Robakis et al., 1991). Thus, nicotinic stimulation may promote the normal secretase as opposed to Aß pathway of APP metabolism. Determination of glutamate and NGF-related activities would be of value in the present series.

In summary, the present data provides additional support for nicotinic therapy in AD (Court and Perry, 1994).

ACKNOWLEDGMENTS

This research was partially funded by Schering, Berlin and the British American Tobacco Company.

REFERENCES

Benwell MEM, Balfour DJK, Anderson JM (1988): Evidence that tobacco smoking increased the density of (-)-[^{3}H] nicotine binding sites in human brain. *J Neurochem* 50: 1243-1247.

Clarke PBS (1992): The Fall and Rise of α -bungarotoxin binding proteins. *Trends Pharmacol Sci* 13:407-413.

Clarke PBS, Schwartz RD, Paul SM, Pert CB, Pert A (1985): Nicotine binding in rat brain: autoradiographic comparison of [^{3}H] acetylcholine, [^{3}H] nicotine and [^{125}I] α-bungarotoxin. *J Neurosci* 5: 1307-1315.

Court JA, Perry EK (1994): CNS nicotinic receptors: therapeutic target in neurodegeneration. *CNS Drugs* 2: 216-233.

Flores CM, Roger SW, Pabreza LA, Wolfe BB, Kellar KJ (1992): A subtype of nicotinic cholinergic receptor in rat brain is composed of $\alpha_4\beta_2$ subunits and is upregulated by chronic nicotine treatment Mol. *Pharmacol* 41: 31-37.

Fowler JS, Volkow ND, Wang GJ, Pappas N, Logan J, MacGregor R, Alexoff D, Shea C, Schlyer D, Wolf AP, Warner D, Zezulkova I, Cilento R (1996): Inhibition of monoamine oxidase B in the brains of smokers. *Nature* 379: 733-736.

Heisenberg CP, Cooper JD, Berke J, Sofroniew MV (1994): NMDA potentiates NGF-induced sprouting of septal cholinergic fibres. *Neuroreport* 5: 413-416.

Lee PN (1994): Smoking and Alzheimer's disease: a review of the epidemiological evidence. *Neuroepidemiol* 13: 131-144.

McGehee DS, Heath MJS, Gelber S, Devay P, Role LW (1995): Nicotine enhancement of fast excitatory synaptic transmission in CNS by presynaptic receptors. *Science* 269: 1692-1696.

Nakaya H, Nakashima T, Kurogochi Y (1991): α_4 is a major acetylcholine binding subunit of cholinergic ligand affinity-purified acetylcholine receptor from rat brains. *Neurosci Lett* 121: 122-124.

Perry EK, Court JA, Johnson M, Piggott MA, Perry RH (1992): Autoradiographic distribution of (^{3}H) nicotine binding in human cortex: relative abundance in subicular complex. *J Chem Neuroanat* 5: 399-405.

Perry EK, Court JA, Lloyd S, Johnson M, Griffiths MH, Spurden D, Piggott MA, Turner J, Perry RH (1996): ß-amyloidosis in normal aging and transmitter signalling in human temporal lobe. *Annals New York Acad Sci* 777: in press.

Perry EK, Morris CM, Court JA, Cheng A, Fairbairn AF, McKeith IG, Irving D, Brown A, Perry RH (1995): Alteration in nicotine binding sites in Parkinson's disease, Lewy body dementia and Alzheimer's disease: possible index of early neuropathology. *Neuroscience* 64: 385-395.

Perry EK, Perry RH (1993): Neurochemical pathology and therapeutic strategies in degenerative dementia. *Int Rev Psychiat* 5: 363-380.

Perry RH, Irving D, Blessed G, Fairbairn AF, Perry EK (1990): Senile dementia of Lewy body type: a clinically and pathologically distinct form of Lewy body dementia in the elderly. *J Neurol Sci* 95: 119-139.

Robakis NK, Anderson JP, Refolo LM, Wallace W (1991): Expression of the Alzheimer amyloid precursor in brain tissue and effects of NGF and EGF on its metabolism. *Clin Neuropharmacol* 14: 515-523.

Ulrich J, Stähelin HB, Johannson G, Orantes M (1994): Does smoking protect from Alzheimer's disease? An epidemiological study based on neuropathology. *Neurobiol Aging* 15: Suppl 1: 546.

Van Duijn CM, Hofman A (1991): Relationship between nicotine intake and Alzheimer's disease. *Brit Med J* 302: 1491-1494.

Warpman U, Nordberg A (1995): Epibatidine and ABT 418 reveal selective losses of $\alpha_4\beta_2$ nicotinic receptors in Alzheimer brains. *Neuroport* 6: 2419-2423.

Alzheimer Disease: From Molecular Biology to Therapy
edited by R. Becker and E. Giacobini

RJR-2403: A CNS-SELECTIVE NICOTINIC AGONIST WITH THERAPEUTIC POTENTIAL

Patrick M. Lippiello, Merouane Bencherif, William S. Caldwell, Sherry R. Arrington, Kathy W. Fowler, M. Elisa Lovette and Leigh K. Reeves
Research and Development, R.J. Reynolds, Winston-Salem, NC 27102

INTRODUCTION

In recent years molecular cloning studies have resulted in the discovery of multiple genes encoding a variety of structural and functional subunits for nicotinic cholinergic receptors (nAChR) (Galzi and Changeux, 1995), 11 having neuronal localization (α2-α9, β2-β4) and 5 expressed in skeletal muscle (α1, β1, γ, δ, ε). This molecular diversity opens up the possibility of complex receptor subtype expression both in the autonomic (PNS) and central (CNS) nervous systems, possibly explaining the complex pharmacological effects of nicotine and other classic nicotinic cholinergic agonists that do not discriminate among the various receptor subtypes expressed in the CNS and PNS. The importance of understanding this complexity is emphasized by the growing body of evidence suggesting that compromised CNS nicotinic cholinergic neurotransmission may play a key role in a variety of CNS and PNS pathologies. In this regard, there is growing interest in the use of nicotinic agonists for the treatment of Alzheimer's disease (AD) (Williams et al., 1994). One consistent feature of this disease is a decline in the function of cholinergic systems. The loss of neurons that release acetylcholine, a key neurotransmitter in learning and memory mechanisms, initially motivated an intense but disappointing search for replacement therapies targeting muscarinic receptors. On the other hand, epidemiological studies have reported an inverse correlation between the incidence of AD and smoking (van Duijn and Hofman, 1991), supporting the notion that nicotinic pharmacology underlies an apparent protective effect. A nicotinic hypothesis for the etiology and treatment of AD is further supported by observations that nicotine treatment of AD patients via subcutaneous injection (Sahakian et al., 1989) or patch (Wilson et al., 1995) improves attention and learning.

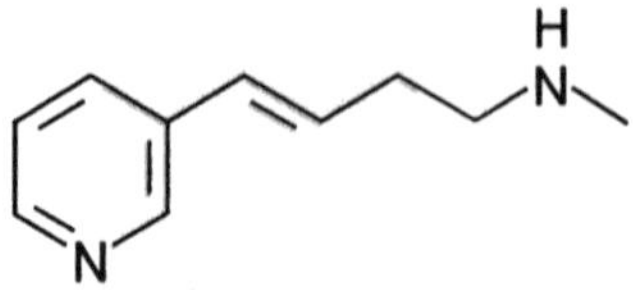

RJR-2403

Because nicotine and most classical nicotinic agonists demonstrate complex CNS and PNS pharmacology, compounds with CNS selectivity might have the beneficial property of acute therapeutic efficacy and reduced peripheral effects. Toward this end, we have synthesized a number of nicotinic compounds and tested their selectivity for various nAChR subtypes. Our results indicate that RJR-2403 [(E)-N-methyl-4-(3-pyridinyl)-3-butene-1-amine fumarate] is a nicotinic agonist with selectivity for a subset of nAChR in the brain. Thus, RJR-2403 could be a very useful research tool to study biological effects mediated by specific nAChR subtypes and a potential lead compound for developing therapeutics to treat AD or other conditions involving compromised nicotinic neurotransmission.

METHODS

Synthesis

The fumarate salt of (E)-N-methyl-4-(3-pyridinyl)-3-butene-1-amine (RJR-2403) was prepared by Dr. Peter Crooks (University of Kentucky).

In Vitro Studies

The following battery of *in vitro* assays was used to evaluate RJR-2403's nAChR selectivity vs nicotine: [^{3}H]-nicotine binding to rat brain membranes (Lippiello and Fernandes, 1986); [^{125}I]-α-bungarotoxin binding to mouse brain membranes (Marks and Collins, 1982); $^{86}Rb^{+}$ efflux from thalamic synaptosomes (Marks et al., 1993); upregulation of high affinity nAChR in M10 cells and primary neuronal cultures (Bencherif et al., 1995a); [^{3}H]-dopamine release from striatal synaptosomes (Grady et al., 1992); cellular $^{86}Rb^{+}$ efflux from PC-12 and TE671 cells (Lukas, 1991); contraction of isolated guinea-pig ileum (Bencherif et al., 1995b).

In Vivo Studies

Analyses of dopamine (DA), acetylcholine (ACh), norepinephrine (NE) and 5-HT release in rat brain cortex were assessed using *in vivo* microdialysis (Summers et al., 1995). Cognitive effects were measured using two paradigms: reversal of scopolamine-induced amnesia in rats using a step through passive avoidance task (Lippiello et al., 1995); improvement of radial arm maze performance in rats with ibotenic acid lesions of the forebrain cholinergic projection system (Hodges et al., 1991). Effects on body

nAchR Subtypes[1]	$\alpha_4\beta_2$				$\alpha_3\beta x$		α_7	$\alpha_3\beta_4$		$\alpha_1\beta_1\gamma\delta$		$\alpha_3\beta_2$
Assays	[^{3}H]-Nicotine Binding	Rb$^+$ Efflux (Synaptosomes)		α4β2 Upregulation	Dopamine Release (Synaptosomes)		[^{125}I] α-BTX Binding	Rb$^+$ Efflux (PC-12 Cells)		Rb$^+$ Efflux (TE671 Cells)		GP Ileum Contraction
	Ki (nM)	BC50 (nM)	Bmax (% Nicotine)	(% Control)	BC50 (nM)	Bmax (% Nicotine)	Ki (uM)	BC50 (uM)	Bmax (% control)	BC50 (uM)	Bmax (% control)	BC30 (uM)
Nicotine	4	591	100	250	100	100	0.6	20	100	60	100	2
RJR-2403	26	732	91	250	938	82	36	>1000	0	> 1000	0	15
RJR-2403 Relative Effect	1/6	1	1	1	1/9	1	1/60	< 1/50	0	< 1/17	0	1/8

TABLE Ia. *In Vitro* receptor selectivity profile of RJR-2403 vs nicotine

temperature (BT), y-maze activity (YM), acoustic startle (AS) and respiration rate (RR) were measured according to Marks et al. (1985). Blood pressure (BP) and heart rate (HR) were determined according to Lippiello et al. (1995).

RESULTS AND DISCUSSION

In Vitro Studies

The *in vitro* effects of RJR-2403 are summarized in Table Ia. Based on its binding to high affinity nAChR in rat brain, RJR-2403 shows very good selectivity for this brain receptor subtype, which contains α4β2 subunits. RJR-2403 binds poorly to ^{125}I-α-bungarotoxin sites in rat brain thought to represent α7-containing nAChRs. Similar to nicotine, RJR-2403 upregulates high affinity (α4β2) receptors in rat brain and M10 cells, seen as an increased B_{max} for [^{3}H]-nicotine binding. RJR-2403 potently activates ^{86}Rb$^+$ efflux from thalamic synaptosomes, an assay proposed to measure the function of α4β2-containing nAChRs (Marks et al., 1993). RJR-2403 is approximately 10-fold less potent than nicotine in eliciting DA release from striatal synaptosomes, believed to reflect interaction with an α3-containing nAChR (Grady et al., 1992). RJR-2403 does not activate human muscle type (α1β1δγ) nAChRs in the TE671 cell line, predicting reduction of potential side effects mediated by receptors at the neuromuscular junction *in vivo.* RJR-2403 does not activate ^{86}Rb$^+$ efflux through nAChRs in PC-12 cells. This cell line expresses α3β4-containing receptors, a subunit pattern similar to that reported for sympathetic ganglia, whose activation *in vivo* may contribute to nicotine's cardiovascular effects. RJR-2403 partially activates nAChRs involved in ileum contraction, being approximately 8 times less potent and 5 times less efficacious than nicotine. This low selectivity for nAChRs in the GI tract suggests that RJR-2403 would minimally stimulate parasympathetic ganglia, thus reducing nicotinically-stimulated GI effects such as nausea and vomiting.

In Vivo Studies

The *in vivo* effects of RJR-2403 are summarized in Table Ib. Similar to nicotine, RJR-2403 increases the release of ACh, DA, NE and 5-HT from 50

Test	In Vivo Microdialysis Peak Release (% of Control)				Passive Avoidance Latency (sec)	Radial Arm Maze (Mean Errors/Trial)				Physiological Test Battery EC50 (umoles/kg) For Decreases					Cardiovascular EC30 (umoles/kg) For Increases	
						Place		Cue								
	Ach	DA	NE	5-HT	+ SCOP	WM	RM	WM	RM	BT	RR	YM Cross	YM Rear	AS	HR	BP
Controls	100	100	100	100	20	0.31	1.76	0.39	1.37	-	-	-	-	-	-	-
Nicotine	165	200	215	ns	135	ns	1.10	ns	ns	6	4	4	5	5	0.7	0.5
RJR-2403	190	150	150	170	125	0.10	1.40	0.05	0.60	130	75	90	90	135	6.0	10.0

TABLE Ib. Physiological and behavioral effects of RJR-2403 and nicotine; ns = not significantly different from controls; all others, $p<0.01$;[1]Putative receptor subtypes based on literature references

to 90% in rat brain cortex. It has been proposed that such nicotinic neuromodulatory effects are mediated by pre-synaptic $\alpha4\beta2$-containing nAChRs (Wonnacott et al., 1990). RJR-2403 (0.1 mg/kg s.c.) is more potent than nicotine in reversing scopolamine-induced amnesia in a step-through passive avoidance test in rats. At this dose of nicotine given subcutaneously, it has been shown that nicotine concentrations in the brain are in the 40 ng/ml (ca. 250 nM) range (Reavill et al., 1990). Thus, based on relative potencies for receptor activation *in vitro* (Table Ia), RJR-2403's effects on this behavior should be mediated primarily by $\alpha4\beta2$-containing receptors. RJR-2403 significantly improves working memory (WM) in lesioned animals, reference memory (RM) in lesioned and control animals, and both WM and RM in control animals at concentrations of 0.1-0.8 mg/kg s.c. Nicotine showed significant effects only for RM. Over this broad concentration range, many CNS nAChR subtypes could be involved. RJR-2403 is significantly less potent than nicotine in causing physiological depression of BT, RR, YM activity and AS, all of which have CNS as well as PNS components. RJR-2403 elicits dose-related increases in both HR and BP, but is significantly less potent than nicotine, consistent with its low efficacy *in vitro* at ganglionic ($\alpha3\beta4$) nAChRs.

CONCLUSIONS

The results of *in vitro* studies indicate that, relative to nicotine, RJR-2403 maintains selectivity for the primary high affinity receptor subtype in the brain ($\alpha4\beta2$), but has significantly lower selectivity for the primary ganglionic ($\alpha3\beta4$), muscle ($\alpha1\beta1\gamma\delta$), and α-bungarotoxin ($\alpha7$) receptor subtypes, as well as those in brain dopaminergic systems ($\alpha3\beta x$) and GI tract ($\alpha3\beta2$). This selectivity profile suggests that the overall rank order potency for RJR-2403 at nAChR subtypes is: CNS ($\alpha4\beta2 > \alpha3\beta2$) >> PNS ($\alpha3\beta4 = \alpha1\beta1\gamma\delta \simeq 0$). RJR-2403's selectivity for CNS vs PNS receptor subtypes correlates well with cognitive improvements seen in amnestic and brain lesion models of impaired memory. RJR-2403's low selectivity for ganglionic and muscle-type nicotinic receptors is predictive of lower cardiovascular effects (i.e., changes in HR, BP) and lower potency in causing physiological depression (e.g. decreased

BT, RR, locomotor activity), compared to nicotine. Based on these findings, RJR-2403 will be a very useful tool for characterizing the biological roles of specific nAChR subtypes and may be a good lead candidate for developing CNS-selective nicotinic agonists with therapeutic potential.

ACKNOWLEDGMENTS

The authors wish to thank Drs. Ezio Giacobini and Kathleen Summers for the microdialysis studies, Dr. Jeffrey Gray et al. for the radial arm maze studies, and Dr. Allan C. Collins for the physiological test battery studies.

REFERENCES

Bencherif M, Fowler K, Lukas RJ and Lippiello PM (1995a): Mechanisms of up-regulation of neuronal nicotinic acetylcholine receptors in clonal cell lines and primary cultures of fetal rat brain. *J Pharmacol Exp Ther* 275: 987-994.

Bencherif M, Lovette ME, Arrington S, Fowler K, Reeves L, Caldwell WS and Lippiello PM (1995b): Metanicotine: a nicotinic agonist with CNS selectivity - *in vitro* characterization. *Abst Soc Neurosci* 21: 605.

Galzi JL and Changeux JP (1995): Neuronal nicotinic receptors; Molecular organization and regulations. *Neuropharmacology* 34: 563-582.

Grady SR, Marks MJ, Wonnacott S and Collins AC (1992): Characterization of nicotinic receptor-mediated [3H]-dopamine release from synaptosomes prepared from mouse striatum. *J Neurochem* 59: 848-856.

Hodges H, Allen Y, Sinden J, Lantos PL and Gray JA (1991): Effects of cholinergic-rich neural grafts on radial maze performance of rats after excitotoxic lesions of the forebrain projection system - II. Cholinergic drugs as probes to investigate lesion-induced deficits and transplant-induced functional recovery. *Neuroscience* 45: 609-623.

Lippiello PM and Fernandes KG (1986): The binding of L-[3H] nicotine to a single class of high affinity sites in rat brain membranes. *Mol Pharmacol* 29: 448-454.

Lippiello PM, Gray JA, Peters S, Grigoryan G, Hodges H, Summers K, Giacobini E and Collins AC (1995): Metanicotine: A nicotinic agonist with CNS selectivity - in vivo characterization. *Abst Soc Neurosci* 21: 605.

Lukas RJ (1991): Effects of chronic nicotinic ligand exposure on functional activity of nicotinic acetylcholine receptors expressed by cells of the PC12 rat pheochromocytoma or the TE671/RD human clonal lines. *J Neurochem* 56: 1134-1145.

Marks MJ and Collins AC (1982): Characterization of nicotine binding in mouse brain and comparison with the binding of α-bungarotoxin and quinuclidinyl benzilate. *Mol Pharmacol* 22: 554-564.

Marks MJ, Romm E, Bealer S, and Collins AC (1985): A test battery for measuring nicotine effects in mice. *Pharmacol Biochem Behav* 23:325-330.

Marks MJ, Farnham DA, Grady SR and Collins AC (1993): Nicotinic receptor function determined by stimulation of rubidium efflux from mouse brain synaptosomes. *J Pharmacol Exp Ther* 264: 542-552.

Reavill C, Walther B, Stolerman, IP and Testa B (1990): Behavioral and pharmacokinetic studies on nicotine, cytisine and lobeline. *Neuropharmacol.* 29: (619-624).

Sahakian B, Jones G, Levy R, Gray JA and Warburton D (1989): The effects of nicotine on attention, information processing, and short-term memory in patients with dementia of the Alzheimer type. *Br J Psychiatry* 154:797-800.

Summers KL, Lippiello PM, Verhulst S and Giacobini E (1995): 5-Fluoronicotine, noranhydroecgonine and pyridyl-methylpyrrolidine release acetylcholine and biogenic amines in rat cortex in vivo. *Neurochem Res* 20: 1089-1094.

van Duijn CM and Hofman A (1991): Relation between nicotine intake and Alzheimer's disease. *Brit Med J* 302: 1491-1494.

Williams M, Sullivan JP and Arneric SP (1994): Neuronal nicotinic acetylcholine receptors. *Drug News & Perspectives* 7: 205-223.

Wilson AL, Langley LK, Monley J, Bauer T, Rottunda S, McFalls E, Kovera C and McCarten JR (1995): Nicotine patches in Alzheimer's disease:pilot study on learning, memory, and safety. *Pharmacol Biochem Behav* 51:509-514.

Wonnacott S, Drasdo A, Sanderson E, and Rowell P (1990): Presynaptic nicotinic receptors and the modulation of transmitter release. In The Biology of Nicotine Dependence. Bock G and Marsh J, eds., John Wiley & Sons, Chichester. pp. 87-101.

Alzheimer Disease: From Molecular Biology to Therapy
edited by R. Becker and E. Giacobini

ABT-089: AN ORALLY EFFECTIVE CHOLINERGIC CHANNEL MODULATOR (ChCM) WITH COGNITIVE ENHANCEMENT AND NEUROPROTECTIVE ACTION

Stephen P. Arnerić, Anthony W. Bannon, Jorge D. Brioni, Clark A. Briggs, Michael W. Decker, Mark W. Holladay, Kennan C. Marsh, Diana Donnelly-Roberts, James P. Sullivan, and Michael Williams
Abbott Laboratories, Neuroscience, D-47W, Pharmaceutical Products Division, Abbott Park, IL, 60064-3500 USA

Jerry J. Buccafusco
Department of Pharmacology and Toxicology, Medical College of Georgia Augusta, GA, 30912-2300 USA

Recent evidence suggests the existence of a diversity of neuronal nicotinic acetylcholine receptor (nAChRs) subtypes with a wide distribution in brain, that each subtype may be involved in mediating specific functions/behaviors, and that these subtypes have a defined pharmacology that may be selectively targeted (Arnerić et al., 1995a). The pharmacologic diversity of neuronal nAChR subtypes suggests the possibility of developing selective compounds which would have more favorable side-effect profiles than existing agents which generally exhibit low selectivity. This broader class of agents, collectively called cholinergic channel modulators (ChCMs) selectively activate some subtypes of nAChRs (i.e. Cholinergic Channel Activators, ChCAs) or inhibit the function of other subtypes (Cholinergic Channel Inhibitors, ChCIs).

Modulation of neuronal nAChR function may be beneficial in the treatment of disorders of attention, especially Alzheimer's disease (AD) (see: Arnerić et al., 1996). ABT-418 is an isoxazole isostere of (-)-nicotine that may represent the first new chemical entity in the general class of ChCMs to be specifically developed for treatment of CNS disorders (Arnerić et al., 1995b). Although encouraging data suggest ABT-418 may have some positive effects on the cognitive symptomology of AD (Newhouse et al., 1996), the preclinical pharmacology of ABT-418 suggest that oral formulation of this compound may not be feasible (Arnerić et al., 1995a).

The focus of this chapter is to summarize the preclinical pharmacology of ABT-089 [3-(2(*S*)-pyrrolidinyl methoxy)-2-methylpyridine], a novel analog of (-)-nicotine under consideration for the treatment of cognitive disorders. ABT-089 is an orally bioavailable ChCM with cognitive enhancement and neuroprotective activity possessing a substantially reduced side-effect profile compared to (-)-nicotine (Lin et al., 1996).

PRECLINICAL PHARMACOLOGY OF ABT-089

ABT-089 potently (K_i = 16 ± 2 nM) interacts with the $\alpha4\beta2$ subtype of nAChRs, the major subtype in rodent brain (Arnerić et al., 1995b). ABT-089 was significantly less potent (K_i > 10,000 nM) against the nicotinic receptors labeled by [^{125}I]α-bungarotoxin in brain. In 28 other receptor/uptake/enzyme binding assays the K_i values for ABT-089 were greater than 10,000 nM.

Functionally, ABT-089 is less potent (estimated EC_{50} = 5 µM) and 34% as efficacious as (-)-nicotine (estimated EC_{50} = 1.5 µM) to stimulate $^{86}Rb^+$ efflux from mouse thalamic synaptosomes which are thought to reflect an interaction with $\alpha4\beta2$ nAChRs. In IMR-32 cells which express ganglionic-like nAChRs distinct from the $\alpha4\beta2$ subtype, ABT-089 was a very weak agonist (EC_{50} = 150 µM; efficacy <20%), which is consistent with the absence of cardiovascular liabilities in dog and monkeys. ABT-089 and (-)-nicotine stimulated the evoked release of [^{3}H]dopamine from rat striatal slices with EC_{50} values of 1000 nM and 40 nM, respectively, while the potency and efficacy to evoke [^{3}H]ACh release was similar, 3000 and 1000 nM, respectively. The decreased potency of ABT-089 on dopamine release compared to (-)-nicotine agrees with *in vivo* drug discrimination studies suggesting that ABT-089 has less dependence potential than (-)-nicotine. Specifically, ABT-089 was 100- fold less potent and 52% as efficacious as (-)-nicotine to generalize with (-)-nicotine. Thus, ABT-089 demonstrates partial agonist activity for some of the major subtypes of brain nAChRs.

ABT-089 also has *in vitro* neuroprotective activity as seen with (-)-nicotine and ABT-418 (Donnelly-Roberts et al., 1996). Exposure of primary rat cortical cultures to glutamate (300 µM) for 15 min results in a significant ($p < 0.05$) increase (3 to 4- fold) in the levels of lactate dehydrogenase (LDH) in the extracellular media 24 hours later. ABT-089 (EC_{50} = 6 µM), ABT-418 (EC_{50} = 6 µM) and (-)-nicotine (EC_{50} = 10 µM) administered 2 hours prior to glutamate were neuroprotective based on the reduction of LDH release into the media. The neuroprotective effects were attenuated by pretreatment with either 100 µM mecamylamine, 1 nM α-BgT or 10 nM MLA suggestive of the possible involvement of the $\alpha7$ subtype. Morphological assessment of the cells confirmed the cytoprotective effects of the ChCMs. Similar results have been obtained in the human differentiated IMR 32 cell line. In contrast to the effects of ABT-418 and ABT-089, tacrine at concentrations up to 100 µM did not protect against glutamate-induced toxicity (data not shown). Not all ChCMs elicit neuroprotective activity. The R-enantiomer of ABT-089, A-94224 (10 μM), was without effect. Compounds structurally related to ABT-089, but with 100-fold greater affinity to bind to nAChRs were also ineffective.

The potential involvement of Aβ in the etiology of (AD) has been well documented and Aβ peptides, most notably $A\beta_{1-40}$ and $A\beta_{1-42}$ have been shown to be neurotoxic *in vitro* as well as *in vivo*. Exposure of either differentiated human IMR 32 cells or rat primary cortical neurons to 10 μM $A\beta_{1-42}$ for 5 days elicits a significant ($p < 0.05$) increase (> 3-fold) in extracellular LDH levels. Preincubation of the cells with either ABT-418 (EC_{50} = 10 μM), or ABT-089 (EC_{50} = 3 μM) protected against the $A\beta_{1-42}$ toxicity, an effect prevented by mecamylamine (100 μM) and α-BgT (1 nM).

ABT-089 has been evaluated in a series of animal test paradigms to assess cognition enhancement. Administered acutely, ABT-089 only marginally improved the spatial discrimination water maze performance of septal lesioned rats. However, more robust improvement (45% error reduction on the last training day) was observed when ABT-089 was administered continuously via subcutaneous osmotic pumps (minimum effective dose: 1.3 μmol/kg/day). Continuous infusion of (-)-nicotine produced comparable improvement in the spatial discrimination water maze performance of septal-lesioned rats, but a 40-fold higher dose of (-)-nicotine was required (62 μmol/kg/day). Continuous infusion of ABT-089 to aged rats enhanced spatial learning in a standard Morris water maze, as indexed by spatial bias exhibited during a probe trial conducted after four days of training, but not after additional training or when they were subsequently trained in a 2-platform spatial discrimination water maze. ABT-089 did not affect performance of either the aged or young rats during inhibitory (passive) avoidance training. In monkeys, acute administration of ABT-089 modestly improved the delayed matching-to-sample performance of mature, adult monkeys (40 μmol/kg, i.m., $p<0.05$) and significantly improved performance in aged monkeys ($p<0.05$). Improved performance in the aged monkeys was restricted to the longest delay intervals and was not accompanied by changes in response latencies. No adverse events were observed with ABT-089 at doses up to 500 μmol/kg following i.m. administration.

While ABT-089 had approximately the same potency as (-)-nicotine in memory tasks, the compound was remarkably less potent than (-)-nicotine in producing EEG activation, hypothermia, seizures, death, and reduction of locomotor activity in rodents. ABT-089 had significantly less pressor liability in dog as compared to (-)-nicotine. Moreover, one of the dose limiting effects of oral (-)-nicotine is significant agitation of gastrointestinal smooth muscle. Thus, development of an oral formulation of ABT-089 would require substantially reduced gastrointestinal activity compared to (-)-nicotine. ABT-089 does not elicit emesis in dogs or monkeys at plasma levels which are 15- to 385-fold greater than the behaviorally effective range (i.e. 2-50 ng/ml).

In rodent, dog and monkey ABT-089 demonstrated substantial oral bioavailability [rat = 32%; dog = 63%; monkey = 77%] when compared to (-)-nicotine and ABT-418. Levels of ABT-089 within the range that elicited cognitive enhancement in the Morris water maze paradigm can be maintained for up to 24 hours following a single oral dose. ABT-089 (100 μM) directly applied to the smooth muscle of guinea pig ileum does not elicit contraction. This is in contrast to the very potent contraction elicited by the muscarinic agonist methylcholine, or (-)-nicotine itself. Furthermore, ABT-089 does not inhibit the ability of other natural constrictive agents such as ACh, histamine, serotonin or K^+ to produce normal responses (n=3; data not shown). Thus, in contrast to many other cholinergic therapies, ABT-089 given at behaviorally relevant concentrations would not be anticipated to affect the natural motility of the gastrointestinal system following oral administration.

ABT-089 is a prototype of a new class of compounds that selectively activates some, but not all, neuronal nAChRs without eliciting the dose-limiting side effects typically observed with (-)-nicotine. ABT-089 may be a safe and effective ChCM for the potential treatment of the cognitive and neurodegenerative components of AD.

ACKNOWLEDGMENTS

The authors thank D.J. Anderson, M.J. Buckley, D. Cox, P. Curzon, D.J.B. Kim, K. L. Gunther, M. Piattonni-Kaplan, D. McKenna, A.B. O'Neill, M. Pendergast, S. Quigley, R.J. Radek, J.L., and J.T. Wasicak for their contributions.

REFERENCES

Arnerić SP, Holladay MW, and Sullivan JP (1996): Cholinergic channel modulators as a novel therapeutic strategy for Alzheimer's disease. *Exp. Opin. Invest. Drugs* 5(1): 79-100.

Arnerić SP, Sullivan JP and Williams M (1995a): Neuronal nicotinic acetylcholine receptors: novel targets for CNS therapeutics. In: *Psychopharmacology: 4th Generation of Progress,* Bloom F.E. and Kupfer DJ, eds., New York, Raven Press; pp. 94-110.

Arnerić SP, Sullivan JP, Briggs CA, Donnelly-Roberts D, Decker MW, Brioni JD, Marsh KC, and Rodrigues AD (1995b): Preclinical pharmacology of ABT-418: A prototypical cholinergic channel activator for the potential treatment of Alzheimer's disease *CNS Drug Rev.* 1: 1-26.

Donnelly-Roberts DL, Xue I, Arnerić SP and Sullivan JP (1996): In vitro neuroprotective properties of the novel cholinergic channel activator (ChCA), ABT-418. *Brain Res.* (In press).

Lin N-H., Gunn DE, Ryther KB, Garvey DS, Donnelly-Roberts DL, Decker MW, Brioni JD, Buckley MJ, Rodrigues AD, Marsh KG, Anderson DJ, Sullivan JP, Williams M, Arnerić SP, and Holladay MW (1996): Structure-Activity Studies on ABT-089: An orally bioavailable 3-pyridyl ether cholinergic channel modulator with cognitive enhancing properties and improved safety profile. *J Med Chem.* (In press).

Newhouse P, Potter A, and Corwin J (1996): Acute administration of the cholinergic channel activator ABT-418 improves learning in Alzheimer's disease. *SRNT 2nd Annual Conference,* Washington, D.C., March 16.

[illegible]

Alzheimer Disease: From Molecular Biology to Therapy
edited by R. Becker and E. Giacobini

BIOCHEMISTRY, PHARMACODYNAMICS AND PHARMACOKINETICS OF CI-1002, A COMBINED ANTICHOLINESTERASE AND MUSCARINIC ANTAGONIST

Mark R. Emmerling, Michael J. Callahan, William J. Lipinski, M. Duff Davis, Leonard Cooke, Howard Bockbrader, Nancy Janiczek, Bill McNally and Juan C. Jaen
Parke-Davis Pharmaceutical Research, Division of Warner-Lambert Company, 2800 Plymouth Road, Ann Arbor, MI 48105 USA

INTRODUCTION

The inhibition of brain acetylcholinesterase (AChE) has proven itself beneficial in treating the dementia associated with Alzheimer's Disease (AD). A six-month, multicenter clinical trial showed that statistically significant improvements in measures of cognitive function (ADAS-Cog), global assessments by clinicians and caregivers, and activities of daily living occur in AD individuals receiving tacrine (Cognex®), an AChE inhibitor (Knapp et al., 1994). The cognitive changes produced by tacrine accompany a delay in the time to nursing home placement (Knopman et al., 1996). This translates into increasing the time AD patients spend in their homes and delays entrance into costly institutionalized care. The greatest benefit occurs at the 160 mg/day dose of tacrine. However, a number of patients (25%) fail to receive the highest dose because of associated cholinergically mediated side effects (Knapp et al., 1994). Thus, eliminating cholinergic side effects should increase the comfort of those receiving the AChE inhibitor while enlarging the number of AD patients who otherwise would not attain the benefits of higher doses.

We have been developing a novel compound that is both an AChE inhibitor and muscarinic antagonist (Emmerling et al., 1994, 1995). CI-1002 (PD 142676) is a potent inhibitor of human AChE (IC_{50} = 40 nM) that also acts as a muscarinic receptor antagonist at higher concentrations both *in vitro* (IC_{50} = 400 nM) and *in vivo*. The compound improves cognitive behavior in rodents and aged rhesus monkeys while producing no peripheral cholinergic side effects at these doses. The separation between doses that benefit cognition and cause side effects may be related to the dual pharmacology of CI-1002. Our preclinical studies indicate that the properties of CI-1002 make it unique among AChE inhibitors. Thus, CI-1002 represents a new class of AChE that may be better tolerated in humans than conventional AChE inhibitors.

RESULTS

Short-term memory often declines with age in animals and man (Bartus et al., 1982). In a delayed match-to-sample task, rhesus monkeys see a symbol that must be correctly identified from a selection of three symbols after a short (1 sec) or long delay (>5 sec) to receive a reward. The ability of rhesus monkeys to respond correctly after a long delay is dramatically reduced with age. CI-1002 significantly improves performance by 50% on tasks with long delays following a CI-1002 dose of 0.1 mg/kg (IM) given to aged rhesus monkeys. This improvement in performance brings the level of correct responses by the aged rhesus to the same as seen with short delay intervals and similar to what is seen with younger animals. Above and below 0.1 mg/kg, the level of performance improves relative to controls but not to a statistically significant level. Tacrine also improves performance on this task at 0.1 mg/kg but to a lesser extent (27%) than CI-1002.

The cognitive responses to CI-1002 in monkeys are related to plasma levels of drug. The 0.1 mg/kg dose of CI-1002 produces an average peak plasma concentration of 20 ng/mL. This value is similar to the plasma levels of tacrine (7 to 20 ng/mL) related to improved cognitive function in AD patients (Ford et al., 1993).

It is assumed that improved cognitive function resulting from treatment with CI-1002 corresponds to changes in central cholinergic activity. To test this hypothesis, the levels of acetylcholine (ACh) were measured by *in vivo* microdialysis in combination with electrochemical detection of ACh in the striatum of squirrel monkeys receiving CI-1002 (Davis and Cooke, 1995). A dose (0.32 mg/kg) was selected that produced a plasma level of CI-1002 at about 20 ng/mL, corresponding to the plasma level associated with improved cognition in the match-to-sample task. At this level of CI-1002, there was a 40% increase of ACh over baseline levels in response to CI-1002. The increase reached its peak at 60 minutes and remained above baseline for up to 3 hours. Higher doses of CI-1002 produced even greater increases in the levels of ACh but these levels were associated with decreased cognitive performance, suggesting that excessive elevation of central ACh leads to behavioral disruption.

Overt parasympathetically mediated cholinergic responses (e.g., emesis) in rhesus occur at higher doses of CI-1002 (3.2 mg/kg), that correspond to an

Compound	Improvement in MWM (optimal dose mg/kg)	Cholinergic Side Effects (mg/kg)	Ratio Side Effects/MWM
CI-1002	3.2	32.0	10
Tacrine	10	3.20	0.32
E2020	10	10.0	1

TABLE I. Comparison of different AChE inhibitors on the separation between the doses that produce cognitive improvement and cholinergic side effects.

average peak plasma concentration of 500 ng/mL. These results indicate about a 20-fold separation (32-fold based on dose) exists between the plasma level of CI-1002 that improves cognitive function and that which results in cholinergically mediated adverse events. This separation between cognitive efficacy and side effects is better for CI-1002 than for tacrine (10-fold based on dose) when tested in the same paradigm.

A separation is also seen in the dose of CI-1002 that causes improved cognitive performance in the Morris Water Maze (MWM) test and the minimum dose at which cholinergic side effects are detected in C57Bl/10SnJ (B10) mice. Direct comparison with tacrine and E2020 to CI-1002 in MWM shows that CI-1002 has a better separation between doses that improve cognitive performance and that cause detectable cholinergic side effects (Table 1). Doses of the different AChE inhibitors that maximally decrease the latency taken to find the hidden platform in MWM were compared to the doses that produce cholinergic side effects (i.e., salivation, tremors, lacrimation) in at least 40% of the animals studied. Using this criteria there was no separation between doses of tacrine and E2020 that produce cholinergic side effects and that improve MWM performance. In contrast, there was about a 10-fold separation for CI-1002. ACh reduce core body temperature in rats. CI-1002 produces hypothermia but consistently only at a high dose (100 mg/kg) and to a lesser extent than other AChE inhibitors (Fig. 1). In contrast, tacrine and E2020 cause a dose-dependent-decrease in core body temperature that is detectable at 3.2 mg/kg and is maximal at a 32 mg/kg dose. The 32 mg/kg doses of tacrine and E2020 cause about a 1.2%C decrease in core body temperature. In this respect, CI-1002 is an atypical cholinesterase inhibitor. However, it is not unheard of for a centrally active cholinomimetic not to lower body temperature. For example, m1-selective muscarinic agonists also produce poor hypothermia.

The departure of CI-1002 pharmacology from other centrally active AChE inhibitors may indicate that sufficient CI-1002 enters the brain at high doses to cause antagonism of muscarinic receptors, thus negating the effects of AChE inhibition. We have observed that the hypothermic effects of the muscarinic agonist CI-979 are antagonized by CI-1002 at doses of 32 mg/kg. However, measurements of cortical EEG activity in rats show no indication

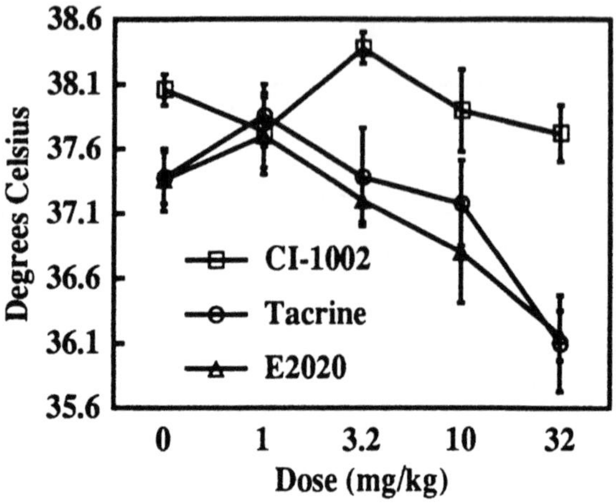

FIGURE 1. Effects of CI-1002, tacrine and E2020 on rat core body temperature

that CI-1002 blocks this parameter of central cholinergic activation (Emmerling et al., 1994). Doses of 3.2 mg/kg to 32.0 mg/kg of CI-1002 in rats reliably decrease cortical EEG activity, an effect common to centrally active AChE inhibitors.

It is difficult to reconcile the lack of effect of CI-1002 to produce hypothermia and its ability to increase cortical EEG activity since both are cholinergically mediated events that require muscarinic receptor activation differentially sensitive to the muscarinic antagonist properties of CI-1002. Alternatively, CI-1002 may be distributed differently in the brain than other AChE inhibitors, resulting in selective activation of central cholinergic systems. Our studies show that after oral administration to rats, CI-1002 is concentrated mostly in the hippocampus followed by the cortex. In comparison, tacrine appears more concentrated in the cortex than the hippocampus. Thus, differences in the regional distribution of CI-1002 may account for differences in its pharmacological effects compared to other AChE inhibitors. Regardless of the explanation, it is clear that decreasing core body temperature by a cholinesterase inhibitor is not critical for improving cognitive function.

The pharmacokinetic properties of CI-1002 have been extensively studied in rats. The systemic bioavailability of CI-1002 is dose dependent. It ranges from 27% to 85% at oral doses of 5 to 30 mg/kg. This non-linearity in plasma levels with increasing doses of CI-1002 suggests a saturable first-pass metabolism of the compound. The half-life of CI-1002 in the plasma is about 2 hrs, which agrees well with the production of cholinergically mediated central effects (e.g., cortical EEG). The major metabolite of CI-1002 in rats is a carboxylic acid derivative that is a poor AChE inhibitor (IC_{50} = 20 &M).

This metabolite is found both in plasma and brain, but it is not suspected of contributing to the pharmacology of CI-1002 at doses that improve cognition.

Although the drug is highly protein bound (92%) in rat serum, studies with radioactive CI-1002 show that it enters the brain rapidly after oral administration. The brain to plasma ratio is on the order of 7 to 1, making CI-1002 comparable to tacrine in its ability to enter the brain.

DISCUSSION

There are about 38 new AChE inhibitors under development by pharmaceutical companies worldwide. The majority of these agents are being developed for the treatment of AD. Of this lot, there are only 4 that are characterized as having combined pharmacological properties. In the case of CI-1002, this combination consists of AChE inhibition and muscarinic receptor antagonism, of which the antagonism is the weaker of the two effects. This combination, along with other pharmacological properties, permits CI-1002 to produce no detectable peripheral cholinergic side effects at doses that improve cognitive performance. This can not be said for other AChE inhibitors, indicating that CI-1002 is a unique second generation AChE inhibitor for the treatment of AD.

REFERENCES

Bartus RT, Dean RL, Beer B and Lippa AS (1982): The cholinergic hypothesis of geriatric memory dysfunction. *Science* 217:408-414.

Davis MD and Cooke LW (1995): Changes in brain acetylcholine overflow during rat and monkey microdialysis following administration of the acetylcholinesterase inhibitor, CI-1002. *Soc Neurosc Abstracts* 21:1975

Emmerling MR, Gregor VL, Schwarz RD, Scholten JD, Callahan MJ, Lee C, Moore CJ, Raby C, Lipinski WJ, and Davis RE (1994): PD 142676 (CI 1002), a novel anticholinesterase and muscarinic antagonist. *Mol Neurobiol.* 9:93-106.

Emmerling MR, Gregor VL, Callahan MJ, Schwarz RD, Scholten JD, Orr EL, Pugsley T, Moore CJ, Raby C, Myers SL, Davis RE and Juan J (1995): CI-1002: a combined acetylcholinesterase inhibitor and muscarinic antagonist. *CNS Drug Rev* 1:27-49.

Ford J, Truman CA, Wilcock, GK, and Roberts CJC (1993): Serum concentrations of tacrine hydrochloride predict its adverse effects in Alzheimer's Disease. *Clin Pharmcol Ther* 53:691-695.

Knapp MJ, Knopman DS, Solomon PR, Pendlebury WW, Davis CS, Gracon SI. (1994): A 30-week randomized controlled trial of high-dose tacrine in patients with Alzheimer's Disease. *JAMA* 271:985-991.

Knopman D, Schneider SK, Davis K, Talwalker S, Smith F, Hoover T, Gracon SI (1996): Long-term tacrine (Cognex) treatment: Effects on nursing home placement and mortality. *Neurology*, in press.

Alzheimer Disease: From Molecular Biology to Therapy
edited by R. Becker and E. Giacobini

MUSCARINIC PARTIAL AGONISTS IN THE SYMPTOMATIC TREATMENT OF ALZHEIMER'S DISEASE

Rajinder Kumar, Eve Cedar, Michael S.G. Clark,
Julia M. Loudon and James P.C. McCafferty*
SmithKline Beecham Pharmaceuticals, New Frontiers Science Park,
Harlow, Essex, UK CM19 5AW and *Upper Providence, PA

INTRODUCTION

In the late 1970s a profound and consistent loss of choline acetyl transferase (ChAT), a presynaptic marker for acetylcholine (ACh), was observed in the brains of Alzheimer disease (AD) patients when compared to age-matched controls (see Bowen, 1983; Perry and Perry, 1980). A correlation was then reported between these. Such post mortem changes were correlated with the degree of clinical dementia measured premortem (Bartus et al., 1982). Losses in ACh receptor density in the cerebral cortex and hippocampus in AD patients correlated with the degree of intellectual impairment (Lehericy et al., 1993; Terry et al., 1991). These receptor losses were predominantly to presynaptic receptors with little or no change in the number of postsynaptic receptors (Mash, 1985).

There is, therefore, a strong link between central cholinergic function and AD. Although not the only consequence of disease pathology - a number of neuronal systems are degenerated - the cholinergic deficit represents the most consistent transmitter depletion, is an early event in the disease process and is highly correlated with cognitive symptoms. Thus, it has provided a major subject for applied clinical pharmacological research. Of the possible mechanisms by which cholinergic function in AD could be restored, two have been the focus for many research efforts in this area.

CHOLINESTERASE INHIBITORS

Cholinesterase inhibitors (AChEIs) have been the most clinically successful to date, with the approval of tacrine in 1993. However, AChEIs are dependent

Limiting Effect	Compound(s)	References
Nausea, vomiting, sweating, salivation, cardiovascular effects, GI effects	Arecoline, RS86, oxotremorine	e.g. Sunderland et al. (1988), Spiegal et al. (1984)
Short duration of action	Arecoline	Tariot et al. (1988)
Limited brain penetration	Bethanechol	Penn et al. (1988)
Depressive symptoms	Oxotremorine	Davis et al. (1987)

TABLE I. Limiting effects during early clinical trials with muscarinic agonists.

on the integrity of presynaptic neurons for their activity and enhance ACh activity at muscarinic and nicotinic sites all over the body, and so have the inherent propensity to induce undesirable peripheral side effects in addition to their cognition-enhancing properties. Details of this approach are widely documented (e.g. Becker and Giacobini, 1988).

DIRECT-ACTING MUSCARINIC AGONISTS

Findings that the postsynaptic muscarinic receptors in the cerebral cortex and hippocampus remained intact in AD led to the development of direct-acting agonists as potential therapies. However, early clinical results were disappointing, as a result of poor bioavailability or lack of separation between cognition enhancement and side effects (Table 1).

DEVELOPMENT OF PARTIAL AGONISTS AS A THERAPY FOR AD

Many limiting side effects in early agonist studies were peripheral in origin, suggesting that compounds selective for central receptors would overcome these problems. Structure-activity studies demonstrated that agonist efficacy of a compound could be predicted from its relative ability to displace the binding of antagonist and agonist ligands (Brown et al., 1988; Freedman et al, 1988). This was followed up by investigating the agonist efficacy required to produce cognition-enhancing effects. Use of molecular cloning techniques then led to description of 5 muscarinic receptor subtypes (Bonner et al., 1988), the functional correlates of which have since been termed M1 (neuronal), M2 (cardiac) and M3 (glandular). An important finding was the demonstration that the M1 subtype was predominant postsynaptically in the cortex and hippocampus; therefore, it was likely to be important in cognitive processing. Functional *in vitro* studies then showed that low efficacy agonists (partial agonists) exhibit apparent selectivity for M1-mediated effects, probably as a result of inadequate receptor reserve in those tissues mediating effects via M2 or M3 receptors (Hargreaves et al., 1992; Shannon et al., 1994).

Review of data obtained from a large number of novel compounds with a range of agonist efficacies led to a clearer understanding of the relationship between muscarinic agonist efficacy and pharmacological profile. This relationship can be exploited to produce efficacious compounds for the treatment of AD which lack unwanted side effects. The high receptor reserve in some tissues (e.g. salivary glands) allows low efficacy (partial) agonists to induce a full response. However, some effects on the cardiovascular system, for example, are avoided by such compounds because of lack of adequate receptor reserve.

SB 202026 [R- (Z) - α - (methoxyimino) -1 - azabicyclo [2.2.2] octane-3-acetonitrile, (Figure 1) is a muscarinic partial agonist with high affinity for all muscarinic receptor subtypes but which lacks the high selectivity for the M2 subtype (up to 100-fold *cf* M1) thought responsible for typical adverse activity of high efficacy agonists (Table 2).

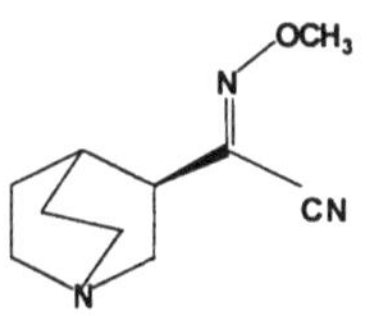

FIGURE 1. Structure of SB 202026

A range of *in vitro* and *in vivo* studies have confirmed that SB 202026 possesses the partial agonist profile predicted from its ligand binding properties. Compared with responses elicited by the full agonists carbachol or OXO-M in functional studies *in vitro*, SB 202026 invokes a full effect at M1 sites (depolarization of rat superior cervical ganglion; 100% of the maximum response induced by a full agonist) while its effectiveness is limited at M2 (electrically-stimulated release of ACh from cortical membranes; 43% of the maximum response) and M3 receptors (isolated guinea-pig ileum; 23% of the maximum response). This functional selectivity of SB 202026 is maintained *in vivo*, resulting in potent cognition-related activity in the absence of cardiovascular side effects (Table II). The finding that cognitive enhancement in animal models can be achieved by SB 202026, in the absence of undesirable side effects, confirms the interpretation that partial agonism is responsible for the observed selectivity with the compound. These findings encouraged the investigation of SB 202026 in the clinic.

EARLY CLINICAL STUDIES

The expectation that the partial agonist profile of SB 202026 would result in better tolerability in the clinic has been confirmed in early studies. The compound was well tolerated by both young and elderly volunteers after

Compound	QNB/ OXO- M ratio	Hm1/ Hm2 ratio	Effective dose (mg/kg iv) RSA (M1)	HR (M2)	BP (M3)	HR/ RSA
OXO-M	1816	100	nd	nd	nd	nd
Carbachol	810	100	nd	nd	nd	nd
Arecoline	225	32.6	0.32	0.02	0.02	0.06
Oxotremorine	188	6.8	0.023	0.007	0.01	0.30
SB 202026	22	0.94	0.018	>0.32	>0.32	>17

TABLE II. Summary of ligand binding data and *in vivo* functional effects of SB 202026 and other muscarinic agonists. QNB/OXO-M ratio predicts agonist efficacy: ratios near or >100 indicate full agonism, those close to unity antagonism, and intermdediate values partial agonism. Hm1/Hm2 ratio indicates relative selectivity for M2 sites. RSA (M1), - induction of rhythmical slow wave activity equivalent to that of a standard dose of arecoline (0.32 mg/kg iv); HR (M2) - bradycardia (dose inducing 50% fall); BP (M3) - hypotension (dose inducing a 50% fall) in the anaesthetised rat following iv administration. HR/RSA ratio indicates the separation between desirable central and undesirable peripheral effects, nd - not determined as these compounds do not penetrate into the CNS.

single and repeat dosage, with the major dose-limiting feature being sweating. Incidence of other side effects was low, and no cardiovascular effects were seen.

In an early efficacy study (14 weeks duration) involving patients with probable AD, there was a statistically significant separation from placebo on the ADAS-Cog scale at all doses tested, with the onset of effect being apparent from 4 week onwards. SB 202026 was again well tolerated, with AE profile at the lowest dose being indistinguishable from placebo. It is noteworthy that this dose was equally efficacious (ADAS-Cog) with higher doses. Further large scale efficacy studies are ongoing.

CONCLUSION

The development of muscarinic partial agonists provides an opportunity to enhance central cholinergic function in the absence of limiting peripheral effects. The hypothesis that such compounds could prove useful as a therapy in AD has been shown to have merit and should result in the availability of safe and effective compounds to palliate cognitive decline in AD.

REFERENCES

Bartus, RT et al. (1982): The cholinergic hypothesis of geriatric memory dysfunction. *Science* 217:408-416.

Becker RE and Giacobini E (1988): Mechanisms of cholinesterase inhibition in senile dementia of the Alzheimer type: Clinical, pharmacological and therapeutic aspects. *Drug Dev Res* 12:163-195.

Bonner TI et al. (1988): Cloning and expression of the human and rat m5 muscarinic acetylcholine receptor genes. *Neuron* 1:403-410

Bowen DM (1983): Biochemical assessment of neurotransmitter and metabolic dysfunction and cerebral atrophy in Alzheimer's disease. In: *Banbury Report* 15, Katzman R, ed. Cold Spring Harbour, pp. 219-232.

Brown F et al. (1988): Variations of muscarinic activities of oxotremorine analogues. *Drug Dev Res* 14:343-347.

Davis KL et al. (1987): Induction of depression with oxotremorine in patients with Alzheimer's disease. *Amer J Psychiatry* 144: 469-471.

Freedman SB et al. (1988): Relative affinities of drugs acting at cholinoceptors in displacing agonist and antagonist radioligands: the NMS/Oxo-M ratio as an index of efficacy at cortical muscarinic receptors. *Br J Pharmacol* 93: 437-445.

Hargreaves RJ et al. (1992): L-689,660, a novel cholinomimetic with functional selectivity for M1and M3 receptors. *Br J Pharmacol* 107:494-501.

Lehericy S et al. (1993): Heterogeneity and selectivity of the degeneration of cholinergic neurons in the basal forebrain of patients with Alzheimer's disease *J Comparative Neurology* 330: 15-31.

Mash DC et al. (1985): Loss of M2 muscarine receptors in the cerebral cortex in Alzheimer's disease and experimental cholinergic denervation. *Science* 228:1115-1117.

Penn RD et al. (1988): Intraventricular bethanechol infusion in Alzheimer's disease: Results of double blind and escalating dose trials. *Neurology* 38:219-222.

Perry EK and Perry RH (1980): The cholinergic system in Alzheimer's disease. In: *Biochemistry of Dementia*, Roberts PJ ed., John Wiley & Sons pp. 135-183

Shannon HE et al. (1994): Xanomeline: a novel muscarinic receptor agonist with functional selectivity for M1 receptors. *J PET* 269(1):271-281.

Spiegal R et al. (1984): First results with RS 86, an orally active muscarinic agonist, in healthy subjects and in patients with dementia. In: *Alzheimer's Disease: Advances in Basic Research and Therapies.* Wurtman RJ, Corkin SH and Growdon JH eds. Boston: Birkhauser, pp. 341.

Sunderland T et al. (1988): Differential responsivity of mood, behaviour and cognition to cholinergic agents in elderly neuropsychiatric populations. *Brain Research Reviews* 13:371-389.

Tariot PN et al. (1988): Multiple-dose arecoline infusion in Alzheimer's disease. *Arch Gen Psychiatry* 45:901-905.

Terry RD et al. (1991): Physical basis of cognitive alterations in Alzheimer's disease - synapse loss is the major correlate of cognitive impairment. *Ann Neurol* 30:572-580.

Alzheimer Disease: From Molecular Biology to Therapy
edited by R. Becker and E. Giacobini

SAFETY AND CLINICAL EFFICACY OF S12024 IN PATIENTS WITH MILD TO MODERATE ALZHEIMER'S DISEASE

Hervé Allain,
Faculté de Médecine, Laboratoire de Neuropharmacologie, Rennes, France

David Guez, Eric Neuman, Muriel Malbezin,
Institut de Recherches Internationales Servier, Courbevoie, France

Jean Lepagnol
Institut de Recherches Servier, Croissy, France

Florence Mahieux
Consultation mémoire, Service de Neurologie, Hôpital Tenon, Paris, France

INTRODUCTION

The selection of S12024 as a candidate for the treatment of core symptoms of Alzheimer's disease (AD) was based upon the central noradrenergic and/or vasopressinergic implication hypothesis of AD (Palmer et al., 1987, Van Ree et al., 1985) and the pharmacological data in mnestic models animals in two areas of memory, i.e. acquisition and recall (Lacroix et al., 1993: Raaijmakers et al., 1994: Carey et al., 1993).

Moreover, the effects on a social memory test totally disappeared in the presence of a specific vasopressin antagonist and in the Brattleboro rat (vasopressin synthesis impaired). S12024 did not reverse scopolamine-induced amnesia (Lepagnol et al., 1994). Recently an impact of the compound on nicotinic receptors has been discovered. This could underly neuroprotective effects. The purpose of the present paper is to summarize the results obtained with S12024, a new memory enhancer drug developed in patients with AD.

PHARMACOLOGICAL DATA

In healthy subjects

A cognitive computerized battery, which involved various cognitive and psychomotor performance tasks, was included in a double-blind randomized placebo controlled sequential dose levels phase I study in elderly volunteers (>60 years). Subjects received from 10 mg/d to 200 mg/d twice daily of S12024 during 7 days (Wesnes et al., 1994).

The results demonstrated that S12024 produced a dose-dependent improvement for the speed of recognition performance at the intermediate dose level (50 and 100 mg twice daily) and to a lesser extent at the 200 mg dose level. The most promising effects were seen at the 100 mg dose ($p<0.05$).

In patients with AD.

Three early clinical phase IIa studies were carried out in patients with probable AD according to DSMIII-R and NINCDS/ADRDA criteria.

The first study (Allain et al., 1994) was a multicenter, randomized, placebo-controlled, parallel group comparison of 3 doses of S12024 (100, 200, 400 mg o.d.) over 28 days. Fifty-three in-patients with AD (mean age: 80.0 +/- 1.0 years, mean MMS score: 17.0 +/- 0.5) were included; 42 completed the study. Six patients withdrew from the study for plausible relationship to drug adverse event reactions. The S12024-200 mg dose produced an enhancement of CDR system measures of recognition memory sensitivity. This effect was significant at day 14 ($p=0.02$) for the combined recognition measure from digit, word and picture tasks; $p=0.13$ at day 28). At day 14, 400 mg/d dose showed trends for slowed speed on various tasks, interpreted as a plausible effect on patient's motivation and attention. The SM5 battery (Lieury et al., 1991) also provided support for enhancement of some word memory tasks at 200 mg. The improvement was near the threshold of significance for the word-recognition task at day 28 ($p=0.07$), the combined word-recognition task at day 28 ($p=0.07$) and the false face-recognition task at day 14 ($p=0.056$).

The second study (Derouesne et al., 1994) was a monocenter, randomized, placebo-controlled study of latin square design, with 3 oral doses of S12024-2 (50, 100, 200 mg o.d.), administered during 4 double-blind treatment periods of 7 days, each separated by a 2-week wash-out period. Twelve patients (mean age: 64 years old +/- 10, mean MMS: 23 +/- 3) completed the study. The acceptability was good with no treatment discontinuation and perfect stability of all clinical, ECG and biological parameters.

On neuropsychological battery, no treatment effect was observed neither for MMS, cognitive battery and ADL scores. A significant period effect was shown at day 1 and day 7 between the second and the 4th period of treatment for some items in favour of a learning effect. CIBIC results showed a non-significant treatment effect in favour of active treatment (100>200>50>0). On quality EEG, significant increase in the relative and absolute b1 activity and a decrease in d activities were observed with the active treatment, predominant at the 200 mg dose ($p=0.01$) in favour of a non-specific stimulation of vigilance. On ERPs, a significant difference ($p<0,05$) was shown between placebo and S12024, whatever the active doses on the amplitude of the processing negativity (PN) and mismatch negativity (MMN) signals. This was in favour of a facilitation of automatic attention processes without dose-effect relationship, indicating a pharmacodynamic central activity.

The third study (Franco-Maside et al., 1995) was a monocenter randomized, placebo-controlled, parallel group comparison of 2 doses of S12024 (100, 300 mg o.d.) over 28 days. Thirty-four patients with AD (mean age: 66 +/- 8, mean MMS score: 20 +/- 3) were included; 31 completed the study. S12024 induced a significant improvement with the 100 mg/d dose at day 28 versus placebo for the global improvement item of CGI (p=0.007). MMS (p=0.67), ADAS-Cog (p=0.45), verbal fluency (p=0.65), and Quality of Life (p=0.144) were not significantly modified but a tendency of improvement was shown for the 100 mg/d dose. Quantitative EEG analysis found statistically significant increase of relative and absolute b1 power in right central and occipital areas and decrease of relative d power at 100 mg/d versus placebo (p<0.05) with opposite effect at high dose. Transcranial Doppler analysis found statistically significant decrease in pulsatility and resistance index scores (p<0.05) and an increase in diastolic velocity (p<0.03) at 100 mg/d dose compared to placebo. This was in favour of enhanced vascular perfusion in middle artery territories.

CLINICAL DATA

Four hundred and four out-patients (mean age: 70 ± 8 years; mean MMS score: 19.5 ± 3.5) diagnosed as suffering from AD received either 100 mg, or 300 mg S12024 or matching placebo o.d. for three months in a European, double-blind randomized parallel group study of the efficacy and safety of S12024.

In the overall analysis, no significant change between treatments were observed, whatever the efficacy outcome measures. A subgroup analysis on moderate AD patients (according to MMS between 14 to 18 at inclusion) showed a statistically significant improvment (*P*<0.05) of ADAS-Cog scores (mean difference of -1.8 point), of MMSE score (mean difference of +1.8 point) and in global performance (mean difference of -0.3 point at CIBIC), in the 100 mg active treatment after 12 weeks versus placebo, while IDDD score did not change. Moreover, a significant interaction (*P*<0.002) between the ApoE e4 allele and the effect of the 100 mg dose of S12024-2 was detected for the MMSE (*P*<0.002) and CIBIC (*P*<0.006). At the active dose of 100 mg/d, the drug was well tolerated.

DISCUSSION

The clinical effects observed at 3 months were convergent whatever the outcome measure (except IDDD) and the 2 types of analyses, in intention to treat and on evaluable patients (per protocol analyses) came to essentially the same conclusions. However, this efficacy was only observed in patients with moderate AD (MMS between 14 and 18 at inclusion). These results could be explained by a more rapidly cognitive deterioration in this population, over a 3-month period treatment compared with mild population (MMS 19-26), according to the tri-linear model of AD evolution (Brooks et al., 1993) and does not mean an inefficacy of S12024-2 in mild population at 6 months or more. We can only assert that patients with moderate AD appear to be more responsive to a therapeutic efficacy.

These results are in concordance with those expected with an anti-Alzheimer's drug, the mechanism of action of which is to show the decline of the disease instead of a symptomatic effect. S 12024-2 seems to slow down AD evolution in patients bearing ApoE e4 allele.

The phenotyping of the ApoE in the population studied shows that dose with the genetic marker were the group in which efficacy was concentrated (Amouyel et al., 1995). This fact is interesting in that it provides some extra rationale for the treatment of AD. The view at the moment is that AD is a multifactorial disorder and that the treatment regimen will probably need to use different agents together for an optimal response. These results lead to open discussion, despite the ethical difficulties involved, or whether it would be preferable to enrich the samples with positive ApoE testing in future studies, which should exclude those with mild dementia. This would also avoid some risk of non-responders being unnecessarily exposed to unwanted effects. The 3-month period was probably too short to demonstrate a significant clinical difference, especially in the mild group (MMS 19-26) in which the decline of cognitive functions is slower.

In spite of the small size of population and the short duration of the treatment which underpowered the results of the study, most of the objectives were reached. The analysis of the overall data raises several methodological questions. The main questions are: i) what is the best duration for phase 2 studies? and ii) how to dwindle the heterogeneity of the population? That is, how to reduce the variance of the main criteria (and lead to a better sensitivity) and is there some identifiable subgroups more sensitive to the studied drug?

The best duration for phase 2 studies depends on both the nature of the expected effect (symptomatic improvement or stabilization) and the amount of knowledge on safety of the drug. It is obvious that it is impossible to directly reveal a stabilization effect on dementia's symptoms in less than 6 months as it is the shortest duration to see a clinically significant deterioration. That is quite long for a phase 2 study and it would be necessary to develop substitutive criteria in order to select the best dose with a shorter exposure of patients to the drug. An alternative or a completion would be the development of bridging studies; that is, studies conducted in AD patients with the sole aim at determining the maximum tolerated dose, as in phase 1 studies. On the contrary, it is possible to expect that a symptomatic drug shows a significant effect in a much shorter duration. However, tacrine's studies clearly demonstrated that a significant placebo effect could be observed up to

3 months. Thus, it could be necessary to extend the duration of a phase 2-A study up to 4 months in order to be given an "off-placebo" effect.

Heterogeneity seems to be inherent in AD. Therefore, the extent of the variance of neuropsychological parameters leads to an increase of the size of sample necessary to demonstrate a significant effect. Reducing this variability, particularly for the main evaluation criteria, would be a considerable goal. Maybe the selection criteria could take into account a neuropsychological range on ADAS-Cog rather than MMS. On the other hand, it is quite possible that a given drug may have an impact on a fraction of the population, given the "responder" concept. Thus, the goal of some phase 2 studies could be to determine the characteristics of a subgroup of best responders among the whole population. Today, it does not seem possible to foresee which subpopulation would be more sensitive to a given drug. However, several independent characteristics could be studied to help determine such a subgroup. Among them, the fourth form of the apolipoprotein E is now well-known as a genetic risk-factor. Moreover, it has been emphasized that E4 non-bearers are more sensitive to tacrine treatment than E4 bearers. Our study with S12024 provide arguments for a differential sensitivity of the two subgroups of patients. Another way is to stratify patients according to their level of deterioration. Two reasons could account for a differential effect. First, it is possible to consider that the extent of the understanding lesions conditions the response to a drug. Second, according to the trilinear model (Brooks et al., 1993) of the evolution of AD seen above, a stabilizing effect could be shorter to demonstrate in a subgroup whose evolution is more rapid.

However, given the difficulties to foresee if (and which) subpopulation will be more sensitive, phase 2 studies have to be conducted in a whole population. But a stratification according to the latter parameters have to be considered and taken into account for the design (and decision) of phase 3 studies.

REFERENCES

Allain H, Wesnes K, Neuman E, Gandon JM., Salzman V, Malbezin M and Guez D (1994): Acceptability and clinical activity of 4 weeks' treatment by S 12024-2 in 53 in-patients with moderate to severe Alzheimer's Disease. *Neurobiol. Aging* 15 (S1):S136-137.

Amouyel P, Neuman E, Dillemann L, Richard F, Barrandon S and Guez D (1995): Characterization of the apolipoprotein E genotypes in a European multicentre trial on Alzheimer's disease with the S-12024. *J Neurol* 242 (S2):S58-S78.

Brooks JO, Kraemer HC, Tanke ED and Yesavage JA (1993): The methodology of study decline in Alzheimer's disease. *JAGS* 41:623-628.

Carey GJ, Domeney AM, Costall B, Brown CL and Detolle-Sarbach S (1993): Effects of S12024-2 on the performance of marmosets in an object discrimination task. *Br J Pharmacol* 108:298.

Franco-Maside A, Vinagre MD, Caamano J, Gomez MJ, Alvarez XA, Fernandez-Novoa L, Zas R, Novo B, Polo E, Garcia M, Castagné I, Neuman E, Guez D and Cacabelos R (1995): Effects of S 12024-2 on brain electrical activity parameters in patients with Alzheimer's disease. *Ann Psychiat.* 5:281-294.

Derouesné C, Renault B, Gueguen B, Van der Linden M, Lacomblez L, Homeyer P, Ouss L, Neuman E, Malbezin M, Barrandon S and Guez D (1994): Clinical and neurophysiological evaluation of 3 doses of S 12024-2 (50, 100, 200 mg o.d.) during repeated oral administration (7 days) in 12 out-patients with mild to moderate Alzheimer's disease *Neurobiol. Aging.* 15, (Suppl 1), S100.

Lacroix P, Reymann JM, Rocher N, Allain H, Castagne I, Detolle-Sarbach S and Guez D (1993): Effects of S12024-2 on memory impairment in aged rats. *Eur Neuropsychopharmacol* 3, 3:419-420.

Lepagnol J and Lestage P (1994): Memory enhancing effects of S 12024-2. Involvement of vasopressinergic neurotransmission. *Third International Springfield Symposium on Advances in Alzheimer Therapy.* Springfield, IL, USA.

Lieury A, Raoul P, Bouton C, Bernoussi M and Allain H (1991): Le neillisseweur des composante de la memoire: analyze factoriolle de 17 scores de memoire. *L'Aunee Psycholog* 91:169-186.

Palmer AM, Francis PT, Bowen DM, Benton JS, Neary D, Mann DM and Snowden JS (1987): Catecholaminergic neurons assessed ante-mortem in Alzheimer's Disease. *Brain Res* 414:365-375.

Raaijmakers W, Prickaerts J, Schffelmeer A and Jolles J (1994): The novel cognition enhancing agent S12024-2 facilitates object and social recognition memory after acute or chronic treatment in rats. *Soc Neurosci Abstr* Miami, FL, USA.

Van Ree JM, Hijman R, Jolles J and DeWied D (1985): Vasopressin and related peptides: animal and human studies. *Progress Neuro-Psychopharmacol & Biol Psychiatr* 9:551-559.

Wesnes K, Neuman E, de Wilde HJG, Malbezin M, Crijns HJM, Jonkman JHG and Guez D (1994): Pharmacodynamic effects of repeated oral administration of 4 different doses of S12024-2 (cognitive enhancer) in 36 healthy elderly volunteers. *Neurobiol Aging 15*, (Suppl.1) S100.

Alzheimer Disease: From Molecular Biology to Therapy
edited by R. Becker and E. Giacobini

PHARMACOLOGICAL CHARACTERIZATION OF PD151832, AN M1 MUSCARINIC RECEPTOR AGONIST

Roy D. Schwarz, Michael J. Callahan, Robert E. Davis, Mark R. Emmerling, Juan C. Jaen, William Lipinski, Thomas A. Pugsley, Charlotte Raby, Carolyn J. Spencer, Katharyn Spiegel, and Haile Tecle
Parke-Davis Pharmaceutical Research, Division of Warner-Lambert Co., Ann Arbor, MI USA

Mark R. Brann
Receptor Technologies, Inc., Winooski, Vermont USA

INTRODUCTION

Alzheimer's disease (AD) is an age-related disorder characterized by progressive neurological impairment. Among neurotransmitters affected, the loss of forebrain cholinergic neurons has been well documented with the results contributing to the idea that therapies based on cholinergic pharmacology would be useful treatments in AD (Bartus et al., 1982). Among those suggested have been acetylcholinesterase (AChE) inhibitors and muscarinic agonists. Clinical studies have shown that the AChE inhibitor tacrine is therapeutically active (Knapp et al., 1994). However, robust and reproducible clinical data with classical muscarinic agonists have been variable and no agonists have been approved to date for AD treatment.

A variety of evidence now exists to suggest that m1 subtype selective muscarinic agonists would be useful for the treatment of AD. Within the CNS m1 receptors appear to be highly localized in the cortex and hippocampus, two brain regions associated with learning and memory (Levey et al., 1991), while in the periphery m2 and m3 receptors are localized in smooth muscle and glandular tissue (Dorje et al., 1991). Unwanted peripheral side effects associated with these subtypes include salivation, sweating, nausea, vomiting and changes in heart rate. Thus, an m1 selective agonist should be clinically efficacious in AD at doses lower than those producing dose-limiting peripheral cholinergic side effects (Davis et al., 1995).

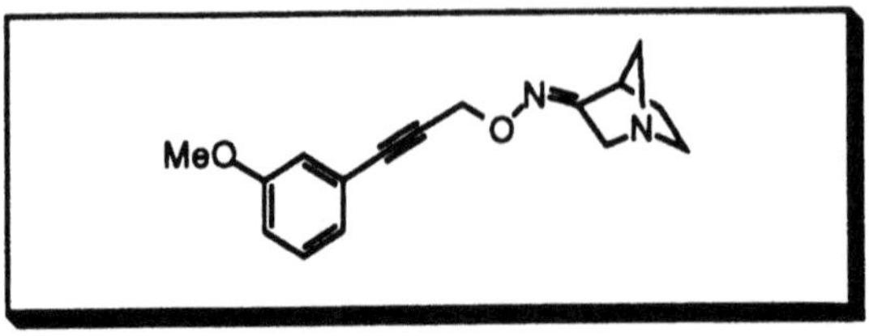

FIGURE 1. Structure of PD151832: (4R)-(Z)-1-Azabicyclo[2.2.1]heptan-3-one,O-[3-(3-methoxyphenyl)-2-propynyl]oxime.

As reported (Tecle et al., 1993; Jaen et al., 1995), a novel series of subtype selective 1-azabicyclo[2.2.1]heptan-3-one oxime agonists were synthesized which *in vitro* showed functional selectivity in a number of models. Pharmacological characterization of one agonist from this series, PD142505, has been previously described (Schwarz et al., 1994). The following report summarizes the biochemical and behavioral properties of PD151832 (Figure 1), the R enantiomer of PD142505.

BIOCHEMICAL CHARACTERIZATION OF PD151832

Receptor binding studies in rat brain tissue show that PD151832 binds to muscarinic receptors in a manner similar to other known agonists having low nM affinity with an agonist ligand and low mM affinity with antagonist ligands. Low mM affinity was also observed across subtypes in transfected CHO cells (TABLE I). Additionally, no binding was observed to a number of other neurotransmitter receptors, indicative of its selectivity for muscarinic receptors.

Since binding in transfected cells underestimates the subtype selectivity of various muscarinic agonists (Schwarz et al., 1993), functional assays were used to address receptor selectivity (TABLE I). PD151832 potently stimulated phosphatidylinositol (PI) hydrolysis in Hm1 CHO cells ($EC_{50} = 1.0$ μM) with little activation of Hm3 or Hm5 CHO cells. There was no inhibition of forskolin-stimulated cAMP accumulation in Hm2 CHO cells, although inhibition was observed in Hm4 cells (1.6 μM). A similar selectivity profile (activity in m1 and m4, but little in m2 or m3) was also observed across subtypes when measuring foci formation in transiently transfected NIH 3T3 cells using beta-galactosidase as a marker in the R-SAT assay (Brauner-Osborne et al., 1995). A clear separation of m1 versus m2 activity was displayed when assessing cellular metabolism in CHO cells with a microphysiometer measuring extracellular pH changes. And finally, in rat cortical slices, PD151832 decreased the K^+-stimulated release of [^{3}H]-ACh in

	EC$_{50}$ (μM)				
	Hm1	Hm2	Hm3	Hm4	Hm5
[^{3}H]-NMS binding[a]	10.2	10.9	13.9	29.6	6.1
PI hydrolysis	1.0	---	NA	---	NA
cAMP accumulation	---	NA	---	1.6	---
Cellular metabolism	0.3	NA	1.5	6.0	NT
Foci formation	0.04	1.3	0.9	0.05	NT

TABLE I. Summary of *in vitro* biochemical results. [a]K_{app} values; NA: not active up to 1 mM; NT: not tested

a dose-dependent manner showing its interaction with presynaptic autoreceptors.

Muscarinic agonists have been shown to stimulate the secretion of soluble amyloid precursor protein (APPs) and decrease the levels of amyloidogenic fragments (Davis et al., 1995). In Hm1 CHO cells, PD151832 was able to increase APPs secreted into the culture during a one hr incubation, with a maximal effect at 10 μM. At this time it is unknown what the direct result of this finding is in humans. However, it is possible that muscarinic agonists could alter disease progression by changing the ratio of soluble to amyloidogenic fragments produced in the brain.

IN VIVO ACTIVITY OF PD151832

CNS activity following oral administration of PD151832 was observed in both rats and mice. Cognitive performance was assessed using C57BL/10SnJ mice in the Morris Water Maze (MWM), a task of spatial working memory. A typical inverted U-shaped dose-response curve was observed with a dose of 1.0 mg/kg (PO) yielding the best improvement in performance. In Long-Evans rats which had forebrain cholinergic neurons destroyed by ibotenic acid lesions, PD151832 improved MWM performance compared to sham-lesioned control animals at 0.1 mg/kg (PO).

Side effects (salivation, lacrimation, diarrhea) in rats or mice are not observed until doses greater than 100 mg/kg (PO). In mice, 160 mg/kg is the minimally effective dose for producing salivation, while no effects on gastrointestinal function (e.g. gastric emptying and intestinal propulsion) were noted in rats up to doses of 178 mg/kg. Thus, consistent with its *in vitro* biochemical profile of M1 subtype selectivity, PD151832 displays behavioral activity in rodents mediated by central activation at doses (0.1-1.0 mg/kg; PO) far lower than those producing peripheral cholinergic side effects ($\geq$160 mg/kg; PO).

In addition to behavioral models, PD151832 was assessed neurochemically in rats for the ability to alter catecholamine metabolism in the CNS following oral administration. Similar to the effects produced by the M1

	Minimally Effective Dose (mg/kg)
Rodents	
MWM performance in mice	1.0
nbM lesioned rat[a]	0.1
Gastrointestinal motility	NE
Side effects (e.g. salivation)	>100.0
Monkeys	
Reversal of scopolamine	0.3
Side effects	>3.2

TABLE II. Summary of *in vivo* activity in rodents and monkeys. [a] nbM: ibotenic acid lesion of nucleus basalis of Meynert. NE: no effect

agonist xanomeline (Shannon et al., 1994), PD151832 increased dopamine synthesis in rat brain striatal and mesolimbic regions. A significant increase indopa accumulation was measured at 10 and 30 mg/kg (PO), in the presence of a decarboxylase inhibitor, NSD 1015, 30 minutes post treatment. Additionally, serotonin synthesis in striatum and cortex was increased at similar doses.

Activity in rhesus monkeys also showed a marked separation between central and peripheral muscarinic activation. At a dose of 0.32 mg/kg (IM), PD151832 significantly reversed the impairment in a continuous performance task produced by the nonselective antagonist scopolamine. However, salivation and emesis occurred in these same monkeys only at doses above 3.2 mg/kg (IM).

SUMMARY

PD151832 is an M_1 subtype selective muscarinic agonist based upon its *in vitro* activity in functional measures of receptor activation. It is orally active, shows strong CNS penetration, and improves cognitive performance in both rodents and monkeys. Based upon the clear separation between behaviorally active doses and those producing peripheral cholinergic side effects, this agonist has a desirable profile for the treatment of AD and at the present time is actively under development.

ACKNOWLEDGMENTS

The authors would like to also acknowledge S. Barrett, P. Doyle, D. Lauffer, L. Lauffer, and D. Moreland for their contributions to this work.

REFERENCES

Bartus RT, Dean RL, Beer B and Lippa AS (1982): The cholinergic hypothesis of geriatric memory dysfuction. *Science* 217:408-416.

Braunner-Osborne H, Ebert B, Brann MR, Falch E, and Krogsgaard-Larsen P (1995): Annulated heterocyclic bioisosteres of norarecoline. Synthesis and molecular pharmacology at five recombinant human acetylcholine receptors. *J Med Chem* 38:2188-2195.

Davis RE, Doyle PD, Carroll RT, Emmerling MW, and Jaen J (1995): Cholinergic therapies for Alzheimer's disease. Palliative or disease altering? *Arzneim Forsh/Drug Res* 45:425-431.

Dorje F, Levey AI, and Brann MR (1991): Immunological detection of muscarinic receptor subtype proteins (m1-m5) in rabbit peripheral tissues. *Mol Pharmacol* 40:459-462.

Jaen J, Barrett S, Brann M, Callahan M, Davis R, Doyle P, Eubanks D, Lauffer D, Lauffer L, Moreland D, Nelson C, Raby C, Schwarz R, Spencer C, and Tecle H (1995): In vitro and in vivo evaluation of the subtype selective muscarinic agonist PD 151832. *Life Sci* 56:845-852.

Knapp MJ, Knopman DS, Solman PR, Pendlebury WW, Davis CS, Gracon SI, Tacrine Study Group (1994): A 30 week randomized controlled trial of high-dose tacrine in patients with Alzheimer's disease. *JAMA* 271:985-991.

Levey AI, Kitt CA, Simonds WF, Price DL and Brann MR (1991): Identification and localization of muscarinic acetylcholine receptor proteins in brain with sybtype-specific antibodies. *J Neurosci* 11:3218-3226.

Schwarz RD, Davis RE, Jaen JC, Spencer CJ, Tecle H and Thomas AJ (1993): Characterization of muscarinic agonists in recombinant cell lines. *Life Sci* 52:465-472.

Schwarz RD, Callahan MJ, Davis RE, Jaen JC, Lipinski W, Raby C, Spencer CJ and Tecle H (1994): Selective muscarinic agonists for Alzheimer disease treatment. In: *Alzheimer Disease: Therapeutic Strategies*, Giacobini E and Becker R, eds. Boston, Birhaser, pp 224-228.

Shannon HE, Bymaster FP, Calligaro DL, Greenwood B, Mitch CH, Sawyer BD, Ward JS, Wong DT, Olesen PH, Sheardown MJ, Swedberg MD, Suzdack PD and Sauerberg P (1994): Xanomeline: A novel muscarinic receptor agonist with fuctional selectivity for M_1 receptors. *J Pharmacol Exp Ther* 269:271-281.

Tecle H, Lauffer DJ, Mirzadegan T, Moos WH, Moreland DW, Pavia MR, Shwarz RD and Davis RE (1993): Synthesis and SAR of bulky 1-azabicyclo[2.2.1]-3-one oximes as muscarinic receptor subtype selective agonists. *Life Sci* 52:505-511.

Alzheimer Disease: From Molecular Biology to Therapy
edited by R. Becker and E. Giacobini

NOVEL M1 AGONISTS: FROM SYMPTOMATIC TREATMENT TOWARDS DELAYING THE PROGRESSION OF ALZHEIMER'S DISEASE

Abraham Fisher, Rachel Haring, Zippora Pittel, David Gurwitz, Yishai Karton, Haim Meshulam, Daniele Marciano, Rachel Brandeis and Eliahu Heldman
Israel Institute for Biological Research,
PO Box 19, Ness-Ziona 74100, ISRAEL

Einat Sadot, Jacob Barg, Leah Behar and Irit Ginzburg
The Weizmann Institute, Rehovot, ISRAEL

INTRODUCTION

Restoration of ACh levels or replacement with an M1 (or m1) muscarinic agonist may be effective in treating at least some of the cognitive symptoms in Alzheimer's disease (AD), (see Court and Perry, 1991; Fisher and Barak, 1994). Treatment approaches for AD have to address abnormalities occurring also along various signal transduction pathways (Harrison et al., 1991). Novel activities associated with m1 muscarinic receptors (m1 mAChR) indicate that m1 agonists may also activate hypofunctional signalling pathways in AD (Fisher and Barak, 1994; Gurwitz et al., 1994).

Recent studies indicate that some apparently different neuropathological damages in the AD brain may be linked. Among these are the relation between the formation of ß-amyloid peptide and amyloid plaques, and the loss of cholinergic function in AD brains (Nitsch et al., 1992). As demonstrated by Nitsch et al. (1992) and later by others labs (Buxbaum et al., 1992; Emmerling et al., 1993; Bowen et al., 1994; Haring et al., 1994; Eckols et al., 1995) cholinergic stimulation of m1 mAChRs can increase the secreted, non-amyloidogenic amyloid precursor proteins (APPs). The increased secretion of APP by cells treated with cholinergic agonists results in decreased synthesis of Aβ (Hung et al.,1993).

Understanding the primary lesions that lead to the development of AD pathology can indicate how to prevent, arrest or delay them. Yet, an effective therapy for AD has to treat also the cognitive disorders of AD patients.

Originally the cholinergic approach was aimed only at treating the symptoms of AD, such as memory loss and cognitive dysfunction. The above findings on activation of m1 mAChRs in conjunction with recent data that activation of such receptors induces neurotrophic-like activities (Pinkas-Kramarski et al., 1992; Gurwitz et al., 1995; Mount et al., 1994; Alberch et al., 1995), decreases *tau* phosphorylation (Sadot et al., 1996) and inhibits apoptosis (Lindenboim et al., 1995) raise the exciting perspective that restoring the cholinergic tone in AD brains may alter the onset and progression of AD dementia.

This presentation is an attempt to address some of these findings and hypotheses together with some novel functionally selective m1 agonists designed by our lab (Fisher and Barak, 1994). Such agonists can be found among certain rigid analogs of ACh, e.g., the *AF series* [project drug AF102B].

RESULTS AND DISCUSSION

Some of the *AF series* compounds are: i) selective ligands for the M1 muscarinic receptor [M1 mAChR] in rat cortex; ii) full agonists in elevating $[Ca^{2+}]i$ in Chinese hamster ovary (CHO) cells stably transfected with cloned m1 mAChR; iii) depending on the compound, partial or full agonists in stimulating phosphoinositides (PI) hydrolysis or arachidonic acid (AA) release; but all are antagonists in elevating cAMP levels [in CHO and in rat pheochromocytoma (PC12) cells stably transfected with m1 mAChR (PC12M1)]. Thus, m1 agonists may activate distinct sets of G-proteins and drug selectivity may extend beyond the ligand recognition site. Selective responses are achieved when the agonist-mAChR complex activates only certain G-proteins, which in turn activate distinct signal transduction pathways (Fisher and Barak, 1994). The *AF series* compounds exhibit such an activity. The notion of *"ligand-mediated selective signaling"* (Gurwitz et al., 1994) e.g., activation of only distinct G-protein subset(s), (but not Gs), might be of clinical significance, since altered signal transduction via Gs might be relevant in the pathophysiology of AD (Harrison et al., 1991; Gurwitz et al., 1994).

m1 Agonists can induce neurotrophic-like effects which are synergistic with NGF in PC12M1 cells (Gurwitz et al., 1995) and can promote survival of cultured primary CNS neurons (e.g., Purkinje and striatal neurons). Cerebellar Purkinje cell survival is under the trophic control of ACh (Mount et al., 1994) and agonists like the *AF series* acting via M1 mAChR (Alberch et al., 1995). NGF potentiated the trophic action of low agonist concentrations and the effect was blocked by pirenzepine (Alberch et al., 1995). m1 mAChR-coupling to phospholipase C and/or phospholipase A2 (e.g., AA release) may underlie these neurotrophic-like effects. However, m1-associated biochemical signals responsible for this synergistic response with NGF in PC12M1 cells did not seem to involve changes in either PI hydrolysis or cAMP levels (Gurwitz et al., 1995). Thus m1 agonists may exert neurotrophic activities in conjunction with some NGF-dependent and -independent signal(s) (Haring et al., 1996). Some of the neurotrophic-like effects induced by such agonists may also involve increased release of APPs following activation of m1 AChR. The mechanisms underlying these neurotrophic-like effects are presently under investigation in our lab. It appears that the synergy between neurotrophines and m1 muscarinic agonists occurs via activation of multiple transduction pathways acting in parallel, which together intensify the signal at their convergence point (Haring et al., 1996). Neurotrophic effects of m1 mAChR stimulation can promote regeneration or cell rescue, and therefore slow down degeneration. Should such effects occur *in vivo*, these might have clinical significance and may constitute a novel treatment for AD. Notably, NGF does not cross the blood-brain-barrier. A more practical approach would involve modulation of the function of endogenous NGF and/or other neurotrophines by a synergistic agent such as an M1 (or m1) agonist.

Stimulation of m1 mAChRs by AF102B in PC12M1 cells enhances secretion of APPs to the culture medium, and lowers the level of membrane associated APPs. The enhanced APPs secretion induced by AF102B is potentiated by NGF and blocked by atropine (Haring et al., 1994, 1995, 1996). Thus, activation of m1 mAChR leads to opposite effects on APPs secretion and Aß production. Consequently, M1 (or m1) agonists may be of value in preventing amyloid formation by selectively promoting the "α-secretase" processing pathway in AD (Nitsch et al., 1992; Buxbaum et al., 1992; Hung et al., 1993; Bowen et al., 1994; Haring et al., 1994). AF102B increased also APPs secretion from rat cortical slices (rich in M1 mAChRs), but not from cerebellar slices (>90% rich in M2 mAChRs), (Pittel et al., 1996). This, together with data reported by Farber et al. (1995), indicate that m1 selective agonists may alter APP processing in cortex and hippocampus where ml (and m3) mAChRs are abundant.

Tau microtubule-associated protein is neuronal specific, and its expression is necessary for neurites outgrowth. Hyperphosphorylated *tau* proteins are the principal fibrous component of the neurofibrillary tangle pathology in AD (Lee et al., 1991). Stimulation of m1 mAChR in PC12M1 cells with CCh or AF102B decreased *tau* phosphorylation as indicated by specific *tau* monoclonal antibodies which recognize phosphorylation dependent epitopes and by alkaline phosphatase treatment (Sadot et al., 1996). In addition, a synergistic effect on *tau* phosphorylation was found between treatments with these muscarinic agonists and NGF (Sadot et al., 1996). The decreased phosphorylation of *tau* protein via m1 mAChR deserves special attention. This suggests for the *first time* a linkage between the muscarinic signal transduction system(s) and the neuronal cytoskeleton, via regulation of phosphorylation of *tau* microtubule-associated protein (Sadot et al., 1996). Moreover, these studies propose a possible correlation between the cholinergic deficiency and *tau* hyperphosphorylation in AD. It can be speculated that activation of m1 mAChRs might provide a novel treatment strategy for AD by modifying *tau* processing in the brain and perhaps delaying the formation and accumulation of overphosphorylated *tau* (Sadot et al., 1996).

Tacrine is the only approved drug for the treatment of AD. Notably AF102B is more effective than tacrine in improving selective attention in monkeys (Siembieda et al., 1995). The favorable results with m1 agonists such as AF102B in rodent models (see Fisher and Barak, 1994), monkeys (Siembieda et al., 1995) and finally in AD patients (Fisher and Barak, 1994) encourage controlled long term clinical trials to evaluate the therapeutic potential of this and other M1 (or m1) agonists in AD. Eventually, clinical studies can be designed to determine whether m1 agonists may reduce Aß and/or *tau* overphosphorylation. Notably, such a hypothesis is now under evaluation. In this trial the effects of physostigmine and AF102B, respectively, are tested in AD patients on CSF, Aß and APP levels, before and after treatment (Nitsch et al., 1996). Thus, it is not inconceivable to assume that M1 (or m1) agonists, in addition to their use as a cholinergic replacement, may have a unique value in delaying the progression of AD.

ACKNOWLEDGMENTS

Supported in part by Snow Brand, Japan.

REFERENCES

Alberch J, Gurwitz D, Fisher A and Mount HTJ (1995): Novel muscarinic M1 receptor agonists promote survival of CNS neurons in primary cell culture. *Soc Neurosci Abstr* 21: 2040.

Bowen DM, Francis PT, Chessell P and Webster MT (1994): Neurotransmission - the link integrating Alzheimer research? *TINS* 17:149-150.

Buxbaum JD, Oishi M, Chen HI, Pinkas-Kramarski R, Jaffe EA, Gandy SE and Greengard P (1992): Cholinergic agonists and interleukin 1 regulate processing and secretion of the Alzheimer ß/A4 amyloid protein precursor. *Proc Natl Acad Sci USA* 89:10075-10078.

Court JA AND Perry EK (1991): Dementia the neurochemical basis of putative transmitter orientated therapy. *Pharmac Ther* 52:423-443.

Eckols K, Bymaster FP, Mitch CH, Shannon HE, Ward JS and DeLapp NW (1995): The muscarinic M1 agonist xanomeline increases soluble amyloid precursor protein release from CHO-m1 cells. *Life Sci* 57:1183-1190.

Emmerling RM, Moore CJ, Doyle PD, Carroll RT and Davis RE (1993): Phospholipase A2 activation influences the processing and secretion of the amyloid precursor protein. *Biochem Biophys Res Commun* 197:292-297.

Farber SA, Nitsch RM, Schulz JG and Wurtman RJ (1995): Regulated secretion of ß-amyloid precursor protein in rat brain. *J Neurosci* 15: 7442-7451.

Fisher A and Barak D (1994): Progress and perspectives in new muscarinic agonists. *Drug News & Perspectives* 7:453-464.

Gurwitz D, Haring R, Heldman E, Fraser CM, Manor D and Fisher A (1994): Discrete activation of transduction pathways associated with acetylcholine m1 receptor by several muscarinic ligands. *Europ J Pharmacol (Mol Pharmacol)* 267:21-31.

Gurwitz D, Haring R, Pinkas-Kramarski R, Stein R, Heldman E, Karton Y and Fisher A (1995): NGF-dependent neurotrophic-like effects of AF102B, an M1 muscarinic agonist, in PC12M1 cells. *Neuro Report* 6:485-488.

Haring R, Gurwitz D, Barg J, Pinkas-Kramarski R, Heldman E, Pittel Z, Wengier A, Meshulam H, Marciano D, Karton Y and Fisher A (1994): Amyloid precursor protein secretion via muscarinic receptors: Reduced desensitization using the M1-selective agonist AF102B. *Biochem Biophys Res Comm* 203:652-658.

Haring R, Gurwitz D, Barg J, Pinkas-Kramarski R, Heldman E, Pittel Z, Danenberg HD, Wengier A, Meshulam H, Marciano D, Karton Y and Fisher A (1995): NGF promotes amyloid precursor protein secretion via muscarinic receptor activation. *Biochem Biophys Res Comm* 213:15-23.

Harrison PJ, Barton AJL, McDonald B and Pearson RCA (1991): Alzheimer's disease: specific increases in a G-protein subunit (Gs) mRNA in hippocampal and cortical neurons. *Mol Brain Res* 10:71-81.

Haring R, Heldman E, Pittel Z, Kloog Y, Marciano D, Danenberg H, Kushnir M and Fisher A (1996): Transduction pathways mediating muscarinic-stimulated amyloid precursor proteins secretion and cell differentiation. *Eur Soc Neurochem,* Groningen, Netherlands, June 15-20.

Hung AY, Haass C, Nitsch RT, Qiu WQ, Citron M, Wurtman RJ, Growdon JH and Selkoe DJ (1993): Activation of protein kinase C inhibits cellular production of the amyloid ß-protein. *J Biolog Chem* 268:22959-22962.

Lee LM-Y, Balin BJ, Otvos L and Trojanowski JQ (1991): A68: A major subunit of paired helical filaments and derivatized forms of normal *tau*. *Science* 251:675-678.

Lindenboim L, Pinkas-Kramarski R, Sokolovsky M and Stein R (1995): Activation of muscarinic receptors inhibits apoptosis in PC12M1 cells. *J Neurochem* 64:2491-2499.

Mount HTJ, Dreyfus CF and Black IB (1994): Muscarinic stimulation promotes cultured Purkinje cell survival: A role for acetylcholine in cerebellar development? *J Neurochem* 63: 2065-2073.

Nitsch RN, Slack BE, Wurtman RG and Growdon JH (1992): Release of Alzheimer amyloid precursor derivatives stimulated by activation of muscarinic acetylcholine receptors. *Science* 258: 304-307.

Nitsch R, Tennis M, Growden J (1996): Regulation of AAP Processing by Stimulation of Muscarinic Acetylcholine Receptors in AD Patients. *Abstract 4th Intl Symposium on Advances in Alzheimer Therapy,* Nice, France, April 10-14.

Pinkas-Kramarski R, Stein R, Lindenboim L and Sokolovsky M (1992): Growth factor-like effects mediated by muscarinic receptors in PC12M1 cells. *J Neurochem* 59:2158-2166.

Pittel Z, Heldman E, Haring R, Barg J and Fisher A (1996): AF102B, an M1 Agonist, potentiates Amyloid Precursor Protein (APP) Secretion from Cerebrocortical but not from Cerebellar Slices. *Abstract 4th Intl Symposium on Advances in Alzheimer Therapy,* Nice, France, April 10-14.

Sadot E, Gurwitz D, Barg J, Behar L, Ginzburg I and Fisher A (1996): Activation of m1-muscarinic acetylcholine receptor regulates *tau* phophorylation in transfected PC12 cells. *J Neurochem* 66:877-880.

Siembieda D, Fitten LJ, O'Neill J, Crawford KC, Halgren E, Fisher A and Perryman KM (1995): Memory-enhancing effects of AF102B and THA in monkeys. *Soc Neurosci Abstr* 21:167.

Alzheimer Disease: From Molecular Biology to Therapy
edited by R. Becker and E. Giacobini

FREE RADICAL SCAVENGERS BLOCK THE ACTIONS OF β-AMYLOID ON NEURONS IN TISSUE CULTURE

J. Steven Richardson and Yan Zhou
Department of Pharmacology and Department of Psychiatry,
University of Saskatchewan, Saskatoon, SK, Canada

INTRODUCTION

The invariable co-localization of ß-amyloid, an insoluble polypeptide comprised of about 40 amino acids, with the neuritic plaques and neurofibrillary tangles that are characteristic of Alzheimer's disease (AD), has focused considerable world-wide research activity on the role of ß-amyloid in the progressive loss of synapses and neurons that is also characteristic of AD. ß-Amyloid is toxic to neurons in tissue culture but generally only after aging for several days in solution when the initially soluble ß-amyloid strands aggregate and form clumps. A fragment of ß-amyloid, the amino acids in positions 25 to 35 of the 40 amino acids making up the ß-amyloid found in the brains of Alzheimer's patients, aggregates very rapidly in aqueous solution (Pike et al., 1993), and is cytotoxic to neurons in tissue culture (Kumar et al., 1994) without the aging process needed to develop the toxicity of ß-amyloid. ß-Amyloid 1 to 40 disrupts the regulation of intracellular calcium (Mattson et al., 1992) and it has been considered that this disruption of cytosolic calcium is responsible for the neurotoxic actions of ß-amyloid.

The ß-amyloid molecule itself is a fragment of a much larger precursor protein of about 700 amino acids. The amyloid precursor protein (APP) appears to be synthesized, inserted into cellular membranes and normally cut within the ß-amyloid region to release a soluble protein (sAPP) that participates in the regulation of cytosolic calcium. The sAPP seems to have neuroprotective and neural growth inducing functions, and reduces the damage to neurons exposed to excitotoxic amino acids or to the neurotoxic forms of ß-amyloid (Schubert and Behl, 1993; Mattson et al., 1993). To synthesize the neuroprotective sAPP, the precursor molecule that is inserted into the cell membrane is cut in such a way that some of the amino acids needed to make ß-amyloid are on the sAPP molecule, and some are on the fragment still in the membrane. In the normal processing of the precursor protein, ß-amyloid is not formed and it appears that the neurotoxic ß-amyloid is produced by aberrant mechanisms. Some of these aberrant processes may involve hereditary abnormalities related to the genes for the amyloid precursor protein or for the enzymes or other biochemical processes involved in the synthesis and degradation of sAPP. However, since most cases of AD do not show evidence of a hereditary link, mechanisms other than errant genes may play a more universal role.

The possible involvement of free radicals and oxidative stress in the pathogenesis of AD has been under consideration for quite some time but little evidence directly supporting this hypothesis has been published (Subbarao et al., 1990; Zhou et al., 1995). However, in recent years evidence that free radicals are involved in the neurotoxic actions of ß-amyloid *in vitro* has been reported by several laboratories. Butterfield and coworkers demonstrated by electron paramagnetic resonance spectroscopy that ß-amyloid 25 to 35 in aqueous solution generates free radical spin adducts (Hensley et al., 1994; Harris et al., 1995), and Mattson and coworkers have shown that the disruption of cytosolic calcium produced by ß-amyloid is accompanied by increased oxidation in primary cultures of hippocampal neurons (Goodman and Mattson, 1994; Mattson and Goodman, 1995). The present study used PC 12 cells, a rat neuroblastoma cell line, as a model of brain neurons to investigate the effects of antioxidant agents on the calcium disrupting and the cytotoxic actions of the ß-amyloid 25 to 35.

MATERIALS AND METHODS

PC 12 cells, grown in collagen-coated flasks, were dislodged from the bottom of the flask by a stream of media ejected from a pipette, dispersed through a 22 gauge hypodermic needle, and used suspended in Krebs-HEPES buffer for cytosolic calcium determination, or plated in the wells of a 96 well tissue culture plate for the cytotoxicity studies. Cytosolic calcium was measured by the Fura-2 method using a Jasco CAF-100 Calcium Analyzer. Cell viability was determined with the MTT assay using a microelisa plate reader.

ß-Amyloid 25 to 35, purchased from Sigma (St Louis, MO), was dissolved in distilled water and was added to the PC 12 cells to give final concentrations as indicated in the Results. The other compounds were dissolved in distilled water or in culture medium and were added to the PC 12 cell preparations prior to the addition of the ß-amyloid 25 to 35.

RESULTS

Basal cytosolic free calcium levels in the PC 12 cells was about 100 nM. ß-Amyloid 25 to 35 (20 μM to 160 μM) produced a concentration-dependent increase in the cytosolic calcium of PC 12 cells that was evident within seconds of adding the ß-amyloid solution. The 40 μM concentration of ß-amyloid increased cytosolic calcium levels by about 60 nM. Pretreating the PC 12 cells with 1 to 40 μM U-83836E, a free radical scavenger drug of the lazaroid family, reduced the effects of 40 μM ß-amyloid 25 to 35 on cytosolic calcium in a concentration-dependent manner, with cytosolic calcium remaining near basal levels at U-83836E concentrations of 10 μM or higher. Pretreatment with 140 μM vitamin E or with 1 mM cholesterol, also shifted the ß-amyloid concentration-response curve to the right. The calcium channel blockers nifedipine (20 μM) and cobalt (1 mM) did not alter the effects of ß-amyloid 25 to 35 on cytosolic calcium. ß-Amyloid did not change the free cytosolic calcium of PC 12 cells in calcium-free buffer.

The viability of PC 12 cells exposed to ß-amyloid 25 to 35 (.01 nM to 100 μM), as determined by the MTT assay performed 24 hours later, was reduced in a concentration-dependent manner. The 40 μM concentration of ß-amyloid 25 to 35 decreased 24 hr cell viability by about 40%. The concentration-response curve of ß-amyloid on cell viability was shifted to the right by 10 μM U-83836E, by 140 μM vitamin E, and by 200 μM cholesterol. At a concentration of 10 μM, U-83836E produced a 25% reduction in the cytotoxic effects of 40 μM ß-amyloid 25 to 35. Nifedipine (1 to 40 μM) and cobalt (25 to 200 μM) did not alter the cytotoxicity of 40 μM ß-amyloid 25 to 35. Moreover, cobalt alone reduced 24 hr cell viability at concentrations above 100 μM, with 150 μM cobalt producing as great an effect as 40 μM ß-amyloid 25 to 35. Cell viability was also reduced by about 20%, at 50 μM concentrations of U-83836E.

DISCUSSION

These results suggest that the activity of free radicals are central to the deleterious effects of ß-amyloid on neurons in tissue culture. Both ß-amyloid 25 to 35 (Hensley et al., 1994) and ß-amyloid 1 to 40 (Harris et al., 1995) have been shown to generate free radical spin adducts in aqueous solutions. And free radical scavenging agents have been shown to block the cytotoxic effects of both ß-amyloid 1 to 40 (Behl et al., 1992; Kumar et al., 1994) and ß-amyloid 25 to 35 (the present study). The free radicals generated by ß-amyloid appear to be necessary for the entry of extracellular calcium into PC 12 cells exposed to ß-amyloid and scavenging these radicals prevents the ß-amyloid-induced elevation of cytosolic calcium. However, even though the cytotoxic actions of ß-amyloid are blocked by antioxidants, the free radicals responsible for reducing cell viability may include radicals from other sources besides those generated by ß-amyloid. One such source is mitochondrial activity. Free radicals are continuously being formed in the mitochondria by various enzymes and by the electron transport chain. The mitochondria are also involved in the homeostatic regulation of cytosolic calcium, and in the presence of excessive calcium there is an increase in the mitochondrial generation of free radicals (Hanukoglu et al., 1993). In general, free radicals are highly reactive and very short-lived and interact only with molecules in the immediate vicinity of their site of formation. There is as yet no evidence that extracellular ß-amyloid enters cells or that ß-amyloid is formed within cells. However, by inducing the mitochondria to generate free radicals in response to elevated cytosolic calcium, ß-amyloid increases intracellular oxidative stress without itself being present intracellularly.

The formation of the cytoprotective sAPP from the membrane bound precursor protein involves the action of an α-secretase enzyme. It has been postulated that the formation of ß-amyloid involves a ß-secretase enzyme. However, the enzymatic nature of the ß-amyloid forming event has not been established nor has it been determined how an enzyme might cleave the precursor protein within the transmembrane domain needed to release a fragment containing all of the amino acids needed to form ß-amyloid. This process need not involve enzymes at all but rather could be mediated by free radicals. A free radical driven system may not be very efficient but by continuing over many decades of a person's life, sufficient product could be formed to account for the pathology of AD. These considerations suggest the following hypothetical scenario. In response to damage to its membrane by free radicals, a cell activates repair processes that include the insertion of amyloid precursor protein into its membrane. Free radicals generated by a peroxidation wave cascading through the lipids of the cell's membrane, cleave some molecules of precursor protein at the transmembrane locus necessary for the release of the amyloidogenic sequence of amino acids. Some of these are cleaved (perhaps by radicals or perhaps by the ß-secretase enzyme postulated by others) to produce the soluble ß-amyloid peptide containing about 40 amino acids. Upon exposure to free radicals in the extracellular fluid, strands of this peptide aggregate, become insoluble, and begin to form additional free radicals which damage nearby cells. These newly damaged cells make precursor protein, some of which ends up as additional molecules of the destructive ß-amyloid rather than the protective sAPP. The ß-amyloid-generated free radicals also elevate cytosolic calcium which increases mitochondrial formation of free radicals which produce intracellular oxidative damage which leads to the death of the cell. When a sufficient number of cells have died such that the surviving cells can no longer maintain the normal functions of the brain, the person begins to experience the symptoms of AD. Although this ß-amyloid/free radical amplification hypothesis is consistent with current data, it remains to be determined if it bears any resemblance to the actual pathophysiology of AD.

ACKNOWLEDGMENTS

Supported in part by a grant from the Scottish Rite Foundation of Canada, and by donations from individuals in Saskatchewan.

REFERENCES

Behl C, Davis J, Cole GM and Schubert D (1992): Vitamin E protects nerve cells from amyloid ß protein toxicity. *Biochem Biophys Res Communs* 186:944-950.

Goodman Y and Mattson MP (1994): Secreted forms of ß-amyloid precursor protein protect hippocampal neurons against amyloid ß-peptide-induced oxidative injury. *Exper Neurol* 128:1-12.

Hanukoglu I, Rapoport R, Weiner L and Sklan D (1993): Electron leakage from the mitochondrial NADPH-adrenodoxin reductase-adrenodoxin-p450scc (cholesterol side chain cleavage) system. *Arch Biochem Biophys* 305:489-498.

Harris ME, Hensley K, Butterfield DA, Leedle RA and Carney JM (1995): Direct evidence of oxidative injury produced by the Alzheimer's ß-amyloid peptide (1-40) in cultured hippocampal neurons. *Exper Neurol* 131:193-202.

Hensley K, Carney JM, Mattson MP, Aksenova M, Harris M, Wu JF, Floyd RA and Butterfield DA (1994): A model for ß-amyloid aggregation and neurotoxicity based on free radical generation by the peptide: relevance to Alzheimer disease. *Proc Natl Acad Sci USA* 91:3270-3274.

Kumar U, Dunlop DM and Richardson JS (1994): The acute neurotoxic effect of ß-amyloid on mature cultures of rat hippocampal neurons is attenuated by the anti-oxidant U-78517F. *Intern J Neurosci* 79:185-190.

Mattson MP, Cheng B, Davis D, Bryant K, Lieberberg I and Rydel RE (1992): ß-amyloid peptides destabilize calcium homeostasis and render human cortical neurons vulnerable to excitotoxicity. *J Neurosci* 12:379-389.

Mattson MP, Cheng B, Culwell AR, Esch FS, Lieberberg I and Rydel RE (1993): Evidence for excitoprotective and intraneuronal calcium-regulating roles for secreted forms of ß-amyloid precursor protein. *Neuron* 10:243-254.

Mattson MP and Goodman Y (1995): Different amyloidogenic peptides share a similar mechanism of neurotoxicity involving reactive oxygen species and calcium. *Brain Res* 676:219-224.

Pike CJ, Burdick D, Walencewicz AJ, Glabe CG and Cotman CW (1993): Neurodegeneration induced by ß-amyloid peptides *in vitro*: the role of peptide assembly state. *J Neurosci* 13:1676-1687.

Schubert D and Behl C (1993): The expression of amyloid ß protein precursor protects nerve cells from ß-amyloid and glutamate toxicity and alters their interaction with the intracellular matrix. *Brain Res* 629:275-282.

Subbarao KV, Richardson JS and Ang LC (1990): Autopsy samples of Alzheimer's cortex show increased peroxidation *in vitro*. *J Neurochem* 55:342-345.

Zhou Y, Richardson JS, Mombourquette MJ and Weil JA (1995): Free radical formation in autopsy samples of Alzheimer and control cortex. *Neurosci Letts* 195:89-92.

Alzheimer Disease: From Molecular Biology to Therapy
edited by R. Becker and E. Giacobini

THERAPEUTIC STRATEGIES IN ALZHEIMER'S DISEASE

M. Flint Beal
Neurology Service, Massachusetts General Hospital, Boston, MA

INTRODUCTION

A substantial amount of evidence has implicated defects in energy metabolism in Alzheimer's disease (AD) (Beal, 1992). Several studies have utilized positron emission tomography (PET) to show that a decrease in glucose metabolism occurs in parietal and temporal cortex early in the course of AD (Duara et al., 1986; Haxby et al., 1986). There is high metabolic rate for oxygen relative to glucose consistent with impaired glucose utilization. Furthermore, a recent study has demonstrated an increase in organic phosphate relative to phosphocreatine using phosphorous nuclear magnetic resonance spectroscopy (Smith et al., 1995).

Parker et al. (1990) and colleagues initially reported reduced cytochrome oxidase activity in platelets of AD patients. Three groups have subsequently documented that there is indeed reduced cytochrome oxidase activity in cerebral cortex of AD patients; the initial report was that of Kish et al. (1992). We subsequently confirmed their results and found a significant decrease of cytochrome oxidase activity of 25-30% in four different regions of cerebral cortex (Mutisya et al., 1994). There are no consistent changes in other electron transport enzymes. A study of highly purified mitochondria from AD patient brains showed that there was a 50% decrease in cytochrome oxidase activity with only small decreases in other electron transport enzymes (Parker et al., 1994). Concentrations of cytochrome aa3, a component of cytochrome oxidase, were normal, arguing for a decrease in catalytic activity rather than a reduction in enzyme concentrations. This raises the possibility that there could be mutations in the cytochrome oxidase subunits encoded by mitochondrial DNA. Consistent with this possibility it has been recently demonstrated that it is possible to transfer cytochrome oxidase defects from platelets of affected patients to mitochondria deficient cell lines (Glasco et al., 1995). This suggests that there is a defect which is encoded on the mitochondrial genome. Other work has demonstrated several point mutations in mitochondrial DNA which seem to be associated with the disease; in particular, a transfer RNA glutamine mutation reported by Shoffner et al. (1993). This observation was recently confirmed (Hutchin and Cortopassi, 1995).

A defect in the electron transport chain at the cytochrome oxidase level may lead to increased free radical generation. Inhibition of cytochrome oxidase with azide leads to increased free radical generation as assessed using electron paramagnetic resonance technology. Numerous reports have demonstrated increased oxidative damage in AD. Protein oxidation products are increased in AD brain samples as compared to young controls (Smith et al., 1991). In addition, a recent novel technique for examining free radical damage to proteins and lipids using specific spin trapping techniques has demonstrated increased oxidative damage in AD postmortem tissue (Hensley et al., 1995). Several studies demonstrate evidence of increased lipid peroxidation in AD cerebral cortex (Subbarao et al., 1990; Hajimohammadreza and Brammer, 1990; Lovell et al., 1995). A reduction in glutamine synthetase activity, which is particularly vulnerable to oxidative stress, has also been reported in AD postmortem tissue (Smith et al., 1991). We recently found the first direct evidence for oxidative damage to DNA in AD postmortem brain tissue (Mecocci et al., 1994). We examined 13 AD patients as compared to 13 age-matched controls. There was a 50% increase in oxidative damage to nuclear DNA in the AD samples; however, there was a 3-fold increase in oxidative damage to mitochondrial DNA as compared with the controls. On a regional basis the most marked changes were in parietal cortex. This is of interest due to observations of decreased glucose metabolism in this brain region as an early and consistent finding in AD patients. Oxidative damage to mitochondrial DNA, however, appeared to be widespread, even involving the cerebellum which shows little pathologic involvement in AD. A previous study found a 2-fold increase in DNA strand breaks in cerebral cortex of AD patients, which is consistent with our finding since strand breaks may be a manifestation of oxidative damage to DNA. More recent studies demonstrate evidence of increased oxidative stress at a cellular level. It has been demonstrated that there is increased staining of senile plaques and neurofibrillary tangles with antibodies to superoxide dismutase or catalase (Furuta et al., 1995), and neurofibrillary tangle containing neurons stained with antibodies to glycation end-products, malondialdehyde, and heme-oxygenase epitopes (Smith et al., 1994a,b; Schipper et al., 1995). These findings, therefore, directly localize oxidative stress at a cellular level. Furthermore, messenger RNA levels for heme-oxygenase have been shown to be significantly elevated in AD postmortem tissue.

Consequences of increased free radical formation may be aggregation and deposition of amyloid, as well as cross linking of tau protein which may contribute to a formation of neurofibrillary tangles (Dyrks et al., 1992). A defect in cytochrome oxidase activity may also lead to increased generation of amyloidogenic fragments as demonstrated in cultured neurons (Gabuzda et al., 1994). A recent study demonstrated that neurons cultured from Down's syndrome fetal tissue show increased cell death by an apoptotic mechanism between 7 and 14 days in culture, whereas normal neurons can be cultured up to 100 days (Busciglio and Yankner, 1995). This increase in an apoptotic cell death is associated with increased free radical generation. This finding, therefore, may be a link between the accelerated course of AD in Down's syndrome patients and free radical generation.

Another potential source for increased free radicals in AD postmortem tissue is activated microglia. Activated microglia have been associated with senile plaques, and are known to generate both superoxide radical and nitric oxide. It has been recently demonstrated that microglia exposed to b-amyloid have increased nitric oxide production (Meda et al., 1995; Goodwin et al., 1995). A consequence of increased generation of nitric oxide and superoxide may be the generation of peroxynitrite, which can then lead to nitration of tyrosine residues. This may again contribute to neurofibrillary tangle formation by impairing the normal interactions of cytoskeletal components.

If increased oxidative damage or protein nitration do indeed occur in AD postmortem tissue, then this suggests that strategies to either block nitric oxide production or to scavenge free radicals might be efficacious in slowing the progress of the disease. We have investigated a number of strategies to treat neurodegenerative diseases with either free radical scavengers or with nitric oxide synthase inhibitors. We initially examined the effects of coenzyme Q_{10}, which is an essential cofactor of the electron transport chain as well as a potent antioxidant compound (Beal et al., 1994). It is a particularly effective antioxidant in preventing lipid peroxidation. It is localized in the mitochondrial membrane, which may make it a particularly effective antioxidant, since the mitochondria appear to be the major source of free radicals in the cell. We first examined the effects of oral administration of coenzyme Q_{10} at doses of 100, 200 and 400 mg/kg for 10 days prior to intrastriatal injections of the mitochondrial toxin malonate. This dosing regimen shows that there is dose dependent neuroprotective effects which appear to be maximal at a dose of 200 mg/kg. We also examined the effects of combining oral treatment with coenzyme Q_{10} with nicotinamide. This combination produced additive neuroprotective effects. Furthermore, coenzyme Q_{10} protected against malonate-induced ATP depletions and malonate-induced increases in lactate concentrations as assessed using magnetic resonance spectroscopy.

Another novel strategy for blocking the generation of free radicals utilizes free radical spin traps - compounds that react with free radicals to form more stable adducts. They can serve as free radical scavengers, effective in attenuating both age related oxidative damage to protein and ischemic lesions *in vivo.* We examined the effects of pretreatment with n-tert-butyl-a-(2-sulphophenyl)-nitrone (S-PBN) (Schulz et al., 1995a), and found that S-PBN significantly attenuated striatal excitotoxic lesions in rats produced by N-methyl-D-aspartate, kainic acid and AMPA. We also investigated the effects of S-PBN and PBN on striatal lesions produced by 3-acetylpyridine, MPP^+ and malonate. 3-Acetylpyridine is a nicotinamide analogue that, after systemic administration, produced selective lesions in the inferior olive and substantia nigra. MPP^+ is the major metabolite of 1-methyl-4-phenyl-1,2,3,6-tetrahydropyridine (MPTP), which produces a Parkinsonism syndrome in both humans and experimental animals. We and others demonstrated that malonate, a competitive inhibitor of succinate dehydrogenase, can produce selective striatal lesions which resemble Huntington's disease. We found that intrastriatal injections of all 3 compounds produced both energy depletion and, secondarily, excitotoxic striatal lesions. Lesions produced by all 3 compounds were significantly attenuated by pretreatment with S-PBN. It was particularly effective against 3-acetylpyridine which may be due to depletion of NADPH, which is required for a regeneration of oxidized glutathione by glutathione reductase.

To verify that the effects of S-PBN were due to free radical scavenging, we used the salicylate method to detect hydroxyl radicals (Schulz et al., 1995a). Both 3-acetylpyridine and malonate striatal injections resulted in an increased production of 2,3 and 2,5-dihydroxybenzoic acid, products of hydroxylation of salicylate. Pretreatment with S-PBN significantly attenuated malonate-induced increases in 2,3-dihydroxybenzoic acid production, consistent with a free radical scavenging mechanism. S-PBN had no significant effects on malonate-induced ATP reductions. We also examined whether S-PBN might have effects on excitatory amino acid receptors *in vivo*. Electrophysiologic studies showed that S-PBN, in contrast to MK-801, had no significant effect on spontaneous activity of striatal neurons.

We examined the effects of free radical spin trapping compounds on toxicity produced by 3-nitropropionic acid. 3-Nitropropionic acid is an irreversible inhibitor of succinate dehydrogenase. It is of particular interest since it can produce selective striatal lesions in man following accidental ingestion. These lesions are associated with an encephalopathy and a delayed onset of dystonia. We demonstrated that chronic administration of 3-nitropropionic acid to both rats and primates produces selective striatal lesions which closely mimic the pathologic and neurochemical features of Huntington's disease. Furthermore, in primates one can reproduce the characteristic movement disorder which accompanies Huntington's disease. We found that continuous administration of DMPO, another spin trapping compound, could significantly attenuate striatal lesions produced by 3-nitropropionic acid.

Since excitotoxicity and oxidative stress may produce neuronal injury by sequential interactive mechanisms, we examined whether combining a free radical spin trap with an NMDA antagonist might produce additive neuroprotective effects (Schulz et al., 1996). The combination of S-PBN with

MK-801 had an additive neuroprotective effect against lesions produced by both malonate and 3-acetylpyridine, which was significantly better protection than MK-801 alone.

Another potential neuroprotective strategy against oxidative damage is to block the production of nitric oxide. The interaction of nitric oxide with superoxide leads to the formation of peroxynitrite. This reaction occurs in an extremely fast rate of 6.7×10^9 M/sec, which is essentially diffusion-limited and does not require transition metals. Therefore, it depends on the concentration of superoxide and nitric oxide in the cell. Peroxynitrite is a highly oxidizing agent which can cause tissue damage through a hydroxyl radical-like mechanism. At physiological pH, peroxynitrite may be able to diffuse over several cell diameters and produce damage by oxidizing lipids, proteins, and DNA. It also can inhibit mitochondrial function.

A relatively selective inhibitor of the neuronal isoform of nitric oxide synthase is 7-nitroindazole (NI). Although it inhibits both the endothelial and the macrophage isoforms *in vitro, in vivo* studies showed no effects on blood pressure or on endothelium dependent blood vessel relaxation or acetylcholine induced blood vessel relaxation. It has antinociceptive effects. We studied the neuroprotective effects of 7-NI against striatal lesions produced by excitotoxins *in vivo* (Schulz et al., 1996). 7-NI significantly attenuated lesions produced by NMDA, but it had no effects on lesions produced by either kainic or AMPA, consistent with results reported *in vitro.* We also investigated the effects of 7-NI in animal models of neurodegenerative diseases. 7-NI dose-dependently protected against MPTP-induced dopamine depletion using 2 different dosing regimens, which produced varying degrees of dopamine depletion (Schulz et al., 1995c). At a dose of 50 mg/kg, 7-NI produced almost complete protection in both dosing regimens. Furthermore, we found that MPTP-induced increases in 3-nitrotyrosine, an indicator of peroxynitrite, were attenuated by 7-NI.

We examined the effects of 7-NI on neurotoxicity produced by striatal injections of the reversible succinate dehydrogenase inhibitor, malonate (Schulz et al., 1995b). 7-NI dose-dependently attenuated striatal malonate lesions and the protection was reversed by L-arginine but not by D-arginine. 7-NI protected against malonate-induced decreases in ATP and increases in striatal lactate as assessed *in vivo* using chemical shift magnetic resonance spectroscopy. 7-NI had no effects on spontaneous electrophysiological activity in the striatum, arguing that its neuroprotective effects were not mediated by an interaction with excitatory amino acid receptors.

We also examined the effects 7-NI on 3-nitropropionic acid toxicity (Schulz et al., 1995b). 7-NI was coadministered with 3-nitropropionic acid for 5 days. Under these conditions all of the control animals developed both behavioral toxicity and striatal lesions, whereas there was complete protection in the 7-NI treated animals. Furthermore, 3-nitropropionic acid administration produced increases in hydroxyl radical and 3-nitrotyrosine levels in the striatum, which were significantly attenuated by 7-NI treatment. These findings, therefore, argue for a role of peroxynitrite in 3-nitropropionic acid neurotoxicity which is blocked by inhibiting neuronal nitric oxide synthase.

As discussed above, there is increasing evidence which implicates oxidative stress in the pathogenesis of AD. Potential sources for oxidative stress are mitochondrial free radical generation as a consequence of defects in cytochrome oxidase, or the generation of free radicals by activated macrophages. Oxidative stress could then play a direct role in the pathogenesis of cell death as well as amyloid deposition in neurofibrillary tangle formation in AD. If this is the case then strategies to scavenge free radicals might prove efficacious in slowing the disease process. Our recent evidence suggests that strategies to either block the generation of nitric oxide using neuronal nitric oxide synthase inhibitors, or free radical scavenging using the spin trapping compounds or coenzyme Q_{10} are both effective strategies for blocking cell death in animal models of neurodegenerative diseases. It is, therefore, possible that similar strategies might be useful in the treatment of AD.

REFERENCES

Beal MF (1992): Does impairment of energy metabolism result in excitotoxic neuronal death in neurodegenerative illnesses? *Ann Neurol* 31:119-130.

Beal MF, Henshaw DR, Jenkins BG, Rosen BR and Schulz JB (1994): Coenzyme Q_{10} and nicotinamide block striatal lesions produced by mitochondrial toxin malonate. *Ann Neurol* 36:882-888.

Busciglio J and Yankner BA (1995): Apoptosis and increased generation of reactive oxygen species in Down's syndrome neurons *in vitro*. *Nature* 378:776-779.

Duara R, Grady C, Haxby J, Sundaram M, Cutler NR, Heston L, Moore A, Schlageter N, Larson S and Rapoport SI (1986): Positron emission tomography in Alzheimer's disease. *Neurology* 36:879-887.

Dyrks T, Dyrks E, Hartmann T, Masters C and Beyreuther K (1992): Amyloidogenicity of bA4 and bA4-bearing amyloid protein precursor fragments by metal-catalyzed oxidation. *J Biol Chem* 267:18210-18217.

Furuta A, Price DL, Pardo CA, Troncoso JC, Xu Z-S, Taniguchi and Martin LJ (1995): Localization of superoxide dismutases in Alzheimer's disease and Down's syndrome neocortex and hippocampus. *Am J Pathol* 146:357-367.

Gabuzda DH, Busciglio J, Chen LB, Matsudaira P and Yankner BA (1994): Inhibition of energy metabolism alters the processing of amyloid precursor protein and induces a potentially amyloidogenic derivative. *J Biol Chem* 6:13623-13628.

Glasco S, Miller SW, Thal LJ and Davis RE (1995): Alzheimer's disease hybrids manifest a cytochrome oxidase defect. *Soc Neurosci* 21:979.

Goodwin JL, Uemura E and Cunnick JE (1995): Microglial release of nitric oxide by the synergistic action of b-amyloid and IFN-g. *Brain Res* 692:207-214.

Hajimohammadreza I and Brammer M (1990): Brain membrane fluidity and lipid peroxidation in Alzheimer's disease. *Neurosci Lett* 112:333-337.

Haxby JV, Grady CL, Duara R, Schlageter NL, Berg G and Rapoport SI (1986): Neocortical metabolic abnormalities precede nonmemory cognitive deficits in early Alzheimer-type dementia. *Arch Neurol* 43:882-885.

Hensley K, Hall N, Subramaniam R, Cole P, Harris M, Aksenov M, Aksenova M, Gabbita SP, Wu JF, Carney JM, Lovell M, Markesbery WR and Butterfield DA (1995): Brain regional correspondence between Alzheimer's disease histopathology and biomarkers of protein oxidation. *J Neurochem* 65:2146-2156.

Hutchin T and Cortopassi G (1995): A mitochondrial DNA clone is associated with increased risk for Alzheimer disease. *Proc Natl Acad Sci USA* 92: 6892-6895.

Kish SJ, Bergeron C, Rajput A, et al. (1992): Brain cytochrome oxidase in Alzheimer's disease. *J Neurochem* 59:776-779.

Lovell MA, Ehmann WD, Butler SM and Markesbery WR (1995): Elevated thiobarbituric acid-reactive substances and antioxidant enzyme activity in the brain in Alzheimer's disease. *Neurology* 45:1594-1601.

Mecocci P, MacGarvey U and Beal MF (1994): Oxidative damage to mitochondrial DNA is increased in Alzheimer's disease. *Ann Neurol* 36:747-751.

Meda L, Cassatella MA, Szendrei GI, Otvos Jr L, Baron P, Villalba M, Ferrari D and Rossi F (1995): Activation of microglial cells by b-amyloid protein and interferon-g. *Nature* 374:647-650.

Mutisya EM, Bowling AC and Beal MF (1994): Cortical cytochrome oxidase activity is reduced in Alzheimer's disease. *J Neurochem* 63:2179-2184.

Parker WD Jr, Filley CM and Parks JK (1990): Cytochrome oxidase deficiency in Alzheimer's disease. *Neurology* 40:1302-1303.

Parker WD Jr, Parks J, Filley CM and Kleinschmidt-DeMasters BK (1994): Electron transport chain defects in Alzheimer's disease brain. *Neurology* 44:1090-1096.

Schipper HM, Cisse S and Stopa EG (1995): Expression of heme oxygenase-1 in the senescent and Alzheimer-diseased brain. *Ann Neurol* 37:758-768.

Schulz JB, Henshaw DR, Siwek D, Jenkins BG, Ferrante RJ, Cipolloni PB, Kowall NW, Rosen BR and Beal MF (1995a): Involvement of free radicals in excitotoxicity in vivo. *J Neurochem* 64:2239-2247.

Schulz JB, Matthews RT, Henshaw DR and Beal MF (1996): Neuroprotective strategies for treatment of lesions produced by mitochondrial toxins: implications for neurodegenerative diseases. *Neuroscience*, in press.

Schulz JB, Matthews RT, Jenkins BG, Ferrante RJ, Siwek D, Henshaw DR, Cipolloni PB, Mecocci P, Kowall NW, Rosen BR and Beal MF (1995b): Blockade of neuronal nitric oxide synthase protects against excitotoxicity in vivo. *J Neurosci* 15:8419-8429.

Schulz JB, Matthews RT, Muqit MMK, Browne SE and Beal MF (1995c): Inhibition of neuronal nitric oxide synthase by 7-nitroindazole protects against MPTP induced neurotoxicity in mice. *J Neurochem* 64:936-939.

Shoffner JM, Brown MD, Torroni A, Lott MT, Cabell P, Mirra SS, Beal MF, Yang C-C, Gearing M, Salvo R, Watts RL, Juncos JL, Hanson LA, Crain BJ, Fayad M and Wallace DC (1993): Mitochondrial DNA mutations associated with Alzheimer's and Parkinson's disease. *Genomics* 17:171-184.

Smith CD, Carney JM, Starke-Reed PE, et al. (1991): Excess brain protein oxidation and enzyme dysfunction in normal aging and Alzheimer disease. *Proc Natl Acad Sci USA* 88:10540-10543.

Smith CD, Pettigrew LC, Avison MJ, Kirsch JE, Tinkhtman AJ, Schmitt FA, Wermeling DP, Wekstein DR and Markesberry WR (1995): Frontal lobe phosphorus metabolism and neuropsychologic function in aging and in Alzheimer's disease. *Ann Neurol* 38:194-201.

Smith MA, Kutty K, Richey PL, Yan S-D, Stern D, Chader GJ, Wiggert B, Petersen RB and Perry G (1994a): Heme oxygenase-1 is associated with the neurofibrillary pathology of Alzheimer's disease. *Am J Pathol* 145:42-47.

Smith MA, Taneda S, Richey PL, Miyata S, Yan S-D, Stern D, Sayre LM, Monnier VM and Perry G (1994b): Advanced maillard reaction end products are associated with Alzheimer disease pathology. *Proc Natl Acad Sci USA* 91:5710-5714.

Subbarao KV, Richardson JS and Ang LC (1990): Autopsy samples of Alzheimer's cortex show increased peroxidation in vitro. *J Neurochem* 55:342-345.

Alzheimer Disease: From Molecular Biology to Therapy
edited by R. Becker and E. Giacobini

NEW SYNTHETIC BIOANTIOXIDANTS - ACETYLCHOLINESTERASE (AChE) INHIBITORS

Faina I. Braginskaya, Elena M. Molochkina, Olga M. Zorina, Irina B. Ozerova, Elena B. Burlakova
Institute of Biochemical Physics, Russian Academy of Sciences, Moscow, Russia

INTRODUCTION

Cholinergic enhancement and preserving by ChE inhibitors is considered the most fruitful approach to the treatment for cognitive and memory disorders as the manifestations of Alzheimer's disease (AD). On the other hand in the past decade some information has appeared concerning the action of antioxidants (AO) - inhibitors of free radical reactions on learning and memory, the effect owing to maintenance the proper composition and structure of membranes through lipid peroxidation (LPO) control (Burlakova, 1994). Some authors consider AO a new group of nootropic drugs (Voronina, 1994).

Earlier we showed the inhibitory action of AO (the derivatives of 3-hydroxypyridine, pyrimidine, benzimidazole and hindered phenols) on membranebound AChE of human erythrocytes (Braginskaya et al., 1992). The effects on AChE were shown to be both direct and mediated with lipid phase. As combining AO and antiAChE properties, AO seems to be promising for AD treatment. However, the antiAChE capacity of AO proved to be not very high: I_{50} (the concentration of 50 % inhibition) was 0.1- 1mM.

Prof. G. A. Nikiforov's group synthesized the compound N3, (Table I) which is an analogue of acetylcholine (ACh), where instead of acetyl fragment there is a radical of so called "phenozan"- β-(4-hydroxy-3,5-ditert. butylphenyl) propionic acid. Some of its derivatives were also synthesized containing "tail" radical R (H, alkyl C_1,C_8,C_{10},C_{12},C_{16}) neighboring to N, compounds NN 2-7 in Table I. Also, the "parent" AO phenozan with power AO effect possesses a number of various biological activities. It is worthwhile to emphasize its effects on learning and memory providing the changes in neuronal membrane lipid phase (Burlakova, 1994).

The main aims of the present study were to examine the action of the above mentioned newly synthesized compounds on the activity of soluble and membranebound AChE, to investigate their AO properties in the biological oxidation model and their effect on structural integrity of membranes. In other words, to explore if those compounds correspond to the criteria needed for practical use, inhibit LPO; exert antiAChE effect, “anchor” in membranes in order to prolong the effects with minimal perturbation of membrane.

METHODS

AChE activity was determined by the colorimetric method (Ellman et al., 1961) and by means of potenciometric titration of CH_3COOH (both methods with permanent registration of the reaction kinetics).

The decrease (in percentage of control value) of LPO rate (tested by accumulation of malondialdehyde type products) in mice brain homogenates in the presence of $2*10^{-5}$ M of AO was used as a measure of antioxidative activity (AOA).

The action of the AO on structural integrity of membranes was assessed by the capacity to induce erythrocyte hemolysis or (at low concentrations) to enhance sensitivity to ultrasonic hemolysis (Braginskaya et al., 1992).

RESULTS AND DISCUSSION

Table I shows AO properties of new compounds not to decrease as compared with phenozan. Incorporation of “anchoring” radicals into AO molecules even increases their AOA. This effect is probably the consequence of prolonged sojourn of AO in the membrane due to a long chain “tail”.

As it can be concluded from Table I, the changes in AO molecules brought about alterations in both inhibition type and inhibition power towards soluble enzyme. Probably the substances may interact with both active and peripheral sites of the enzyme in various modes, depending on the structure of the compounds. The replacement of tertiary N atom by quaternary one resulted in pronounced increase of enzyme-inhibitor association. Besides, incorporation into AO molecules of hydrophobic hydrocarbon radicals led to 10-fold decrease of inhibitory constant K_i, thus enhancing the inhibition. The percent of inhibition at the definite concentration of substrate showed the same effect. The “phenozan part” of AO molecule proved to be of no importance for the action on soluble enzyme. It was shown by using a precursor without phenozan moiety. Apparently, the long chain “tail” enhances the inhibitory effect of the drug, providing its interaction with appropriate hydrophobic effector sites of the enzyme. It also follows, from the analysis of Table I, that “tail” length (a number of C atoms in R) does not influence the capacity of the compounds to inhibit soluble AChE.

Compounds NN	AOA arbitrary units	Soluble enzyme inhibition type	inhibitory constants, $M*10^4$	inhibition % at $S_0=1.7*10^{-4}$ M ATCh[1] $AO\approx3*10^{-5}$M	Membrane enzyme I_{50},mM*10^2	inhibition type	inhibitory constants, $M*10^4$	inhibition % at $S_0=1.7*10^{-4}$ M ATCh $AO\approx3*10^{-5}$ M
0.	42±8							
1[2].	30±10	noncompet.	K_i=2.8	15	30±3	noncompet.	K_i=3.0	
2.	67±8	incompet.	K_i=5.0	16	10±1	noncompet.	K_i=1.0	
3.	40±10	mixed	K_i=0.51	43	25±2	noncompet.	K_i=2.5	
4.	80±5	compet.	K_i=0.025	81	5±0.5	mixed non	trivial	80 (0 at $S_0=5*10^{-5}$
5.	70±5	mixed	K_i=0.042	80	7±0.7			
6.	76±8				8±0.8			
7.	50±10	mixed	K_i=0.027	81	15±2	noncompet.	K_i=1.6	24
8. α-tocoph.	25±3							

TABLE I. Antioxidative and antiAChE effects of newly synthesized AO compounds (without preincubation). [1]acetylthiocholine [2]contains tertiary N atom and "phenolic" but not phenosan moiety.

It thus appears that the newly synthesized compounds, while possessing profound AOA, reveal an expressed antiAChE effect on the purified soluble enzyme.

The data in Table I show that the AO exert appreciable inhibitory action on membrane erythrocyte AChE. Incorporation of hydrophobic hydrocarbon radical into AO molecule resulted in expressed enhancement of inhibition capacity (the decrease of I_{50}). Types and parameters of inhibition of membrane AChE were determined for the compounds NN 1-4 and N 7. The drugs, NN 1 and 2, that contain a tertiary N atom showed the inhibitory constants of similar values both for soluble and for membranebound enzyme. Apparently the changes in the enzyme conformation, because of its including in membrane, does not make any important contribution to the drugs-AChE interaction. The compounds, quaternary ammonium salts (NN 3-7), proved to have some differences both in types and parameters of inhibition for soluble and membranebound AChE. Thus, it can be concluded that the inhibitory effect on membranebound form of enzyme is lower, except for the compound N 4. Compound N 4 exerts the same percent of inhibition at the definite concentration of substrate for both forms of AChE. However, the effect of this compound is dual. Either inhibition or activation can take place because of nontrivial mixed inhibition type at different substrate concentrations. The distinctions between action of the AO on soluble and membranebound AChE should be attributed to the membrane microenvironment that mediates drug-enzyme interaction. The influence of AO on membrane structure is supported by the following observations. Elongation of "tail" radical does not affect the inhibition of soluble AChE while inhibition of membrane AChE does weaken with the elongation of carbon chain. Figure 1 shows that the increase of "tail length" parallels the effect on structural integrity of erythrocyte membranes. For the compound N 4, kinetic parameters of the enzyme were determined after 30-min preincubation of original "concentrated" erythrocyte suspension with ethanolic solution of the AO. It can be seen from Table II that the mode of action on the enzyme depends on AO concentration. When AO content in reaction medium (after dilution of original suspension for AChE assay) was the same that had affected AChE in the case of AO adding directly at the moment of the reaction start, the similar effect took place with and without preincubation. As compared with the case of soluble AChE a non-competitive component arose. At the lower AO concentration it was apparently the changes in membrane that resulted in the opposite effect (see Table II) on kinetic parameters.

Thus, it appears that the synthetic AO which contain quaternary N, phenozan fragment and "anchoring" radical, can influence membrane AChE both through interaction with the enzyme (its active and/or peripheral centers) and through modifying membrane state, the direction of the modification depending on the AO concentration.

AO	concentration	K_M,rel. units	V_{max}	V_{max}/K_M
in original suspension, M	final in reaction medium, M	exp./contr.	rel. units	rel. units
A.				
0	0	1	1	1
$1.3*10^{-5}$	$0.9*10^{-7}$	0.3	0.35	1.25
$1.3*10^{-4}$	$0.9*10^{-6}$	2.5	1.3	0.5
B.				
0	$5*10^{-7}$	1	1	1
0	$5*10^{-6}$	3.7	1.4	0.4

TABLE II. The action of the AO N 4 on membranebound erythrocyte AChE with (A) and without (B) preincubation to provide or not AO entering membrane lipids.

It is of interest that the inhibitory capacity of the investigated AO (tertiary and quaternary ammonium salts) correlates with their AOA. A connection that exists between I_{50} of the compounds NN 1-7 toward membrane AChE and their AOA can be expressed with regression equation: $y=-0.127+0.146x$, where $y=I_{50}$, $x=1/AOA$; correlation coeff. $r=0.9925$ ($p<0.001$). This connection might be due to the role of membrane: since both AOA and antiAChE properties are dependent on binding to the membrane and on the duration of AO sojourn. For both suggestions, the charge on N and "tail length" are important. However, the same compounds exert such a correlation (the more AOA, the more inhibition capacity) toward soluble form of the

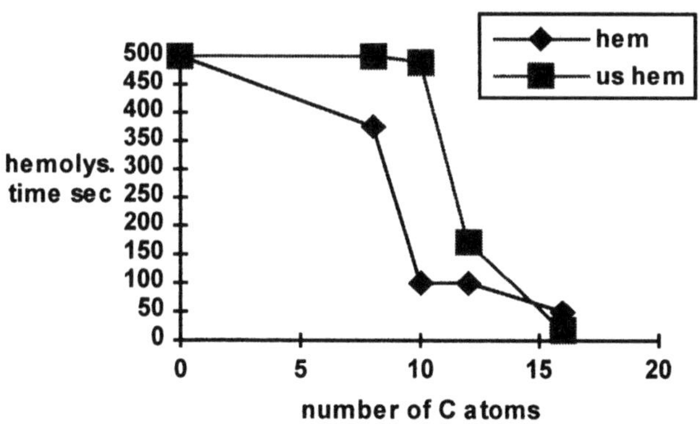

FIGURE 1. The influence of AO on the structural integrity of erythrocyte membranes. **hem** - induced, **us hem** - ultrasonic hemolysis in the presence of $2*10^{-4}$ and $2*10^{-5}$ M AO correspondingly.

AChE: $y=-2.97-3.3x$, where $y=\log KI$, $x=AOA$; $r=-0.7619$ ($0.1 \geq p > 0.05$). Apparently, the connection between the above mentioned properties exists also at the level of the enzyme itself, and might be due to electron density distribution in hybrid molecule.

From the presented data it can be concluded that the synthesized compounds correspond to the proposed criteria *in vitro*. It is worthwhile to examine some of these drugs *in vivo* to reveal the possibility of their use in AD therapy.

REFERENCES

Braginskaya FI, Zorina OM, Molochkina EM, Nikiforov GA, Burlakova EB (1992): Synthetic bioantioxidants: inhibitors of acetylcholinesterase activity. *Izvestiya Akademii Nauk Ser.biol.* 5:690-698.

Burlakova EB (1994): Antioxidant drugs as neuroprotective agents. In: *Alzheimer Disease: Therapeutic Strategies*, Giacobini E and Becker R eds. Birkhauser: Boston, p.p. 313-317.

Ellman GL, Courtney KD, Andres VJ, and Featherstone RM (1961): *Biochem Pharmacol* 7:88-96.

Voronina TA (1994): Nootropic drugs in Alzheimer disease treatment. New pharmacological strategies. In: *Alzheimer Disease: Therapeutic Strategies*, Giacobini E and Becker R eds. Birkhauser: Boston, p.p. 265-269.

Alzheimer Disease: From Molecular Biology to Therapy
edited by R. Becker and E. Giacobini

RATIONALE TO TREAT ALZHEIMER'S DISEASE WITH SELEGILINE - CAN WE PREVENT THE PROGRESSION OF THE DISEASE?

Paavo J. Riekkinen Sr., Keijo J. Koivisto, Eeva-Liisa Helkala
Department of Neurology, University of Kuopio, Kuopio, Finland

Kari J. Reinikainen, Olavi Kilkku, Esa Heinonen
Orion Corporation, Orion Pharma International, Clinical R&D, CNS Drugs, Turku, Finland

INTRODUCTION

Alzheimer's disease (AD) is a major cause of dementing illness in the developed countries and accounts for a disproportionate share of medical resource utilization and health care expenditures (Hay and Ernst, 1987). Unless an effective treatment regime or preventive measures are found for AD, the magnitude of the problem will continue to increase.

There are several experimental and clinical lines of evidence, suggesting that selegiline, a selective monoamine oxidase-B (MAO-B) inhibitor, might have therapeutic efficacy in the treatment of AD (Reinikainen, 1994). Animal and human studies have suggested that oxidative stress and increased deamination, leading to excessive formation of toxic endogenous or even exogenous by-products, may lead to damage of membranes and neuronal death in AD. Thus, inhibition of MAO-B leading to decreased production of free radicals and other possible neurotoxins can reduce these events as indicated by experimental work, where selegiline has shown neuroprotective effects in many different models (MPTP, DSP-4, 6-OHDA). Recent experimental evidence indicate that selegiline may have not only neuroprotective effect but also neuronal rescue effects (Tatton, 1993). Furthermore, selegiline treated aged rats have shown better cognitive performance than their controls, suggesting that selegiline may slow down age-related cognitive deterioration (Heinonen et al., 1995). Finally, 16-17 out of 19 available reports of clinical studies suggest that selegiline may have therapeutic effects in patients with AD.

We initiated in 1990 a randomized, double-blind, placebo-controlled, parallel group, three-year follow-up study to investigate the long-term efficacy and safety of selegiline in the treatment of AD. The evaluation of possible effect of selegiline in slowing down the progression of AD is the primary aim of this study.

PATIENTS AND METHODS

A total of 80 AD patients (40 in the selegiline group, 40 in the placebo group) with possible or probable AD according to the NINCDS-ADRDA criteria (McKhann et al., 1984) diagnosed at the Department of Neurology, Kuopio University Hospital, were included in the study. Patients had to fulfill the following inclusion criteria: 1) age 40-90 years; 2) Mini-Mental State Examination (MMSE) score of between 11 and 25; 3) Hachinski Ischemic Score as modified by Rosen of less than or equal to 4; 4) written informed consent signed by the patient or the legal guardian.

Patients were randomized to receive either selegiline, two 5 mg tablets, or a placebo, given each morning in conjunction with breakfast. Randomization was carried out using stratification into three classes according to the MMSE score: 1) mild dementia, MMSE 21-25; 2) moderate dementia, MMSE 16-20; and 3) severe dementia, MMSE 11-15.

The efficacy of selegiline was assessed by short cognitive tests (MMSE, Blessed's Memory-Information-Concentration Test (BIMC), with global behavioral or psychiatric scales (Clinical Global Impression Scale (CGI), Instrumental Activities of Daily Living (IADL), Brief Psychiatric Rating Scale (BPRS), Hamilton's Depression Rating Scale) and with brief neuropsychological tests (Buschke Selective Reminding Test (BSR), Verbal Fluency Test (VFT) on letters and category, Trail Making Test (TMT) and Russell's Adaptation of the Visual Reproduction Test (VRT) at baseline, at 6, 12, 18, 24, 30 and 36 months. Furthermore, more comprehensive neuropsychological tests evaluating various cognitive abilities were assessed at baseline at 18 and 36 months.

Electrocardiography and the laboratory tests for safety were taken at baseline and thereafter at the one-year interval. Heart rate and blood pressure in supine and standing positions were measured at baseline and thereafter during each follow-up visit. Adverse events reported by the patient or the relative or observed by the investigator were recorded on case record forms.

The data was analyzed by intention-to-treat method, including all randomized patients with at least one subsequent visit after baseline during three-year study period. The comparison of the neuropsychological tests were accomplished by applying analysis of variance (ANOVA) model appropriate for an unbalanced repeated measurements with two grouping factors (stratification group, study drug) and one within factor (follow-up visit). The changes of the follow-up values from baseline value were used in the analysis.

For CGI (data on seven point ordinal scale) the proportional odds regression model, based on the 36 months and the last individual follow-up visit, was applied. Difference in proportions of patients with adverse events between the treatment groups were compared by using Fisher's exact test. A p-value of <0.05 was considered statistically significant.

RESULTS

There were no statistically significant differences between the selegiline and placebo groups at baseline in age, sex, concomitant diseases or in neuropsychological or psychiatric tests. At baseline 34 AD patients had a score in the MMSE between 21 and 25, 31 scored between 16 and 20 and 15 scored between 11 and 15. Altogether 54 patients (67.5 %) completed the 3-year follow-up. The reasons for the premature discontinuation of the study were death (6 patients), severe concomitant disease (11 patients), severe progression of the AD (1 patient), noncompliance (2 patients), and the personal unwillingness to continue the trial (6 patients).

The severity of dementia as assessed by the MMSE test deteriorated significantly more rapidly in the placebo group than in the selegiline group ($p=0.038$) during 3 years follow-up. At 36 months the total score of the MMSE decreased 7.8 points (from 20.8 ± 3.0 to 13.0 ± 8.4; Mean$\pm$SD) in the placebo group and 6.2 points (from 18.6 ± 3.3 to 12.4 ± 6.7) in the selegiline group (patients with complete data). The deterioration of dementia as evaluated by the MMSE was more rapid in the placebo group than in the selegiline group in AD patients having advanced disease (MMSE below or equal to 20 at baseline, $p=0.027$), but not in patients with less severe dementia at baseline (MMSE 21-25). There were also some statistically significant findings ($p<0.05$) in other assessments favoring selegiline: in relatives' CGI, copying of VRT, in the number of false and missing numbers of TMT version A, in mistakes of VFT and in the Digit Span. There were also some trends ($p<0.10$) supporting these findings in patients with moderate or severe dementia (MMSE 11-20) in the Hamilton's Depression Rating Scale, and in the number of total words and recognition in the BSR.

Selegiline was well tolerated. Adverse events reported by the patient, his/her caregivers or noticed by the investigator included dizziness, confusion, depression, insomnia, obstipation, cardiac failure, pneumonia, chest pain, headache and drowsiness. Dizziness, confusion and depression were significantly more often reported or noticed in the selegiline group than in the placebo group. The laboratory values did not reveal any marked toxic effect due to selegiline when compared to baseline values or values during placebo treatment. Neither were any remarkable differences found in blood pressure or heart rate between the selegiline and the placebo groups.

DISCUSSION

The theoretical background for the use of selegiline in the long-term treatment of AD is strong. Our study shows a decline in the MMSE total scores for all patients studied during the three years follow-up, but this is more pronounced in the placebo-treated patients than in those taking selegiline. The difference between treatment groups was more pronounced in AD patients with advanced dementia (MMSE score below 20 at baseline). Furthermore, some other assessments also supported the findings found in the MMSE scores. These results suggest that the progression of AD may be slowed down by the selegiline treatment.

Our study together with the study published by Burke et al. (1993) are so far the only studies which have evaluated the long-term effects of selegiline in AD. Multicenter studies with larger numbers of patients are, however, in progress. In contrast to our study, Burke and coworkers could detect no statistically significant group difference in the total score of the MMSE. They concluded that they were unable to rule out that, over a longer time frame or with more subjects, selegiline could demonstrate a slowing of the rate of disease progression. Lower number of patients and inclusion of only mildly demented patients in their study may explain the difference compared to our findings.

The significant slowing of the progression rate of AD in our study as assessed by the MMSE was only partly reflected in other assessments. Interestingly, the relatives' CGI was significantly in favor of selegiline, but this finding was not confirmed by other scales assessing the patient's ability to deal with the practical aspects of everyday life. This, together with the finding that only a proportion of other neuropsychological tests differed significantly between treatment groups diminish the importance of our results. This could be interpreted to suggest that selegiline is not an effective agent, or that the neuropsychological tests we assessed displayed severely impaired performance at entry, allowing little opportunity for further decline in either the placebo group or the selegiline group. Indeed, this floor effect could explain the unchanged follow-up scores in the BSR test assessing verbal memory and in the VRT assessing visual memory. Similar findings concerning assessments of delayed verbal memory in the long-term follow-up of AD have been published by Morris et al. (1993).

In the present study, selegiline did not induce severe or frequent adverse events, confirming the results achieved in other investigations (Tariot et al., 1987; Mangoni et al., 1991). The frequency of CNS-related symptoms seemed, however, to be more frequent in the selegiline group as compared to placebo group, but not concerning other body systems. The present study further emphasizes that selegiline is a safe and well-tolerated drug also in the long-term treatment of AD, as has been similarly shown in the long-term treatment of Parkinson's disease (The Parkinson Study Group, 1993).

The results of our long-term study suggest that the progression of the AD may be slowed down by the selegiline treatment, particularly in patients with advanced AD. However, our results are not conclusive. We have to wait for the results of ongoing, multicenter trials with larger populations to confirm our preliminary findings.

REFERENCES

Burke WJ. Roccaforte WH, Wengel SP, Bauer BL, Ranno AE and Willcockson NK (1993): L-deprenyl in the treatment of mild dementia of the Alzheimer type: Results of a 15-month trial. *J Am Geriatr Soc* 41:1219-1225.

Hay JW and Ernst RL (1987): The economic cost of Alzheimer's disease. *Am J Public Health* 77:1169-75.

Heinonen EH, Haapalinna A, Suhonen J and Hervonen A (1995): Selegiline may slow down age-related cognitive deterioration. *Eur J Neurol* 2 (suppl. 1):130.

Mangoni A, Grassi MP, Frattola L, Piolti R, Bassi S, Motta A, Marcone A and Smirne S (1991): Effects of a MAO-B inhibitor in the treatment of Alzheimer's disease. *Eur Neurol* 31:100-107.

McKhann G, Drachman D, Folstein M, Katzman R, Price D and Stadlan EM (1984): Clinical diagnosis of Alzheimer's disease: Report of the NINCDS-ADRDA work group under the auspices of the Department of Health and Human Task Force on Alzheimer's disease. *Neurology* 34:939-944.

Morris JC, Edland S, Clark C, Galasko D, Koss E, Mohs R, van Belle G, Fillenbaum G and Heyman A (1993): The Consortium to establish a registry for Alzheimer's disease (CERAD). Part IV. Rates of cognitive change in the longitudinal assessments of probable Alzheimer's disease. *Neurology* 43:2457-2465.

Reinikainen KJ (1994): Rationale for the use of selegiline in the treatment of Alzheimer's disease. *Neurobiol Aging* 15:s66.

Tatton WG (1993): Selegiline can mediate neuronal rescue rather than neuronal protection. *Mov Disord* 8 (suppl 1):20-30.

The Parkinson Study Group (1993): Effects of tocopherol and deprenyl on the progression of disability in early Parkinson's disease. *N Engl J Med* 328:176-183.

Alzheimer Disease: From Molecular Biology to Therapy
edited by R. Becker and E. Giacobini

INFLAMMATORY PROCESSES - ANTI-INFLAMMATORY THERAPY

Paul S. Aisen, Deborah B. Marin and Kenneth L. Davis
Department of Psychiatry, Mount Sinai Medical Center,
New York, NY 10029

INTRODUCTION

Histopathologic studies suggest that inflammatory processes are active at sites of neurodegeneration in Alzheimer's disease (AD) (see Aisen and Davis, 1994). Acute phase proteins such as a-1 antichymotrypsin (ACT), a-2 macroglobulin and C-reactive protein (CRP) are components of plaques. Interleukins 1 and 6, inflammatory cytokines with the capacity to stimulate expression of acute phase proteins, are present in increased amounts. In addition, early components of the complement cascade are found in the region of neuritic plaques. The cellular component of the inflammatory reaction in the AD brain appears to be the activated microglial cell. These cells bear HLA surface markers, and appear to be closely related to, perhaps derived from, blood-borne mononuclear phagocytes. Activated microglia seem to be involved in the maturation of neuritic plaques.

A number of studies suggest that inflammatory cytokines, complement proteins and microglial cells are contributing to neurodegeneration rather than merely reacting to the degenerative process. A transgenic mouse model demonstrates that overexpression of IL-6 in the brain causes neurodegeneration (Campbell et al., 1993). IL-1 increases the expression of the amyloid precursor protein (APP) in cell culture systems (Buxbaum et al., 1992). In rat pheochromocytoma (PC12) cells, IL-1 augments the toxicity of $Aß_{1-42}$ (Fagarasan and Aisen, 1996). Similarly, complement proteins may augment Aß cytotoxicity; C1q interacts with Aß, shifting the peptide to a more aggregated, toxic state (Webster et al., 1994). Stimulated mononuclear inflammatory cells release neurotoxins (Giulian et al., 1994), suggesting that microglial activation may contribute to neuronal cell loss in AD.

If inflammatory mediators contribute to neurodegeneration as is suggested by the animal and cell culture studies, they represent targets for therapeutic intervention. The selection of anti-inflammatory agents for testing in AD should be based on consideration of the efficacy of candidate drugs in the suppression of these specific mechanisms, and should draw upon the extensive experience with anti-inflammatory/immunosuppressive drugs in the treatment of rheumatic diseases.

GLUCOCORTICOID THERAPY

Glucocorticoids are the most effective and widely used drugs for the pharmacologic suppression of inflammation and autoimmunity, and therefore the appropriate first choice to test the inflammatory hypothesis of AD. However, an important question must be addressed: Does glucocorticoid therapy cause hippocampal damage in animals and humans?

For almost two decades, it has been suggested that hippocampal damage in aging and AD might be related to glucocorticoid effects. Indeed, glucocorticoids can induce hippocampal damage, and increase susceptibility of hippocampal neurons to other toxic insults (Sapolsky et al., 1990). However, it is becoming evident that glucocorticoid effects are complex, and may be protective (McEwen et al., 1992). For example, recent studies indicate that while prolonged stress causes hippocampal dendritic atrophy in the rat, an effect also seen with oral corticosterone therapy alone, steroid therapy protects against stress-induced hippocampal atrophy (Magarinos et al., 1995).

Before proceeding with a therapeutic trial of glucocorticoid treatment in AD, we conducted pilot studies to determine whether a regimen of prednisone which suppressed peripheral markers of inflammation would cause deleterious cognitive or behavioral effects in AD patients (Aisen et al., 1996). In the initial pilot study, 10 patients were treated with an initial dose of prednisone 10 mg daily for four weeks, tapered over an additional three weeks. This regimen was well tolerated, but failed to suppress ACT and CRP levels.

In the second study, the initial dose was 20 mg for two weeks, tapered over an additional five weeks. This regimen was also well tolerated, with no change in mean cognitive or behavioral assessments associated with treatment. ACT and CRP levels fell with this treatment; ACT levels remained suppressed during taper, while CRP levels rose when the dose dropped below 10 mg daily. Plasma C3a levels, an indicator of complement activation, also fell with treatment.

Based on these results, a multicenter placebo-controlled trial of prednisone has been initiated by the Alzheimer's Disease Cooperative Study consortium. A total of 150 patients will be randomized to receive prednisone or matching placebo at an initial daily dose of 20 mg, tapered to a maintenance dose of 10 mg. The primary outcome measures are the one year change in the cognitive component of the Alzheimer's Disease Assessment Scale and the Clinical Dementia Rating scale sum of boxes. The relationship of serum ACT and CRP levels to clinical progression will also be examined.

To date, over one hundred subjects have been enrolled at twenty sites around the country. There have been no serious adverse events attributed to study medication. The study will be completed in late 1997.

COLCHICINE AND HYDROXYCHLOROQUINE

Colchicine and hydroxychloroquine are anti-inflammatory drugs which target mononuclear/microglial cell activation, APP processing, and in the case of colchicine, amyloidogenesis. Both drugs are well-tolerated with chronic use, and are candidates for the treatment of AD.

Colchicine binds tubulin, inhibiting microtubule function; this is presumed to be the mechanism of its anti-inflammatory effects, which includes inhibition of mononuclear cell accumulation (Wallace, 1974). Colchicine blocks the internalization of membrane proteins to lysosomes via endosomes, which could block the processing of the amyloid precursor protein into amyloidogenic fragments. Colchicine is dramatically effective in the prevention and treatment of amyloidosis associated with familial Mediterranean fever (FMF) (Zemer et al., 1992), and in animal models of amyloidosis (Shirahama and Cohen, 1974).

Hydroxychloroquine has broader anti-inflammatory/immunosuppressive efficacy than colchicine; it has been proven effective in rheumatoid arthritis (it is classified as a disease-modifying anti-rheumatic drug, or DMARD) and in systemic lupus erythematosus. Its anti-inflammatory activity may be related to its lysosomotropic properties: it raises lysosomal Ph, inhibiting acid proteases. A lysosomotropic drug might have a favorable effect on APP processing (Haass et al., 1992). Hydroxychloroquine inhibits IL-1 and IL-6 release by stimulated mononuclear inflammatory cells (Sperber et al., 1993).

A recently completed pilot study of colchicine in AD indicates that it is safe, with no adverse effect of cognitive or behavioral measures. At a dose of 0.6 mg twice daily (a typical regimen used in the treatment of crystal arthropathy or FMF), mild diarrhea occurred in three of ten subjects, suggesting that this may be the maximal tolerated dose for patients with AD. A pilot study of hydroxychloroquine is still in progress. A multicenter therapeutic trial comparing the effect on cognitive and clinical measures of one year of treatment with colchicine, hydroxychloroquine or placebo is planned.

OTHER AGENTS

A number of investigators have suggested that nonsteroidal anti-inflammatory drugs (NSAIDs) should be studied. In support of this class of drug, retrospective studies have demonstrated a negative association between reported past use of NSAIDs and the expression of AD (Breitner et al., 1994). A pilot study suggests that indomethacin slows the rate of progression (Rogers et al., 1993). However, the retrospective studies are subject to recall bias, and the pilot study is inconclusive.

NSAIDs have been most useful in the suppression of acute, neutrophil-mediated inflammation. In rheumatoid arthritis, they relieve inflammation, but do not suppress the acute phase response and (unlike drugs such as hydroxychloroquine) do not modify joint damage. However, new NSAIDs such as tenidap have additional effects, such as inhibition of IL-1 release that could prove useful in the treatment of AD.

Other classes of drugs that seem appropriate for consideration based on their use in rheumatic and neurologic diseases are methotrexate, sulfasalazine and ß-interferon. Unfortunately, clinical trials are very time-consuming and expensive. Screening of agents with cell culture and animal models of the inflammatory component of AD will improve the likelihood that the anti-inflammatory strategy will succeed.

ACKNOWLEDGMENTS

This work was supported in part by grants (AG05138, AG02219, AG10483) from the National Institute on Aging, and by a grant (5 MO1 RR00071) for the Mount Sinai General Clinical Research Center from the National Center for Research Resources, National Institutes of Health.

REFERENCES

Aisen PS, Marin D, Altstiel L, Goodwin C, Baruch B, Jacobson R, Ryan T and Davis KL (1996): A pilot study of prednisone in Alzheimer's disease. *Dementia* (in press).

Aisen PS and Davis KL (1994): Inflammatory mechanisms in Alzheimer's disease: Implications for therapy. *Am J Psychiatry* 151:1105-1113.

Breitner JC, Gau BA, Welsh KA, Plassman BL, McDonald WM, Helms MJ and Anthony JC (1994): Inverse association of anti-inflammatory treatments and Alzheimer's disease: initial results of a co-twin control study. *Neurology* 44:227-232.

Buxbaum JD, Oishi M, Chen HI, Pinkas-Kramarski R, Jaffe EA, Gandy SE and Greengard P (1992): Cholinergic agonists and interleukin 1 regulate processing and secretion of the Alzheimer b/A4 amyloid protein precursor. *Proc Natl Acad Sci USA* 89: 10075-10078.

Campbell IL, Abraham CR, Masliah E, Kemper P, Inglis JD, Oldstone MBA and Mucke L (1993): Neurologic disease induced in transgenic mice by cerebral overexpression of interleukin 6. *Proc Natl Acad Sci USA* 90: 10061-10065.

Fagarasan MO and Aisen PS (1996): IL-1 and anti-inflammatory drugs modulate Ab cytotoxicity in PC12 cells. *Brain Res* (in press).

Giulian D, Li J, Leara B and Keenen C (1994): Phagocytic microglia release cytokines and cytotoxins that regulate the survival of astrocytes and neurons in culture. *Neurochem Int* 25:227-233.

Haass C, Koo EH, Mellon A, Hung AY and Selkoe DJ (1992): Targeting of cell-surface β-amyloid precursor protein to lysosomes: alternative processing into amyloid-bearing fragments. *Nature* 357:500-503.

Magarinos AM, Orchinik M and McEwen BS (1995): Oral administration of corticosterone mimics effects of stress on hippocampal CA3c dendritic structure. *Soc for Neurosci Abstr* 21:1948.

McEwen BS, Angulo J, Cameron H, Chao HM, Daniels D, Gannon MN, Gould E, Mendelson S, Sakai R, Spencer R and Woolley C (1992): Paradoxical effects of adrenal steroids on the brain: protection versus degeneration. *Biol Psychiatry* 31:177-199.

Rogers J, Kirby LC, Hempelman SR, Berry DL, McGeer PL, Kaszniak AW, Zalinski J, Cofield M, Mansukhani L, Willson P and Kogan F (1993): Clinical trial of indomethacin in Alzheimer's disease. *Neurology* 43:1609-1611.

Sapolsky RM, Uno H, Rebert CS et al. (1990): Hippocampal damage associated with prolonged glucocorticoid exposure in primates. *J Neurosci* 10:1897-1902.

Shirahama T and Cohen AS (1974): Blockage of amyloid induction by colchicine in an animal model. *J Exp Med* 140:1102-1107.

Sperber K, Quraishi H, Kalb TH, Panja A, Stecher V and Mayer L (1993): Selective regulation of cytokine secretion by hydroxychloroquine: Inhibition of interleukin 1 alpha (IL-1-a) and IL-6 in human monocytes and T cells. *J Rheumatol* 20:803-808.

Wallace SL (1974): Colchicine. *Seminars Arthritis Rheum* 3:369-381.

Webster S, O'Barr S and Rogers J (1994): Enhanced aggregation and b structure of amyloid b peptide after coincubation with C1Q. *J Neurosci Res* 39: 448-456.

Zemer D, Livneh A and Langevitz P (1992): Reversal of the nephrotic syndrome by colchicine in amyloidosis of familial Mediterranean fever. *Ann Intern Med* 116:426.

[illegible] and Selkoe DJ (1992). [illegible] surface β-amyloid precursor protein to lysosomes: alternative processing into amyloid-bearing fragments. Nature 357:500-503

[illegible]

Alzheimer Disease: From Molecular Biology to Therapy
edited by R. Becker and E. Giacobini

PROPENTOFYLLINE - PRECLINICAL DATA

Karl A. Rudolphi
Hoechst Marion Roussel, Frankfurt/Main, Germany

INTRODUCTION

Dementia belongs to the most serious sociomedical problems of the aging society in the industrialized countries; however, despite enormous worldwide research efforts, still no satisfactory pharmacotherapy exists. Many of the compounds under investigation provide a symptomatic treatment by enhancing vigilance and cognition. These drugs, classified as cognition enhancers or nootropics, are often criticized for insufficient clinical proof of efficacy and/or limited tolerability. Another approach is to interfere with the neurodegenerative process and to reduce brain tissue damage. One of these neuroprotective drugs is the novel xanthine derivative propentofylline [1-(5′-oxohexyl)-3-methyl-7-propylxanthine]. It improves cognitive functions in experimental models of vascular dementia and/or Alzheimer′s Disease (AD), inhibits inflammatory processes such as excessive microglia activation, formation of free radicals, cytokines and abnormal amyloid precursor proteins, stimulates the synthesis and release of nerve growth factor and reduces ischemic brain damage. Propentofylline is in advanced clinical development in ameliorated cognitive functions, global functions and activities of daily living in patients with AD and vascular dementia (Rother et al., 1996; Saletu et al., 1991). This chapter reviews the data from animal studies conducted to profile the pharmacological activities of the drug.

LEARNING AND MEMORY, CHOLINERGIC SYSTEM

The predominant cognitive symptom in dementia is the progressive loss of memory aquisition and retrieval. Two experimental animal models were selected to test antidementic effects of propentofylline: aged spontaneously hypertensive rats (Goto et al., 1987) and rats with basal forebrain lesions by local injection of ibotenic acid (Fuji et al., 1993a,b). In both cases the animals develop severe deficiencies in learning and memory. Aged spontaneously hypertensive rats show cerebrovascular abnormalities (Nordborg and

Johansson, 1980) and may thus resemble the situation of vascular dementia, whereas the basal forebrain lesions may mimic the degeneration and loss of cholinergic neurons of the basal forebrain nucleus of Meynert in Alzheimer's dementia patients (Etienne et al., 1986). In the study with SHR (Goto et al., 1987), the long-term oral administration of propentofylline (25 mg/kg once daily) for 15 days improved the learning and memory performance of the rats in an active avoidance paradigm (shuttle box). Similar results were obtained in rats with basal forebrain lesion (Fuji et al., 1993a). Propentofylline given orally at doses of 10 or 25 mg/kg once daily for 28 days significantly improved the performance of the basal forebrain lesioned rats in three learning and memory paradigms (one-trial dark avoidance, water maze and habituation test). It is unlikely that the beneficial effects of propentofylline in the basal forebrain lesioned rats was due to a reduction of the primary brain lesion because the drug treatment was started 8 days after the lesion.

One of the most prominent pathophysiological changes in AD is the loss and dysfunction of cholinergic neuronal systems including changes of muscarinic receptors. Propentofylline (given in daily single oral doses of 25 mg/kg over 28 days starting 8 days after the induction of the lesion) significantly reversed the compensatory receptor upregulation of muscarinic binding sites and binding affinity in the frontal and parietal cortex and the hippocampus of basal forebrain lesioned rats (Fuji et al., 1993b). In a separate study in basal nucleus-lesioned rats, propentofylline prevented the decrease of choline acetyltransferase activity in the hippocampus but not in the cortical areas (Fuji et al., 1993a). Thus, propentofylline improved experimentally induced alterations of the cholinergic nervous system, which resemble those seen in the brains of Alzheimer patients.

Hippocampal long term potentiation (LTP), a long lasting increase in synaptic transmission following brief high frequency stimulation, is widely regarded as an electrophysiological correlate of learning and memory processes. Propentofylline enhanced LTP in guinea pig hippocampal slices with effective concentrations of 10 μM (Tanaka et al., 1990) or 10 nM (Yamada et al., 1994).

MICROGLIAL CELLS

Propentofylline inhibits potentially neurotoxic functions of pathologically activated microglial cells. In *in vitro* experiments in primary cultures of microglial cells from rat brain propentofylline (20-50 μM) significantly decreased the formation of free oxygen radicals (Banati et al., 1994) or TNFα (Suzumura et al., 1995) and inhibited the phorbolester-stimulated microglia proliferation with an IC_{50} of 3μM (Si et al., 1996). In an *in vivo* immunohistochemical study in gerbils with 5 min global forebrain ischemia, the chronic treatment with propentofylline (14 successive single daily doses of 10 mg/kg i.p. starting one day after the induction of ischemia) inhibited the ischemia-induced expression of cell antigens which are indicative of a potentially cytotoxic phenotype of activated microglial cells. In addition, the enhanced immunostaining for an amyloid precursor protein was strongly suppressed by the propentofylline treatment (McRae et al., 1994). This study shows that propentofylline can inhibit an ongoing and potentially neurotoxic process of microglia activation in the postischemic brain. It also gives evidence for a therapeutic potential of the drug in Alzheimer's dementia in which microglia activation and an increased amyloid formation represent important pathomechanisms (Banati et al., 1993).

BRAIN ISCHEMIA

Propentofylline given before or up to 1 hr after induction of focal or global ischemia in gerbils or rats reduced ischemic nerve cell death, mortality, neurological symptoms, edema, intraneuronal calcium accumulation, extracellular glutamate concentration, reactive astrocytosis, prevented the loss of hippocampal M_1 muscarinic binding sites and increased energy-rich phosphates (see Rudolphi et al., 1992). Specifically, in a clinically relevant model of stroke, the permanent occlusion of the left middle cerebral artery in rats, a continuous i.v. infusion over 24 hrs of propentofylline (0.01, 0.05 or 0.1 mg/kg/min) starting 15 min after the induction of ischemia significantly improved neurological symptoms and reduced infarct volume by 39 % (Park and Rudolphi, 1994). In conclusion, propentofylline showed a tissue saving neuroprotective effect in different experimental models of brain ischemia, significantly improved the neurological status and reduced mortality.

MOLECULAR MECHANISM OF ACTION

Although the drug is a derivative of the same heterocyclic ring system (1,3,7-trisubstituted xanthines), its molecular mechanisms of action differ considerably from those of the classical methylxanthines theophylline and caffeine (Parkinson et al., 1994). Propentofylline blocks Na-independent adenosine transporters and acts simultaneously as an inhibitor of cyclic nucleotide phosphodiesterases (IC_{50} 8 - 30 μM). It thereby enhances the active concentrations of the homeostatic cell modulator adenosine and the second messenger cyclic AMP (cAMP) at the sites of action, i.e. extracellular adenosine A_1 and A_2 receptors or intracellular cAMP-regulated systems, and reinforces their multiple neuroprotective actions (Rudolphi et al., 1992). These include the inhibition of a series of pathophysiological key processes, which appear to be instrumental for both vascular and Alzheimer's type dementia, i.e. microglia activation, excessive release of glutamate, free radical formation, accumulation of free intracellular calcium, and enhanced expression of ß-amyloid precursor proteins.

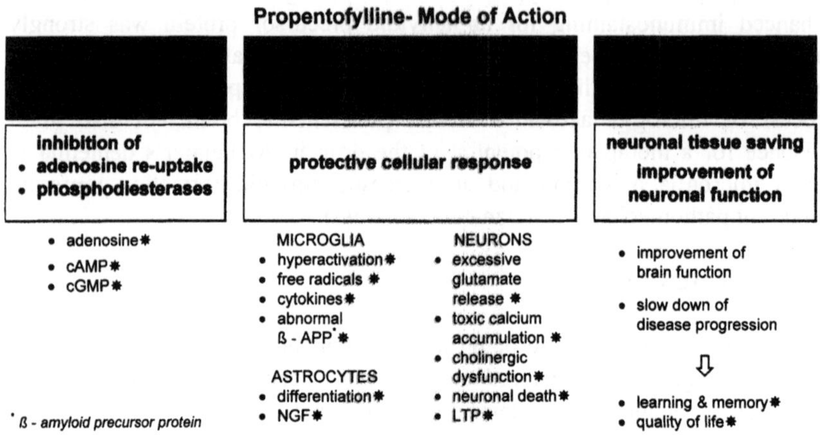

Drugs which increase intracellular cAMP are also known to stimulate the synthesis/release of nerve growth factor (NGF), an endogenous neurotrophin, which is required to maintain neuronal structure and function, and which stimulates restorative processes following brain injury (Kromer, 1987). Propentofylline stimulated the synthesis and release of NGF in mouse astrocyte cultures (Shinoda et al., 1990), reversed the age-associated decrease of NGF concentration in the cortex of aged rats (Nabeshima et al., 1993) and counteracted the learning and memory deficits in rats induced by intraseptal infusion of anti-NGF antibodies (Nitta et al., 1996). The drug may therefore support endogenous restorative processes in the brain by enhancing NGF formation or by preventing the age-related decrease of NGF synthesis.

CONCLUSION

In conclusion, our preclinical data indicate that propentofylline is a novel antidementic drug which modulates glial and neuronal functions by reinforcing multifarious actions of endogenous adenosine and cAMP. The outcome of the ongoing clinical studies with propentofylline will be proof as to whether the drug may offer an efficacious and safe therapy of dementia.

REFERENCES

Banati RB, Gehrmann J, Schubert P and Kreutzberg GW (1993): Cytotoxicity of microglia. *Glia* 7:111-118.

Banati RB, Schubert P, Rothe G, Gehrmann J, Rudolphi K, Valet G and Kreutzberg GW (1994): Modulation of intracellular formation of reactive oxygen intermediates in peritoneal macrophages and microglia/brain macrophages by propentofylline. *J Cereb Blood Flow Metab* 14:145-149.

Etienne P, Robtaille Y, Wood P, Gauthier S, Nair NPV and Quirion R (1986): Nucleus basalis neuronal loss, neuritic plaques and choline acetyltransferase activity in advanced Alzheimer's disease. *Neuroscience* 19:1279-1291.

Fuji K, Hiramatsu M, Kameyama T and Nabeshima T (1993a): Effects of repeated administration of propentofylline on memory impairment produced by basal forebrain lesion in rats. *Eur J Pharmacol* 236:411-417.

Fuji K, Hiramatsu M, Hayashi S, Kameyama T and Nabeshima T (1993b): Effects of propentofylline, a NGF synthesis stimulator, on alterations in muscarinic cholinergic receptors induced by basal forebrain lesion in rats. *Neurosci Lett* 99-102.

Goto M, Demura N and Sakaguchi T (1987): Effects of propentofylline on disorder of learning and memory in rodents. *Japan J Pharmacol* 45:373-378.

Kromer LF (1987) Nerve growth factor treatment after brain injury prevents neuronal death. *Science* 235:214-216.

McRae A, Rudolphi K and Schubert P (1994): Propentofylline depresses amyloid and Alzheimer's CSF microglial antigens after ischemia. *NeuroReport* 5:1193-1196.

Nabeshima T, Nitta A and Hasegawa T (1993): Impairment of learning and memory and the accessory symptom in aged rat as senile dementia model (3): oral administration of propentofylline produces recovery of reduced NGF content in the brain of aged rats. *Jpn J Psychopharmacol* 13:89-95.

Nitta A, Ogihara Y, Onishi J, Hasegawa T, Furukawa S and Nabeshima T (1996): Propentofylline prevents neuronal dysfunction induced by infusion of anti-nerve growth factor antibody into the rat septum. *Eur J Pharmacol* (in press).

Nordborg C and Johansson BB (1980): Morphometric study on cerebral vessels in spontaneously hypertensive rats. *Stroke* 11:266-270.

Park CK and Rudolphi K (1994): Antiischemic effects of propentofylline (HWA 285) against focal cerebral infarction in rats. *Neurosci Lett* 178:235-238.

Parkinson FE, Rudolphi KA and Fredholm BB (1994): Propentofylline: a nucleoside transport inhibitor with neuroprotective effects in cerebral ischemia. *Gen Pharmac* 25:1053-1058.

Rother M, Kittner B, Rudolphi K, Rößner M and Labs KH (1996): HWA 285 (Propentofylline) - a new compound for the treatment of both vascular dementia and dementia of the Alzheimer type. *Ann New York Acad Sci* 777:404-409

Rudolphi KA, Schubert P, Parkinson FE and Fredholm BB (1992): Adenosine and brain ischemia. *Cerebrovasc Brain Metab Rev* 4:346-369.

Saletu B, Möller HJ, Grünberger J, Deutsch H and Rössner M (1990-91): Propentofylline in adult-onset cognitive disorders: double-blind, placebo-controlled, clinical, psychometric and brain mapping studies. *Neuropsychobiology* 24:173-184.

Shinoda I, Furukawa Y and Furukawa S (1990): Stimulation of nerve growth factor synthesis/secretion by propentofylline in cultured mouse astroglial cells. *Biochem Pharmacol* 39:1813-1816.

Si QS, Nakamura Y, Schubert P, Rudolphi K and Kataoka K (1996): Propentofylline, a xanthine derivative, inhibits proliferation of microglia in culture *Experimental Neurology* (in press)

Suzumura A, Sawada M and Marunouchi T (1995): Propentofylline suppresses TNF-alpha production by glial cells. *Neuroimmunology* 3:76-77.

Tanaka Y, Sakurai M, Goto M and Hayashi S (1990): Effect of xanthine derivatives on hippocampal long-term potentiation. *Brain Res* 522:63-68.

Yamada Y, Nakamura H and Okada Y (1994): Propentofylline enhances the formation of long-term potentiation in guinea pig hippocampal slices. *Neurosci Lett* 176:189-192.

Alzheimer Disease: From Molecular Biology to Therapy
edited by R. Becker and E. Giacobini

PROPENTOFYLLINE (HWA 285): A SUBGROUP ANALYSIS OF PHASE III CLINICAL STUDIES IN ALZHEIMER'S DISEASE AND VASCULAR DEMENTIA

Barbara Kittner for the European Propentofylline Study Group
Hoechst-Marion-Roussel, TD Cardiovascular Agents,
Werk Kalle-Albert, D-65174 Wiesbaden, Germany

INTRODUCTION

Dementia is a syndrome characterized by progressive deterioration in intellectual capacity and other mental functions. The most common form of dementia is Alzheimer's disease (AD). Vascular dementia (VaD) is generally considered to be the second-most common form of dementia.

There are no definitive markers or methods for diagnosing AD or VaD. Many patients with AD may also suffer from atherosclerosis or be at risk of cerebrovascular disease. In addition, AD is often associated with vascular changes caused by amyloid deposits (similar to senile plaques) in cerebral blood vessels (Munoz, 1991). Conversely, most patients with VaD have evidence of cortical senile plaques and neurofibrillary tangles. Neurotransmitter abnormalities have been described in both AD and VaD (Gottfries et al., 1991). Furthermore, similar cellular mechanisms are thought to be involved in both degenerative and ischemic neuronal damage, including glutamate 'excitotoxicity', intracellular calcium overload, damage induced by free radicals and particularly overactivation of microglial cells (Rudolphi. 1996).

Since AD and VaD share key elements in their pathophysiology, it follows that both may respond to treatments which address these common disease processes. Propentofylline (HWA 285), a xanthine derivative developed by Hoechst-Marion-Roussel, acts to reduce glutamate 'excitotoxicity', intracellular calcium overload, damage induced by free radicals and overactivation of microglial cells.

A 3-month, Phase II clinical study in patients with probable dementia suggested that propentofylline has a beneficial clinical effect in both VaD and AD (Saletu et al., 1990). Here, we report results from the first Phase III trials with propentofylline, and focus on the question of whether propentofylline is effective in both VaD and AD.

PHASE III STUDIES AND SUBGROUP ANALYSIS

Two placebo-controlled, double-blind studies comparing propentofylline with placebo in patients with dementia were conducted at centers in 7 European countries. The two studies were identical in terms of design, entry criteria and evaluation criteria, differing only in duration (6 and 12 months respectively).

The principal inclusion criterion for both studies was a DSM-III-R diagnosis of dementia. Forms of dementia other than VaD, AD or mixed AD/VaD were excluded. The individual studies did not stratify patients according to type of dementia since the Phase II study had indicated equal efficacy for propentofylline in both AD and VaD (Saletu et al., 1990). However, all patients were classified at baseline on the basis of DSM-III-R criteria and/or Hachinski ischemia score as having either VaD, AD or mixed AD/VaD, so that a subgroup analysis of the results could be carried out.

Patients received 300 mg propentofylline three times daily (or placebo equivalent) orally 1 hour before meals. The primary efficacy variables were the SKT (Syndrom Kurztest) total score for cognitive function (Erzigkeit, 1986) and GBS (Gottfries-Brane-Steen) (Gottfries et al., 1982) and CGI (Clinical Global Impressions) Item II for global function. Secondary variables included CGI Item I (severity), NAB (Nürnberger-Alters Beobachtungsskala) total score for Activities of Daily Living (Oswald and Fleischmann, 1986), and MMSE (Mini-Mental State Examination) total score. Patients were evaluated at baseline and every 3 months (although at every month for CGI scores) until endpoint. For inclusion in the analysis, all patients had to have at least one 3-month evaluation after entry into the study.

RESULTS

In total, 293 patients entered the 6-month study and 260 entered the 12-month study (Table I). In both studies, patients receiving propentofylline or placebo were comparable in terms of baseline demographic and background data, medical history (including severity of dementia as assessed by baseline variables) and medication.

Only a small number of patients had a mixed etiology of dementia and these are not discussed further. The AD and VaD subgroups were comparable (for both propentofylline and placebo) in all baseline criteria with the exception of the Hachinski ischemia score, which, as would be expected, was higher in the VaD subgroups.

	6-month study		12-month study	
	HWA 285	Placebo	HWA 285	Placebo
No. of patients entered	146	147	129	131
- with AD	83	75	81	89
- with VaD	45	48	25	26
- with mixed AD/VaD	18	24	23	16
No. of patient drop-outs	25	20	43	30
Mean age (yrs) of patients	70.7±8.6	71.1±8.2	71.9±7.3	72.9±7.1
Mean MMSE score	21.9±2.9	20.7±2.8	20.8±3.0	20.7±3.1
Mean Hachinski score	5.4±4.1	5.6±4.1	4.0±3.2	4.2±3.4

TABLE I. Patient populations entered in 2 studies comparing propentofylline (HWA 285) with placebo. (Key: AD = Alzheimer's disease, VaD = Vascular dementia)

Pooled Analysis

The results of the individual 6- and 12-mo studies confirmed the finding of the Phase II trial that propentofylline was indeed effective in the treatment of dementia. In the individual studies, propentofylline efficacy was demonstrated statistically for most of the primary assessment criteria, and approached statistical significance in the remainder. When the data sets from the two studies were pooled, the combined results were statistically significant for all variables (Table II) and clearly demonstrated that propentofylline is effective in treating dementia. Data from the 12-mo study suggests that the progression of the disease in patients treated with propentofylline may be delayed by one year, as evaluated by assessments of cognitive and global functioning and activities of daily living.

Subgroup Analysis

The individual 6- and 12-month studies were not designed to assess the efficacy of propentofylline in specific etiological subgroups. However, by pooling data from the two studies, sufficient statistical power was obtained to carry out a post-hoc subgroup analysis of the effects of propentofylline on AD

	P-values for treatment comparison		
	6-month study (n= 263)	12-month study (n=229)	Pooled analysis (n=492)
Primary efficacy variables			
GBS total score	0.011	0.002	<0.001
SKT total score	0.088	0.004	0.001
CGI item II	0.029	0.064	0.001
Secondary efficacy variables			
CGI item I (severity)	0.309	0.005	0.008
NAB total score	0.687	0.012	0.044
MMSE total score	0.450	0.012	0.023

TABLE II. Propentofylline versus placebo in the treatment of dementia: statistical significance of results in 6- and 12-month studies and pooled analysis at endpoint.

	P-values for treatment comparison	
	AD patients (n= 288)	VaD patients (n=132)
Primary efficacy variables		
GBS total score	0.028	0.009
SKT total score	0.019	0.023
CGI item II	0.134	0.003
Secondary efficacy variables		
CGI item I (severity)	0.178	0.029
NAB total score	0.013	0.560
MMSE total score	0.126	0.016

TABLE III. Propentofylline versus placebo in AD and VaD subgroups: Statistical significance of results. (Key: AD = Alzheimer's disease, VaD = Vascular dementia)

and VaD (Table III). In both VaD and AD, propentofylline was associated with statistically significant improvements relative to placebo according to all the primary efficacy criteria, with the exception of CGI score for AD. Among the secondary efficacy variables, a statistically significant response was observed for NAB total score for patients with AD, and on CGI item I and MMSE total score for VaD patients.

Safety and Tolerability

In both studies, propentofylline was well tolerated with a good safety profile with respect to adverse events, vital signs, ECG and laboratory variables.

CONCLUSIONS

The most important conclusion of these results is that propentofylline is clinically effective in the treatment of both AD and VaD. Clinical relevance is considered to be established if significant improvements (relative to placebo) in cognition are observed and are supported by global improvements assessed by an independent clinician. An assessment of activities of daily living should also be made to evaluate the impact of treatment on the patient's lifestyle.

In the studies reported here, cognitive function was assessed by the SKT test. Compared with placebo, a statistically significant benefit was observed in both AD and VaD subgroups receiving propentofylline. Global improvements were assessed by the GBS and CGI Item II scales (the GBS test is considerably more detailed than the CGI). Significant improvements in both AD and VaD were recorded using the GBS test, but global improvements measured by CGI Item II only reached significance in the VaD group. This reflected the fact that there was considerable variability of response within both the AD and VaD subgroups. This prevented the attainment of statistical significance in the AD subgroup, where the mean response, although marked, was smaller than in the VaD subgroup. Variability in the rate of progression of disease was particularly apparent in the AD subgroup, a factor which could not be quantified at baseline but became apparent during the course of the study. These considerations call into question the sensitivity of CGI in the assessment of AD in extended trials.

In terms of activities of daily living, the NAB test revealed statistically significant differences in the AD, but not the VaD, patients treated with propentofylline. This may be because many VaD patients are physically handicapped by their stroke and consequently are unable to respond significantly in terms of daily living assessment criteria, despite improved cognitive and global functioning.

The results suggest that among the patients treated with propentofylline in these studies, the smaller group who were suffering from VaD experienced a clinical benefit which was greater than that experienced by the AD patients.

It is likely, however, that propentofylline would achieve greater efficacy in AD during longer trials, and especially if it is used at an earlier stage in the disease. In the studies described here, neuronal damage due to disease progression was probably already severe in many patients. In our current studies, we are attempting to address this issue by limiting the enrollment of AD patients to those with mild disease or who are in the upper range of moderately severe disease (i.e. a minimum MMSE score of 18 points).

Finally, as propentofylline's mode of action is quite distinct from that of other drugs for AD, combination therapies of propentofylline and other drug classes (e.g. cholinergic agents) might prove effective in the future. At present, however, it is clear that propentofylline as monotherapy offers significant clinical benefits to patients with both AD and VaD.

REFERENCES

Erzigkeit H (1986): SKT: Ein Kurztest zur Erfassung von Gedächtnis- und Aufmerksamkeitsstörungen. Weinheim: Beltz Test GmbH.

Gottfries CG, Bråne G, Gullberg B, Steen G (1982): A new rating scale for dementia syndromes. *Arch Gerontol Geriatr* 1:311-330.

Gottfries CG, Alafuzoff I, Carlsson A et al. (1991): Neurochemical changes in brains from patients with vascular dementia. In: *Cerebral ischemia and dementia*, Hartman A, Kuschinsky W, Hoyer S, eds. Berlin: Springer-Verlag, pp.94-102.

Munoz DG (1991): The pathophysiological basis of multi-infarct dementia. *Alzheimer Dis Assoc Discord* 5:77-90.

Oswald WD, Fleischmann UM (1986): Nürnberger Altersinventar NAI. Nürnberg: Universität Erlangen-Nürnberg.

Saletu B, Möller HJ, Grunberger J et al. (1990): Propentofylline in adult-onset cognitive disorders: Double-blind, placebo-controlled, clinical, psychometric and brain mapping studies. *Neuropsychobiology* 24:173-184.

Alzheimer Disease: From Molecular Biology to Therapy
edited by R. Becker and E. Giacobini

ANAPSOS: NEW THERAPEUTIC STRATEGIES FOR NEURODEGENERATION AND BRAIN AGING WITH NEUROIMMUNOTROPHIC FACTORS

X. Antón Alvarez, Raquel Zas, Raquel Lagares, Lucía Fernández-Novoa, Andrés Franco-Maside, José J. Miguel-Hidalgo and Ramón Cacabelos
EuroEspes Biomedical Research Center, Bergondo-A Coruña, Spain.

Joaquín Díaz and José M. Sempere
A.S.A.C. Pharmaceutical International, Alicante, Spain

INTRODUCTION

Alzheimer's disease (AD) is a neurodegenerative disorder characterized by an heterogeneous etiopathogenesis on a genetic basis. The most prominent neuropathological findings in AD are senile plaques, neurofibrillary tangles (NFT), amyloid deposition in neural tissues and vessels, synaptic loss and subsequent neuronal death (Cacabelos et al., 1993, 1995; Hardy and Allsop, 1991; Terry et al., 1991). However, other neurochemical mechanisms may also account for cell death and neurodegeneration in AD, including neuroimmune dysfunction, free radical formation, neurotransmitter deficits and alterations in brain calcium homeostasis (Cacabelos et al., 1994, 1995; Richardson, 1993). So, new approaches to AD treatment are oriented to search for agents acting on etiopathogenic events and pleiotropic compounds displaying multifactorial effects.

Anapsos is an extract obtained from the rhizomes of the fern *polypodium leucotomos* (Horvath et al., 1967). Initially, it has been found that anapsos increased survival in patients with advanced neoplasms. Anapsos exerts immunosuppressive effects in healthy volunteers and in patients with atopic dermatitis or psoriasis (Alvarez et al., 1995, Vargas et al., 1983), and prolongs the allograft survival in rodents with skin transplantations (Tuominen et al., 1991).

In this report we summarize the effects of anapsos on: (1) learning and brain interleukin-1ß (IL-1ß) levels in intact rats; (2) learning and IL-1ß production in rats with neurotoxic lesions in the nucleus basalis of Meynert (nbM); (3) longevity, behavior, and brain histamine (HA) and IL-1ß concentrations in old mice; and (4) learning, psychomotor coordination (PMC), brain superoxide dismutase (SOD) activity and IL-1ß levels, and neuronal degeneration in rats with ß-amyloid implants in the hippocampus.

RESULTS

1. Effects of anapsos on learning and brain IL-1ß levels in intact rats. Female Sprague-Dawley rats (N=6 rats/group) received acute or daily i.p. injections of anapsos (25 mg/kg). In a passive avoidance behavior (PAB) task, in which the animals had to learn to stay in a neutral platform in order avoid a 0.25 mA electric footshock in the surrounding area, anapsos (25 mg/kg/day; 5 days; i.p.) increased the number of avoidances during ten trials (Anapsos=6.12±1.64 Av; Control=4.00±1.30 Av, $p<0.02$) and mean trial latency (Anapsos=21.06±3.12 sec; Control=17.11±3.95 sec, $p<0.05$). Anapsos (25 mg/kg; i.p.; 15, 30, 60 and 120 minutes before sacrifice) induced a time-dependent decrease in the concentrations of IL-1ß in the brain of intact rats, with maximum effects 120 minutes after injection in the hippocampus (58% reduction) and cortex (51% reduction).

2. Effects of anapsos on learning and brain IL-1ß in nbM-lesioned rats. Female Sprague-Dawley rats were injected with 2.5 µg of ibotenic acid in 2 µl saline (L) or vehicle alone (S) into the right and left nbM (coordinates: -0.8 mm posterior, 2.6 mm lateral, 8.2 mm ventral). After 7 days for recovery, learning performance in a PAB task and brain IL-1ß levels were evaluated in lesioned animals receiving acute i.p. injections of anapsos (LA; 100 mg/kg) or saline (LS), and in sham-operated animals treated with vehicle (SS). Lesioned rats tested in a PAB paradigm after acute treatment with anapsos showed a significant increase in the number of avoidances (LS=1.57±1.27 Av; LA=3.85±1.21** Av; SS=3.57±1.39** Av; ** $p<0.01$ vs LS) and in mean trial latency (LS=11.79±4.6 sec; LA=18.98±2.9** sec; SS=16.88±2.74* sec; *$p<0.05$ and **$p<0.01$ vs LS). Rats with nbM lesions showed increased IL-1ß immunoreactivity with respect to sham animals in the hypothalamus (66.6% increase; $p<0.01$), hippocampus (29.5% increase; $p<0.05$) and brain cortex (70.1% increase; $p<0.01$). Acute i.p. injections of anapsos produced significant decreases in hippocampal (46.3%; $p<0.05$) and cortical (60.5%; $p<0.01$) IL-1ß levels in lesioned rats.

3. Effects of anapsos on longevity, behavior and brain neuroimmune function in mice.

The effects of the daily administration of anapsos (25 and 100 mg/kg/day; s.c.) on the survival rate, psychomotor activity (PMA), PMC and learning in a PAB task, as well as on brain HA and IL-1ß levels were evaluated in mice. Eleven month old mice received anapsos or saline (C) injections during another 11-month period before testing. Anapsos-treated mice showed a significant

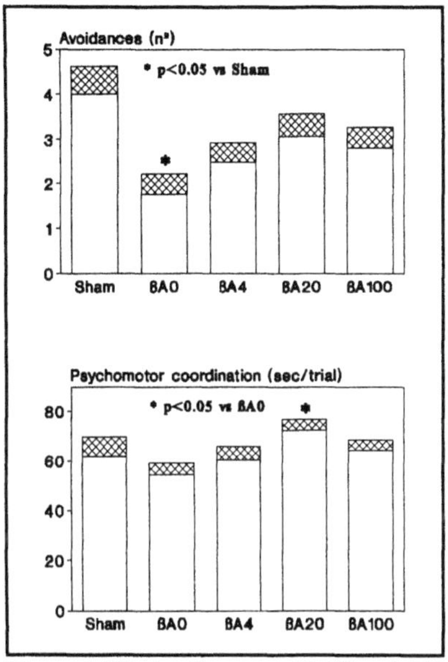

Figure 1. Effects of anapsos (4, 20 and 100 mg/kg/day; 7 days; i.p.) on learning and psychomotor coordination in rats with hippocampal ß-amyloid implants. X±SD.

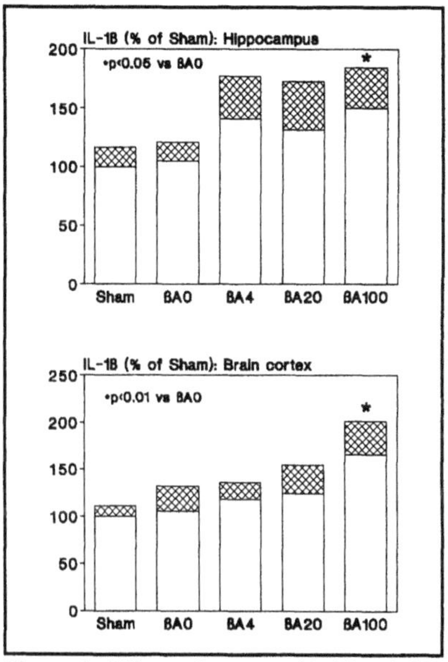

Figure 2. Effects of anapsos (4, 20 and 100 mg/kg/day; 7 days; i.p.) on brain IL-1ß levels in rats with ß-amyloid implants into the hippocampus. X±SD.

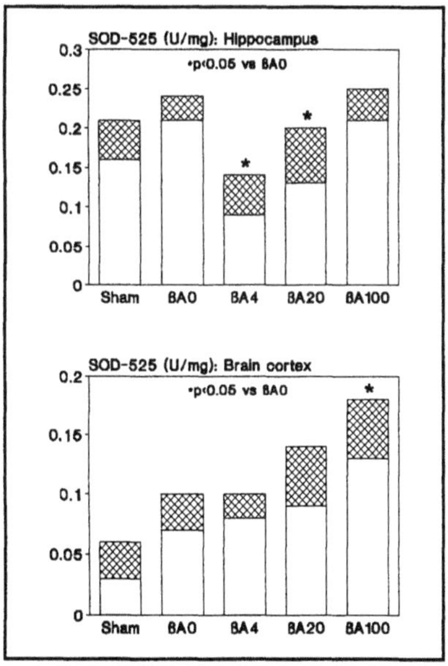

Figure 3. Effects of anapsos (4, 20 and 100 mg/kg/day; 7 days; i.p.) on brain SOD activity in rats with ß-amyloid implants into the hippocampus. X±SD.

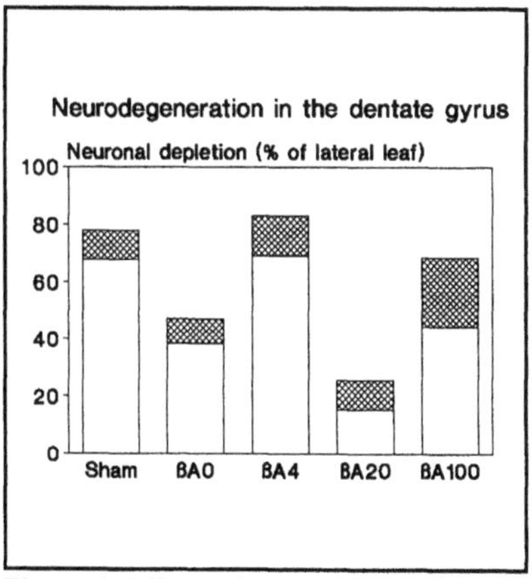

Figure 4. Effects of anapsos (4, 20 and 100 mg/kg/day; 7 days; i.p.) on neuronal degeneration in the dentate gyrus of rats with ß-amyloid implants. X±SD.

increase in PMA (C=670±239 I/O; A25=1088±318** I/O; A100=945±283* I/O; *p<0.05 and **p<0.01 vs C) and PMC scores (C=22±17 sec; A25=38±17* sec; A100=27±21 sec; *p<0.05 vs C) with respect to untreated rats. HA levels tended to be higher in the posterior hypothalamus (C=0.67±0.23 nmol/mg; A25=1.14±0.83 nmol/mg) and lower in the hippocampus (C=0.6± 0.6 nmol/mg; A25=0.25±0.1 nmol/mg) of animals receiving 25 mg/kg/day of anapsos. In addition, anapsos increased cortical IL-1ß levels in a dose-dependent manner (C=0.65±0.26 pg/mg; A25=0.91±0.23 pg/mg; A100=1.61±0.74 pg/mg, p<0.01 vs C), without altering them in the hippocampus. Learning and survival rates were similar in mice treated or not with anapsos.

4. Neuroprotective effects of anapsos in rats with ß-amyloid implants into the hippocampus.

Sprague Dawley rats received ß-amyloid protein (ßA) or water (sham) injections into the hippocampus and were treated with saline (sham, ßA0) or anapsos (4, 20 and 100 mg/kg/day) during 7 days after neurosurgery. Then, PAB learning, PMC, SOD activity, IL-1ß levels, and the extent of hippocampal neurodegeneration were assessed. Animals treated with anapsos showed a dose-dependent increase in PMC values (Sham=61.8±30.6 sec; ßA0=54.4±18.3 sec; ßA4=60.2±22 sec; ßA20=72.2±18* sec; ßA100=64±16.8 sec; *p<0.05 vs ßA0) and in the number of avoidances in a PAB task (Sham=4.0±2.1* Av; ßA0=1.7±1.6 Av; ßA4=2.4±1.7 Av; ßA20=3.1±1.9 Av; ßA100=2.8±1.8 Av; *p<0.05 vs ßA0) (Figure 1). Anapsos also increased IL-1ß concentrations, exerting maximum effects with the 100 mg/kg dose in the hippocampus (43% increase) and cortex (65% increase) (Figure 2). On the other hand, anapsos reduced SOD activity in the hippocampus (Sham=0.16±0.05 U/mg; ßA0=0.21±0.03 U/mg; ßA4=0.09±0.05* U/mg; ßA20=0.13±0.07* U/mg; ßA100=0.21±0.04 U/mg; *p<0.05 vs ßA0), increasing it in brain cortex (Sham=0.03±0.03 U/mg; ßA0=0.07±0.03 U/mg; ßA4=0.08±0.02 U/mg; ßA20=0.09±0.05 U/mg; ßA100=0.13±0.05* U/mg; *p<0.05 vs ßA0) (Figure 3). Finally, anapsos reduced the extent of hippocampal neurodegeneration, expressed as the percentage of the lateral leaf of the gyrus dentatus devoid of neurons (Sham=67%; ßA0=38%; ßA4=69%; ßA20=15%; ßA100=44%) (Figure 4). Most of the positive effects of anapsos were observed with 20 mg/kg.

CONCLUSIONS

According to the present results we can conclude:

1. Anapsos improves learning and suppresses brain IL-1ß production in intact rats, indicating that anapsos has procognitive and immunomodulatory activities.

2. Anapsos reverses learning impairment and brain IL-1ß overexpression in nbM-lesioned rats, suggesting that this vegetal extract might influence learning and neuroimmune functions by acting directly on the brain cholinergic system or by exerting a neuroprotective action.

3. The chronic administration of anapsos is well tolerated, reverses impairment of motor functioning, influences brain histamine levels and stimulates cortical cells producing IL-1ß without altering learning and survival in old mice.

4. Anapsos reduces behavioral deficits and hippocampal neurodegeneration in rats with ß-amyloid implants in the hippocampus, increasing IL-1ß production and modulating SOD activity in the brain.

5. According to these results, it appears that anapsos is a neuroprotective compound that improves learning and protects against neurodegeneration in rats. It is possible that procognitive and neuroprotective effects of anapsos might be mediated, at least in part, by its actions on brain IL-1ß production and SOD activity. It is suggested that anapsos represents a novel pleiotropic strategy for the treatment of AD and other neurodegenerative disorders.

ACKNOWLEDGMENTS

This research was supported by A.S.A.C. Pharmaceutical International and the EuroEspes Foundation.

REFERENCES

Alvarez XA, Zas R, Lagares R, Franco A, Maneiro E, Miguel-Hidalgo JJ, Lao JI, Fernández-Novoa L, Díaz J and Cacabelos R (1995): Neuroinmunomodulatory and neurotrophic activity of anapsos: Studies with laboratory animals. *Ann Psychiatry* 5:267-280.

Cacabelos R, Alvarez XA, Fernández-Novoa L, Franco A, Mangues R, Pellicer A and Nishimura T (1994): Brain Interleukin-1ß in Alzheimer's disease and vascular dementia. *Meth Find Exp Clin Pharmacol* 16:141-151.

Cacabelos R, Hofman A, Mullan M, Ferris SH, Portera A, Frey H, Swaab DF, de Leon M, Terry R, Masters CL, Iqbal K, Wisniewski HM, Takeda N, Giacobini E, Nordberg A and Winblad B (1993): Alzheimer's disease: Scientific progress for future trends. *DN&P* 6:242-244.

Cacabelos R, Nordberg A, Caamaño J, Franco-Maside A, Fernández-Novoa L, Gómez MJ, Alvarez XA, Takeda M, Prous JJr, Nishimura T and Winblad B (1995): Therapeutic strategies in Alzheimer's disease. *Ann Psychiatry* 4: 221-279.

Hardy J and Allsop D (1991): Amyloid deposition as the central event in the aetiology of Alzheimer's disease. *Trend Pharmacol Sci* 12:383-388.

Horvath A, Alvarado F, Szöcs J, De Alvarado ZN and Padilla G (1967): Metabolic effects of Calagualine, an antitumoral saponine of Polypodium Leucotomos. *Nature* 214:1256-1258.

Richardson JS (1993): Free radicals in the genesis of Alzheimer's disease. *Ann NY Acad Sci* 695:73-76.

Terry RD, Masliah E, Salmon DP, Butters N, De Teresa R, Hill R, Hansen LA and Katzman R (1991): Physical basis of cognitive alterations in Alzheimer's disease: Synapse loss is the major correlate of cognitive impairment. *Ann Neurol* 30:572-80.

Tuominen M, Bohlin L, Lindbom L-O and Rolfsen W (1991): Enhancing effect of Calaguala on the prevention of rejection on skin transplants in mice. *Phytotherapy Res* 5:234-236.

Vargas J, Muñoz C, Osorio C and García-Olivares E (1983): Anapsos, an antipsoriatic drug which increases the proportion of suppressor cells in human peripheral blood. *Ann Immunol* 134:393-400.

Part IV

DEVELOPMENT OF NEW THERAPIES IN ALZHEIMER DISEASE

Design of Clinical Trials
Advances in Assessment
SPECT and PET
Potential Markers of Disease Progression and Drug Efficacy in Alzheimer's Disease
Treatment of Behavioral and Psychosocial Disturbances

Alzheimer Disease: From Molecular Biology to Therapy
edited by R. Becker and E. Giacobini

CLINICAL TRIALS TO PREVENT ALZHEIMER'S DISEASE IN A POPULATION AT-RISK

Michael Grundman and Leon J. Thal
Department of Neurosciences and Alzheimer's Disease
Cooperative Study Unit, University of California, San Diego

INTRODUCTION

A recent cost analysis indicates that Alzheimer's disease (AD) is the nation's third most costly disease after cancer and heart disease. Direct medical and social service costs are calculated at $40,000 per patient per year. Total annual direct costs for AD care in the United States for 1994 is estimated at $83 billion (Ernst and Hay, 1994). This enormous financial burden suggests that attempts to delay the onset of disease appearance are needed to reduce both social and financial consequences to the nation.

The incidence of AD is strongly age-dependent and best estimates based on available epidemiological data suggest that the age-specific incidence of AD is approximately:

INCIDENT AD

Age	Incidence (%)	Cases per thousand/year
65-70	0.2	2
70-75	0.6	6
75-80	1.2	12
80-85	2.4	24
85-90	3.0	30

Thus, to carry out a primary prevention trial, approximately 5,000 individuals over the age of 70 with normal cognition at entrance would need to be enrolled and followed for 5 years in order to detect a 30% decrease in disease incidence with an 80% probability. The total cost of such a trial would be approximately $25 million. Given the limited observational data suggesting that any agent might alter the incidence of AD, it is unlikely that such a trial will be initiated in the very near future.

METHODOLOGICAL CONSIDERATIONS FOR A TRIAL OF INDIVIDUALS WITH MILD COGNITIVE IMPAIRMENT (MCI).

The concept of MCI is that it might represent a transitional phase between cognitive changes of normal changing and incipient dementia. These individuals are likely to progress in a relatively rapid and identifiable fashion towards dementia. Clinicians have utilized a wide variety of terms to describe these subjects including MCI, at risk for dementia and cognitive impairment without functional impairment. Using the Clinical Dementia Rating Scale (CDR) these individuals have a CDR of 0.5. A sample of 67 individuals with mild cognitive impairment and a CDR rating of 0.5 was shown to convert to frank dementia at a rate of approximately 44% after three years of follow up (Petersen et al., 1995).

The Alzheimer's Disease Cooperative Study (ADCS) therefore postulated that a trial to prevent the conversion of individuals with MCI to dementia could be carried out and would require fewer subjects and cost significantly less than a primary prevention trial. We collected follow up data on subjects with MCI or evaluated prospectively at Alzheimer Disease Centers associated with the ADCS. Data collected included age at presentation, gender, time to reach diagnosis of dementia defined as a CDR of 1, or meeting DSMIII-R or NINCDS-ADRDA criteria for dementia, and follow up time if dementia was not reached. Individuals with MCI were defined as:

- having a memory complaint
- demonstrating mild memory impairment on examination
- having absence or minimal impairment in other cognitive domains
- independence in activities of daily living
- meeting criteria for CDR = 0.5
- failing to meet NINCDS-ADRDA or DSMIII-R criteria for dementia

Seventeen centers submitted data on 687 subjects meeting these criteria; 55% were women, 45% men, and the mean age was 72 ± 10. Conversion to dementia occurred at the rate of:

INCIDENCE OF DEMENTIA IN SUBJECTS WITH MCI

Years of Follow Up	Cumulative Incidence (%)
0	0
1	9
2	31
3	44

Thus, with three years of follow up, approximately 45% of individuals or 15% per annum will make the transition from MCI to dementia. Older individuals progress at a more rapid rate, while gender did not affect the rate of conversion. Other possible predictors of dementia such as Apolipoprotein E status and the presence of hippocampal atrophy or delayed memory impairment were not examined in this data set.

POTENTIAL AGENTS FOR A DEMENTIA PREVENTION TRIAL IN SUBJECTS WITH MCI

In a trial of MCI, the agent administered will be given to both those individuals who will and will not develop dementia. Thus, any agent chosen will need to meet several requirements including: preliminary data supporting efficacy such as observational or epidemiological data, safety, and biological plausibility. Many agents currently meet these criteria and four will be mentioned as exemplars.

Selegiline is an MAOB inhibitor. Upregulation of MAOB can lead to excessive deamination of biogenic amines resulting in hydrogen peroxide and free radical formation. This may in turn lead to lipid peroxidation and neural membrane breakdown. Selegiline inhibits oxidative deamination and may reduce production of hydrogen peroxide. Selegiline was examined in a 15 month study in a small sample of mildly impaired AD patients. Although no statistically significant differences occurred between the treated and placebo group, there was a strong trend towards slower deterioration in the treated group on cognitive measures (Burke et al., 1993).

Observational data available from the East Boston study suggests that the use of Vitamin E is associated with a lower risk of AD (Evans, personal communication). Reactive oxygen species are produced in the central nervous system and may damage membranes due to their high lipid concentration. Vitamin E is a lipid soluble vitamin capable of breaking the chain reaction of lipid peroxidation (Halliwell & Gutteridge, 1985).

Vitamin C is an efficient antioxidant in the aqueous phase (Beyer, 1994). Vitamin C also markedly enhances the survival-promoting effects of Vitamin E on cultured hippocampal and striatal neurons (Sato et al., 1993). Unpublished data from the East Boston study suggests that the use of Vitamin C supplements is associated with a lower risk of AD (Evans et al., personal communication).

Co-enzyme Q protects against lipid peroxidation and functions as an essential cofactor in the electron transport chain and as a free radical scavenger (Stocker et al., 1991; Mukai et al., 1990.) Little clinical data is available demonstrating the effects of co-enzyme Q in patients with AD.

MCI TRIAL DESIGN

The overall design of an MCI trial would involve the enrollment of subjects diagnosed with MCI in a standardized fashion. These individuals would meet the following criteria: memory complaint, normal activities of daily living, normal general cognitive function, abnormal memory function, CDR = 0.5, not demented. The study design would be a parallel multiple group design with three years of follow up. Treatment groups could be exposed to: different doses of one antioxidant, more than one antioxidant, or combinations of two antioxidants in a 2 X 2 factorial design, The endpoint for the study would be dementia defined as a CDR of 1. Sample size calculations would depend on the degree of reduction in conversion to dementia sought, and the number of groups studied. For example, a 33% reduction in dementia could be reached if the active treatment reduced the conversion to dementia to 10% per annum for a total of 30% conversion to dementia after three years. This would be compared to the placebo-treated condition with 15% conversion to dementia per annum or 45% after three years. A two arm parallel study seeking this degree of reduction in dementia would require 360 subjects with an alpha level of 0.05 and a power of 0.80.

DISCUSSION

A trial examining conversion from MCI to AD is clearly both feasible and desirable. Overall costs for such a trial would be approximately $8 million, far less than the costs associated with a primary prevention trial. Preventing progression from MCI to frank dementia would have a profound public health impact on AD. In reality, it is likely that individuals with MCI who progress to AD during a trial and normal individuals who convert to dementia during a primary prevention trial both already have the pathological changes characteristic of AD. This is supported by the neuropathological findings of Braak and Braak (1994) whose work suggest that the neuropathology of AD begins 10-30 years prior to clinical presentation. Use of an MCI population represents an enriched group of individuals likely to develop dementia at a high rate during a short period of follow up. There are many other strategies that might be employed to enrich a sample. This strategy has the advantage of not excluding subjects based on biological variables and should therefore be generalizable to other individuals with MCI who have similar clinical characteristics.

ACKNOWLEDGMENT

This work is supported by AGO 10483.

REFERENCES

Beyer RE (1994): The role of ascorbate in antioxidant protection of biomembranes: interaction with vitamin E and coenzyme Q. *Journal of Bioenergetics and Biomembranes* 26:349-358.

Braak H and Braak E (1994): Diagnosis of Alzheimer's disease: development of cytoskeletal changes and staging of Alzheimer-related intraneuronal pathology. *Neurobiology of Aging* 15:5141.

Burke WJ, Roccaforte WH, Wengel SP, Bayer BL, Ranno AE, Willcockson NK (1993): L-deprenyl in the treatment of mild dementia of the Alzheimer type: results of a 15-month trial. *Journal of the American Geriatrics Society* 41:1219-1225.

Ernst RL and Hay JW (1994): The US economic and social costs of Alzheimer's disease revisited. *Am J Public Health* 84:1261-1264.

Halliwell B and Gutteridge JMC (1985): Oxygen radicals in the nervous system. *Trends in Neuroscience* 8:22-26.

Mukai K, Kikucki S, Urano 5 (1990): Stopped-flow kinetic study of the regeneration reaction of tecopheroxyl radical by reduced uniquinone- 10 in solution. *Biochimica et Biophysica Acta* 1035:77-82.

Petersen RC, Smith GE, Ivnik 'U (1995): Apolipoprotein E status as a predictor of the development of Alzheimer's disease in memory impaired individuals. *JAMA* 273:1274-1278.

Sato K, Saito H, Katsuki H (1993). Synergism of tocopherol and ascorbate of the survival of cultured brain neurons. *Neuroreport* 4:1179-1182.

Stocker R, Bowry VW, Frei B (1991): Ubiquinol-lO protects human low density lipoprotein more efficiently against lipid peroxidation than does atocopherol. *Proc Natl Acad Sci USA* 88:1646-1650.

Alzheimer Disease: From Molecular Biology to Therapy
edited by R. Becker and E. Giacobini

EFFECTS ON DECLINE OR DETERIORATION

Serge Gauthier and Judes Poirier
McGill Centre for Studies in Aging and
Douglas Hospital Research Centre
Montreal, Quebec, Canada

Julian Gray
F. Hoffmann-La Roche Ltd, Basel, Switzerland

INTRODUCTION

There is a great interest in slowing down the progression of symptoms in Alzheimer's disease (AD), particularly in early stages when people are fully autonomous. Such a stabilization strategy would logically target primary or secondary pathophysiological events of AD (Gray and Gauthier, 1996; Gauthier and Poirier, 1996). On the other hand there is a possibility of sustained symptomatic therapeutic effects that would lead to delay of progression, comparable to levodopa in Parkinson's disease (PD). Finally, some agents may have mixed stabilization and symptomatic effects, such as selegiline in PD. Our discussion will not eliminate any of these mechanistic approaches, since all could have effects on decline or deterioration. We will rather critically examine the methodological issues relevant to proof of efficacy for a clinically meaningful stabilization in early, intermediate as well as late stages of AD.

NATURAL HISTORY OF AD

There are a number of AD patients cohorts studied longitudinally and demonstrating a number of facts relevant to drug trial design. The standardization of diagnosis of probable AD by DSM-III and NINCDS-ADRDA criteria allow a 85% accuracy at autopsy (Galasko et al., 1994). There is a great individual variability in rate of progression, but most patients follow a hierarchic loss of domain specific (cognitive, functional and social) skills, culminating in a late stage loss of automatic motor skills, well described in instruments such as the FAST (Sclan and Reisberg, 1992).

In regard to rate of cognitive deterioration, there is good evidence that neither the age of onset nor the place of residence play a significant role, at least for ages 50 to 95 (Katzman et al., 1988). Cases with early onset familial AD related to point mutations are rare and can be excluded from stabilization studies, as well as very old subjects who have a higher mortality rate (Becker et al., 1994).

There are a number of disease milestones potentially useful as end points with face validity, such as loss of instrumental ADL, basic ADL, incontinence, institutionalization, and death (Galasko et al., 1995). Finally, there are clinical predictors of more rapid progression towards institutionalization, such as early appearance of psychiatric disturbances and extrapyramidal signs (Panisset and Stern, 1996).

OUTCOME VARIABLES RELEVANT TO PROGRESSION

Global assessment of severity has been measured using scales such as the CDR, FRS and BCRS (Rockwood and Morris, 1996). The CDR has the advantages of hierarchy and concordance, with six domains explored by a semi-structured interview with patients and caregivers. The CDR has been developed and used for early to intermediate stages of AD. The FRS has been proposed as an expanded CDR, incorporating the additional domains of language and mood. The BCRS takes into account eleven domains and is applicable to later stages of AD. Berg et al. (1992) have discussed the use of the slope of CDR as an outcome measure for evaluating disease progression. These authors point out how sensitivity of this measure may be enhanced by adding data points at the beginning and end of the study period.

There is a large body of data on performance-based cognitive scales such as the ADAS-Cog, Blessed IMC, MMSE. There is evidence for non-linearity over time because of slower loss of cognitive abilities in early AD, a steeper measurable loss in an intermediate phase, and a floor effect (Morris et al., 1993).

Functional scales such as IADL and PSMS have also shown a curvilinear pattern of deterioration (Mortimer et al., 1992). The time to emergence of behavioral symptoms, described extensively by Reisberg et al. (1987), has only recently been incorporated in drug trial design.

Time to institutionalization has been studied systematically because of its potential pharmaco-economic relevance, and recent reports suggest ways to resolve confounding factors (Brodaty et al., 1993; Stern et al., 1996).

TRIAL DESIGN AND ANALYSIS

Longitudinal studies of AD suggest that a minimum of one year and two outcome variables are required to have clinically meaningful data, including a global assessment and a domain specific measure of severity. Variations on the traditional parallel groups, placebo-controlled design have been proposed by Leber (1994), including the blinded withdrawal and follow-up, and the randomized start design. Time to clinically relevant end points has also been tested by the ADSCU.

Estimates of the number of patients required to show efficacy have been made, using the sum of boxes on the CDR (Berg et al., 1992). Homogeneity of the patients can be enhanced by stratification for severity, favouring intermediate stage patients (CDR 1 or 2) whose cognitive deterioration falls in the linear portion of the curve (Holmes et al., 1995). There remains the option of enrichment of a study population with clinical characteristics predictive of fast progression (e.g., aphasia, extrapyramidal signs, psychosis), leading to shorter study time. Recent evidence suggest that apoE 4 allele load influence cholinomimetic response in patients with sporadic AD (Poirier et al., 1995). Stratification based on apoE genotype prior to randomization should be considered, especially since progression appears to be modulated by gender and apoE allele load (Poirier et al., unpublished data).

Different methods of analysis are available such as the mean change score between baseline and the last observation; slope analysis using linear, trilinear or curvilinear models; and finally, survival analysis or time to reach a critical endpoint, using the Kaplan-Meier life tables method and the Cox proportional hazard model (Drachman et al., 1990).

REGULATORY GUIDELINES

The US draft guidelines and ongoing harmonization discussions suggest that an absolute global severity instrument (such as the CDR) and a domain specific measure (cognitive or functional), in a long enough study will provide proof of stabilization effects. A blinded withdrawal and an adequate duration follow-up period would provide additional evidence of a disease process modifying action. The Canadian published (Mohr et al., 1995) and European draft guidelines concur, with additional emphasis on pharmaco-economically relevant measures and ADL changes, respectively.

ETHICAL CONSIDERATIONS

Ethics of research in patients with AD must always be an essential consideration in the planning and implementation of long studies required for proof of stabilization effects. Relevant ethical issues range from the prolonged use of placebo to washout from active treatment, loss of competency as the disease is progressing, need for repeated consent from patients and proxy. The use of advance directives for research would be highly recommended for such studies. The current paucity of research protocols in late stages of AD could be overcome by trial designs using time to emergence of neurological signs and loss of autonomy (Franssen et al., 1993).

CONCLUSIONS

There is a need for preclinical models of AD that could lead academia and industry towards etiology-driven therapies. Clinical testing will require short term (3-6 months) feasibility studies in order to establish safety and tolerability, followed by studies of at least one year in parallel groups for determination of slopes using a global severity measure and a domain specific scale. Even longer studies would be needed to establish the effect of stabilization on delaying the time to major endpoints, or to determine whether such therapy can delay onset of frank dementia when administered in the presymptomatic stage.

Stabilization therapy is not incompatible with symptomatic treatment, and it is likely that cholinesterase inhibitors and receptor agonists will be used in combination with agents primarily targeted at stabilization.

ACKNOWLEDGMENTS

The research of Serge Gauthier and Judes Poirier is supported by the Alzheimer Disease Society of Canada, the " Fonds de la recherche en santé du Québec", the Medical Research Council of Canada, the National Health Research and Development Program. We thank Ms. Christina Kyriakou for her expert secretarial assistance.

REFERENCES

Becker JT, Boller F, Lopez OL, Saxton J et al. (1994): The natural history of Alzheimer's disease. Description of study cohort and accuracy of diagnosis. *Arch Neurol* 51:585-594.

Berg L, Miller JP, Bary J, Rubin EH et al. (1992): Mild senile dementia of the Alzheimer type. Evaluation of intervention. *Ann Neurol* 31:242-249.

Brodaty H, McGilchrist C, Harris L and Peters KE (1993): Time until institutionalization and death in patients with dementia. *Arch Neurol* 50:643-650.

Drachman DA, O'Donnell BF, Lew RA and Swearer JM (1990): The prognosis in Alzheimer's disease. "How far" rather than "how fast" best predict the course. *Arch Neurol* 47:851-856.

Franssen EH, Kluger A, Torossian CL and Reisberg B (1993): The neurologic syndrome of severe AD. Relationship to functional decline. *Arch Neurol* 50:1029-1039.

Galasko D, Hansen LA, Katzman R et al. (1994): Clinical-neuropathological correlations in Alzheimer's disease and related dementias. *Arch Neurol* 51:888-895.

Galasko D, Edland SD, Morris JC, Clark C et al. (1995): CERAD. Part XI. Clinical milestones in patients with Alzheimer's disease followed over 3 years. *Neurology* 45:1451-1455.

Gauthier S and Poirier J (1996): Neuroprotection in Alzheimer's disease. In: *Neuroprotection: fundamental and clinical aspects*, Bär PR and Beal MF, eds. Marcel Dekker Inc. (in press)

Gray J and Gauthier S (1996): Stabilization approaches to Alzheimer's disease. In: *Clinical diagnosis and management of* Alzheimer's disease, Gauthier S, ed. London: Martin Dunitz. (in press)

Holmes E, Merskey H, Fox H, Fry RN et al. (1995): Patterns of deterioration in seniledementia of the Alzheimer type. *Arch Neurol* 52:306-310.

Katzman R, Brown T, Thal LJ et al. (1988): Comparison of rate of annual change of mental status score in four independent studies of patients with Alzheimer's disease. *Ann Neurol* 24:384-389.

Leber P (1994): Some observations and suggestions on antidementia drug development. Presented at the NIA Symposium on Presymptomatic Strategies in Alzheimer's disease, Washington, DC.

Mohr E, Feldman H and Gauthier S (1995): Canadian guidelines for the development of antidementia therapies: a conceptual summary. *Can J Neurol Sci* 22:62-71.

Morris JC, Edland S, Clark C, Galasko D et al. (1993): CERAD. Part IV. Rates of cognitive change in the longitudinal assessment of probable Alzheimer's disease. *Neurology* 43:2457-2465.

Mortimer JA, Ebbitt B, Jun SP and Finch MD (1992): Predictors of cognitive and functional progression in patients with probable Alzheimer's disease. *Neurology* 42:1689-1696.

Panisset M and Stern Y (1996): Prognosis factors. In: *Clinical diagnosis and management of Alzheimer's disease*. Gauthier S, ed. London: Martin Dunitz. (in press).

Poirier J, Delisle MC, Quirion R, Aubert I et al. (1995): Apolipoprotein E4 allele as a predictor of cholinergic deficits and treatment outcome in Alzheimer's disease. *Proc Natl Acad Sci* 92:12260-12264.

Reisberg B, Borenstein J, Salob SP et al. (1987): Behavioral symptoms in Alzheimer's disease: phenomenology and treatment. *J Clin Psychiatry* 48 (suppl):9-15.

Rockwood K and Morris JC (1996): Global staging measures in dementia. In: *Clinical diagnosis and management of Alzheimer's disease*, Gauthier S, ed. London: Martin Dunitz (in press).

Sclan SG and Reisberg B (1992): Functional Assessment Staging (FAST) in Alzheimer's disease: reliability, validity and ordinality. *International Psychogeriatrics* 4 (suppl 1): 55-69.

Stern Y, Tang MX, Albert M, Brandt J et al. (1996): Algorithms for predicting time to nursing home admission and death in individuals with Alzheimer's disease. *Neurology* 46 (suppl): A434.

[illegible] S and [illegible] L (1990) [illegible] In: [illegible] Dat PR and [illegible] eds. Marcel Dekker Inc. (in press).

[illegible] and Gauthier S (1990) [illegible] In: [illegible] (in press).

[illegible] Pattern of [illegible]

[illegible] comparison of [illegible] of [illegible] patients [illegible]

[illegible]

[illegible]

[illegible]

[illegible]

[illegible]

[illegible]

[illegible]

Alzheimer Disease: From Molecular Biology to Therapy
edited by R. Becker and E. Giacobini

THE BRIDGING STUDY: OPTIMIZING THE DOSE FOR PHASE II/III

Neal R. Cutler and John J. Sramek
California Clinical Trials, Beverly Hills, CA

INTRODUCTION TO THE BRIDGING STUDY

Treatment of Alzheimer's disease (AD) has recently focused on strategies which target deficits in the cholinergic system of the brain. Such focus is reasonable, given the loss of cholinergic neurons and associated choline acetyltransferase seen in the AD brain (Cutler et al., 1994). However, treatment with cholinergic compounds has proven complicated. All compounds thus far tested, including tacrine, appear to have limited efficacy, and treatment is often accompanied by significant adverse events. Many can affect plasma cholinesterases and have short half-lives as a result of rapid metabolism. More research is needed to find selective compounds with safer adverse event profiles.

One neglected area of research has been examination of the impact of cholinergic deficits on the pharmacodynamic profiles of potential AD compounds. Our clinical experience with "bridging studies," safety/tolerability studies in the patient population, have indicated that there can be considerable variation in the maximum tolerated doses (MTDs) of the healthy elderly and AD populations (Cutler and Sramek, 1995a,b). Table I summarizes the maximum tolerated doses (MTDs) determined in a number of safety/tolerability studies of cholinergic compounds for comparison of these values between healthy volunteers and AD patients. For the acetylcholinesterase inhibitor velnacrine, the MTD in elderly normals was 100 mg tid, while AD patients could only tolerate 75 mg tid (Puri et al., 1988, 1989, 1990; Cutler et al., 1990). For another acetylcholinesterase (AChE) inhibitor, eptastigmine, whereas 32 mg was the single-dose MTD in healthy elderly subjects, the MTD in AD patients was significantly higher at 48 mg tid, or 144 mg/day (Goldberg et al., 1991; Sramek et al., 1994). For the muscarinic agonist CI-979, the normal MTD was defined as 1 mg q6h, while for AD patients the MTD was found to be twice as high, 2 mg q6h (Reece et al., 1992; Sramek et al., 1995b). Similarly, for xanomeline tartrate, the healthy elderly MTD was 50 mg tid, or 150 mg/day, compared to an MTD of 100 mg tid, or 300 mg/day, for AD patients (Sramek et al., 1995a).

These results point to two major advantages of bridging studies: Where the MTD is higher in patients than in healthy normals, higher doses can then be used in Phase II efficacy studies to maximize the probability that efficacy can be determined. For example, in a recent efficacy trial of three doses of xanomeline (25, 50, and 75 mg tid) in 300 patients with AD, only the highest dose (75 mg tid) showed superiority over placebo (Tollefson, 1994). This dose would not have been investigated had we not previously conducted a bridging study that demonstrated tolerability of doses above the healthy elderly MTD of 50 mg tid. In such cases,bridging prevents potentially useful agents from appearing ineffective. In cases where the MTD is lower in patients than in healthy normals, as with velnacrine, the performance of a bridging study prevents use of a dose which might result in unacceptably high levels of clinically significant adverse events in large-scale efficacy studies.

CONDUCT OF A BRIDGING STUDY

A major difficulty in evaluating Phase I results is that different investigators use different definitions of the MTD. We determine the MTD by reaching the minimal intolerated dose (MID), defined as the dose at which a

	Velnacrine	Eptastigmine	CI-979	Xanomeline	Bespiridine
Young normals single-dose	MTD not reached; 200 mg well tolerated	MTD not reached; 20 mg well tolerated	NP*	MTD not reached; 150 mg well tolerated	MTD not reached; 30 mg well tolerated
Young normals multiple-dose	NP	MTD = 32 mg tid	MTD = 1 mg q6h	MTD not reached; 75 mg bid well tolerated	NP
Elderly normals single-dose	MTD not defined; 300 mg well tolerated	MTD = 32 mg/day	NP	NP	NP
Elderly normals multiple-dose	MTD = 100 mg tid	NP	in progress	MTD = 50mg bid	MTD = 100mg bid
AD patients "bridging study"	MTD = 75 mg tid	MTD = 48 mg tid	MTD = 2 mg q6h	MTD = 100 mg tid	MTD = 50 mg bid

TABLE I. Review of Phase I safety/tolerance studies of cholinergic compounds: Comparison of maximum tolerated doses (MTDs) in normals and patients (for first four compounds, see text for appropriate references; for besipirdine see Sramek et al, 1995c). *NP = not performed.

majority (50%or more) of the subjects receiving active drug experience severe or multiple moderate adverse events, or the dose at which one serious adverse event occurs in one patient. The highest dose tolerated before reaching the MID is then designated as the MTD.

The MTD determined in healthy volunteers serves as the basis for designing a dosing schedule in a bridging study. We recommend a panel design beginning with a panel tested at a dose 50% below the MTD of healthy normals. Each subsequent panel is tested at a higher dose, moving on to 25% below the normal MTD, the MTD, 25% above the MTD, 50% above the MTD, and so on in equivalent intervals until an MID is reached. The highest dose tolerated before this dose is then designated the MTD. A second panel at this dose may be tested using new patients to increase confidence in the safety of this dose.

The use of equal dose increments is preferable to a logarithmic design because it better characterizes the part of the dose-response curve with the greatest slope, allowing for a more precise definition of the MTD (Cutler and Sramek, 1995a,b). Increments of approximately 25%, which can be adjusted based on the pharmacokinetics of a particular compound, permit the plasma concentration curves for each dose to overlap somewhat but remain easily distinguishable.

Furthermore, we believe that there are several advantages to determining a fixed-dose as opposed to a dose titration MTD. In a dose titration trial, safety assessments are more difficult to make because drug accumulation effects and increased tolerance due to acclimation can obscure the true safety profile (Geller, 1984). Moreover, the MTD is likely to depend on the rate of dose ascension, for which there is no clear standard. Thus, a dose titration MTD provides little information about the dose level than can be tolerated without prior exposure to the drug. Dose titration can be useful with some drugs in order to reach therapeutic doses; however, it is prudent to determine a fixed-dose MTD first in order to establish safety guidelines for higher dose ranges and provide information for caregivers about symptoms of intolerance due to poor compliance. The identification of such symptoms through bridging studies contributes significantly to the safety of Phase II clinical trials.

Exploration of a new compound's toxicology before initiating a bridging study is important. In particular, toxicity studies in animals should investigate dosages well beyond any anticipated human dosage in order to provide a clear profile of potential adverse events. This is especially significant since the patient population may tolerate doses much higher than expected. If there is inadequate preclinical toxicology data, a bridging study may have to be terminated before an MTD has been reached.

ACKNOWLEDGMENTS

The research from our laboratory reviewed here was supported by Hoechst-Roussel Pharmaceuticals, Somerville, New Jersey and Montreal, Quebec; Merck Research Laboratories, West Point, Pennsylvania; Parke-Davis Pharmaceuticals, Ann Arbor, Michigan; and Eli Lilly Laboratories, Indianapolis, Indiana. We thank Giedra Miller for preparing the manuscript.

REFERENCES

Cutler NR, Murphy MJ, Nash RJ, Prior PL (1990): Clinical safety, tolerance, and plasma levels of HP 029 in Alzheimer's disease. *J Clin Pharmacol* 30:556-561.

Cutler NR, Sramek JJ (1995a): Scientific and ethical concerns in clinical trials in Alzheimer's patients: the bridging study. *Eur J Clin Pharmacol* 48:421-428.

Cutler NR, Sramek JJ (1995b): The target population in phase I clinical trials of cholinergic compounds in Alzheimer's disease: the role of the "bridging study." *Alzheimer Dis Assoc Disord* 9:139-145.

Cutler NR, Sramek JJ, Veroff AE (1994): Alzheimer's Disease: Optimizing Drug Development Strategies. Sussex England: John Wiley & Sons, Ltd. Geller NL (1984): Design of phase I and II clinical trials in cancer: a statistician's view. *Cancer Invest* 2:483-491.

Goldberg MR, Barchowsky A, McCrea J, et al (1991): Heptylphysostigmine (L-693,487): safety and cholinesterase inhibition in a placebo-controlled rising-dose healthy volunteer study. Abstr 2nd Int Symposium Advances in Alzheimer Therapy.

Puri SK, Ho I, Hsu R, Lassman HB (1990): Multiple dose pharmacokinetics, safety, and tolerance of velnacrine (HP029) in healthy elderly subjects: a potential therapeutic agent for Alzheimer's disease. *J Clin Pharmacol* 30:948-955.

Puri SK, Hsu R, Ho I, Lassman HB (1988): Single-dose safety, tolerance, and pharmacokinetics of HP029 in elderly men: a potential Alzheimer agent. *Curr Ther Res* 44:766-780.

Puri SK, Hsu RS, Ho I, Lassman HB (1989): Single dose safety, tolerance and pharmacokinetics of HP029 in healthy young men: A potential Alzheimeragent. *J Clin Pharmacol* 29:278-284.

Reece PA, Bockbrader H, Sedman AJ (1992): Safety, pharmacodynamics, and pharmacokinetics of a new muscarinic agonist, CI-979. *Clin Exp Pharmacol Physiol* 21 (Suppl):58.

Sramek JJ, Block GA, Reines SA, Sawin SF, Barchowsky A, Cutler NR (1994): A multiple-dose safety trial of eptastigmine in Alzheimer's disease, with pharmacodynamic observations of red blood cell cholinesterase. *Life Sci* 56:319-326.

Sramek JJ, Hurley DJ, Wardle TS, Satterwhite JH, Hourani J, Dies F, Cutler NR (1995a) The safety and tolerance of xanomeline tartrate in patients with Alzheimer's disease. *J Clin Pharmacol* 35:800-806.

Sramek JJ, Sedman, AJ, Reece PA, Underwood B, Seifert RD, Bockbrader H, Cutler NR (1995b): Safety and tolerability of CI-979 in patients with Alzheimer's disease. *Life Sci* 57:503-510.

Sramek JJ, Viereck C, Huff FJ, Wardle T, Hourani J, Stewart JA, Cutler NR (1995c): A "bridging" (safety/tolerance) study of besipirdine hydrochloride in patients with Alzheimer's disease. *Life Sci* 57:1241-1248.

Tollefson GD, Bodick NC, Shannon HC (1994): Xanomeline: a potent and specific M1 agonist in the treatment of Alzheimer's disease. Abstr American College Neuropsychopharmacology, December 12-16.

Stamek [illegible] [illegible] *Wa*[illegible] [illegible]
[illegible]
[illegible]
[illegible]
[illegible]
[illegible]

Alzheimer Disease: From Molecular Biology to Therapy
edited by R. Becker and E. Giacobini

ASSESSMENT OF THERAPEUTIC STRATEGIES FOR SLOWING PROGRESSION OF ALZHEIMER'S DISEASE

Michael J Pontecorvo
Janssen Research Foundation, Titusville, N.J.

Wim Parys
Janssen Research Foundation, Beerse, Belgium

The first drugs for treatment of Alzheimer's disease (AD) were designed to symptomatically improve cognition by augmenting cholinergic transmission. A second generation of compounds, now entering clinical trials, may not directly enhance cognition, but could slow the progression of cognitive deterioration. Compounds like antioxidants (deprenyl, vitamin E), and NMDA antagonists have shown potential as neuroprotective agents in other disease conditions. Alternatively agents that stabilize cytoskeletal function or block abnormal processing, deposition or toxic effects of amyloid may provide disease-specific neuroprotective effects. Although the methods used to evaluate these drugs will be generally similar to those used in evaluation of symptomatic treatments, the unique challenges of studying disease progression may require refinements in study design, selection criteria and assessment instruments.

Drugs intended to slow progression of AD may not produce an acute improvement in cognitive performance. Thus, significant differences between placebo and drug-treated patients may only be seen when the placebo group has been followed for an adequate amount of time. Longitudinal studies suggest that the sample size required to detect a reduction in rate of cognitive decline decreases as the length of the trial increases (Stern et al., 1994). Together with observations that the rate of deterioration in clinical trials is likely to be slower than in natural observation (Knopman and Gracon, 1994), these data suggest that treatments proposed to slow deterioration will need to be evaluated several times over a period of at least one year. If the difference between the drug-treated and the placebo groups increases as a function of treatment duration (e.g., 12 vs 6 months) then it may be argued that the drug has altered the progression of the cognitive deterioration.

An alternative approach (Thal, 1994) utilizes a survival analysis. The main efficacy variable is time to some clinically relevant endpoints (nursing home placement, loss of basic skills, etc.). Time to a clinical endpoint can be readily translated into an estimate of rate of progression and used to estimate the social and economic impact of the treatment. The difficulty is in choosing endpoints that are specific surrogates for disease state. For example, time to nursing home placement seems, on the surface, an appropriate indicator of disease progression. However, this endpoint may be influenced by caregiver variables (support network, attitude toward patient) and patient variables, such as behavioral disturbances, that are not monotonically related to disease progression (Mittelman et al., 1993).

A finding that patients receiving active-treatment deteriorate more slowly than placebo-treated patients supports the inference that the compound slows the clinical progression of AD. However, inferences regarding a compound's effects on the AD pathophysiology are more problematic. One strategy is to evaluate the effects of treatment withdrawal. If a drug slows the progression of neurodegeneration, then the advantage gained by drug treatment should be maintained after treatment is withdrawn (patients withdrawn from drug vs placebo should deteriorate at similar rates). Alternatively, if a compound acts symptomatically to mask progression of pathopysiology, then symptoms will rapidly reappear upon withdrawal (patients withdrawn from drug should deteriorate more rapidly). Unfortunately this design has both ethical problems (withdrawal of an effective medication) and practical limitations (pharmacological rebound and psychological withdrawal effects can complicate interpretation; moreover, for a positive outcome this design requires a statistical demonstration of equivalence, and thus requires more subjects than the original proof of efficacy).

Long-term studies may provide an alternative to a withdrawal design. If the between-treatment continued to increase over two or more years, it would be difficult to argue that the compound did not slow progression of the disease process. Such clinical results could also be bolstered by demonstration that biomarkers of disease progression are altered by treatment. Our understanding of the role of processes such as amyloidogenesis, tau phosphorylation and/or neurofibrillary tangle formation, may soon develop to the point that biomarkers such as amyloid burden in CSF will be used in clinical studies. Until then, imaging techniques (e.g., hippocampal volume, as measured by CT or MRI, Jobst et al., 1994) may be the most useful biomarkers of disease progression.

Compounds that delay disease progression may also delay onset of dementia. This endpoint (development of clinically diagnosed dementia) is both clear-cut and clinically/societally relevant. The required sample size and trial duration are not prohibitive. For example, Petersen et al. (1995) reported that approximately 40 to 45% of subjects with mild cognitive impairment (Clinical Dementia Rating-CDR=0.5) progressed to clinically diagnosed dementia (CDR=1.0 or greater) within 3 years. Assuming that diagnosis will be less consistent and progression more variable in a multicenter trial, and making allowances for a high dropout rate over the three year period, the sample size necessary to detect a 30% decrease in percentage of patients becoming demented in 3 years would be approximately 500. Nevertheless, this design still requires a substantial investment of resources for an extended period of time, and may not be advisable until a compound has shown efficacy in more traditional trials. Second, it will be important to carefully define both the subject population and the clinical endpoint (e.g., CDR 0.5 and 1.0). Third, until symptomatic treatments have been tested in this type of design it will be difficult to rule out the possibility that an experimental agent can mask onset of dementia by symptomatically improving cognition. On the other hand, any positive outcome (delay of onset), of reasonable magnitude, would have an enormous consequence for public health. Thus, these kinds of trials should be considered with compounds that have otherwise shown evidence of efficacy.

Trial designs must also take into account factors such as disease stage/severity at baseline, presence/absence of psychiatric symptoms or extrapyramidal side-effects, concurrent medications (particularly vitamins, non-steroidal anti-inflammatory agents and hormone treatments) and apolipoprotein-E (APO-E) genotype, all of which may influence the rate of cognitive deterioration. The effects of baseline level of severity appear to be particularly robust. For example, Stern et al. (1994) reported that the yearly rate of progression, as assessed by the ADAS-Cog and Blessed dementia scale is greater for patients in moderate to severe stages of AD than for patients with milder disease. It is unclear at this point whether this is due to differential sensitivity of these assessment scales or whether there is a real difference in rate of disease progression. In either case, it follows that it may be easier to demonstrate treatment effects in a moderately demented population than in a mildly demented population. However, a balance must be struck between selecting a relatively homogeneous cohort of patients that is likely to show substantial progression over the treatment period, and including a sufficient range of patients to show generality of effect. Additionally, it is advisable to plan sub-group analyses to evaluate the contribution of factors such as those mentioned above to the treatment effect.

Finally, assessment variables for studies of treatment effects on AD progression must be chosen both to meet current regulatory and scientific standards and to optimize assessment of progression of cognitive deterioration. Over the past several years a consensus has emerged that trial assessments should include a psychometric/performance measure covering the major domains of intellectual impairment seen in AD (e.g., memory, orientation, judgement, language and praxis) and a measure of clinical relevance (e.g., a global measure and/or an activities of daily living scale). It remains to be seen whether different criteria will emerge as we come to better understand the currently available assessment scales and develop new drugs that slow disease progression as well as treat symptoms. In the meanwhile, experimenters need to consider how the currently available measures might perform in trials aimed at evaluation of effects on disease progression.

The Alzheimer's disease assessment scale (ADAS) is perhaps the most well know and widely accepted of the cognitive psychometrics used in AD assessment. It consists of two major subscales -- a cognitive subscale and a behavioral subscale. The ADAS behavioral subscale may not be ideally suited for use in clinical trials. Other more comprehensive scales are available for evaluation of changes in behavior, and adverse event reporting provides a more frequent and timely picture of behavioral change. The ADAS cognitive subscale (ADAS-Cog) has been shown to distinguish AD patients vs normal elderly, mild vs moderate vs severe AD patients and to change measurably over a one year period (Kramer-Ginsberg et al., 1988; Stern et al., 1994). Although there exist numerous alternatives to the ADAS, most of these are batteries composed of unrelated tests that are not amenable or validated for combination into a single overall test score for a summary evaluation of treatment effects. Moreover, the ADAS has a history of use across a significant number of clinical trials allowing at least rough qualitative comparisons across studies.

Clinical relevance of a treatment effect is commonly evaluated using a clinician's interview based global evaluation of change (CIBIC). The primary advantage of a CIBIC is that it provides an independent evaluation, uncontaminated by formal evaluation with psychometric scales. Several published trials have now shown that a CIBIC or CIBIC-plus (caregiver input considered) can detect symptomatic improvements produced by cholinesterase inhibitors (e.g., tacrine). The utility of this assessment in long-term studies of disease progression may be more problematic. Dahlke et al. (1992) argued that inter-rater and test-retest reliability for global evaluations of change in dementia patients may be low, due to differing concepts of disease change and difficulties in reconstructing/remembering the condition of a patient at baseline. Additionally, it must be noted that the 7-item change scale was constructed to be sensitive to improvement in global condition. It is unclear whether this rating would be sufficiently sensitive to detect a slowing of rate of deterioration. Use of semistructured interview techniques, comprehensive notes or taping of baseline sessions and use of a broader change scale (e.g., 11 points) could contribute to improved accuracy and sensitivity of the CIBIC.

Activities of daily living (ADL) scales may be used as an alternative or supplement to global evaluations. There currently exists little consensus as to choice of an ADL scale. The issues seem more related to how to measure ADLs than what to measure. For example, some scales utilize caregiver report whereas others rely on direct observance of performance in the clinician's office. The former are subject to caregiver bias, whereas the latter may be distorted by the unfamiliar experimental situation. Additionally, many scales have not been studied in sufficient numbers of patients or for sufficient periods of time to predict their utility in studies of disease progression. We have recently employed the Disability Assessment for Dementia (DAD) scale (Gauthier et al., 1993) in two one-year placebo-controlled studies (SAB-INT-12, SAB-USA-25, Janssen data on file). In both studies, the baseline DAD score was correlated with the baseline ADAS-Cog, and change over the one-year period was significant and correlated (r=0.4-0.5) both with change in the ADAS-Cog and with CIBIC score. Thus, the DAD scale could be useful in evaluating the effects of pharmacologic treatments on progression of functional deterioration in patients with AD.

ACKNOWLEDGEMENTS

The authors appreciate the statistical support of Hong-Lin Su and Minerva Cortens and the efforts of the SAB-INT-12 and SAB-USA-25 investigators.

REFERENCES

Dahlke F, Lohaus A, and Gutzman H (1992): Reliability and clinical concepts underlying global judgments in dementia: Implications for clinical research. *Psychopharmacol Bull* 28:425-432.

Gauthier L, Gauthier S, Gelinas I, McIntyre M and Wood-Dauphinee S (1993): Functional assessment in Alzheimer's Disease. *Abstracts of the 16th Annual Meeting of the Canadian College of Neuropharmacology and the British Association for Psychopharmacology*, Montreal, Canada, S4.5.

Jobst KA, Hindley NJ, King E and Smith AD (1994): The Diagnosis of Alzheimer's Disease: A Question of Image? *J Clin Psychiatry* 55::22-31.

Knopman DS and Gracon S (1994): Observations on the short-term 'natural history' of probable Alzheimer's disease in a controlled clinical trial. *Neurology* 44:260-265.

Kramer-Ginsberg E, Mohs RC, Aryan M, Lobel D, Silverman J, Davidson M and Davis KL (1988): Clinical predictors of course for Alzheimer patients in a longitudinal study: A preliminary report. *Psychoharmacol Bull* 24:458-462.

Mittelman MS, Ferris SH, Steinberg G, Shulman E, Mackell JA, Ambinder A and Cohen J (1993): An intervention that delays institutionalization of Alzheimer's disease patients: Treatment of spouse-caregivers. *Gerontologist* 33:730-740.

Petersen RC, Smith GE, Ivnik RJ, Tangalos EG, Schaid DJ, Thibodeau SN, Kokmen E, Waring SC and Kurland LT (1995): Apolipoprotein E status as a predictor of the development of Alzheimer's disease in memory impaired individuals. *J Am Med Assoc* 273:1274-1278.

Stern RG, Mohs RC, Davidson M, Schmeidler J, Silverman J, Kramer-Ginsberg E, Searcy T, Bierer L, and Davis KL (1994): A longitudinal study of Alzheimer's disease: Measurement, rate, predictors of cognitive deterioration. *Am J Psychiatry* 151:390-396.

Thal LJ (1994): Clinical testing of new drugs for efficacy in Alzheimer's disease. In: *Alzheimer's Disease: Therapeutic Strategies*, Giacobini E and Becker R, eds. Boston, Birkhauser, pp. 436-440.

Alzheimer Disease: From Molecular Biology to Therapy
edited by R. Becker and E. Giacobini

POTENTIAL FOR PROGRESS IN THE THERAPEUTICS OF ALZHEIMER'S DISEASE; UNANSWERED QUESTIONS

Kenneth L. Davis
Department of Psychiatry, Mount Sinai Medical Center
New York, NY 10029

INTRODUCTION

Palpable progress has been made in recent years in the therapeutics of Alzheimer's disease (AD). The approval of tacrine in a number of countries throughout the world has brought to the clinician, for the first time, a drug with some efficacy in the palliative of treatment of symptoms in AD. Although in many patients these effects are not robust, and the subgroup of responsive patients is not as large as may have been desired, benefit has been established in measures of cognition, activities of daily living, and global performance (Davis et al., 1992; Farlow et al., 1992; Knapp et al., 1994). However, perhaps the most meaningful reflection of the perceived efficacy of the treatment of the symptoms of AD with drugs that increase cholinergic activity are the large number of compounds in development that are poised to compete with tacrine.

A series of cholinesterase inhibitors including E2020, ENA713, galanthamine, metrifonate, and long acting physostigmine are all in either late-phase II or phase III development. In fact, E2020 awaits regulatory approval in the United States. Cholinergic agonists all about to enter Phase III studies, include 5B202026, PD979-14/16, and xenomoline. Nicotinic agonism is also being investigated for its therapeutic potential. Taken together it is likely that the armamentarium of cholinomimetic compounds available to the clinician will substantially increase in the next five years.

The purpose of this paper is to delineate, and then elaborate, a number of unanswered questions that ultimately need to be addressed as cholinomimetics move forward in clinical trials.

CHOLINOMIMETICS

It has been well recognized since some of the earliest clinical trials with cholinesterase inhibitors (Mohs et al., 1985, 1985a; Thal et al., 1983), and continuing into the large scale studies (Davis et al., 1992; Farlow et al., 1992; Knapp et al., 1994), that not all patients with AD respond to cholinesterase inhibitors. Thus, a marker to identify treatment-responsive patients, prior to therapy, would be quite useful.

Theoretical reasons for nonresponsivity are multiple. Surely the efficacy of a cholinesterase inhibitor is dependent upon the integrity of the cholinergic neuron, as that neuron must produce some acetylcholine. However, studies with biopsy tissue from end-stage Alzheimer's patients indicate that even in the most severe form of this disease, it is possible to produce comparable levels of brain acetylcholine following releasing agents and a cholinesterase inhibitor, as exist in normal brain (Nilsson et al., 1986). These results, suggest alternate explanations for the nonresponsivity of many patients to cholinesterase inhibitors. Indeed, a very plausible alternative explanation is that AD is much more than simply a cholinergic deficiency. The multiple other neurotransmitters and neuropeptide deficits in AD can be expected to influence the responsivity to cholinesterase therapies, as well as cholinomimetics in general. Support for this position derives from a series of experiments indicating that the addition of a noradrenergic deficiency to a cholinergic deficiency all but eliminates the ability of a cholinergic drug to reverse the behavioral consequences of the cholinergic deficiency. Hence, a relatively intact noradrenergic innervation seems to be a prerequisite to the efficacy of a cholinergic compound in improving performance on learning tasks (Haroutunian et al., 1990). Potential markers of responsivity to cholinomimetics based on indexes of noradrenergic neurotransmission should be tested.

Recently, it has been suggested that the presence of the type 4 isoform of Apolipoprotein E (ApoE4) is associated with diminished responsivity to the cholinesterase inhibitor tacrine (Poirier et al., 1995). This is a potentially important finding whose replication is awaited from ongoing large trials of cholinomimetic compounds. Explanations of this result include the possibility that ApoE4 status is associated with increased neuropathology, and as a consequence either a more damaged cholinergic innervation or more widespread disease affects multiple neurotransmitters and neuropeptides.

Clearly, it would substantially aid the clinician to have a pretreatment marker of cholinergic responsivity in general, or cholinesterase responsivity in particular. Such a marker would eliminate the need for prolonged exposure of patients to drugs in which only equivocal responsivity is being detected. Conversely, patients with a high likelihood of responsivity would be encouraged to tolerate some of the bothersome, but not threatening, side effects of these agents until signs of efficacy were achieved. Whether this marker of responsivity is thought to be ApoE4, a measure of noradrenergic transmission, or any number of other possible parameters, the search for a predictive test of responsivity should be given a high priority.

Another central question for the cholinergic approach is if patients nonresponsive to cholinesterase inhibitors prove responsive to cholinergic agonists? Indeed, a fundamental rationale for the development of M_1 agonists is that patients whose disease has progressed to the point that cholinergic innervation is no longer adequate to support the efficacy of a cholinesterase inhibitor will prove responsive to an agonist. Although this argument has a reasonable theoretical base, it has not been established in the clinic. Furthermore, studies comparing the efficacy of a muscarinic agonist to a cholinesterase inhibitor are not ongoing. A relatively efficient way to test this hypothesis would be to randomize patients who are nonresponsive to a cholinesterase inhibitor, to a parallel-designed study in which they are randomly assigned to either an alternative cholinesterase inhibitor or a cholinergic agonist. However, only with the answer to this question will the limits of cholinergic responsivity be defined.

Clearly, this issue of whether cholinomimetic agents alter the course of AD need to be addressed. Cholinomimetic compounds were initially conceived as palliative therapies. However, an enlarging data base has suggested that altering cholinergic transmission, either with M_1 agonist or cholinesterase inhibitors, can affect the processing of amyloid precursor protein (Buxbaum et al., 1992; Nitsch et al., 1992; Farber et al., 1995; Haroutunian et al., 1996; Haroutunian et al., 1995; Lahiri et al., 1994; Lahiri et al., 1994a; Lahiri et al., 1992). Following an interruption in cholinergic neurotransmission, either through a neurotoxic lesion, the administration of an anesthetic, or scopolamine, levels of messenger RNA for APP, APP protein, and secreted forms of APP, are all elevated. Some, but not all, cholinesterase inhibitors can normalize this process in these model systems, and return APP levels as well as secreted APP products to pre-insult levels (Haroutunian et al., 1996). It is possible to extrapolate from these data that the use of some cholinesterase inhibitors could alter the processing of APP in a way that might alter the course of AD.

These pre-clinical results strongly encourage long-term studies with cholinomimetic compounds, designed to determine if the course of the disease is altered. Although such studies have not yet been conducted, a retrospective analysis of one of the multi-center tacrine studies has been completed (Knopman et al., 1996). Patients in this study received either 80, 120, or 160 milligrams of tacrine for approximately 26 weeks. At the conclusion of the double blind portion of this study, patients were given the opportunity to continue on tacrine treatment. A retrospective analysis, of patients who continued on tacrine treatment for the next two years, indicated that those patients receiving either 120 or 160 milligrams of tacrine were significantly less likely to enter a nursing home than patients who received 80 milligrams or less of the drug. Similarly, patients who were taking 160 milligrams of tacrine were less likely to die than those patients taking a lower dose of tacrine. Data analysis included only those patients taking tacrine at the time of death, nursing home placement, or for the entire 2-year period of follow up.

The persuasiveness of this analysis is necessarily limited by its retrospective nature, and the absence of any parallel control groups. Nonetheless, they are consistent with some of the preclinical work indicating an effect of cholinesterase inhibitors to diminish secreted APP products (Haroutunian et al., 1995). However, it is equally plausible that patients on high-dose tacrine, experiencing meaningful symptom reduction, were less likely to be placed in a nursing home simply because they were more cognitively intact. Whatever the cause of these putatively favorable long-term results, the necessity to conduct a long-term study on the effects of a cholinesterase inhibitor to alter the outcome of Alzheimer's patients, and particularly such outcomes as nursing home placement and/or death, must be performed.

The range of psychometric changes reported with tacrine have, to many clinicians, been less robust than desired. A frequently asked question is, what can be done to enhance the magnitude of drug effect on the ADAS? It is possible that subsequent cholinesterase inhibitors, or cholinergic agonists, differing somewhat in their mechanism of action (e.g., competitive vs. noncompetitive inhibitors) may produce a larger effect on the ADAS than has so far been obtained. However, alternative approaches to enhancing the efficacy of cholinomimetic compounds should be considered.

An extremely intriguing result that derives from the tacrine studies, is the retrospective analysis that indicates that those patients receiving estrogen had a larger response than patients who were not receiving estrogen. The availability of compounds possessing many of the properties of estrogen, but without the feminizing characteristics, encourages the conduct of combination therapies with estrogen and estrogen-related compounds.

The importance of noradrenergic neurotransmission to the efficacy of cholinomimetic therapies provides a convincing rationale for combining cholinomimetic compounds with drugs that will increase noradrenergic activity. Indeed, the addition of relatively high doses of deprenyl to cholinesterase inhibitors proved more efficacious than cholinesterase inhibitors alone (Schneider et al., 1993).

Finally, the possibility that combining cholinomimetic compounds with differing mechanisms of action may produce enhanced central cholinergic neurotransmission with fewer side effects needs to be considered. This possibility has already been suggested by preclinical data (Flood et al., 1983; 1985). A synergistic interaction between M_1 agonists and cholinesterase inhibitors, or releasing agents such as 4-aminopyridine and cholinesterase inhibitors, should be tested. Combination therapy is the standard practice in the treatment of hypertension. In the future, it is possible that clinicians treating Alzheimer's patients will use a similar paradigm.

Implicit in the search for more substantial cognitive effects with future cholinesterase inhibitors is the assumption that all cholinesterase inhibitors are not equivalent. In fact, there are important differences among cholinesterase inhibitors, but whether those differences have any clinical relevance still needs to be determined. For example, some cholinesterase inhibitors have no butyrylcholinesterase inhibiting activity. Interestingly, to date, those cholinesterase inhibitors without butyrylcholinesterase activity have not reported neutropenia, in contrast to the aminoacradines and heptylphysostigmine.

Patients often report nausea and vomiting, and sometimes diarrhea at the initiation of therapy with cholinesterase inhibitors. It is not clear if this is an inevitable adverse event that is equivalent across all cholinesterase inhibitors, or whether some cholinesterase inhibitors are more prone to these adverse events than others. Although a component of the gastrointestinal adverse event may reside in the gastrointestinal tract, undoubtedly, some component of the nausea and vomiting experienced by some patients is of central origin. It is certainly possible to reduce nausea and vomiting by developing drugs that have more specific central nervous system effects, and petition preferentially into the brain, nonetheless, some gastrointestinal events persist even in relatively brain-specific compounds. The rate of absorption (reflected in the T_{max}, of the cholinesterase inhibitors) may be a predictor of the severity and frequency of nausea and vomiting. Hence, drugs with a more gradual absorption and penetrance into the brain seem to be associated with less problems than drugs that are rapidly absorbed. For this reason, nausea and vomiting has been diminished by administering cholinesterase inhibitors with meals.

Cholinesterase inhibitors can differ markedly in their interaction with the cholinesterase enzyme. For example, diisoflurophosphate is an irreversible cholinesterase inhibitor. Clearly, this is not a characteristic that is desirable in drugs administered to Alzheimer's patients. The rate at which the cholinesterase inhibitor dissociates from cholinesterase is another important property. Drugs with a prolonged inhibiting capacity can become problematic unless dosing frequency can be carefully tailored to guarantee no accumulation of cholinesterase inhibition over repeated administration. Since the most elderly patients generally metabolize drugs particularly slowly, drugs with half lives suggesting twice-a-day dosing in relatively healthy 73-year olds, may require only daily dosing in more frail 85-year olds. Similarly, once-a-day dosing in younger elderly, may be quite excessive, and even dangerous, in frail very elderly.

Just as the interaction between cholinesterase inhibitors and the cholinesterase enzyme can be irreversible or non-irreversible, it can be competitive or noncompetitive. From the perspective of desiring a drug that is least likely to suppress cholinergic neurotransmission, it can be argued that a competitive inhibitor would offer some advantages over a noncompetitive inhibitor.

Geriatric patients are often taking multiple drugs for a host of indications. Consequently, drug interactions must be of concern. Cholinesterase inhibitors differ in their metabolism, and their degree of protein binding. Obviously, the ideal cholinesterase inhibitor is one that does not alter the metabolism of commonly used drugs, nor compete for binding sites or plasma proteins.

There are also potentially important differences among M_1 agonists, differences that impact on their very rationale for development. For example, it has been thought that development of a drug that is an M_1 agonist would be relatively devoid of cardiovascular interactions, perhaps not associated with gastrointestinal adverse events, and relatively specific for enhancing cognitive performance in the Alzheimer's patient. Although this premise behind the development of M_1 agonists in AD remains hypothetical, available compounds differing in their activity at the various muscarinic sites, and are not yet as specific for the M_1 receptor as might be desired. Binding affinities derived from *in vitro* studies present differing results depending on whether the muscarinic receptor profile being obtained is from cloned human muscarinic receptors, or available rat receptors. Furthermore, *in vivo* potencies at these receptors may not be congruent with even receptor binding profiles from cell lines transfected with human muscarinic receptors. Hence, it will truly be in the clinic that the selectivity of these drugs, and their relative specificity to augment cognition, will be proven.

Other important assumptions underlie the use of cholinergic agonists, particularly M_1 specific agonists. One assumption is that the tonic activity that will be engendered by an agonist will have positive therapeutic consequences. It is possible that if some of the behaviors for which the putative cognitive enhancers were designed are mediated through phasic neurotransmission, tonic stimulation will be ineffective, or even countertherapeutic. Furthermore, the commitment to muscarinic stimulation assumes that nicotinic activity is relatively unimportant therapeutically. Nicotinic compounds have been administered to Alzheimer's patients with some benefit (Newhouse et al., 1988). Alternatively, the presynaptic location of some nicotinic receptors, and their loss in AD, may make the stimulation of these receptors a moot issue.

Receptor mismatches, which is to say the presence in the brain of receptors for which there are no current innervation, poses a particular dilemma for the agonist strategy (Herkenham, 1994). The administration of agonists will stimulate receptors that under basal circumstances are not stimulated. Such receptors may be the consequence of reorganization. Their stimulation, from the perspective of the agonist strategy, is hopefully irrelevant, but possibly associated with undesired behavioral side affects (Davis et al., 1978).

Despite these theoretical arguments, one of the most intriguing results obtained with M_1 agonists is the diminution of non-behavioral symptoms encountered with xanomeline (Tollefson et al., 1994). The possibility that an M_1 agonist may be particularly efficacious in diminishing noncognitive symptoms is intriguing, and would be a result not to date reported with cholinesterase inhibitors. This might derive from interactions between cholinergic and other neurotransmitter systems, initiated by cholinergic stimulation, or it could be through stimulation of cholinergic receptors otherwise not being stimulated by cholinesterase inhibitors. Whatever the result, the efficacy of M_1 receptors in AD, and their profile of activity, will be eagerly awaited.

CONCLUDING REMARKS

The knowledge base regarding the biology of AD is increasing almost expedentially. The application of molecular and cell biological methodologies to elucidating the pathophysiology of AD has had enormous benefits. As a consequence, there are multiple, credible hypotheses around which therapeutic agents can be developed. This scientific opportunity, combined with the magnitude of the Alzheimer's problem, has encouraged an enormous effort from the pharmaceutical community. From small biotech startup companies to multi-national giants, therapeutic programs in AD have proliferated. Thus, the therapeutic nihilism that surrounded this field is no longer appropriate. It would be indeed surprising if the next generation of Alzheimer's patients do not have a far better outcome than our last generation of patients.

REFERENCES

Buxbaum JD, Oishi M, Chen HI, Pinkas-Kramarski R, Jaffe EA, Gandy SE and Greengard P (1992): Cholinergic agonists and interleukin-1 regulate processing and secretion of the Alzheimer's β/A4 amyloid protein precursor. *Proc Natl Acad Sci USA* 89:10075-10078.

Davis KL, Hollister LE and Tepper J (1978): Cholinergic inhibition of methylphenidate induced steriotypy: Oxotremorine. *Psychopharmacology* 56:1-4.

Davis KL, Thal U, Gamzu ER et al. (1992): A double-blind, placebo controlled multicenter study of Tacrine for Alzheimer's disease. *N Eng. J Med.* 327:1253-1259.

Eikelenboom Farber S, Nitsch R, Schulz J and Wurtman R (1995): Regulated secretion of β amyloid precursor protein in rat brain. *The Journal of Neuroscience* 15:7442-7451.

Farlow M, Gracon SI, Hershey LA, Lewis KW, Sadowsky CH, Dolan-Ureno I (1992): A controlled trial of Tacrine in Alzheimer's disease. The Tacrine Study Group. *JAMA* 268(1 8):2523-2529.

Flood JF, Smith GE and Cherkin A (1983): Memory Retention: Potentiation of cholinergic drug combination drug in mice. *Neurobiology of Aging* 4:37-43.

Flood JF, Smith GE and Cherkin A (1985): Memory enhancement: Supra-additive effect of subcutaneous cholinergic drug combinations in mice. *Psychopharmacology* D25-D3 1.

Haroutunain V, Greig N, Davis KL and Wallace WC (1996): Pharmacological modulation of Alzheimer's βamyloid precursor protein levels in the CSF of rats with forebrain cholinergic lesions. *Mol Brain Research,* submitted.

Haroutunian V, Greig N, Gluck R, Fiber E, Davis KL and Wallace W (1995): Selective attenuation of lesion-induced increases in secreted B-APP by acetyleholinesterase inhibitors. *Society for Neuroscience* Abst. 208.1.

Haroutunian V, Kanof PD, Tsuboyama G and Davis KL (1990): Restoration of cholinomimetic activity by clonidine in cholinergic plus noradrenergic lesioned rats. *Brain Res* 507:261-266.

Herkenham M (1994): Mismatches between receptor and transmitter localizations in the opiate system: implications for nonsynaptic opioid actions. In *Autoradiology and Correlative Imaging for Cells and Tissues* (Stumpf WE and Solomon HF eds.) New York, Raven Press, in press.

Knapp MJ, Knopman DS, Solomon PR et al. (1994): A 30-week randomized controlled trial of high-dose tacrine in patients with Alzheimer's disease. *JAMA* 271(13):985-991.

Knopman D, Schneider L, Davis K, Talwalker S, Smith F, Hoover T and Gracon 5 (1996): Long-term tacrine (Cognex^{R}) treatment: effects on nursing home placement and mortality. *Neurology,* in press.

Alzheimer Disease: From Molecular Biology to Therapy
edited by R. Becker and E. Giacobini

THE ALZHEIMER'S DISEASE ASSESSMENT SCALE: MODIFICATIONS THAT CAN ENHANCE ITS USE IN FUTURE CLINICAL TRIALS

Richard C. Mohs, Deborah Marin,
Cynthia R. Green, and Kenneth L. Davis
Mount Sinai School of Medicine, Psychiatry Service,
VA Medical Center, Bronx, New York

INTRODUCTION

The Alzheimer's Disease Assessment Scale (ADAS) is widely used in clinical trials of potential treatments for Alzheimer's disease (AD). The cognitive portion of the ADAS (ADAS-Cog) consists of 11 items designed to assess the severity of memory, language, praxis and orientation impairments. The noncognitive portion of the ADAS (ADAS-Non-Cog) consists of 10 clinician rated items assessing the severity of depressive symptoms, psychosis, agitation, attention and tremors. These items are described in previous publications (Mohs et al., 1982; Rosen et al., 1984), and longitudinal data have demonstrated that the cognitive portion of the scale measures progressive cognitive impairment across a broad range of dementia severity (Stern et al., 1994). The ADAS-Cog has been used as the principal measure of cognitive outcome in several multicenter clinical trials of potential treatments for AD, including the trials of tacrine (Farlow et al., 1992; Knapp et al., 1994).

In its current form the ADAS-Cog has been shown to be a reliable, valid and useful measure of the effects of potential antidementia treatments in patients with mild to moderately severe AD. Like all psychological measuring instruments, however, the ADAS has limitations and those limitations have become more well defined as the number and diversity of treatment trials for patients with dementia has increased. One limitation is that the ADAS-Cog may not be sufficiently sensitive to cognitive change in patients with very mild dementia or in persons with mild cognitive impairment who are at high risk for dementia. This limitation is evident from longitudinal data indicating that even in mild AD the annual rate of change on the ADAS-Cog is quite small (Stern et al., 1994). Secondly, the ADAS-Cog may not include items to assess every aspect of cognitive dysfunction in AD. In particular, the scale has no items specifically designed to assess attention and concentration, nonverbal memory, planning and executive functions, or the full range of praxis functions often assessed in a full neuropsychological exam. A final limitation of the ADAS-Cog is that it is not particularly useful in severely demented patients since they cannot perform most of the test items.

The noncognitive portion of the ADAS has not been used nearly as widely as the ADAS-Cog, in part because few treatment trials are designed to evaluate effects on noncognitive symptoms. In comparison with the amount of information available about the cognitive aspects of AD, information about noncognitive aspects of AD is both less plentiful and less precise. The evaluation of drug effects on noncognitive aspects of AD would be greatly facilitated by more precise data concerning the prevalence, natural history and clinical impact of noncognitive symptoms . Longitudinal studies using both cognitive and noncognitive measures are needed to determine the extent to which the ADAS-Noncog or other measures are useful for assessing treatment effects on the noncognitive aspects of AD.

The present paper summarizes recent data which suggest possible modifications of the ADAS that could make it more useful for certain kinds of studies. The data summarized in this paper are from two sources. Some were obtained as part of the Instrument Development Protocol of the Alzheimer's Disease Cooperative Study (ADCS), a multisite consortium funded by the U.S. National Institute on Aging to conduct clinical trials of new treatments for AD and to develop improved assessments instruments for those trials (Ferris et al., 1996). The data from the cognitive instruments portion of the ADCS study were designed to determine whether certain cognitive tests might provide reliable and valid additions to the ADAS-Cog. The data on the ADAS-Non-Cog were collected as part of ongoing longitudinal studies of AD at the Memory Disorders Clinics of the Mount Sinai Medical Center and the Bronx VA Medical Center. These data were analyzed to determine the natural history of noncognitive symptoms assessed by the ADAS-NonCog and their relationship to cognitive and functional impairment.

THE COGNITIVE INSTRUMENT STUDY OF THE ADCS

Details of the ADCS Instrument Development Protocol can be found in Ferris et al. (1996). Briefly, this multisite protocol involved 50 nondemented elderly subjects and 50 AD patients in each of three severity groups defined by screening scores on the Mini Mental State Exam (MMSE); the AD groups were mild (MMSE 21-26), moderate (MMSE 16-20) and severe (MMSE 10-15). This range of severity was selected in order to determine the extent to which each test could measure differences in severity across the range usually included in clinical trials of antidementia drugs. All subjects were tested at baseline, 6 and 12 months in order to determine the extent to which each test was sensitive to change in severity of symptom over a time period in which patients, on average are known, to show increasing cognitive impairment (Stern et al., 1994). A subset of subjects in each severity group was retested at 1 and 2 months to determine the effect of repeated exposure to the test materials and to evaluate retest reliability.

The tests selected for the study were designed to assess aspects of cognition that are not explicitly tested by the 11 items included in the ADAS-Cog as it is currently used. Those areas are listed in Table 1 along with the tests investigated and a summary of the conclusions from the reliability and validity analyses. Details of the findings are presented in Mohs et al. (1996). The test of delayed recall was found to be reliable and a very sensitive test for differentiating mild AD from normals but, since performance on this test is very poor even in mild AD (Welsh et al., 1991), it did not discriminate across the range of dementia severity and was not very sensitive to longitudinal change. The test is likely to be useful as a measure of treatment effects only in very mild AD or in patients at high risk for AD. The facial memory test was neither sensitive to group differences or to differences over time. The praxis measures drawn from the Boston Diagnostic Aphasia Exam were completed correctly by all but the most demented patients and thus would not be a useful addition in most treatment protocols. The mazes task (Christensen et al., 1991) designed to measure executive or planning functions is available in seven forms of increasing complexity. Even the most simple forms of this task were so difficult that some AD patients could not complete the task making it impossible to use the usual time scores. The simplest forms of this task might be useful in studies of treatment effects in very mild patients or at risk subjects but not in the great majority of AD treatment trials. Finally, the

Test	**Cognitive Domain Assessed**	**Reliability**	**Sensitivity to Differences Across All Groups**	**Sensitivity to Longitudinal Change**
Delayed Recall	Memory	Good	Only Mild vs. Normal	No
Face Memory	Nonverbal Memory	Good	No	No
Praxis	Praxis	Good	No	No
Mazes (Time)	Planning/ Executive Function	Good	Only Mild vs. Normal	No
Visual Search/ Multiple Forms	Attention/ Concentration	Varies, Good for some forms	Yes	Yes

TABLE 1. Description of tests included in the ADCS Cognitive Instruments Study and summary of the results of the reliability and validity studies.

study evaluated several forms of a visual search task as possible measures of attention and concentration. The forms varied in terms of reliability and most were sensitive to change across a fairly broad range of dementia severity. One form, in which subjects searched a page of digits and crossed out all occurrences of two digits (e.g. "2" and "8") seemed to have the best combination of reliability and sensitivity across a broad range of dementia severity. This item is brief and could be a useful addition to the ADAS-Cog in many drug treatment studies.

THE LONGITUDINAL STUDY OF THE ADAS-NON-COG

Results of the longitudinal study of the ADAS-Cog have been published (Stern et al., 1994), and the present summary of longitudinal data on the ADAS-Non-Cog is drawn from a somewhat larger sample of patients followed with the same protocol. Briefly, patients with clinically diagnosed AD are evaluated every six months with the full ADAS, other cognitive tests and the Instrumental Activities of Daily Living Scale (IADLS) and the Physical and Self Maintenance Scale (PSMS) of Lawton and Brodie (1969). Patients are included in the longitudinal analysis only if they have two evaluations at least one year apart. At baseline, all patients are living in the community but they continue to be followed until they die, refuse to participate or move away so that staff cannot evaluate them even with home or nursing home visits. At the time of the present analyses, longitudinal data were available for 147 patients followed for an average of 36.1 (±22) months. Details of the analysis are presented in Marin et al. (1996).

Ten clinician rated items, each scored from 0 (symptom not present) to 5 (very severe), are included in the ADAS-Non-Cog. There are two items to rate depression (depressed mood, tearfulness), 3 items to rate agitation (pacing, excess motor activity, uncooperativeness), 2 items to rate psychosis (delusions, hallucinations), and items to rate concentration, tremors, and appetite change. Analyses were conducted to determine for each item and for summed totals of items: 1) Do scores increase over time as does the ADAS-Cog? 2) Do scores correlate with the severity of cognitive impairment?, and 3) Do scores correlate with functional impairment measured by the IADLS and PSMS? Only the attention/concentration item worsened monotonically over the follow-up period and it was the only item to have a high correlation ($r = 0.83$) with the ADAS-Cog, suggesting that this item actually measures a cognitive disturbance in AD. For the remaining items, scores did not systematically increase during follow-up and visits on which patients had high scores tended to be followed by visits with lower scores suggesting that these noncognitive symptoms are episodic rather than relentlessly progressive. Functional impairment measured by the IADLS and PSMS was highly correlated with scores on the ADAS-Cog ($r = 0.86$ and $r = 0.80$, respectively). After partialing out the effect of ADAS-Cog scores, none of the ADAS-Non-

Cog items correlated significantly with functional impairment except for the attention item; again, this supports the view that attentional disturbances in AD are a part of the cognitive impairment.

To use the ADAS-Non-Cog appropriately, the attention/concentration items would have to be moved to the cognitive portion or, alternatively, eliminated and a psychometric item added to the ADAS-Cog. The addition would be one such as the cancellation item described above. Scores on the ADAS-Non-Cog should also not be expected to progress in untreated or placebo treated patients since these symptoms are episodic. Finally, the fact that functional impairments are highly related to cognitive deficits but only modestly related to depressed mood, agitation and psychosis suggests that only by improving cognition will major clinical improvements in function be achieved.

REFERENCES

Christensen KJ, Mutchu KS, Norstrom S et al. (1991): A cognitive battery for dementia. Development and measurement characteristics. *Psychol Assessment* 3:168-174.

Ferris SH, Mackell J, Mohs RC, et al. (1996): A multicenter evaluation of new treatment efficacy instruments for AD clinical trials. *Alz Dis and Assoc Disorders* (in press).

Farlow M, Gracon SI, Hershey LA, et al. (1992): A controlled trial of tacrine in Alzheimer's disease. *J Am Med Assn* 268:2523-2529.

Knapp MJ, et al. (1994): 30-week randomized controlled trial of high-dose tacrine in patients with Alzheimer's disease. *J Am Med Assn* 271:985-991.

Lawton MP, Brody EM (1969): Assessment of older people: Self-maintaining and instrumental activities of daily living. *Gerontologist* 9:179-186.

Marin DB, Green CR, Schmeidler J, et al. (1996): Noncognitive disturbances in Alzheimer's disease. *J Am Geriat Soc* (in press).

Mohs RC, Knopman D, Petersen RC, et al. (1996): Development of cognitive Instruments for use in clinical trials of antidementia drugs: Additions to the Alzheimer's Disease Assessment Scale that broaden its scope. *Alz Dis and Assoc Disorders* (in press).

Mohs RC, Rosen WG, and Davis KL (1982): Defining treatment efficacy in patients with Alzheimer's disease. In: *Alzheimer's Disease: A Report of Progress in Research*, Corkin S, Davis KL, Growden JH, Usdin E and Wurtman RJ, eds. New York: Raven Press, pp. 351-356.

Rosen WG, Mohs RC, Davis KL (1984): A new rating scale for Alzheimer's disease. *Am J Psychiat* 141:1356-1364.

Stern RG, Mohs RC, Davidson M, et al. (1994): A longitudinal study of Alzheimer's disease: Measurement, rate and predictors of cognitive deterioration. *Am J Psychiatry* 151:390-396.

Welsh KA, Butters N, Hughes J, Mohs RC, Heyman A (1991): Detection of abnormal memory decline in mild cases of Alzheimer's disease: use of the neuropsychological measures developed for the Consortium to Establish a Registry for Alzheimer's Disease (CERAD).*Archs Neurol* 48:278-281.

Alzheimer Disease: From Molecular Biology to Therapy
edited by R. Becker and E. Giacobini

AN ITEM POOL TO ASSESS ACTIVITIES OF DAILY LIVING IN ALZHEIMER'S DISEASE

Douglas R. Galasko
Department of Neurosciences (Neurology), University of California, San Diego, CA

David Bennett
Department of Neurology, Rush-Presbyterian Medical Center
Chicago, IL

The Alzheimer's Disease Cooperative Study

INTRODUCTION

The evaluation of activities of daily living (ADL) is an underdeveloped area in clinical trials for Alzheimer's Disease (AD). If an anti-dementia drug improves cognitive function or slows the rate of progression of AD, ideally this effect should extrapolate to a change in a patient's ADL if it is clinically appreciable. Consequently, ADL assessment was identified as an important initiative by the Alzheimer's Disease Cooperative Study (ADCS), a multi-center consortium formed to carry out clinical trials in AD.

ADL scales were originally developed to describe performance of basic ADL abilities such as walking, eating and bathing needed for self-care (Katz et al., 1963). Basic ADL can be affected by physical or intellectual disability, and in AD are typically compromised only late in the course of dementia (Spector et al., 1978). Another group of more complex ADL, designed to assess an individual's ability to live independently and to interact with members of a community, is called Instrumental ADL (IADL). Scales for IADL (Lawton and Brody, 1969; Blessed et al., 1968; Pfeffer et al., 1982) were not designed to fit the needs of multi-center clinical trials or to track change in patients over short to medium intervals. Some of the items in these scales are imprecisely worded, gender-biased, or infrequently performed and show substantial floor effects, i.e. mildly demented AD patients do not attempt many IADL. More detailed scales, such as the Record of Independent Living (Weintraub, 1986) have not been examined for reliability and have not undergone a detailed item analysis.

An ADCS subcommittee was charged with developing an ADL scale for clinical trials in AD. We developed and tested a large pool of ADL items and analyzed their metric properties to identify those most suitable for use in AD clinical trials. Performance ADL scales, in which patients are asked to carry out simulated ADL using props, were thought to be too time-consuming for clinical trial use and do not rate patients' functional abilities in their own living situations. We decided that a questionnaire for informants of patients with AD was the most appropriate type of instrument for multi-center trials. Several important properties of ADL questions were outlined to guide the development and analysis. First, ADL should be performed regularly by most elderly individuals, and attempted by patients with mild, moderate or severe dementia, without a strong influence of age or gender. Ratings of ADL should be reliable, with little short-term (test-retest) variation. Finally, items should detect change over a period relevant to clinical trials, for example 1 year.

TESTING THE ADL INVENTORY

ADL were selected from existing scales or proposed based on clinical experience, and covered a wide range of activities necessary for personal care, communicating and interacting with other people, maintaining a household, conducting hobbies and interests, and making judgments and decisions. Questions were pre-tested at three sites and modified to enhance clarity. All ADL questions had a similar format. First, the informant was asked whether the patient performed the ADL during the past four weeks or not. If the answer was "yes", the informant next selected the single most accurate descriptor of the patient's usual performance of that ADL from a number of choices. For some ADL, the descriptors were "independently" (the highest level), "with supervision" (needing verbal instructions during ADL performance, an intermediate level) or "with physical help" (a lower level). If a subject needed reminders to get started but performed an ADL independently and successfully, that was rated as independently. For other ADL, descriptors of different ways of accomplishing the ADL were offered, arranged in presumed decreasing order of difficulty. For example, for "writing things down," the descriptors were: wrote letters or long notes that other people understood; wrote short notes, lists or messages that other people understood; and wrote his/her signature or name.

A pool of 45 items, the ADL Inventory (Galasko et al., 1996), was administered as part of the ADCS Instrument Development Protocol, a multi-center study of assessment scales covering cognitive and non-cognitive domains. Subjects aged 60 or older were recruited from 27 participating sites throughout the USA. They included both elderly controls (n = 64, 24 men and 40 women) and subjects who met NINCDS-ADRDA criteria for probable AD (n = 242, 94 men and 148 women). The recruitment of AD patients was stratified according to five levels of Mini-Mental State Examination (MMSE) scores: 0-5; 6-10; 11-15; 16-20; > 20. All subjects were assessed at baseline, 6 and 12 months, and half of the subjects at 1 and 2 months. Informants needed to have direct contact with the subject for several hours per week, spread over at least 2 days. The Inventory was administered in person, or by telephone when informants were unable to travel to a clinic site. Psychometrists or clinicians experienced in the assessment of patients with AD carried out ADL interviews. A representative from each Center attended a training session for the Inventory, and each site received a procedural manual describing each question.

TOWARDS A SCALE

To reduce the pool of items, we began by analyzing each item in terms of the criteria discussed above. For clinical trials, broad applicability is important: an ADL should be attempted or should have been regularly performed in the past by most subjects. We found that several ADL were attempted by fewer than 90% of controls or AD patients at baseline or premorbidly. These mainly showed a gender effect (men do fewer ADL) and included managing a checkbook, listening to the radio, cleaning up spills, watering plants, cleaning a room, doing laundry, setting a table, taking medications, attending a group event, making a bed, and listening to the radio. The first four of these ADL had the highest rates of non-performance. A different reason for not attempting ADL is the severity of dementia. More difficult ADL are often completely abandoned by patients with moderate or severe dementia, or are taken over by caregivers. Although an overall ADL scale should contain items attempted by most patients with AD, it is important to incorporate some items that are difficult for less severely demented patients. Managing a checkbook or personal finances was the only ADL not attempted even by mildly demented patients. Although this question discriminated AD patients from elderly controls, it would not be useful in a rating scale for clinical trials. At the opposite end of the spectrum, ADL such as turning a faucet on and off, eating, or following simple 1-step instructions were attempted by virtually all patients but did not show change at 12 months. The remainder of the ADL were attempted by 30 - 97% of patients with AD at baseline.

We designed each ADL item to consist of a number of descriptions of levels of performance to increase sensitivity. We first eliminated any decriptors that rarely or never were recorded for patients with AD or controls. We next examined whether descriptions of different ways of performing each ADL behaved as levels by examining MMSE scores among patients who performed each ADL at each presumed level. If patients who performed an ADL at two "levels" did not differ in their MMSE scores, we considered collapsing those levels into one. This was necessary mainly for descriptors of low levels of ability, such as performing ADL "with physical help" compared to "not at all." As an overall test of the relationship between these assumed levels of functional ability and cognitive ability, we calculated Spearman correlation coefficients for each ADL's levels of performance and MMSE scores; for most ADL these ranged from 0.5 - 0.75. Therefore, the levels of performance appeared to behave as a hierarchy, since these cross-sectional analyses indicate that ADL performance progressively decreases with cognition.

We next examined short-term (test-retest) reliability, using kappa statistics to assess the extent to which ratings of levels of performance of each ADL item agreed at baseline and at 1 month. In general, kappas fell into the 0.4 to 0.75 range, indicating satisfactory to good agreement. The kappas from month 1 to month 2 were similar to those for baseline to 1 month. Several items, namely cleaning up spills, making conversation and listening to a radio had low kappas (< 0.4) from baseline to month 1, but all were rated more reliably from month 1 to month 2. ADL items that were dichotomous and the traditional basic ADL had the highest kappas. This indicates the trade-off in defining ADL performance as a number of levels rather than a dichotomy: describing several intermediate levels enhances sensitivity to detect small increments of ADL change, but introduces more variability into informant reports.

We analyzed change for each ADL from baseline to 12 months as any amount of decline or improvement of performance. At 12 months, for virtually all ADL, the percentage of patients with AD who declined significantly exceeded those who improved. As a first step towards extracting an ADL scale from the Inventory, we selected those ADL which showed wide applicability, good reliability and declined at 12 months in at least 25% of patients with AD in one or more of the baseline MMSE strata. This procedure retained 30 ADL. The maximum obtainable score for each ADL was weighted equally, then each patient's total ADL score was calculated and converted to a percentage; 100% thus represented optimal ADL function and 0% total impairment. These total ADL scores are shown in Table 1.

Controls performed ADL optimally at baseline and at 12 months, reflecting the selection of healthy elderly individuals for this study. At baseline, ADL scores differed significantly between controls and patients with AD, and between AD patients with different strata of MMSE scores. At 12 months, patients with AD declined in ADL. Patients with mild dementia (MMSE > 20) declined less than did those with moderate or severe dementia.

Group	Baseline ADL score	Change at 12 months
Controls	98 ± 3.3	-1.1 ± 12.6
AD, MMSE > 20	84 ± 9.5	-8.4 ± 20.3
AD, MMSE 10 - 20	67 ± 14.0	-19.3 ± 19.8
AD, MMSE 0 - 9	40 ± 20.1	-19.7 ± 15.3

TABLE 1. Total ADL scores in controls and patients with AD.

ADL whose performance declined at 12 months in 25% or more of these AD subjects with mild dementia were: reading, handling mail, discussing current events, shopping, using a telephone, remembering details of television shows, making a snack or meal, using a household appliance, traveling beyond home, carrying out a pastime or hobby, choosing clothes to wear, taking medications, writing a note or letter, and ability to be left alone.

CONCLUSIONS

The ADL Inventory contains items that show good scaling, applicability to patients with AD regardless of gender, adequate test-retest reliability, and sensitivity to detect decline over 12 months. A scale constructed from the best items in the Inventory should be suitable for use in typical clinical trials for AD, which recruit patients in an MMSE range of about 10 - 26. However, for studies of very mild AD or of severely demented patients, customized scales emphasizing ADL most likely to show change in the target population would be more sensitive. The assessment of ADL in patients with mild cognitive impairment who are at risk of developing AD will need further development.

ACKNOWLEDGMENTS

Supported by Grant AGO 5161 from the National Institute on Aging. We thank ADCS Centers for making these studies possible.

REFERENCES

Blessed G, Tomlinson BE, Roth M (1968): The association between quantitative measures of dementia and of senile changes in the cerebral grey matter of elderly subjects. *Br J Psychol* 225:797-811.

Galasko D, Bennett D, Sano M et al., (1996): Development of a pool of items to assess activities of daily living in clinical trials for Alzheimer's disease. *Neurology* 46:A248 (abstr).

Katz S, Ford AB, Moskowitz RW et al., (1963): The index of ADL: A standardized measure of biological and psychosocial function. *JAMA* 185:914-919.

Lawton MP, Brody EM (1969): Assessment of older people: self-maintaining and instrumental activities of daily living. *Gerontologist* 9:179-186.

Pfeffer RI, Kurosaki TT, Harrah CH, Chance JM, Filos S (1982): Measurement of functional activities in older adults in the community. *J Gerontol* 37:323-329.

Spector WD, Katz S, Murphy JB, Fulton JP (1978): The hierarchical relationship between activities of daily living and instrumental activities of daily living. *J Chronic Dis* 40:481-489.

Weintraub S (1986): The record of independent living: an informant-completed measure of activities of daily living and behavior in elderly patients with cognitive impairment. *Am J Alzheimer Care* 1:35-39.

Alzheimer Disease: From Molecular Biology to Therapy
edited by R. Becker and E. Giacobini

SEVERE IMPAIRMENT BATTERY: A POTENTIAL MEASURE FOR AD CLINICAL TRIALS

Frederick A. Schmitt and J. Wesson Ashford
Sanders-Brown Center on Aging, University of Kentucky, Lexington, KY

Steven Ferris and Joan Mackell
Department of Psychiatry, NYU Medical Center, New York, NY

Judith Saxton
Department of Psychiatry, University of Pittsburgh, Pittsburgh, PA

Lon Schneider
Department of Psychiatry, University of Southern California, School of Medicine, Los Angeles, CA

Christopher Clark
Department of Psychiatry, University of Pennsylvania, Philadelphia, PA

Chris Ernesto, Kimberly Schafer, and Leon Thal
Alzheimer's Disease Cooperative Study (ADCS), University of California at San Diego, San Diego, CA

The Alzheimer's Disease Cooperative Study Units and the ADCS Instrument and Severe Impairment Committees

Objective measures of cognitive change as an index of drug efficacy are an integral part of Alzheimer's disease (AD) clinical trials. As potential AD therapies are evaluated, objective measures of cognition are needed if more advanced AD patients are enrolled. As AD progresses, many objective assessments of patient functioning become limited in their ability to measure cognitive functions given the continued dissolution of expressive and receptive abilities in the moderate to severe stages of AD.

One approach to the measurement of mental abilities in advanced stages of AD has been to use observer-based rating scales including the Clinical Dementia Rating Scale (CDR) (Hughes et al., 1982), Global Deterioration

Scale (GDS) (Reisberg et al., 1982), and the Functional Assessment Staging (FAST) scale (Reisberg, 1988). While providing similar schemes with varying detail for staging the severity of AD, these severity rating scales rely on informant reports and clinical observation of the presence or absence of behaviors and symptoms characteristic of the later stages of AD.

Recent efforts to circumvent the floor effects of cognitive tests that occur during the later stages of AD have focused on neurological procedures such as the Glasgow Coma Scale (Benesch et al., 1993), the presence of neurological signs and symptoms observed in conjunction with deteriorating cognitive and functional skills (Volicer et al., 1994), and ratings of disease severity based on nursing home medical records (Hartmaier et al., 1994). Although these approaches reliably assess dementia severity, they provide a limited performance-based evaluation of the AD patient's abilities.

An important goal from the standpoint of clinical trials is to measure changes in mental functions in the moderate to severe range with precision similar to the more frequently assessed mild to moderate levels of AD (Ashford et al., 1995; Reisberg et al., 1994). Further, as effective treatments of AD are developed, objective measures may be needed if advanced AD patients are to be evaluated for therapeutic efficacy. For example, while the AD Assessment Scale (ADAS) (Rosen et al., 1984) is a standard index of cognition in clinical trials of mild-to-moderate AD patients, a similar measure is needed if more severe patients are to be studied. To this end, three procedures have recently appeared in the literature for use in the evaluation of severe impairment in AD. They are: (1) the Test for Severe Impairment (TSI) (Albert & Cohen, 1992), (2) the Modified Ordinal Scales of Psychological Development (M-OSPD) (Auer et al., 1994), and (3) the Severe Impairment Battery (SIB) (Saxton et al., 1990; Panisset et al., 1994). Longitudinal data from these three measures are currently limited, however, 26 AD patients have been evaluated with the SIB over an average of roughly 35 weeks (Panisset, et al., 1994). Given the perceived need for a performance-based measure of treatment efficacy in moderate to severe AD, the SIB was examined in an ADCS protocol for its ability to track change over 12 months across a range of AD severity.

METHODS

Preliminary data. Prior to the initiation of the ADCS *Instrument Protocol,* which was designed to evaluate the ability of several measures to track change across a spectrum of AD severity, the SIB was given to 44 AD patients residing in a nursing home in order to gain experience with this measure. These patients were being followed by the ADRC at the University of Kentucky and included 12 men and 32 women with a mean age of 76.8 years (SD=7.3). The average MMSE for this group was 6.4 (SD=5.6) and their median rating on the FAST was 6.5 (moderate to severe AD). As shown in

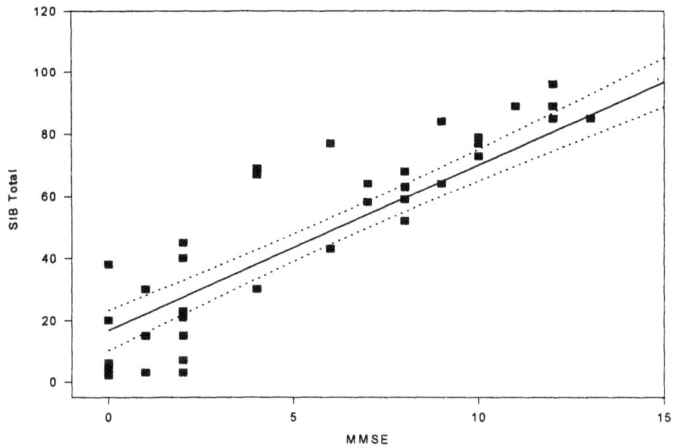

FIGURE 1. Association between SIB and MMSE in a nursing home sample.

Figure 1, there was a significant association between SIB and MMSE scores (r = 0.90), but more importantly, patients with MMSE scores of 0 could still perform on the SIB.

Given these pilot data, the SIB was included in the ADCS Instrument Protocol. Patients with AD were enrolled and their cognitive function was stratified by MMSE (groups: MMSE = 0-4, 5-9, 10-15, 16-20). Of the initial 192 patients who met criteria, 175 (91.1%) completed all of the baseline, 6 and 12 month assessments. For the change analyses there were 61 (34.9%) men and 114 (65.1%) women with a mean age of 71.8 (SD = 9.2) years while the number of years of educational attainment was 13.02 (SD = 2.7).

Procedure

All participants were administered the SIB in addition to cognitive, global, activities of daily living, and behavioral scales. The SIB was completed by all patients at baseline, six, and twelve month intervals and a subset of 88 subjects were also tested at one and two months after baseline.

RESULTS

Concurrent (construct) validity of the SIB was evaluated by comparing each patient's baseline scores on the CDR, GDS, FAST, and MMSE with their SIB total score. The Pearson product-moment correlations between the SIB and other measures were: CDR r = -0.64 ($p < .01$); MMSE r = 0.82 ($p < .01$); GDS r = -0.67 ($p < .01$) and FAST 4= -0.59 ($p < .01$). Additional correlations showed that the SIB score was not related to patient age at the time of

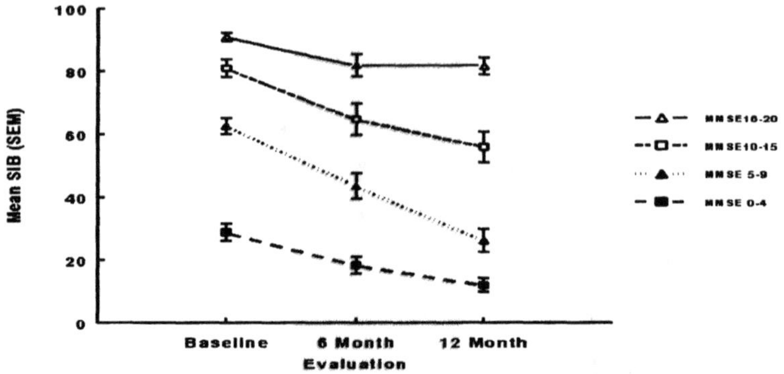

FIGURE 2. Mean SIB over time by baseline MMSE severity group.

assessment (r = 0.06) or years of education (r = 0.01). Data from those 88 subjects who were tested at baseline and again at one and two months allowed for the assessment of test-retest reliability. The correlation between the baseline and one month SIB was 0.90 ($p<.001$) and the correlation between baseline and two month SIB scores was 0.86 ($p < .001$).

Analysis of change

Analyses of covariance (ANCOVA) showed that change in SIB at twelve months was not influenced by testing schedule (F(1,172) = 3.43, p = 0.07) or an interaction between testing schedule and baseline SIB score ($F(1,173) = 0.00$, $p = 0.99$). Mean SIB scores at baseline, 6 and 12 months are shown in Figure 2.

CONCLUSIONS

While there are a few objective measures being developed for the evaluation of mental functioning in the more advanced stages of AD, the data evaluating the ability of these scales to index change over time is limited. The results of this study suggest that the SIB is a reliable, valid, and useful instrument for evaluating change in AD patients, particularly for those persons in the moderate to severe (but not end-stage) range of functioning. More importantly, these data show that in more severe cases of AD (MMSE below 15) the SIB provides a broad range of assessment that should allow for a more careful evaluation of changes in patient functioning over time as well as their response to potential therapeutic interventions. Further, the SIB may be useful as a secondary outcome measure in extended clinical trials where patients might progress to moderate or severe AD during the trial.

ACKNOWLEDGMENT

This work was supported by NIA grant U01-AG10483.

REFERENCES

Albert M and Cohen C (1992): The Test for Severe Impairment: an instrument for the assessment of patients with severe cognitive dysfunction. *J Am Geriatr Soc* 40:449-453.

Ashford JW, Shan M, Butler S, Rajesekar A and Schmitt FA (1995): Temporal quantification of Alzheimer's disease severity: 'Time Index' model. *Dementia* 6:269-280.

Auer SR, Sclan SG, Yaffee RA and Reisberg B (1994): The neglected half of Alzheimer disease: cognitive and functional concomitants of severe dementia. *J Am Geriatr Soc* 42:1266-1272.

Benesch CG, McDaniel KD, Cox C and Hamill RW (1993): End-stage Alzheimer's disease: Glasgow Coma Scale and the neurologic examination. *Arch Neurol* 50:1309-1315.

Hartmaier SL, Sloane PD, Guess HA and Koch GG (1994): The MDS Cognition Scale: a valid instrument for identifying and staging nursing home residents with dementia using the Minimum Data Set. *J Am Geriatr Soc* 42:1173-1179.

Hughes CP, Berg L, Danziger WL, Coben LA and Martin RL (1982): A new clinical scale for the staging of dementia. *Br J Psychiatry* 140:566-72.

Panisset M, Roudier M, Saxton J and Boller F (1994): Severe Impairment Battery: a neuropsychological test for severely demented patients. *Arch Neurol* 51:41-45.

Reisberg B (1988) Functional Assessment Staging (FAST). *Psychopharmacol Bull* 24:653-659.

Reisberg B, Ferris SH, de Leon MJ and Crook T (1982): The global deterioration scale for assessment of primary degenerative dementia. *Am J Psychiatry* 139:1136-1139.

Reisberg B, Sclan SG, Franssen E, Kluger A and Ferris S (1994): Dementia staging in chronic care populations. *Alzheimer Dis Assoc Disord* 8:S188-205.

Rosen WG, Mohs RC and Davis KL (1984): A new rating scale for Alzheimer's disease. *Am J Psychiatry* 141:1356-1364.

Saxton J, McGoingle-Gibson K, Swihart A, Miller M and Boller F (1990). Assessment of the severely impaired patient: description and validation of a new neuropsychological test battery. *Psychol Assess* 2:298-303.

Volicer L, Hurley AC, Lathi DC and Kowall NW (1994): Measurement of severity in advanced Alzheimer's disease. *J Gerontol Med Sci* 49:M223-226.

Alzheimer Disease: From Molecular Biology to Therapy
edited by R. Becker and E. Giacobini

VALIDITY AND RELIABILITY OF THE ALZHEIMER'S DISEASE COOPERATIVE STUDY-CLINICAL GLOBAL IMPRESSION OF CHANGE (ADCS-CGIC)

Lon S. Schneider[1,2], Jason T. Olin[1]
Department of Psychiatry[1], Department of Neurology[2], School of Gerontology, University of Southern California, Los Angeles, CA

Rachelle S. Doody
Department of Neurology, Baylor College of Medicine, Waco, TX

Christopher M. Clark
Department of Neurology, University of Pennsylvania School of Medicine, Philadelphia, PA

John C. Morris
Department of Neurology, Washington University School of Medicine, St. Louis, MO

Barry Reisberg and Steven H. Ferris
Department of Psychiatry, New York University Medical Center, NY, NY

Frederick A. Schmitt
Department of Neurology, University of Kentucky, Lexington, KY

Michael Grundman and Ronald G. Thomas
University of California at San Diego, CA

INTRODUCTION

Clinical global impressions of change (CGICs) are important measures of efficacy in clinical trials. CGIC scales have been used extensively as primary outcome criteria in psychopharmacological trials and in early clinical trials for antidementia drugs (e.g., Schneider and Olin, 1994). CGICs have been reported to be the most sensitive index of change in 14 of 17 dementia trials, when compared to other measures (Lehmann, 1984).

The assumption behind their use is that a skilled clinician should easily be able to see the clinical effect of a treatment after a short interview. CGICs have been used successfully in trials of other neuropsychiatric treatments, usually with the completely unstructured Clinical Global Impression (CGI, Guy, 1976). However, the CGI has shown little sensitivity to the treatment effects of putative medications to treat symptoms of Alzheimer's disease (AD). More recently designed CGICs include structured formats, and ask for documentation by the clinician. Yet, considering that current medications to treat AD will likely have modest to moderate effects, the methodology behind the design of CGICs still requires great scrutiny. To address these concerns, the instrumentation committee of the National Institute on Aging's Alzheimer's Disease Cooperative Study (ADCS) developed a new method for clinically assessing change in AD trials (the ADCS-CGIC; Schneider et al., 1996). Along with other measures, the reliability and validity of the ADCS-CGIC was evaluated in a 12-month multicenter trial.

METHODS

The multicenter trial was a 52-week study to assess new clinical efficacy measures designed by the instrumentation committee of the ADCS. Patients were randomly assigned to one of two visit schedules: Baseline, 1-month, 2-month, 6-month, and 12-month; or Baseline, 6-month and 12-month. Briefly, subjects were men and women (N=242) who met National Institute of Neurological and Communicative Disorders and Stroke-Alzheimer's Disease and Related Disorders Association criteria for AD (McKhann et al., 1984). Patients had no significant concomitant medical conditions, were not taking psychotropic medications, and provided informed consent. A non-demented control group was also enrolled (N=64). The ADCS-CGIC was administered by a clinician who had at least one year's experience making global ratings in clinical trials.

ADCS-CGIC

The ADCS-CGIC consists of three parts: a semi-structured baseline interview administered to the patient and an informant, a follow-up interview administered to the patient and informant, and a clinician's rating review. The baseline interviews serve as a reference for future ratings. At the follow-up visit, clinicians make two assessments of change, one after each interview. Half of the subjects were randomly assigned to be interviewed first, followed by the informant; the other half were interviewed in the reverse order. This permitted an assessment of the informant's contribution and patient's contribution to a CGIC rating. The second rating is used as the global rating. The global rating is made on a 7-point scale (1=very much improved, 2=much

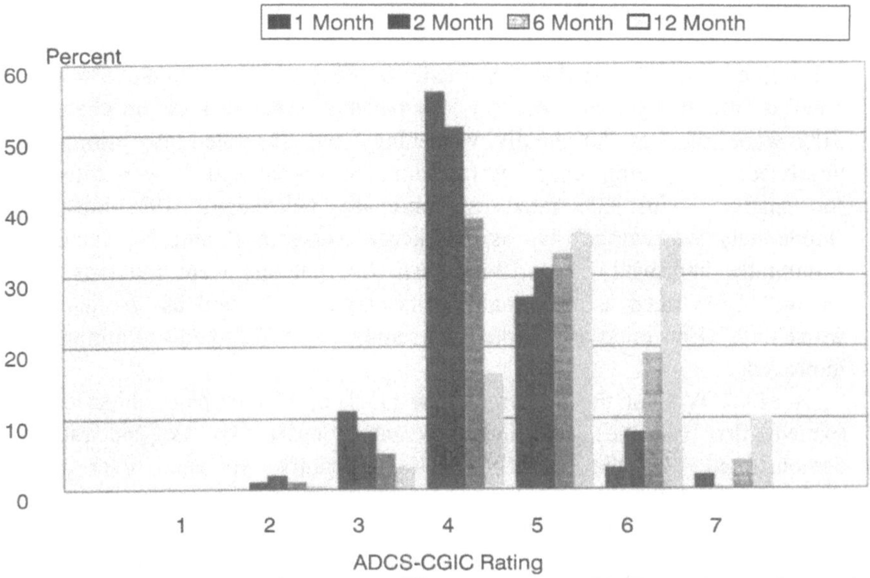

FIGURE 1. Frequency distribution of ADCS-CGIC ratings at each time interval for AD Subjects. The figure shows the frequency distribution of AD subjects' ADCS-CGIC ratings for each of the four assessment periods, with the first column on the left representing the 1-month visit, the second column representing the 2-month visit and so on. Note that over time, more AD subjects received scores indicating worsening. 1=marked improvement, 2=moderate improvement, 3=mild improvement, 4=no change, 5=mild worsening, 6=moderate worsening, 7=marked worsening.

improved, 3=minimally improved, 4=no change, 5=minimally worsening and so on). The third part, the clinician's rating review, was intended for clinicians to provide information about the basis of their ratings.

The ADCS-CGIC assesses 15 areas under the domains of cognitive, behavior, and social and daily functioning. Under each area is a list of sample probes along with space to take notes. Clinicians are expected to use the forms as a guideline. To provide minimal structure, there are few requirements for the interviews: an assessment of mental status must be made; clinicians are not permitted to ask about side effects or treatment effects, nor to discuss the patient's functioning with other staff members. The creation of the ADCS-CGIC is described in detail elsewhere (Olin et al., 1996; Schneider and Olin, 1996; Schneider et al., 1996).

RESULTS

There were 119 males and 187 females with AD entered into the trial. Their mean age was 72 years (SD=9.0) at the time of baseline. Patients had a mean of 13.3 (2.9) years of education.

Regarding the stability of the ADCS-CGIC, at 1-month the majority of patients (56%) were rated as "no change," with 27% rated as "minimally worsening," and 11% rated as "minimally improved." At 2-months, 8% were rated as "minimally improved," 51% of patients were rated as "no change," 31% were rated as "minimally worsening" and 8% rated as "minimally improved." Regarding sensitivity to change, at 6-months, 38% were rated as "no change," with 33% rated as "minimally worsening," 19% rated as "moderately worsening," 4% as "markedly worsening," and 5% rated as "minimally improved." At 12-months, 16% patients were rated as "no change," 36% rated as "minimally worsening," 35% rated as "moderately worsening," 10% rated as "marked worsening," and 3% rated as "minimally improved."

A MANOVA for the factors of time (1, 2, 6, 12-months), subject order (patient first vs. 2nd), and interview rating order (1st vs. 2nd rating) demonstrated that the ADCS-CGIC significantly worsened over time ($F(3,240)=69$, $p<0.0001$).

There was an interaction between subject order (patient 1st vs. 2nd) and interview rating order (1st vs. 2nd rating) ($F(1,88)=2.46$, $p=0.01$). When patients were interviewed first, the ADCS-CGIC rating showed *less* worsening than the informant interview. When informants were interviewed first, the ADCS-CGIC rating showed *greater* worsening than the patient interview.

Concurrent validity for the ADCS-CGIC was evaluated. 12-month ADCS-CGIC ratings correlated with change scores on the Clinical Dementia Rating Morris, 1993; CDR; $r=0.23$, p 0.001), Mini-Mental State Examination (Folstein et al., 1975; MMSE; $r=-0.32$, p 0.001), Global Deterioration Scale (Reisberg et al., 1982; GDS; $r=0.21$, $p \leq 0.001$), and Functional Assessment Staging (Reisberg, 1988; FAST; $r=0.12$, $p \leq 0.05$), . When MMSE score was controlled, the ADCS-CGIC continued to correlate with the CDR ($r=0.15$, $p=0.074$) and GDS ($r=0.15$, $p \leq 0.05$).

DISCUSSION

The ADCS-CGIC, unlike the completely unstructured CGIC (Guy, 1976) used in most neuropsychopharmacological research, includes an organized, but nevertheless unstructured format with which a clinician may address clinically relevant overall change in clinical trials. Overall, we found the ADCS-CGIC to be a stable measure of global function in AD over short intervals, and sensitive to decline over longer ones. In addition, we found support for the predictive validity of the instrument, from correlations with other related measures. Finally, our results suggest that the ADCS-CGIC is a comprehensive measure that captures different information than mental status exams or global measures of severity. In summary, the ADCS-CGIC appears to be a valid, reliable, and easily operationalized instrument for use in clinical trials.

ACKNOWLEDGMENTS

This report was supported, in part, by the United States Public Health Service, National Institute on Aging, Alzheimer's Disease Cooperative Study (U01 AG10483). We wish to acknowledge the additional support of Joan A. Mackell, Ph.D., at New York University and Kimberly Schafer, M.A., at University of California, San Diego.

REFERENCES

Clinical Global Impressions (CGI) (1976): In: *ECDEU Assessment Manual for Psychopharmacology*, Guy W, ed. Rockville, MD: U.S. Department of Health and Human Services, Public Health Service, Alcohol Drug Abuse and Mental Health Administration, NIMH Psychopharmacology Research Branch, pp. 218-222.

Folstein MF, Folstein SE, McHugh PR (1975): "Mini-Mental State": A practical method for grading the cognitive state of patients for the clinician. *J Psychiatr Res* 12:189-198.

Leber P. (November, 1991): Unpublished letter to pharmaceutical companies [Issues affecting the implementation of the Clinician's Global Assessment].

Lehmann E (1984): Practicable and valid approach to evaluate the efficacy of nootropic drugs by means of rating scales. *Pharmacopsychiat* 17:71-75.

McKhann G, Drachman D, Folstein M, et al. (1984): Clinical diagnosis of Alzheimer's disease: Report of the NINCDS-ADRDA Work Group under the auspices of the Department of Health and Human Services Task Force on Alzheimer's Disease. *Neurology* 34:939-944.

Morris JC (1993): The Clinical Dementia Rating (CDR): Current version and scoring rules. *Neurology* 43:2412-2414.

Olin JT, Clark C, Doody R, Schmitt F, Ferris S, Morris J, Reisberg B, Schneider LS (1996): Reliability and validity of a new clinical global rating scale for Alzheimer's clinical trials. *J Int Neuropsychol Soc 2*:4.

Reisberg B (1988): Functional assessment staging (FAST). *Psychopharmacol Bull* 24:653-659.

Reisberg B, Ferris SH, deLeon MJ, Crook T (1982): The global deterioration scale for assessment of primary degenerative dementia. *Am J Psychiatry* 139:1136-1139.

Schneider LS, Olin JT (1994): Quantitative overview of the efficacy of Hydergine. *Arch Neurol* 51:787-798.

Schneider LS, Olin JT (1996): Clinical Global Impressions in Clinical Trials. *Int. Psychogeriat* (In Press)

Schneider LS, Olin JT, Doody RS, Clark CM, Morris JC, Reisberg B, Schmitt FA, Grundman M, Thomas RG, Ferris SH (1996): Validity and reliability of the Alzheimer's Disease Study-Clinical Global Impression of Change (ADCS-CGIC). *Alzheimer's Disease and Associated Disorders.* (In Press)

Alzheimer Disease: From Molecular Biology to Therapy
edited by R. Becker and E. Giacobini

ADVANTAGES OF THE "TIME-INDEX" METHOD FOR MEASUREMENT OF ALZHEIMER DEMENTIA: ASSESSMENT OF METRIFONATE BENEFIT

J. Wesson Ashford[1], Frederick A. Schmitt[1], and Daniel Wermeling[2]
Departments of Psychiatry[1], Neurology[1] and School of Pharmacy[2]
University of Kentucky, Lexington, KY

Florian Bieber, John Orazem, and Barbara Gulanski
Bayer Pharmaceutical Corp., New Haven, CT

INTRODUCTION

A central issue in the study of the patient with Alzheimer's disease (AD) is the measurement of the severity of the associated dementia. Determination of the degree of impairment of cognitive function is essential not just for diagnosis, but for monitoring the progression of the disease, for confirmation of diagnosis, and assisting with caregiver planning, as well as for research, including the assessment of the benefit of pharmacologic agents. Numerous scales have been developed for measuring dementia severity (see Ashford et al., 1995). However, all such scales are composites of a diverse variety of queries, which give numeric values that are added together, some using complex algorithms, to give a score which represents the severity of the dementia. While some scales have a relationship with neuropathological features, all of these scales are ordinal in nature, with the numeric values having no direct relationship with any physical factors.

Recently, Ashford et al. (1995) have suggested that the physical factor, "time", can be used as the absolute dimension along which AD severity can be measured. AD is well known to be a gradually progressive disease, with the duration of illness lasting from two to twenty years, and a mean of eight years from first symptoms until death. Therefore, the mean rate of progression of a population of AD patients can be used as a standard for measuring the dementia severity and progression of individual patients. The patient's level of severity is thus quantified with respect to the overall time-course of the illness, and the severity measure is a "time-index", using "day-units" or "year-units" according to the needed precision. The mean rate of progression is defined as "one" across the full range of the illness; a rate of "zero" would be considered non-progression, and a rate of "two" would be considered to be twice the standard rate of progression.

An important application of the "TIME-INDEX" method is assessment of effects of pharmacologic agents on dementia severity and AD rate of progression. The commonly used ordinal scales have traditionally been treated as arithmetic measurements and studied with simple statistics such as means and standard deviations. However, the application of these calculations to non-interval scales is problematic. The "TIME-INDEX" scale has equal intervals and is more amenable to statistical analysis. Further, the "TIME-INDEX" units can be clearly interpreted to understand the degree of benefit of the drug , and rates of change can be followed to determine whether the agent is slowing disease course, preventing deterioration, or resulting in improvement over time.

METHODS

The "TIME-INDEX" method was applied to the Bayer Pharmaceutical directed phase 2B study of the long-acting anticholinesterase, metrifonate (Bieber, 1996). In this study, 441 of 480 patients completed Mini-Mental State Exams (MMSE) (Folstein et al., 1975) on 6 occasions; screening, baseline, and weeks 2, 4, 8, and 12. The study examined four conditions:

panel 0:	48m,66f	- placebo
panel 1:	48m,64f	- 0.5 mg/kg x 14d, 0.2 mg/kg
panel 2:	46m,64f	- 0.9 mg/kg x 14d, 0.3 mg/kg
panel 3:	37m,68f	- 2.0 mg/kg x 14d, 0.65 mg/kg

MMSE scores at each time point were translated into "TIME-INDEX" values using the translation function described previously (Ashford et al., 1995):

$$X = 1.45 * (29 - MMSE)$$

$$TIME\text{-}INDEX = 156.61*X - 3.9928*X2 + 0.049654*X3$$

RESULTS

Analysis of change in "TIME-INDEX" values between baseline and the 12 week examination according to "TIME-INDEX" severity at baseline for each panel showed a learning effect which was progressively stronger for more mildly affected patients. However, linear regression showed that metrifonate has a benefit which was parallel across conditions, indicating that the positive effect occurred regardless of severity for this sample.

Analysis of MMSE scores across visits emphasized the small differences in baselines between the groups, suggesting that the drug benefit could be a learning effect, though this was not borne out by examination of the regression lines for each panel. The change in the "TIME-INDEX" showed a similar pattern. However, correcting for baseline variation, a mathematically precise calculation, clearly demonstrated the benefit of the drug, with the maximum effect for the highest dose. At 12 weeks (84 days), the maximum drug dose showed a benefit relative to placebo of 134 days. This effect may be interpreted as indicating that at the end of the study, the patients in the high dose panel were relatively 50 "days" better than they were at the beginning of the study.

	Baseline MMS	12th Week (Time-Index)	Change (+ = Improvement)
panel 0	17.5 (715)	17.4 (690)	+ 25
panel 1	18.9 (574)	19.8 (433)	+ 141
panel 2	19.0 (561)	19.4 (459)	+ 102
panel 3	18.2 (648)	19.6 (460)	+ 159

Examining the rate of change in the "TIME-INDEX" values between visits allows examination of the drug's progressive effects:

	Standard	Baseline-Week 2	Week 8-Week 12
panel 0	-1	+ 4.1	- 1.8
panel 1	-1	+ 1.6	+ 2.2
panel 2	-1	+ 4.0	+ 0.8
panel 3	-1	+ 7.1	+ 1.4

Placebo and all drug conditions showed a positive rate indicative of improvement between the baseline and the first visit, explained partly by the learning effect. However, the placebo rate dropped back to near zero at the next visit and showed a rebound deterioration between the 8 and 12 week visits. Conversely, the drug effect was consistently above the mean rate of deterioration (-1), and all panels showed a rate indicating continued improvement between the 8 and 12 week visits. While the calculation appears to project a continued benefit of the drug, a longer term study is needed to demonstrate whether this drug will continue to benefit patients beyond the 12 week study period.

DISCUSSION

The "TIME-INDEX" method provides a mathematically sound method of drug effect assessment which also makes a clear demonstration of long-term benefits, and describes the effects in practical terms. A learning effect of test-retest design is clearly demonstrated in this study, with more learning occurring for more mildly affected patients. This is consistent with the perspective that AD is primarily a disease of learning mechanisms, which are most severely destroyed early in the disease course. Metrifonate showed a benefit for patients at 12 weeks relative to baseline, and this effect is clearly shown to be independent of dementia severity. The MMSE scores themselves do not show the striking benefit of metrifonate. Translation of MMSE scores into "TIME-INDEX" values shows a pattern similar to the MMSE scores, but correction for baseline gives an indication of the benefit of this medication in a numerical form which can be clearly explained and compared with other studies. Rate analysis shows that metrifonate is actually continuing to benefit patients after 8 weeks.

REFERENCES

Ashford JW, Shan M, Butler S, Rajasekar A and Schmitt FA (1995): Temporal quantification of Alzheimer's disease severity: 'Time Index' model. *Dementia* 6:269-280.

Bieber F, Mas J, Orazem J, Gulanski B (1966): Results of a Dose Finding Study with Metrifonate in Patients with Alzheimer's Disease (AD). Abstract 4th Intnl Sympsium on Advances in Alzheimer Therapy, Nice, France April 10-14.

Folstein MF, Folstein SE and McHugh PR (1975): 'Mini-Mental State'. A practical method for grading the cognitive state of patients for the clinician. *J Psychiatr Res* 12:189-198.

Alzheimer Disease: From Molecular Biology to Therapy
edited by R. Becker and E. Giacobini

PET-STUDIES USING 11C-TZTP DERIVATIVES FOR THE VISUALIZATION OF MUSCARINIC RECEPTORS IN THE HUMAN BRAIN

Henrik Nybäck, Christer Halldin, Per Karlsson, Yoshifumi Nakashima and Lars Farde
Department of Clinical Neuroscience,
Karolinska Hospital, Stockholm, Sweden

Per Sauerberg
Novo Nordisk A/S, Måløv, Denmark

Harlan E. Shannon and Frank P. Bymaster
Eli Lilly and Company, Indianapolis, IN, USA

INTRODUCTION

The cholinergic hypothesis of memory disorders (Bartus et al., 1982) provides a rationale for the development of drugs for Alzheimer's disease (AD). A decline in nicotinic (Nordberg and Winblad, 1986) and muscarinic (Mash et al., 1985) cholinergic receptors has been described in studies of post-mortem brain material from AD patients.

Using positron emission tomography (PET) radioligand binding to transmitter receptor populations in the human brain can be visualized *in vivo*. We have investigated the distribution and binding of 11C-labelled 3-(3-substituted -1,2,5,-thiadiazol-4-yl)-1,2,5,6-tetrahydro-1-methylpyridines (substituted TZTP) which have high affinity for the muscarine M1-receptor (Sauerberg et al., 1992; Shannon et al., 1994; Nybäck et al., 1995). The compounds have been developed as potential agents for treatment of AD patients and one of them, the hexyloxy-TZTP or xanomeline, has shown promising results in clinical trials (Bodick et al., 1994).

The aims of our study was to evaluate [11C]xanomeline and [11C]butylthio-TZTP as tools for visualization of muscarinic receptors in the living human brain.

EXPERIMENTS IN MONKEYS

Initial studies in Cynomolgus monkeys, using the Scanditronix PC-2048-15B PET-system, showed a rapid uptake of [11C]xanomeline and [11C]butylthio-TZTP into the brain with a high concentration in the frontal cortex and basal ganglia but low levels in the cerebellum. The obtained regional distribution was compatible with the anatomy of muscarinic receptors as found in post-mortem autoradiographic studies (Dewey et al., 1990). A pharmacological characterization of the uptake and binding of was done (Farde et al., 1995) by pretreating the animals with drugs (xanomeline, scopolamine, biperiden, trihexyphenidyl, haloperidol and DuP 734) known for their affinity to muscarinic, dopamine-D2 and sigma binding sites (Culp et al., 1992).

The results indicate that the binding of [11C]butylthio-TZTP is saturable and reversible and may be displaced by xanomeline by about 70 %. Also, the muscarinic antagonist, biperiden, reduced the uptake in the cortex and the striatum but not in the cerebellum. However, haloperidol and DuP 734 reduced the uptake of [11C]butylthio-TZTP also in cerebellum indicating that the radioactivity in this region represents binding to sigma recognition sites. The affinity of the TZTP derivatives for sigma sites has recently been confirmed in studies on guinea pig brains (Farde et al., 1996).

PET EXPERIMENTS IN HUMANS

Three healthy male volunteers, 19-22 yr, were given tracer doses of [11C]xanomeline or [11C]butylthio-TZTP (about 300 Mbq in saline) intravenously and the uptake and distribution in the brain was measured during 50 minutes. Regions of interest were drawn on the reconstructed PET-images with the guidance of CT-images which were obtained with the same head fixation device.

As in the monkeys [11C]xanomeline and [11C]butylthio-TZTP were rapidly taken up into the brain with more than 5 % of the given radioactivity appearing within 5 minutes after the injection. The uptake of [11C]xanomeline showed an earlier and lower peak as compared to [11C]butylthio-TZTP which is consistent with a lower affinity for the M1 receptor as measured *in vitro*. The cortical/cerebellar ratio of both tracers was lower in man than in monkeys indicating a relatively high proportion of sigma binding.

OCCUPANCY STUDIES

Two male and one female elderly but healthy volunteers, 58, 59 and 68 years, participated in a study of receptor occupancy following a pharmacologically active dose of xanomeline. [11C]Butylthio-TZTP was used as a tracer in

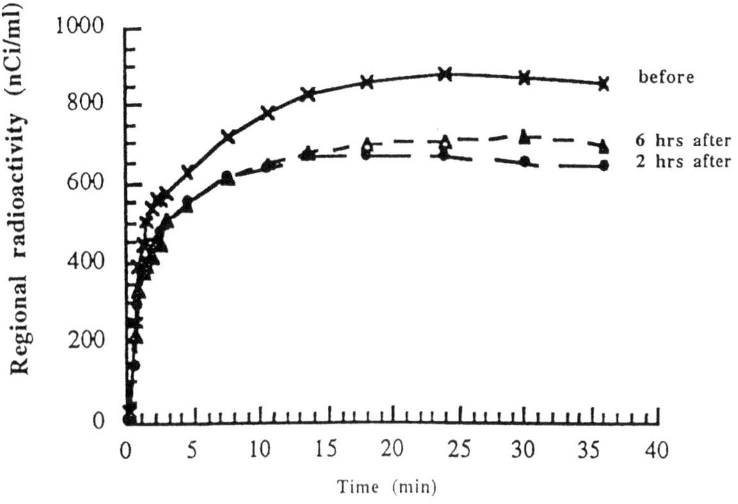

FIGURE 1. Effect of xanomeline, 75 mg p.o., on uptake of [11C]butylthio-TZTP in the human frontal cortex (n = 3).

three consecutive PET examinations, one before and two at 2 and 6 hours after an oral dose of 75 mg of xanomeline. Receptor occupancy was calculated by graphical analysis following correction for the formation of metabolites which were measured by high pressure liquid chromatography.

As in the young volunteers [11C]butylthio-TZTP radioactivity increased continuously during 20 minutes in all brain regions. The uptake of the tracer was significantly lower at 2 hours following the oral xanomeline and remained still lowered at 6 hours (Figure 1). The occupancy of M1-receptors in the living human brain by 75 mg of xnomeline was calculated to be 35-55% at 2 hours and 10-45% at 6 hours following drug ingestion. The subject with the highest occupancy was the only of the three subjects who experienced a pharmacological effect in the form of adverse reactions, i.e. nausea, drowsiness and headache.

CONCLUSIONS

The substituted TZTPs represent a class of compounds suitable for studies of cholinergic mechanisms in cognitive disorders. [11C]-Labelled TZTP-derivatives may be used as tools in PET experiments of muscarinic receptor occupancy, thus enabling the development of new drugs for symptom relief in AD.

REFERENCES

Bartus RT, Dean RL, Beer B et al. (1982): The cholinergic hypothesis of geriatric memory dysfunction. *Science* 217:408-416.

Bodick NC, DeLong AF, Bonate PL et al. (1994): Xanomeline, a specific M1 agonist: Early clinical studies. In: *Alzheimer's Disease: Therapeutic strategies*, Giacobini E and Becker R, eds. Birkhäuser Boston, pp. 234-238.

Culp S, Rominger D, Tam W et al. (1992): [3H]DuP 734: A receptor binding profile of a high affinity novel sigma receptor ligand in Guinea pig brain. *J Pharmacol Exp Ther* 263:1175-1187.

Dewey S, MacGregor R, Brodie J et al. (1990): Mapping muscarinic receptors in human and baboon brain using [N11C-Methyl]-Benztropine. *Synapse* 5:213-223.

Farde L, Suhara T, Halldin C et al. (1995): New radioligands for PET-examination of muscarinic receptors. *Adv Behavioral Biol* 44:499-505.

Farde L, Suhara T, Halldin C et al. (1996): PET-study of the M1-agonists [11C]-Xanomeline and [11C]butylthio-TZTP in monkey and man. *Dementia* (in press).

Nordberg A and Winblad B (1986): Reduced number of (3H)-nicotine and (3H)-acetylcholine binding sites in the frontal cortex of Alzheimer brains. *Neurosci Lett 72*:115-119.

Nybäck H, Halldin C, Karlsson P et al. (1995): Positron emission tomography (PET) studies with ligands for cholinergic receptors in the human brain. *Adv Behavioral Biol* 44:245-249.

Mash D, Flynn D and Potter L (1985): Loss of M2 muscarine receptors in the cerebral cortex in Alzheimer's disease and experimental cholinergic denervation. *Science* 228:1115-1117.

Sauerberg P, Olesen P, Nielsen S et al. (1992): Novel functional M1 selective muscarinic aginists. Synthesis and structure-activity relationships of 3.(1,2,5-Thiadiazolyl)-1,2,5,6-tetrahydro-1-methylpyridines. *J Med Chem* 35: 2274-2283.

Shannon H, Bymaster F, Calligaro et al. (1994): Xanomeline: A novel muscarinic receptor agonist with functional selectivity for M1 receptors. *J Pharm Exp Ther* 269:271-281.

Alzheimer Disease: From Molecular Biology to Therapy
edited by R. Becker and E. Giacobini

PET IMAGING OF NICOTINIC RECEPTORS IN ALZHEIMER'S DISEASE - IMPLICATION WITH DIAGNOSIS AND DRUG TREATMENT

Agneta Nordberg
Dept of Clinical Neuroscience and Family Medicine, Div of Nicotine Res., Karolinska institutet, Huddinge University Hospital, Huddinge, Sweden.

INTRODUCTION

The knowledge about the nicotinic acetylcholine receptors (nAChR) in human brain and their involvement in higher functions such as learning and memory are relatively new phenonema. The nAChRs are candidates for transducing cell surface interactions not only for acetylcholine but also for other neurotransmitters such as dopamine, noradrenaline, GABA and glutamate. The nAChRs are relatively abundant in brain. Three types of nAChRs binding sites, a super-high, a high and a low-affinity nAChRs site have been identified in the human brain (Nordberg et al., 1988a,b). The nAChRs belong to a family of ligand-gated ion channel receptors. Multiple α and ß nAChR subunits have been isolated which in a combination of a given pentamer determine the functional properties of the nAChR. So far the α2-α5, α7, ß2 and ß4 subunits have been cloned in the human brain. Receptor autoradiography, *in situ* hybridization, Northern blot techniques will allow the regional nAChRs and their transcripts to be mapped and quantified *in vitro* in human brain tissue.

Marked losses of nAChRs have been found in cortical autopsy brain tissue from patients suffering from Alzheimer's disease (AD) (Nordberg and Winblad, 1986). The α4ß2 nAChR subtype is vulnerable in AD (Warpman and Nordberg, 1995). Stimulation of remaining nAChRs by receptor selective drugs might be of clinical significance for alleviating some of the cognitive impairments (Sahakian et al., 1989; Newhouse et al., 1988) in AD patients. A drawback of neurochemical studies in brain autopsy material is that the neurochemical analysis represents the final stage of the disease. It is therefore a challenge to explore whether imaging techniques such as positron emission tomography (PET) can be used to reveal selective deficits in nAChRs in AD patients. PET studies of nAChRs disturbances in the early course of AD might be of diagnostic value and provide an early initation of drug treatment in the patient.

VISUALIZATION OF NICOTINIC RECEPTORS BY ^{11}C-NICOTINE

The use of ^{11}C-nicotine as radioligand in PET studies for nAChRs has recently been reviewed (Mazière and Delforge, 1995). Quantification of ^{11}C-nicotine binding in the brain is complex since the uptake of ^{11}C-nicotine to the brain is strongly dependent on the cereberal blood flow (Yokoi et al., 1993) and nicotine is rapidly associated/dissociated from nAChR binding sites (Larsson and Nordberg 1985; Hall et al., 1993). A fraction of ^{11}C-nicotine binds non-specifically in brain dependent on physiological parameters such as pH of the tissue. We initially analyzed the ^{11}C-nicotine binding in human brain using a two-compartment model and the rate constants k_1 and k_2 (Nordberg et al., 1990). We later developed a new dual tracer model where ^{15}O-water is administered in close succession to ^{11}C-(S)-nicotine to compensate for the influence of the cerebral blood flow (CBF) on the rate constants (Nordberg et al., 1995). The rate constant $k_2^* = k_2/$ CBF gives a quantitative measure of ^{11}C-nicotine binding in human brain (Nordberg et al., 1995). The proposed kinetic model has been evaluated by PET studies performed in Rhesus monkeys (Lundtqvist et al., 1996). The k2* rate constants were measured under normal and high flow conditions (changes in pCO2) and found to be independent of the CBF. When ^{11}C-(S)-nicotine was administered to monkeys together with different amounts of unlabelled nicotine (0.010- 0.015 mg/kg) to change the specific radioactivity, the k2* value increased indicating less binding of ^{11}C-(S)-nicotine in brain (Lundqvist et al.,1996).

NICOTINIC RECEPTOR DEFICITS IN AD

Significantly higher k2* values for ^{11}C-(S)-nicotine, e.g. lower nicotine binding, have been observed in the frontal and temporal cortices and hippocampus of AD patients compared to age-matched healthy volunteers (Nordberg et al., 1995). No significant changes in nicotine binding (k2* values) were observed in the thalamus, caudate nucleus, putamen or cerebellum of AD patients compared to controls.

A significant negative correlation between cognitive function (mini-mental-state-examination score, MMSE) and k2* for ^{11}C-nicotine has been found in the temporal cortex of AD patients and age-matched controls (p< 001) (Fig 1a). A negative correlation between glucose metabolism and k2* for ^{11}C-nicotine has also been noticed in the parietal cortex of AD patients.

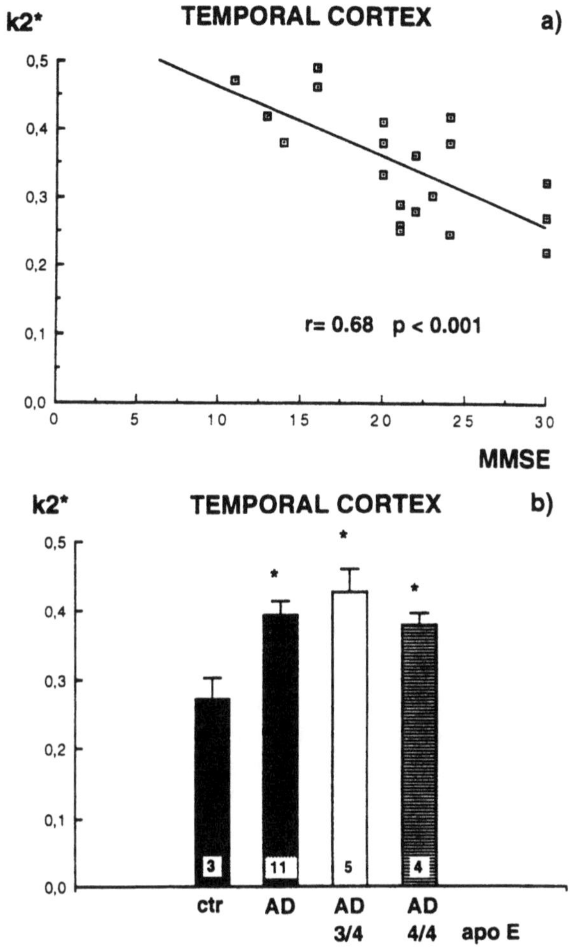

FIGURE 1 a. Correlation between cognition (MMSE) and k2* for ^{11}C-nicotine in the temporal cortex. b. Number of nicotinic receptors (expressed as k2* rate constant) in the temporal cortex AD patients carrying apoE ε3/4 and ε4/4 alleles respectively. * $p < 0.02$ compared to age-matched controls.

APO E GENOTYPE AND NICOTINIC RECEPTORS

The apolipoprotein E (apoE) genotype is an established risk factor for AD. The presence of the apoE ε4 allele results in an earlier clinical manifestation of AD (Corder et al., 1993). Cognitively intact subjects homozygous for ε4 has recently been reported to show reduced glucose metabolism in cortical brain areas compared to non-carriers of ε4 (Reiman et al., 1996). We observed no difference in glucose metabolism in brain between AD patients carrying the ε3/3, ε3/4 or ε4/4 apoE alleles respectively (Corder et al., 1996). In addition, we found no difference in number of nicotinic receptors (expressed as k2*) in the temporal cortex of AD patients carrying 1 compared to 2 ε4 alleles (Figure 1b). Furthermore, analysis of nicotinic receptor deficits in autopsy AD tissues showed no relationship with apoE ε4 dosage (Nordberg et al., in preparation) opposite to the finding of Poirier et al. (1995). Since a greater accumulation of ß-amyloid has been reported in brain of AD subjects with ε4 allele compared to those without (Polvikoski et al., 1995) the PET findings might suggest that the amyloid deposition is not closely related to the functional activity of the brain in AD patients. The data support the assumption that the apoE genotype is more involved in the process of developing AD than in the progression of the disease.

REGULATION OF NICOTINIC RECEPTORS BY DRUG TREATMENT

The cholinergic hypothesis has so far attracted most interest in treatment stategies in AD. PET studies have shown that the nicotinic receptors can respond differently to drug treatment in AD. Ondansetron is a $5HT_3$ antagonist theoretically assumed to increase the cortical acetylcholine release and therefore tested as potential drug in AD. Six AD patients treated with 10 m g/bw ondansetron for 3 months showed a significant increase in k2* (decrease in nicotine binding) in the frontal association cortex. The effect on nicotinic receptors was opposite to that observed for AD patients treated with the cholinesterase inhibitor tacrine. Tacrine in long-term treatment restores the nicotinic receptors in cortical brain regions of AD patients (Nordberg et al., 1992) and a significant correlation is observed between improvement in cognitive tests and nicotinic receptors (decrease in k2*) (Nordberg, 1995). Treatment with nerve growth factor (NGF) intraventricularly has so far been tested in three AD patients. Two AD patients showed improvement in nicotinic receptors in cortical brain regions which persisted after withdrawal of the NGF treatment. In a third AD patient receiving a 1/25 times lower dose of NGF than the other two AD patients, no improvement in nicotinic receptors was observed except for the hippocampus. Generally, the nicotinic receptors seem to respond quicker to drug treatment than the glucose metabolism in brain. ApoE has recently been discussed as a predictor of treatment outcome with cholinergic drugs in AD (Poirier et al., 1995). When AD patients, as a part of a 1-year treatment study, were tested for their response to a single dose of oral tacrine (40 mg) we were unable to observe any difference in neuropsychological response among carriers with different apoE genotype (Almkvist et al., 1996).

ACKNOWLEDGMENTS

The research from our laboratory was supported by the Swedish Medical Research Council, Loo and Hans Osterman's foundation, Stiftelsen för Gamla Tjänarinnor, Swedish Tobacco Research Council, KI foundations, Golje's foundation.

REFERENCES

Almkvist O, Amberla K, Jolic V, Meurling L, Winblad B and Nordberg A (1996): Cognitive Functions and EEG After a Single Dose of Tacrine. *Abstract 4th Intl Symposium on Advances in Alzheimer Therapy*, Nice, France, April 10-14.

Corder EH, Jelic V, Basun H, Lannfelt L, Valind S, Winblad B, Nordberg A (1996): No difference in cerebral glucose metabolism in Alzheimer Patients with differing apolipoprotein E genotypes. Submitted.

Corder EH, Saunders AM, Strittmatter WJ, Schmechel DE, Gaskell PC, Small GW, Roses AD, Pericak-Vance MA (1993): Gene dose of apolipoprotein E type 4 allele and the risk of Alzheimer's disease in late onset families. *Science* 261: 921-923.

Hall M, Zerbe L, Leonard S and Freedman R (1993): Characterization of 3H-cytisine binding to human brain membrane preparations. *Brain Res* 600: 127-133.

Larsson C and Nordberg A (1985): Comparative analysis of nicotine-like receptor-ligand interaction in rodent brain homogenate. *J Neurochem* 45: 24-31.

Lundqvist H, Nordberg A, Hartvid P, Lilja A, Långström B (1996): (S)-(-)-[^{11}C] Nicotine binding in rhesus monkey brain measured with PET - A model approach. Submitted.

Mazière M and Delforge J (1995): PET imaging of ^{11}C-nicotine: Historical aspects. In: *Brain imaging of nicotine and tobacco smoking*. Domino EF, ed. Ann Arbor: NPB Books, pp. 13-28.

Newhouse PA, Sunderland T, Tariot PN, Blumgardt CL, Mellow A, Murphy DL (1988): Intravenous nicotine in Alzheimer's disease: a pilot study. *Psychopharm* 95:171-175.

Nordberg A. (1995): Long-term treatment with tacrine: effects on progression of Alzheimer's disease as determined by functional brain studies. In: *Research Advances in Alzheimer's Disease and Related Disorders*, Iqbal K, Mortimer JA, Winblad B, Wiesniewski HM, eds. Chichester: John Wiley & Sons, pp. 293-298.

Nordberg A, Adem A, Hardy J and Winblad B (1988a): Changes in nicotinic receptor subtypes in temporal cortex of Alzheimer brains. *Neurosci Lett* 86: 317-321.

Nordberg A, Adem A, Nilsson L, Romanelli L and Zhang X (1988b): Heterogenous cholinegic nicotinic receptors in the CNS. In: *Nicotinic acetylcholine receptors in the nervous system,* Clementi F, Gotti C and Sher E, eds. New York: Springer Verlag, pp. 331-350.

Nordberg A, Hartvig P, Lilja A, Viitanen M, Amberla K, Lundqvist H, Andersson Y, Ulin J, Winblad B and Långström B (1990): Decreased uptake and binding of ^{11}C-nicotine in brain of Alzheimer patients as visualized by positron emission tomography. *J Neural Transm* (P-D sect) 2: 215-224.

Nordberg A, Lilja A, Lundqvist H, Hartvig P, Amberla K, Viitanen M, Warpman U, Johansson M, Hellström-Lindahl E, Bjurling P, Fasth KJ, Långström B and

Winblad B (1992): Tacrine restores cholinergic nicotinic receptors and glucose metabolism in Alzheimer patients as visualized by positron emission tomography. *Neurobiol Aging* 13: 747-758.

Nordberg A, Lundqvist H, Hartvig P, Lilja A and Långsröm B (1995): Kinetic analysis of regional (S)(-)^{11}C-nicotine binding in normal and Alzheimer brains- in vivo assessment using positron emission tomography. *AD and Assoc Disord* 9: 21-27.

Nordberg A and Winblad B (1986): Reduced number of 3H-nicotine and 3H-acetylcholine binding sites in the frontal cortex of Alzheimer brains. *Neurosci Lett* 72: 115-119.

Poirier J, Marie-Claude D, Quirion R, Aubert I, Farlow M, Lahiri D, Hui S, Bertrand P, Nalbantoglu J, Gilfix BM and Gauthier S (1995): Apolipoprotein E4 allele as a predictor of cholinergic deficits and treatment outcome in Alzheimer disease. *Proc Natl Acad Sci,* USA 92: 12260-12264.

Polvikoski T, Sulkava R, Haltia M, Kainulainen K, Vuorio A, Verkkoniemi A, Niinistö L, Halonen P, Kontula K. (1995): Apolipoproteint E, dementia, and cortical deposition of ß-amyloid protein. *N Engl J Med* 333: 1242-1247.

Reiman EM, Caselli RJ, Yun LS, Chen K, Bandy D, Minoshima S, Thibodeau SN and Osborne D (1996): Preclinical evidence of Alzheimer's disease in persons homogenous for the e4 allele for apoliprotein E. *N Engl J Med* 334: 752-758.

Sahakian B, Jones G, Levy R, Gray J, Warburton D (1989): The effects of nicotine on attention, information processing and short-term memory in patients with dementia of Alzheimer type. *Br J Psychiatry* 154:797-800.

Yokoi F, Komiyama T, Ito T, Hayashi T, Lio M and Hara T (1993): Application of carbon-11 labelled nicotine in the measurement of human cerebral blood flow and other physiological parameters. *Eur J Nucl Med* 20: 46-52.

Warpman U and Nordberg A (1995): Epibatidine and ABT 418 reveal selective losses of α4ß2 nicotinic receptors in Alzheimer brains. *NeuroReport* 6: 2419-2423.

Alzheimer Disease: From Molecular Biology to Therapy
edited by R. Becker and E. Giacobini

CEREBRAL SPECT IMAGING: ADVANCES IN RADIO PHARMACEUTICALS AND QUANTITATIVE ANALYSIS

Daniel O. Slosman
Division of Nuclear Medicine, Hôpital Cantonal Universitaire, Geneva, Switzerland.

Pierre J. Magistretti
Institut de Physiologie et Service de Neurologie du CHUV, Faculté de Médecine, Lausanne, Switzerland.

INTRODUCTION

For more than a decade, nuclear medicine has been recognized as the unique tool for performing functional brain imaging and for quantifying non-invasively regional cerebral blood flow (rCBF). It required high-cost technology, such as positron emission tomography (PET), and a dedicated research structure.

Today, in addition to the technological improvements in image quality (triple-head gamma-camera and fan-beam collimators), the introduction of ^{99m}Tc-labeled brain tracers contributes largely to the recent widespread use of cerebral perfusion SPECT (single photon emission computed tomography) imaging in a clinical setting (Waldemar, 1995).

These developments in radiochemistry are aimed at determining agents that are 1) easy to label (kit formula), 2) suitable for Tc-99m labeling (taking into account the ready availability of $^{99m}TcO_4^-$, its short half-life and its efficiency in external imaging), 3) stable *in vitro* and *in vivo*, as labeled products, and 4) biologically able to reflect, by their uptake, cerebral functions such as stress or activation studies (short arterial input and high extraction fraction during a single pass).

To reflect accurately rCBF, a tracer should fulfill the following criteria: 1) to be neutral and to have a high degree of lipophilicity, 2) to have a small size and to cross the intact blood-brain barrier (BBB) freely and be independent of blood cell-plasma protein repartition, 3) to show a good linearity over a wide range of cerebral flow, 4) to have the best first-pass brain extraction, 5) to show no redistribution or back-diffusion and 6) to show a tissue retention independent of metabolic dysfunction.

RADIOPHARMACEUTICALS AND BIOLOGICAL BEHAVIOR

Xenon-133 is a noble, freely diffusible inert gas that passes through cell membranes and has free exchange between blood and tissue. It can be inhaled or injected intravenously. Xe-133 is the first rCBF tracer used for the quantitative measurement by SPECT (Lassen et al, 1983). But, its characteristics lead to the use of dedicated instruments with a high degree of sensitivity and a dynamic mode of acquisition. It has a low gamma photon energy (80 keV) resulting in a poor spatial resolution, which is due to significant absorption and scattering (Table I).

I-123, isopropyl-*p* -iodoamphetamine (IMP) becomes the first tracer of a new class of agents used in a clinical setting : the "metabolic microspheres" tracers. It is a lipophilic substance that crosses BBB, and has a 90% first-pass extraction efficiency (Table I). Biological mechanisms involved in its brain tissue retention are not fully understood but, experimentally, it is related to the serotonergic uptake system and is inhibited by imipramine. The brain uptake is less than 10% of the injected dose achieved within one hour. It is considered to reflect accurately rCBF, but for some authors, Xenon-133 remained the best tracer. Limitations are related to the use of iodine-123 (a cyclotron product not always available) and the presence of a significant redistribution over time.

Tc-99m, hexamethylpropyleneamineoxime (HMPAO) was the first Tc-99m labeled brain tracer to be used in a clinical setting. It is a lipophilic substance that crosses BBB, and has a 65% extraction on first pass (Table I). Biological mechanisms involved in its cell retention include rapid hydrophilic conversion by interaction with the intra-cellular glutathione. Our recent observations indicate that the cellular retention of HMPAO is determined by the redox state in a variety of cell types, including a human astrocyte-derived cell line (Jacquier-Sarlin et al, 1996a). The brain uptake achieves 6-7% of the injected dose within five minutes. It is considered to reflect rCBF, but not for high flow values that become underestimated. Limitations are related to its chemical *in vitro* instability, the underestimation of rCBF at high flow and its dependency on the redox state of the tissue. This latest aspect could become valuable if one no longer considers HMPAO as a rCBF tracer, but as a metabolic tracer. Reduced glutathione is important for scavenging oxygen free radicals (implicated in neuronal death). The amount of reduced glutathion is directly related to the redox state of a cell, particularly to its NADP/NADPH ratio. Thus, changes in the levels of NADPH, which is notably produced as a by-product of glucose metabolism through the pentose phosphate pathway, will influence the redox state of a cell, a condition that will impact on the uptake of HMPAO. Therefore, HMPAO SPECT may become a useful handle to elucidate, in particular, certain pathophysiological mechanisms involved in neurodegeneration.

Tc-99m ethylcysteinate dimer (ECD) is a newer Tc-99m labeled brain tracer that has been already introduced in a clinical setting. It is also a lipophilic substance that crosses BBB and has characteristics very similar to HMPAO (Table I). But biological mechanisms involved in its cell retention include rapid hydrophilic conversion by interaction with esterase. Our recent observations indicate that membranar esterasic activity plays an important role in the subsequent cellular trapping (Jacquier-Sarlin et al, 1996b). If intracelllular esterasic activity allows conversion to hydrophilic complexes, the membranar esterasic activity induces extracellular conversion, therefore reducing the amount of lipophilic compound that can enter into the cell.

ECD is considered to be similar to HMPAO in numerous biological or technical aspects with identical limitations in accuracy for high flow values that become underestimated. Advantages over HMPAO are related to its chemical *in vitro* stability and the absence of dependency on the redox state of the tissue. This latest aspect has already been identified: several investigators have reported discrepancies between brain SPECT with ECD and other rCBF agents in different conditions, particularly in the presence of focal increased HMPAO while ECD remains iso-intense. In our experience, 10 out of 18 patients with both ECD and HMPAO brain SPECT showed discrepant results,

	Xenon-133	IMP I-123	ECD-Tc	HMPAO-Tc
Physical T1/2	5.3 days	13.0 hrs	6.03 hrs	6.03 hrs
Energy (keV)	81	159	140	140
Administered dose (MBq)				
Intra/Vein.	350-750	200	350-600	350-600
Intra/Art.	110			
Gas	100/liter			
Absorbed dose (10^{-2} mGy/MBq)				
Brain	0.12	2.9	0.8	0.5
Total body	0.1	0.9	0.4	0.2
Retention				
- whole-body (24h)		70%	15%	65%
- brain (1h)		6%	5.2%	5.6%
- brain T1/2 wash-out		> 52 hrs	> 7 hrs	> 24 hrs
- blood clearance T1		70 sec (52%)	35 sec	90sec (59%)
- blood clearance T2		375 sec (48%)	8 min	4 hr (41%)
Circulating activity at 60 min		1%	5%	12%
First pass extraction		90%	77%	65%

TABLE I. Physical and biological characteristics of the radiopharmaceuticals used for brain imaging.

namely focal increased HMPAO retention and iso-intense ECD (6 neoplasia, 2 vascular diseases, 1 abscess and granuloma).

RADIOPHARMACEUTICALS AND CELLULAR LOCUS

Recently, astrocytes and not neurons have been identified as the target of fluorodeoxyglucose cellular retention, as previously thought (Pellerin and Magistretti, 1994). One can wonder which population of cells does retain SPECT brain imaging agents administered intra-venously (endothelial cells, astrocytes or neurons) if there is any specificity.

Unlike other tracers, Xenon-133 does not behave as radiolabeled "chemical microsphere", and cell specificity is unlikely to be of importance. Moreover, for the other markers, the cellular target should be definitively identified in order to discriminate between flow-dependent to metabolism-dependent effects. Based on the observations made with FDG, one could consider astrocytes as a cellular target for ECD and HMPAO retention.

RADIOPHARMACEUTICALS AND CLINICAL USE

As shown in Table II, one can no longer consider as identical all these tracers' ability to perform brain SPECT for the assessment of rCBF. The adequacy of the choice of rCBF tracer will depend on the question asked and on the knowledge of the biological mechanisms related to its brain tissue retention. For instance, taking into account the recent observations regarding ECD and HMPAO, as mentioned previously, we could conclude that the recommended

	Xenon-133	IMP I-123	ECD-Tc	HMPAO-Tc
Labeling				
- procedure	none	none	kit	kit
- stability	++++	++	++	+/-
Image contrast	+/-	+	+++	++
SPECT quality	+/-	+	+++	++
Accuracy (rCBF)				
- normal state	+++	++	+	+
- pathology	+++	++	+/-	+
Indications				
- Neurodegeneration	+	+	+	++
- Cerebral oncology	-	-	+/-	++
- Vascular Disease	+	+	+	+
- Activation test	+/-	+/-	+	+

TABLE II. Clinical use of the radiopharmaceuticals dedicated to brain imaging.

marker to image rCBF (investigations of vascular diseases or to perform stress-tests/activation-tasks) should be ECD, while HMPAO could be used to image the metabolic properties of the tissues specific to neurodegeneration or neoplasia.

RADIOPHARMACEUTICALS AND QUANTITATIVE MEASUREMENT

Besides the use of a radiotracer, which reflects rCBF proportionally to its tissue retention in the brain, it is necessary to develop techniques to transform radioactive events detected by voxel units in a concentration of tracer (microcurie/ml). Then to assess the accuracy of such a method, to provide normative data and to validate its linearity in healthy subjects as well as in populations of patients with neurological abnormalites (in particular with vascular disease or neurodegenration). It should be recognized that true meas urement using a SPECT device is, at first, related to the development of a technique for scatter correction and attenuation correction. We are currently evaluating such an approach and to date, preliminary observations suggest that such hardware/software improve, at least, the image quality. We already know that ECD or HMPAO underestimate high flow but this would not exclude their use to quantify rCBF, if appropriate models are used and a normative database is set up for each marker.

Because of the combined development of the technology and the radiopharmaceuticals, rCBF SPECT has become a promising tool for the clinical assessment of neurodegeneration. In addition, in taking into account the limit in accuracy, the ability to quantify rCBF with good reproducibility will allow its use for the longitudinal assessment of a new therapeutic approach.

ACKNOWLEDGMENTS

The research from our laboratory reviewed here was supported by a grant from the Swiss National Foundation for Scientific Research (FNSRS N° 31-33781.92).

REFERENCES

Jacquier-Sarlin M, Polla B, Slosman DO (1996a): Oxido-reductive state: the major determinant for cellular retention of HMPAO-^{99m}Tc. *J Nucl Med* In Press.

Jacquier-Sarlin M, Polla B, Slosman DO (1996b): The cellular basis of ECD brain retention. *J Nucl Med* In press.

Lassen NA, Henriksen S, Holm S, Barry DI, Paulson OB, Vorstrup S, Rapin J, Le Poncin-Lafitte M, Moretti JL, Askienazy S, Raynaud C (1983): Cerebral blood-flow tomography: Xenon-133 compared with isopropylamphetamine-iodine-123. *J Nucl Med* 24:17-21.

Pellerin L, Magistretti PJ (1994): Glutamate uptake into astrocytes stimulates aerobic glycolysis : a mechanism coupling neuronal activity to glucose utilization.. *Proc Natl Acad Sci USA* 91: 10625-10629.

Waldemar G (1995): Functional brain imaging with SPECT in normal aging and dementia. *Cerebrovascular and brain metabolism reviews* 89-130.

Alzheimer Disease: From Molecular Biology to Therapy
edited by R. Becker and E. Giacobini

IN VIVO IMAGING OF ANTICHOLINESTERASE DRUGS USED IN ALZHEIMER'S DISEASE.

B. Tavitian, S. Pappatà, F. Branly, A. Jobert, A. Dalger, E. Dumont, F. Simonnet, J. Grassi, C. Crouzel and L. DiGiamberardino
INSERM U334, Service Hospitalier Frédéric Joliot, Direction des Sciences du Vivant, CEA, 4 place du général Leclerc, 91401 Orsay, France

There is a hope that Positron Emission Tomography (PET), a non-invasive *in vivo* technique, will become a useful tool for the study of patients with neurodegenerative disorders in which dysfunction and/or damage of the central cholinergic system is suspected (Foster, 1994). New ligands capable of imaging cholinergic pathways could explore the chemical and functional impairments found in Alzheimer's disease (AD) (Perry, 1987) and the therapeutic effects of potential drugs on these impairments.

Among the key components of cholinergic neurotransmission, there is an interest for the *in vivo* quantification of brain acetylcholinesterase (AChE, the enzyme hydrolyzing acetylcholine) because of its known changes in AD (Davies, 1979), and because AChE is the target of a number of inhibitors designed as therapeutic drugs for the enhancement of cholinergic transmission (Giacobini and Becker, 1994).

PET images show the distribution of tissular radioactivity concentrations after the injection of a positron-emitter-labeled compound. Interpretation of the images thus requires a quantitative link to be established between radioactivity and ligand-target interaction, a task which is far from obvious due to the limitations of *in vivo* experiments. In this paper, I will report animal data obtained in non-human primates with two [^{11}C]-labeled inhibitors of AChE, [^{11}C]-physostigmine ([^{11}C]-PHY) and [^{11}C]-heptylphysostigmine ([^{11}C]-HEP).

METHODS

Male adult baboons (Papio papio) under isoflurane anesthesia received a bolus injection of [^{11}C]-PHY or [^{11}C]-HEP. Arterial blood samples and sequential PET scans were collected during 90 min. Plasma radioactivity was separated into soluble and bound fractions with TCA, and the soluble fraction was further analyzed with HPLC. PET scans were acquired with a ECAT 953B camera in the orbito-meatal plane and corrected for ^{11}C decay and attenuation. Regions of interest were traced manually according to MRI scans obtained in the same position. All radioactive concentrations were expressed in percent of the injected dose per liter of tissue (%ID/l).

[^{11}C]-PHYSOSTIGMINE

Physostigmine (PHY) has been labeled with ^{11}C (Bonnot et al., 1993) on the carboxylic carbon of its carbamic group. Previous autoradiographic studies in rats (Planas et al., 1994) and PET imaging in baboons (Tavitian et al., 1993) showed similar cerebral distributions for [^{11}C]-PHY and AChE. After a bolus injection of [^{11}C]-PHY, a peak of blood radioactivity was observed during the first minute followed by a rapid decline (8.5% ID/1) at 60 min; $t_{½} = 0.58 \pm 0.33$ min). Blood radioactivity was essentially plasmatic and found in three forms: a) unbound (free) radioactivity, which declined very rapidly after injection to account for 10% or less of total plasmatic radioactivity after 30 min; this fraction was identified by HPLC as essentially unmetabolized [^{11}C]-PHY, thus representing [^{11}C]-PHY available for cerebral uptake; b) radioactivity bound to plasmatic proteins remained constant at 60% of total plasmatic radioactivity after the initial peak; selective immunoprecipitation with a monoclonal antibody demonstrated that 50% of this fraction was bound to plasmatic butyrylcholinesterase (BChE); c) volatile radioactivity in the form of [^{11}C]-CO_2, which direct quantitation in respiratory gasses confirmed as the major metabolite of [^{11}C]-PHY.

After a bolus i.v. injection of 15-20 mCi [^{11}C]-HEP, the maximal uptake of radioactivity in the brain was reached at 2-3 min, and was followed by a gradual decrease. Peak uptake was 33% of the injected dose per liter (IDL) in the putamen, 28% IDL in the caudate nucleus, 27% IDL in the cerebellum, and 24% IDL in the cerebral cortex. The washout of radioactivity was slower in the caudate and putamen (half-time 38 min) than in the cortex (20 min) and in the cerebellum (15 min). Competition with cold PHY reduced radioactivity concentrations in a dose and region-dependent manner. The inhibition of uptake was not modified by increasing the dose of cold PHY from 300 to 500 μg/kg/2 hr. Thus, the difference between the control and the 500 μg/kg/2 hr dose competition experiments was used to estimate specific [^{11}C]-PHY binding, which was then normalized by free plasmatic radioactivity (Figure 1). At 25 min, normalized specific binding was 6.9 in the putamen, 6.3 in the caudate nucleus, 2.5 in the cerebellum, and 1.0 in the frontal cortex and white matter, values that reflect the known relative regional concentrations of AChE.

Injection of iso-OMPA, a specific inhibitor of BChE, increased the overall cerebral uptake of [^{11}C]-PHY but had no effect on the striatum to cortex ratio (Figure 2). Since BChE concentrations are higher than AChE concentrations

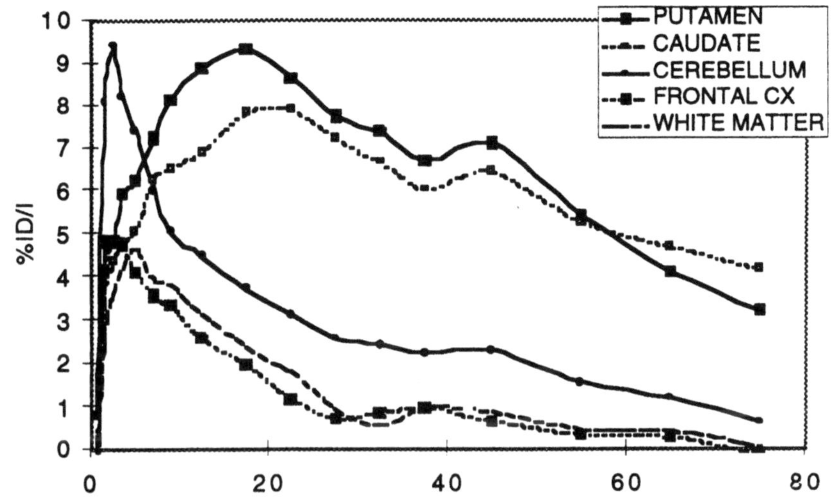

FIGURE 1. Kinetics of [^{11}C] PHY binding in brain regions: "specific," i.e. radioactivity concentrations in control minus saturation experiments. RIGHT: "normalized" binding, i.e. "specific" binding normalized by free plasmatic radioactivity.

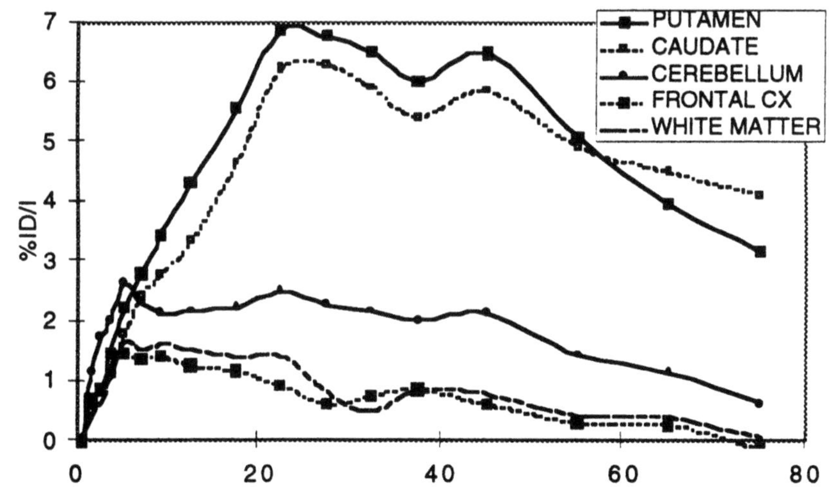

FIGURE 2. Kinetics of [^{11}C] PHY binding in brain regions: "normalized" binding, i.e. "specific" binding normalized by free plasmatic radioactivity.

in the cortex but much lower in the striatum, it is likely that [^{11}C]-PHY binding to BChE *in vivo* is negligible compared to its binding to AChE.

In contrast, pre-injection with tacrine (DOSE) reduced the uptake of [^{11}C]PHY to levels similar to those seen under saturation with high doses of cold PHY, suggesting that tacrine has an inhibitory effect on AChE *in vivo*.

[^{11}C]-HEPTYLPHYSOSTIGMINE

Heptylphysostigmine (HEP) is a cholinesterase (ChE) inhibitor currently under clinical trials for the therapy of Alzheimer's disease (Sramek et al., 1995).

Like PHY, HEP inhibits ChE by carbamylation of the active site, but decarbamylation of ChE is slower than with PHY. IC50 of HEP for AChE is higher than that for BChE, while PHY inhibits preferentially AChE (Pacheco et al., 1995). HEP was labeled with ^{11}C in the same position as PHY (Crouzel et al., 1995).

After a bolus i.v. injection of 15-20 mCi [^{11}C]HEP, the maximal uptake of radioactivity was reached at 10-15 min depending on the brain regions, and was followed by a gradual decrease. Peak uptake was 33% of the injected dose per liter (IDL) in the striatum, 31% IDL in the thalamic area, 30% IDL in the cerebellum, and 24% IDL in the cerebral cortex. The decrease in radioactivity concentrations was slower in the cortex (half-time 85 min) than in the striatum (63 min) and in the cerebellum (60 min).

Iso-OMPA, a specific inhibitor of BChE, increased the striatum to cortex ratio by 25% for peak uptake of [^{11}C]-HEP (Figure 3). This can be interpreted as a consequence of a higher BChE to AChE ratio in the cortex than in the striatum. In contrast, the muscarinic antagonist dexetimide had no effect on [^{11}C]-HEP cerebral binding, suggesting that HEP binding to cerebral muscarinic receptors is negligible *in vivo*.

These results suggest that [^{11}C]-HEP is a valuable PET tracer of ChE, both AChE and BChE, in the primate brain. Its higher lipophilicity and longer duration of residence in cerebral tissue may offer an advantage over

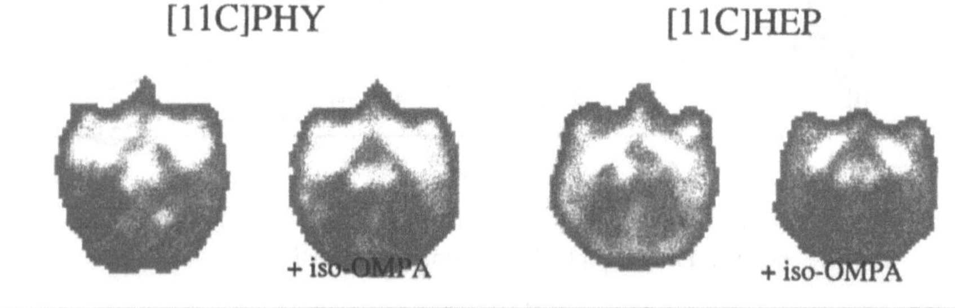

FIGURE 3. Comparison of selective BChE inhibition on the cerebral uptake of [^{11}C]PHY and [^{11}C]HEP, showing a clear reduction of cortical radioactivity under iso-OMPA for [^{11}C]HEP but not for [^{11}C]PHY.

[^{11}C]-PHY for *in vivo* imaging of these enzymes, especially in situations where the involvement of BChE is suspected, such as AD (Mesulam, 1994).

CONCLUSION

[^{11}C]PHY's binding characteristics in non-human primates provide a basis for *in vivo* measurement of brain AChE active sites. [^{11}C]HEP appears to localize both to BChE and AChE active sites. [^{11}C]PHY and [^{11}C]HEP could be useful to study the changes in cholinesterase activity during the course of neurodegenerative disorders such as Alzheimer's disease, or after pharmacological treatment with anticholinesterasic drugs, such as tacrine.

REFERENCES

Bonnot S, Crouzel C, Prenant C and Hinnen F (1993): Carbon-11 labelling of an inhibitor of acetylcholinesterase: [^{11}C] physostigmine. *J. Label. Comp. Radiopharm.* 33(4):277-284.

Crouzel C, Hinnen F and Maitre E (1995): Radiosynthesis of methyl and heptyl [^{11}C]isocyanates from [^{11}C] phosgene, application to the synthesis of carbamates: [^{11}C]physostigmine and [^{11}C]heptylphysostigmine. *Appl. Rad. Isotop.* 46:167-170.

Davies P (1979): Neurotransmitter-related enzymes in senile dementia of the Alzheimer type. *Brain Res.* 171:319-327.

Foster N (1994): PET imaging. In: *Alzheimer Disease*, Terry RD, Katzman R and Bick KL, eds. New York: Raven Press, pp. 87-103.

Giacobini E and Becker R (1994): Development of drugs for Alzheimer therapy: a decade of progress, in: *Alzheimer disease: therapeutic strategies*, eds. E. Giacobini and R. Becker (Birkhäuser, Boston) p. 1-7.

Main AR and Hastings FL (1966): Carbamylation and binding constants for the inhibition of acetylcholinesterase by physostigmine (eserine). *Science* 154:400-402.

Mesulam MM (1994): Butyrylcholinesterase in Alzheimer's disease, in: *Alzheimer disease: therapeutic strategies*, eds. E. Giacobini and R. Becker (Birkhäuser, Boston) p. 79-83.

Pacheco G, Palacios-Esquivel R and Moss D (1995): Cholinesterase inhibitors proposed for treating dementia in Alzheimer's disease: selectivity toward human brain acetylcholinesterase compared with butyrylcholinesterase. *J. Pharmacol. Exp. Therap.* 274:767-770.

Perry EK (1987): Cortical neurotransmitter chemistry in Alzheimer's disease. In: *Psychopharmacology: the Third Generation of Progress*, Meltzer HY, ed. New York: Raven Press, pp. 887-895.

Planas AM, Crouzel C, Hinnen F, Jobert A, Né F, DiGiamberardino L and Tavitian B (1994): Rat brain acetylcholinesterase visualized with [^{11}C]physostigmine. *NeuroImage* 1:173-180.

Sramek JJ,. Block GA, Reines SA, Sawin SF, Barchowsky A and Cutler NA, (1995): A multiple-dose safety trial of eptastigmine in Alzheimer's disease, with pharmacodynamic observation of red blood cell cholinesterase. *Life Sci.* 56:319-

326.

Tavitian B, Pappata S, Planas AM, Jobert A, Bonnot-Lours S, Crouzel C and Di Giamberardino L (1993): *In vivo* visualization of acetylcholinesterase with positron emission tomography. *NeuroReport* 4(5):535-538.

Alzheimer Disease: From Molecular Biology to Therapy
edited by R. Becker and E. Giacobini

SPECT SCAN AND EFFICACY OF THERAPY IN ALZHEIMER'S DISEASE

Jacques Darcourt, Octave Migneco
Department of Nuclear Medicine, Centre Antoine Lacassagne, Université de Nice-Sophia Antipolis

Philippe Robert, Michel Benoit
Department of Psychiatry, Université de Nice-Sophia Antipolis

INTRODUCTION

Recent pharmacological developments in Alzheimer's disease (AD) imply early diagnosis and require tools to assess their efficacy. Beside conventional evaluations, neuroimaging should play a role in providing objective quantitative data for both early diagnosis and patient follow-up. Structural imaging techniques such as MRI have been proposed as sensitive techniques measuring hippocampal atrophy. However, atrophy is related to neuronal loss and is most likely a late feature of the disease. Moreover, the Consortium to Establish a Registry for Alzheimer's Disease (CERAD) did not find a satisfactory interrater agreement for interpreting MRI findings in elderly patients despite careful methodology and readings by experienced neuroradiologists (Davis et al., 1992). Emission tomography imaging has the advantage of providing information on cortical dysfunction which precedes neuronal death and is more likely to correspond to the targets of new drugs. Positron emission tomography (PET) provides information on regional cerebral blood flow (rCBF), metabolism and receptor densities. However, the use of PET is restricted to few centers due to its high cost. Single photon emission computed tomography (SPECT) is widely available and can be used for a large scale patient evaluation. Although various types of tracers have been developed, perfusion tracers are most commonly available and used as routine clinical tools. This paper focuses on the use of perfusion SPECT tracers for the assessment of treatment efficacy on AD.

Several studies have been published using perfusion SPECT with this goal and are analyzed in Robert et al., (1994). With various designs and methodologies, they show a measurable effect of drugs on perfusion SPECT scans which tend to prove that this is an appropriate tool for evaluation of treatment efficacy in this disease (Battistin et al., 1989; Geaney et al., 1990; Hunter et al., 1991; Agnoli et al., 1992; Ebmeir et al., 1992; Minthon et al., 1993). This paper reviews the basic assumptions underlying perfusion SPECT studies from this perspective in order to clarify the advantages and limitations of the technique.

TRACER UPTAKE AND rCBF

The first assumption is that the regional measurements of regional tracer concentrations are a measurement of rCBF. This needs to be discussed with respect to the uptake mechanisms of these tracers. All of them have first to cross the normal blood brain barrier (BBB). Due to the endothelial cell tight junctions, only lipophilic molecules (low molecular weight < 500 Dalton and neutral) can cross the BBB from the blood towards the brain tissue. No modification occurs within the brain in the case of a diffusible tracer like Xenon-133 and it crosses back to the venous blood according to rCBF. Microsphere-like tracers like technetium-99m exametazime (HMPAO, Ceretec®, Amersham) or technetium-99m bicisate (ECD, Neurolite®, Du Pont Pharma) undergo metabolic changes in which they lose their lipophilicity. Therefore, back-diffusion through the BBB is impossible and they remain in the brain tissue. The question is whether or not the uptake of these tracers depends only on rCBF. In theory, it depends on both rCBF and extraction fraction (E), and E depends on the permeability surface product (PS product) which accounts for the membrane interference in the diffusion process. However, in practice the PS product is high for normal BBB, the extraction fraction of highly lipophilic molecules is 100% and the tracer uptake only depends on rCBF. Furthermore this remains true in AD where no BBB abnormalities were found. Therefore, Xenon-133 diffusion rate through the brain is directly related to blood flow and is the basis of the reference method for absolute rCBF measurement when one has the capability of very fast dynamic SPECT acquisitions.

Concerning microsphere-like tracers, the trapping mechanism should also be considered since the delayed brain uptake would reflect brain perfusion at the moment of injection only if no back-diffusion occurred in the meantime. This is the case at low flow rates, but significant back-diffusion occurs above 50 ml/min/100g and the brain uptake is no longer linearly proportional to rCBF with underestimation of high flows. This phenomenon is, however, not significant for AD studies where rCBF is diminished. Although the mechanisms of brain uptake are complex, it is fair to say that local perfusion SPECT tracer concentrations in AD patients are proportional to rCBF. However, only Xenon-133 dynamic studies lead to **absolute** measures in ml/min/100g. Microsphere-like tracers labelled with technetium-99m have the

	Changes in ml/min/100g				Corticocerebellar ratios changes			
ROIs	m	sd	r	p	m	sd	r	p
rGTI	3,82	5,10	0,53	**	0,04	0,09	0,49	**
lGTI	3,82	4,86	0,34	NS	0,03	0,10	0,20	NS
rGTS	3,95	5,08	0,57	**	0,03	0,10	0,46	**
lGTS	3,52	4,28	0,55	**	0,02	0,09	0,38	*
r19	3,38	3,81	0,44	*	0,03	0,09	0,22	NS
l19	4,18	3,33	0,34	NS	0,04	0,08	0,12	NS
rTP	3,66	5,07	0,51	**	0,03	0,12	0,32	NS
lTP	3,95	4,04	0,34	NS	0,03	0,11	0,15	NS
rPrF	4,01	5,23	0,36	NS	0,03	0,12	0,21	NS
lPrF	4,16	5,11	0,34	NS	0,03	0,10	0,24	NS
All ROIs	3,67	3,85	0,47	*	0,02	0,09	0,29	NS

TABLE I. Changes during one year measured by SPECT in 22 AD patients (average MMS worsening: 4.2 (sd=4.8)): comparison between corticocerebellar ratios and absolute measurements in ml/min/100g. r: right; l: left; GTI: internal temporal gyrus; GTS: superior temporal gyrus; 19: Brodmann's area 19; TP: temporoparietal region; PrF: prefrontal cortex; m: mean; sd: standard deviation; r: correlation coefficient with MMSE changes; p: NS=not significant, * $p<0.05$, ** $p<0.01$.

advantage of a stable brain concentration which allows injection apart from the camera room and long acquisitions producing high resolution images, but have the drawback of producing only a **relative** evaluation of rCBF. Therefore, image reading as well as image quantification has to be done by comparison to an internal reference where rCBF is supposed to be normal (the cerebellum or the calcarine cortex are most often used in AD). This becomes a major drawback when follow-up studies are compared since one has to assume that the reference area's rCBF remained unchanged. We have tested a methodology to convert the measured exametazine concentrations using the data of a Xenon-133 SPECT acquired simultaneously into absolute values of rCBF in ml/min/100g (Darcourt et al., 1993a). This methodology was applied to study 22 AD patients twice at a one year interval. The clinical evolution was measured by the MMSE score and compared to the SPECT evolution. Classical corticocerebellar ratios were significantly less correlated than absolute rCBF changes with MMSE evolution. Absolute measurement of rCBF is probably more appropriate for AD SPECT follow-up and drug efficacy evaluation; however, it implies a close relationship between rCBF and disease progression.

CORTICAL rCBF AND DISEASE PROGRESSION

There are two aspects to this question: (1) is rCBF related to brain metabolism and (2) how is brain metabolism related to the progression of the disease? The first question can be answered positively since various data confirm that rCBF is regulated by regional metabolic activity resulting in a perfusion-metabolism coupling (Raichle et al., 1976). This coupling persists in AD (Frackowiack et al., 1981). The second question is more difficult. Patterns of brain metabolism in AD patients are correlated with cognitive deficits, but the rates of change of these metabolic and cognitive patterns are not so robustly correlated (Rapoport, 1991). Fewer than expected correlations may be due to the slowness of metabolic changes (approximately 1 ml/min/100g drop in rCBF for a one point drop in MMSE in our experience), to non-linear time courses, to compensatory phenomenon («metabolic reserve», plasticity) and/or to the fact that the cortical territories evaluated are only secondarily affected by the disease mainly active in deeper structures such as the hippocampal regions. The relatively low sensitivity to evolution of the technique can probably be improved by a better evaluation of hippocampal region along with the use of cognitive stimulation to test the «brain reserve».

PERFUSION SPECT IMPROVEMENTS

As mentioned earlier, the first improvement will come from the possibility of absolute quantification by merging the information obtained by the two types of tracers: diffusible and stable. The hippocampal regions are deeply located and tracer's concentrations are difficult to measure precisely in these areas due to decrease in spatial resolution, non-uniform attenuation and compton scattering. However, recent developments in tomographic reconstruction reduce the consequences of these physical artefacts on SPECT quantification. The mechanism by which the AD process affects brain metabolism and cognition early in the disease is not agreed upon. It can be hypothesized that synaptic transmission is progressively altered and is temporarily compensated by the engagement of a («metabolic reserve», plasticity). The metabolic changes observed on «resting» SPECT images correspond to areas where this reserve is exhausted, while the territories in jeopardy have a preserved rCBF at rest and can only be detected in conditions testing the reserve («stress conditions»). We studied 7 AD patients and 5 controls under visuospatial stimulation; we showed all AD patients failed to activate the right Brodmann's area 19 which was activated by an average of 14% in controls in favor of a limited cognitive reserve in the AD patients (Darcourt et al., 1993b). We suggest that cognitive stimulation SPECT studies might improve the sensitivity of the technique and give a better evaluation of the drug efficacy in that disease.

CONCLUSION

Methodological cautions are necessary to apply conventional brain perfusion SPECT to drug efficacy evaluation due to the complex relationship between tracer uptake, brain metabolism and the degenerative process. However, previous studies already have shown convincing results and technical developments in SPECT quantification are at hand. We suggest to develop a cognitive stimulation test along with an improved quantitative assessment of the hippocampal region as a tool for early diagnosis and drug efficacy evaluation in AD.

REFERENCES

Agnoli A, Fabbrini G, Fioravanti M and Martucci N (1992): CBF and cognitive evaluation of Alzheimer type patients before and after IMAO-B treatment: a pilot study. *Eur Neuropsychopharmacol* 2: 31-35.

Battistin L, Pizzolato G, Dam M et al. (1989): Single Photon emission computed tomography studies with 99mTc-Hexamethylpropyleneamine Oxime in dementia: effects of acute administration of L-Acetylcarnitine. *Eur Neurol* 29:261-265.

Davis PC, Gray L, Albert M et al. (1992): The Consortium to Establish a Registry for Alzheimer's Disease (CERAD). Reliability of a standardized MRI evaluation of Alzheimer's disease. *Neurology* 42: 1676-1680.

Darcourt J, Cauvin JC, Miller B, and Mena I (1993a): Absolute rCBF calibration of HMPAO SPECT using xenon-133. *Clin Nucl Med* p 928.

Darcourt J, Robert P, Migneco O et al. (1993b): Response to visual cognitive stimulation assessed by tc-99mHMPAO in Alzheimer type dementia. *J Nucl Med* 34:5.

Ebmeier KP, Hunter R, Curran SM et al. (1992): Effects of a single dose of the acetylcholinesterase inhibitor velnacrine on recognition memory and regional cerebral blood flow in Alzheimer's disease. *Psychopharmacol* 108:103-109.

Frackowiack R, Pozzilli C, Legg N et al. (1981): Regional cerebral oxygen supply and utilization in dementia. *Brain* 104:753-778.

Geaney DP, Soper N, Shepstone BJ and Cowen PJ (1990): Effect of central cholinergic stimulation on regional cerebral blood flow in Alzheimer disease. *The Lancet* 335:1484-1487.

Hunter R, Wyper DJ, Patterson J et al. (1991): Cerebral pharmacodynamics of physostigmine in Alzheimer's disase investigated using single-photon computerized tomography. *Br J of Psychiat* 158: 351-357.

Minthon L, Gustafson L, Dalfelt G et al. (1993): Oral tetrahydroamino-acridine treatment of Alzheimer's disease evaluated clinically and by regional cerebral blood flow and EEG. *Dementia* 4:32-42.

Raichle ME, Grubb RL, Eichling JO and Ter-Pogossian MM (1976): Correlation between regional cerebral blood flow and oxidative metabolism: in vivo studies in man. *Arch Neurol* 33: 523-526.

Rapoport S (1991): Positron emission tomography in Alzheimer's disease in relation to pathogenesis: a critical review. *Cerebrovascular and Brain Metabolism Rev* 3: 297-335.

Robert P, Benoit M, Darcourt G et al. (1994): Use of SPECT in early diagnosis and to monitor the effect of drugs in Alzheimer disease. In: *Alzheimer Disease: Therapeutic Strategies*. Giacobini E and Becker R eds. Boston: Birhauser, pp 417-423.

previous studies already have shown convincing results and technical developments in SPECT quantification are at hand. We suggest to develop a cognitive stimulation test along with an improved quantitative assessment of the hippocampal region as a tool for early diagnosis and drug efficacy evaluation in AD.

REFERENCES

[illegible]

[illegible]

[illegible]

[illegible]

[illegible]

[illegible]

[illegible]

[illegible]

[illegible]

[illegible]

[illegible]

Alzheimer Disease: From Molecular Biology to Therapy
edited by R. Becker and E. Giacobini

MRI AND COGNITIVE MARKERS OF PROGRESSION AND RISK OF ALZHEIMER DISEASE

Steven H. Ferris and Mony J. de Leon
Aging and Dementia Research Center,
New York University Medical Center, New York, NY

INTRODUCTION

Antemortem, Alzheimer disease (AD) remains a *clinical* diagnosis because no fully validated biologic markers are available. Consequently, case selection for clinical trials is based on clinical symptoms (impaired cognition, functioning and behavior), coupled with a full history and medical evaluation. Similarly, the demonstration of treatment efficacy in clinical trials requires effects on *clinical* symptoms or progression. While validated cognitive assessments are accepted markers of disease severity and progression, certain magnetic resonance imaging (MRI) measures (e.g., hippocampal atrophy) correlate well with cognitive assessments in aging and AD, and thus potentially could serve as biologic markers of disease progression. Furthermore, both cognitive tests (particularly decline in delayed recall) and hippocampal atrophy indicate greatly increased risk of AD in nondemented elderly individuals. MRI, therefore, may provide an important adjunct to both selection of "at risk" cases and for tracking disease progression in early AD. In this brief review, these conclusions are supported primarily by results from our Center, but they are consistent with reports from a number of other laboratories.

POTENTIAL UTILITY OF MRI FOR MEASURING DISEASE PROGRESSION

A major research goal is to develop therapeutic strategies to slow the progression or delay the onset of AD. However, a critical hurdle in proving a pharmacological effect on progression is the inherent ambiguity in using clinical measures to distinguish between a disease altering effect and a purely symptomatic effect. One way to overcome this hurdle would be to employ a biologic correlate of disease severity that is not subject to a symptomatic effect. While biochemical, electrophysiological, and functional neuroimaging measures might serve as sensitive markers of pharmacologic effects, results using these techniques would also be ambiguous with respect to differentiating symptomatic from progression altering effects. Thus *structural* neuroimaging measures are most promising as biologic markers of disease severity and progression because they are least likely to show pharmacologic reversal.

To be useful as an outcome measure in AD clinical trials, an MRI measure must meet the same criteria required for clinical evaluations. Thus *validity* and *reliability* are essential requirements. To be valid, an MRI measure must be a proven feature or correlate of AD. It must discriminate well between AD patients and nondemented controls, and ideally, be validated neuropathologically. MRI measures of hippocampal atrophy clearly provide a valid AD marker and considerable postmortem evidence confirms the consistent and early vulnerability of the hippocampal region in AD (see de Leon et al., 1996). Reliability of measurement studies (test-retest and inter-observer) at individual research centers have yielded good results (de Leon et al., 1993). Reliability across sites, as would be necessary in a multi-center clinical trial, has not been examined. However, good inter-site reliability is feasible, particularly if all scan evaluations are done by the same research team.

Proven *sensitivity to disease severity and progression* are also essential requirements for a useful marker of AD progression. Cross-sectionally, neuropathologic and neuroimaging evidence clearly demonstrates progressive atrophy of the hippocampus in AD (see de Leon et al., 1995), without "ceiling" or "floor" effects that would limit sensitivity range in clinical trials. Atrophy is measurable even in the earliest stages of AD (de Leon et al., 1989b, 1995) and continues to progress (at least as measured postmortem) during the very late, end stages of the disease (Bobinski et al., 1995).

Sensitivity to longitudinal change is another critical requirement for a marker of progression. If a measure does not show significant change over the interval of the trial, it cannot be sensitive to potential slowing of disease course produced by the treatment. Furthermore, data on change (mean and SD) are needed for power and sample size estimates when designing the trial. Related longitudinal data have been reported for ventricular volume (de Leon et al., 1989a) and width of the medial temporal lobe (Jobst et al., 1994), but not specifically for volume of the hippocampus. Thus, additional longitudinal data is the only information needed in orderfor hippocampal atrophy to be a useful marker of AD progression in clinical trials. Several groups, including the Neuroimaging Laboratory at our Center, are currently obtaining this important information.

EARLY DIAGNOSTIC MARKERS OF AD

It is likely that AD begins years before the underlying pathology is sufficient to produce overt symptoms. Therefore, drugs that slow the progression of AD might also be effective in delaying the onset of clinical symptoms in nondemented elderly destined to develop AD. In designing clinical trials of treatments that may delay conversion to AD, it will be essential to have methods for selecting nondemented subjects who are at high risk for conversion to AD over reasonable study intervals of 1-3 years. An initial group of interest will be elderly cases with mild cognitive impairment (MCI), who have significant memory and related cognitive deficits but lack the degree of functional impairment required to meet current diagnostic criteria for dementia (Smith et al., 1996; Ferris and Kluger, 1996). These individuals have a global staging score of 3 on the Global Deterioration Scale (GDS) or 0.5 on the Clinical Dementia Rating (CDR). Most of these cases convert to dementia within several years (Flicker et al.1992; Ferris et al., 1993). Those who have come to autopsy all have met neuropathologic criteria for AD (Morris et al., 1991). Early diagnostic markers are needed both for selecting "at risk" cases and for measuring decline in these subjects over the study. Recent studies suggest that certain cognitive measures such as verbal delayed recall, and MRI measures such as hippocampal atrophy may serve this purpose. These measures may also prove useful for identification of "at risk" *normal* elderly, a group for which the conversion rate over several years generally is quite low.

Cognitive markers of early AD

Sensitivity to early AD was first suggested by cross-sectional studies by several groups comparing early AD with normal elderly subjects. Tests of recent memory, language, psychomotor speed and orientation discriminated these groups, but delayed recall was the most effective. Similar cross-sectional results were found when comparing early AD to MCI cases, but there was considerable overlap between groups and heterogeneity among the MCI cases. Most compelling have been longitudinal studies of nondemented elderly that show verbal memory, psychomotor and language measures to be the best predictors (Flicker et al., 1991; Ferris et al., 1993; Masur et al., 1994), and high conversion rates (up to 70%) over several years for MCI cases. We found that a cut off score on a test of delayed paragraph recall produced an overall prediction accuracy greater than 90% for conversion from MCI to AD (N = 57, mean follow-up period = 3.8 years). These results suggest that delayed story recall could be used for selecting an "enriched" sample of subjects for clinical trials who are very likely to convert to AD over a three-year study period.

Hippocampal markers of early AD.

Since postmortem studies show that the hippocampus is among the first brain regions affected in AD, it is not surprising that hippocampal atrophy as seen with CT or MRI is an early diagnostic marker of AD. Both subjective lucency ratings (reflecting CSF accumulation in perihippocampal fissures) and quantitative volumetric measurements show that hippocampal atrophy discriminates mild AD patients from normal elderly controls (de Leon et al., 1996). Individuals with MCI are heterogeneous with respect to hippocampal atrophy and presence of this marker in MCI individuals predicts decline to AD within several years with greater than 90% overall accuracy (de Leon et al., 1989b; 1993). Furthermore, the early atrophic changes are anatomically specific to the hippocampus, since atrophy of neighboring regions found in early AD is not present in MCI cases (Convit et al., 1993; de Leon et al., 1985). This result suggests that hippocampal atrophy in MCI cases precedes the more widespread atrophy that accompanies clinically evident dementia.

Prediction of decline in normal aging

Since early hippocampal atrophy is present in about 30% of normal elderly subjects (de Leon et al., 1996), it may presage subsequent cognitive decline. Normal elderly subjects generally do not convert to dementia over follow-up intervals of several years (Flicker et al., 1991; 1993), but those with subjective or volumetric MRI evidence for hippocampal atrophy have lower delayed recall scores (particularly delayed paragraph recall) than subjects without hippocampal atrophy (Golomb et al., 1993; 1994). This MRI correlation with delayed recall was specific to the hippocampus. Furthermore, smaller hippocampal size predicted greater longitudinal decline in delayed paragraph recall after a 3.8 year mean follow-up interval (Golomb, et al., 1996). Additional longitudinal research is needed to determine whether early hippocampal atrophy in normal elderly individuals predicts eventual conversion to dementia.

CONCLUSIONS

MRI measurement of hippocampal atrophy in AD patients should be considered for use as a marker of disease progression in AD clinical trials. Measures of hippocampal atrophy and verbal delayed recall also may be useful for identifying nondemented cases at risk for AD, and for measuring change in clinical trials designed to delay progression to dementia.

ACKNOWLEDGMENTS

Supported in part by NIH grants P30 AG08051, R01 AG12101, RO1 AG13616 and RO1 AG0305. We acknowledge the important contributions of the following investigators: Antonio Convit, Susan De Santi, Charles Flicker,

Ajax George, James Golomb, Alan Kluger, Barry Reisberg, Henry Rusinek and Chaim Tarshish.

REFERENCES

Bobinski M, Wegiel J, Wisniewski HM, Tarnawski M, Reisberg B, Mlodzik B, de Leon MJ and Miller DC (1995): Atrophy of hippocampal formation subdivisions correlates with stage and duration of Alzheimer's disease. *Dementia* 6:205-210.

Convit A, de Leon MJ, Golomb J, George AE, Tarshish CY, Bobinski M, Tsui W, De Santi S, Wegiel J and Wisniewski H (1993): Hippocampal atrophy in early Alzheimer's disease: Anatomic specificity and validation. *Psychiatric Quart* 64:371-387.

de Leon MJ, Convit A, De Santi S, Golomb J, Tarshish C, Rusinek H, Bobinski M, Ince C, Miller D, Wisniewski H and George AE (1995): The hippocampus in aging and Alzheimer's disease. In: *Neuroimaging Clinics of North America*, George AE ed. Philadelphia: W.B. Saunders, pp. 1-17.

de Leon MJ, Convit A, George AE, Golomb J, De Santi S, Tarshish C, Rusinek H, Bobinski M, Ince C, Miller D and Wisniewski H (1996): In vivo structural studies of the hippocampus in normal aging and Alzheimer's Disease. In: *Neurobiology of Alzheimer's Disease*, Wurtman RJ, Corkin S, Growdon JH, Nitsch RM eds. New York: New York Academy of Science, pp. 1-13.

de Leon MJ, George AE, Reisberg B, Ferris SH, Kluger A, Stylopoulos LA, Miller JD, La Regina ME, Chen C and Cohen J (1989a): Alzheimer's disease: Longitudinal CT studies of ventricular change. *Am J Neuroradiol* 10:371-376.

de Leon MJ, George AE, Stylopoulos LA, Smith G and Miller DC (1989b): Early marker for Alzheimer's disease: The atrophic hippocampus. *Lancet* 2: 672-673.

de Leon MJ, Golomb J, George AE, Convit A, Tarshish CY, McRae T, De Santi S, Smith G, Ferris SH, Noz M and Rusinek H (1993): The radiologic prediction of Alzheimer's disease: the atrophic hippocampal formation. *Am J Neuroradiol* 14:897-906.

Ferris SH and Kluger A (1996): Commentary on age-associated memory impairment, age-related cognitive decline and mild cognitive impairment. *Aging and Cognition* 3: in press.

Ferris SH, Kluger A, Golomb de Leon MJ, Flicker C and Reisberg B (1993): Assessment in early detection of age-associated cognitive decline. In: *Proceedings of the Sixth Congress of the International Psychogeriatric Association* Berlin: International Psychogeriatric Association, p. 46.

Flicker C, Ferris SH and Reisberg B (1991): Mild cognitive impairment in the elderly: predictors of dementia. *Neurol* 41:1006-1009.

Flicker C, Ferris SH and Reisberg BA (1993): A two-year longitudinal study of cognitive function in normal aging and Alzheimer's disease. *J Geriat Psychiatry Neurol* 6:84-96.

Golomb J, de Leon MJ, George AE, Kluger A, Convit A, Rusinek H, De Santi S, Foo SH, Litt A and Ferris SH. (1994): Hippocampal atrophy correlates with severe cognitive impairment in elderly patients with suspected normal pressure hydrocephalus. *J Neurol Neurosurg Psychiatry* 57:590-593.

Golomb J, de Leon MJ, Kluger A, George AE, Tarshish C and Ferris SH (1993): Hippocampal atrophy in normal aging: an association with recent memory impairment. *Arch Neurol* 50:967-976.

Golomb J, Kluger A, de Leon MJ, Ferris SH, Mittelman M, Cohen J and George AE (1996): Hippocampal formation size predicts declining memory performance in normal aging. *Neurol*, in press.

Jobst KA, Smith AD, Szatmari M, Esiri MM, Jaskowski A, Hindley N, McDonald B and Molyneux AJ (1994): Rapidly progressing atrophy of medial temporal lobe in Alzheimer's Disease. *Lancet* 343: 829-830.

Masur DM, Sliwinski M, Lipton RB, Blau AD and Crystal HA (1994): Neuropsychological prediction of dementia and the absence of dementia in healthy elderly persons. *Neurol* 44:1427-1432.

Morris JC, McKeel DW, Storandt M, Rubin EH, Price JL, Grant EA, Ball MJ and Berg L (1991): Very mild Alzheimer's disease: Informant-based clinical, psychometric, and pathological distinction from normal aging. *Neurol* 41:469-478.

Smith G, Peterson R and Ivnik R (1996): Course and predictors of outcome of mild cognitive impairment. *Aging and Cognition* 3: in press.

Alzheimer Disease: From Molecular Biology to Therapy
edited by R. Becker and E. Giacobini

BRAIN MAPPING AND TRANSCRANIAL DOPPLER ULTRASONOGRAPHY IN ALZHEIMER DISEASE DRUG MONITORING

Ramón Cacabelos, José Caamaño, Dolores Vinagre,
José I. Lao, Katrin Beyer, Antón Alvarez
EuroEspes Biomedical Research Center, Institute for CNS Disorders,
15166-Bergondo, La Coruña, Spain.

INTRODUCTION

International guidelines recommend for the early diagnosis of Alzheimer disease (AD) to proceed according to the following steps: 1) diagnostic criteria (DSM-IV/NINCDS-ADRDA); 2) clinical evaluation (general, neurologic, psychiatric); 3) biochemical study (blood, urine); 4) neuropsychological assessment (cognitive, functional, behavioral); 5) neurophysiological studies (EEG, brain mapping, evoked potentials); 6) neuroimaging (static (CT-Scan, MRI), dynamic (SPECT, PET); 7) cerebrovascular evaluation (BBF, Transcranial Doppler Ultrasonography); 8) neurochemical studies with potential biological markers; and 9) genetic studies (APOE, STM2, 182s, APP21m); relying the confirmation of the disease on 10) neuropathological studies at a postmortem level (Cacabelos et al., 1995a; 1996). The complete administration of this protocol in specialized centers yields a highly accurate diagnostic rate (>95%) in early stages. Despite the high cost of this procedure and its unlikely use in conventional clinical settings, it seems essential its implementation when trying to include AD patients in clinical protocols to investigate a particular aspect of the disease or when we recruit patients for a drug clinical trial. In terms of clinical assessment for the daily practice, the following steps would be sufficient: (a) clinical assessment, (b) basal biochemical screening, (c) psychometric assessment, and (d) static neuroimaging. However, in the selection of patients for drug clinical trials, genetic testing and functional instrumentation have proved to be very useful, since the therapeutic response of most patients varies according to their genotype, and because biological and functional parameters, together with psychometric measurements, have shown to be more reliable and precise than psychometric parameters alone. Furthermore, in the near future, new drug development strategies should consider to design novel therapeutic interventions at a preventive level in subjects with a demonstrable genetic risk (Cacabelos et al., 1994; 1996); and some compounds in the new generations of anti-dementia drugs would act on the CNS prior to the onset of clinical symptoms. These new types of drugs under development in some laboratories also require a new design in clinical trials for drug evaluation, since most protocols at present are basically useless for studying drug efficacy in early stages of the disease or prior to disease-onset.

In this paper we show the utility of brain mapping (BM) and Transcranial Dopppler Ultrasonography (TDS) as a diagnostic aid, as well as their potential use for drug monitoring in clinical trials with AD patients.

BRAIN MAPPING

Computerized EEG spectral analysis and topographic brain mapping are still not widely used for diagnosis and therapeutic assessment in clinical trials. Brain mapping allows noninvasive, harmless and continuous monitoring of brain functional conditions with high sensitivity for changes of conditions such as psychopathology, medication, several types of spontaneous thought, mental task execution or induced perception (Franco-Maside et al., 1995a). In AD patients, brain mapping shows a reduction in alpha activity, diffuse slowing, outline disorganization, and increase in theta and delta activity (Vinagre et al., 1994). It helps in the differential diagnosis of AD and vascular dementia (VD) where computerized EEG changes consist of a focal increase in theta and delta waves and a decrease in alpha activity in damaged brain areas in a patchy fashion (Vinagre et al., 1994).

We have investigated the potential utility of brain mapping as a diagnostic aid in differentiating AD (N=20; age: 64.01±7.78 yrs; range: 53-83 yrs) from VD (N=20; age: 76.70±7.54 yrs; range: 64-90 yrs), and found that theta activity significantly increased in most brain regions. The relative power of delta activity was found increased in occipital, frontal and temporal regions. AD patients, in general, exhibit an overall decrease in fast-frequency bands, with beta activity decreased in frontal, central, parietal, and temporal areas. Basal alpha attenuation is also observed in most regions. Demented patients showed a significant increase in relative delta and theta power, but not significant differences were observed in delta activity between AD and VD. Patients with VD showed a significant increase in theta activity with respect to AD in frontal, central, and temporal regions. In the fast-frequency bands, alpha activity did not differ in demented patients, whereas beta activity was higher in AD patients in T5 (Vinagre et al., 1994).

We have also studied brain mapping changes associated with disease staging and found a progressive increase in relative delta and theta power from GDS-3 to GDS-7 in parallel with a progressive decrease in relative alpha and beta power, as shown in Table 1.

BRAIN MAPPING IN DEMENTIA
Brain Bioelectrical Activity in Alzheimer Disease by Staging

STAGE	Delta	Theta	Alpha	Beta
GDS-3	9.4(1.8)	18.8(1.4)	52.8(4.6	15.2(3.4)
GDS-4	20.0(4.1	33.9(3.4)	19.3(4.5)	10.1(0.8)
GDS-5	35.1(11.6)	24.6(3.0)	15.1(2.9)	4.8(1.6)
GDS-6	51.0(11.9)	29.6(6.8)	6.3(1.4)	3.6(1.7)
GDS-7	41.9(7.7)	37.9(4.9)	7.5(0.8)	3.1(0.8)

TABLE 1. Brain mapping in AD according to disease progression as evaluated by the Global Deterioration Scale (GDS) (N=10/group). Units: Relative Power (%) expressed as mean(±SD).

In AD patients with the most frequent APOE genotypes in FAST-4 staging we found no differences in relative delta, theta, alpha and beta power among the APOE-3/3, APOE-3/4, and APOE-4/4 groups. In any case, brain bioelectrical activity was never worse in patients with APOE-4/4 than in patients with APOE-3/4 or APOE-3/3 with a similar age or mental deterioration pattern (Table 2).

TRANSCRANIAL DOPPLER ULTRASONOGRAPHY

Transcranial Doppler Ultrasonography (TDS) is a non-invasive technique recently introduced in medicine as a diagnostic and monitoring tool for the evaluation of brain hemodynamic parameters (Caamaño et al., 1993,1994b,1995). Blood flow parameters currently measured with TDS include the mean velocity (Mv), the dyastolic velocity (Dv), the pulsatility index of Gosling (PI=[Sv-Dv/Mv]), the resistance index of Pourcelot (RI=[Sv-Dv/Sv]), the effective pulsatility range of Ries (EPR=Mv-(Sv-Dv)), the pulsatility transmission index (PTI), and the interhemispheric asymmetry index (AI) (Caamaño et al., 1995). We have studied TDS parameters in AD and VD (Caamaño et al.,1992,1993,1994b,1995). In both types of dementia

APOE-RELATED BRAIN MAPPING
Brain Mapping Activity in APOE-Related Alzheimer Disease

Activity	APOE-3/3	APOE-3/4	APOE-4/4
Delta	37.0 (6.8)	31.4(8.8)	36.6(11.3)
Theta	26.5(3.4)	27.8(3.5)	21.7(2.7)
Alpha	18.6(5.3)	20.5(6.6)	25.0(8.7)
Beta	6.0(1.4)	6.6(1.5)	6.1(1.4)

TABLE II. Brain mapping in APOE genotype-related Alzheimer disease (N=10/group). Units: Relative Power (%) expressed as mean(±SD)

BRAIN HEMODYNAMICS

Brain Hemodynamic Parameters in Alzheimer Disease by Staging

Parameters	GDS-3	GDS-4	GDS-5
Mv (cm/sec)	47.38±11.0	43.07±6.6	38.81±8.5
Sv (cm/sec)	72.83±15.1	66.42±9.9	65.54±12.2
Dv (cm/sec)	30.77±8.3	27.65±4.8	23.18±6.6
PI (x±sd)	0.90±0.13	0.90±0.16	1.11±0.25
RI (x±sd)	0.58±0.05	0.57±0.04	0.63±0.08
EPR (x±sd)	5.33±6.7	4.30±6.4	-3.54±10.1
Age (x±sd)	64.56±7.61	68.44±7.40	68.12±9.03
Age range (Yrs)	52-79	56-82	51-85
N	18	26	11

TABLE III. Brain hemodynamic parameters in AD according to disease staging

we have found a significant decrease in blood flow velocities and an increase in peripheral resistances. By individual spectral analysis, 50% of VD patients show disorders in blood flow rhythm, with no major changes in AD (Caamaño et al., 1995). When brain hemodynamic parameters are analyzed by TDS in different AD stages, a clear reduction in Mv, Sv, Dv, and EPR is observed in parallel with an increase in PI and RI from GDS-3 to GDS-5 (Table 3).

CONCLUSIONS

From our own studies concerning brain mapping (Vinagre et al., 1994; Franco et al., 1995a) and TDS (Caamaño et al., 1992, 1993, 1994b, 1994) we can conclude the following: (1) Brain mapping and TDS used independently are useful tools as a diagnostic aid in senile dementia; (2) when used in combination, both brain mapping and TDS help in the differential diagnosis of AD and VD; (3) brain bioelectrical and hemodynamic parameters correlate with psychometric assessment and mental deterioration; (4) these low-cost, non-invasive techniques are sensitive enough to establish phenotypic profiles in specific AD genotypes (Cacabelos, 1996); and (5) both procedures are reliable for drug monitoring in clinical trials (Cacabelos et al., 1995a,b; Franco-Maside et al., 1994,1995b; Caamaño et al., 1994a).

REFERENCES

Caamaño J, Valle-Inclán F, Cacabelos R (1992): Middle cerebral artery blood flow velocity in patients with vascular dementia: Assessment by transcranial Doppler ultrasonography. *Ann Psychiat* 3:273-280.

Caamaño J, Gómez MJ, Cacabelos R (1993): Transcranial Doppler ultrasonography in senile dementia: Neuropsychological correlations. *Meth Find Exp Clin Pharmacol* 15:193-199.

Caamaño J, Gómez MJ, Franco A, Cacabelos R (1994a): Effects of CDP-choline on cognition and cerebral hemodynamics in patients with Alzheimer's disease. *Meth Find Exp Clin Pharmacol* 16:211-218.

Caamaño J, Gómez MJ, Vinagre D, Franco-Maside A, Cacabelos R (1994b): Transcranial Doppler ultrasonography in senile dementia. *Drugs Today* 30:283-293.

Caamaño J, Gómez MJ, Vinagre D, Franco A, Fernández-Novoa L, Alvarez XA, Novo B, Zas R, Cacabelos R (1995): Transcranial Doppler ultrasonography in neurogeriatric disorders. *Neurogerontol Neurogeriat* 1:141-157.

Cacabelos R (1996): Diagnosis of Alzheimer's disease: defining genetic profiles (genotype vs phenotype). *Acta Neurol Scand* (in press).

Cacabelos R, Caamaño J, Alvarez XA, Gómez MJ, Franco-Maside A, Vinagre D, Fernández-Novoa L, Lao JI, Beyer K, Recuero M, Sastre I, Bellido MJ, Valdivieso F, Vigo-Pelfrey C (1995a): Diagnostic criteria and genotyping in Alzheimer's disease. *Neurogerontol Neurogeriat* 1:329-353.

Cacabelos R, Lao JI, Beyer K, Alvarez XA, Franco-Maside A (1996): Genetic testing in Alzheimer's disease: ApoE genotyping and etiopathogenic factors. *Meth Find Exp Clin Pharmacol* 18 (Suppl. A):161-179.

Cacabelos R, Norberg A, Caamaño J, Franco-Maside A, Fernández-Novoa L, Gómez MJ, Alvarez XA, Takeda M, Prous J, Nishimura T, Winblad B (1994): Molecular strategies for the first generations of anti-dementia drugs (I) Tacrine and related compounds. *Drugs Today* 30:295-337.

Cacabelos R, Caamaño J, Gómez MJ, Fernández-Novoa L, Franco-Maside A, Vinagre D, Novo B, Zas R, Alvarez XA (1995b): Treatment of Alzheimer's disease with CDP-choline: Effects on mental performance, brain electrical activity, cerebrovascular parameters and cytokine production. *Ann Psychiat* 5:295-315.

Franco-Maside A, Caamaño J, Gómez MJ, Cacabelos R (1994): Brain mapping activity and mental performance after chronic treatment with CDP-choline in Alzheimer's disease. *Meth Find Exp Clin Pharmacol* 16:597-607.

Franco-Maside A, Vinagre D, Caamaño J, Alvarez XA, Fernández-Novoa L, Novo B, Zas R, Cacabelos R (1995a): Brain mapping in neurogeriatrics. *Neurogerontol Neurogeriat* 1:119-139.

Franco-Maside A, Vinagre D, Caamaño J, Gómez MJ, Alvarez XA, Fernández-Novoa L, Zas R, Novo B, Polo E, García M, Castagne I, Newman E, Guez D, Cacabelos R (1995b): Effects of S12024 on brain electrical activity parameters in patients with Alzheimer's disease. *Ann Psychiat* 5:281-294.

Vinagre D, Franco-Maside A, Gómez MJ, Caamaño J, Cacabelos R (1994): Brain electrical activity mapping in cerebrovascular disorders and senile dementia. *Drugs Today* 30:275-281.

Caamaño J, Gómez MJ, Franco A, Cacabelos R (1994a) Effects of CDP-choline on cognition and cerebral hemodynamics in patients with Alzheimer's disease. Meth Find Exp Clin Pharmacol 16:211-218.
Caamaño J, Gómez MJ, Vinuegre D, Franco-Maside A, Cacabelos R (1994b) Transcranial Doppler ultrasonography in senile dementia. [illegible]
[illegible]

Alzheimer Disease: From Molecular Biology to Therapy
edited by R. Becker and E. Giacobini

APOE GENOTYPE AND MRI VOLUMETRY: IMPLICATION FOR THERAPY

Hilkka S. Soininen, Maarit Lehtovirta, Mikko P. Laakso, Kaarina Partanen, Paavo Riekkinen Jr, Merja Hallikainen, Tuomo Hänninen, Keijo Koivisto and Paavo J. Riekkinen Sr.
Departments of Neurology and Radiology, Kuopio University and University Hospital, Kuopio, Finland

INTRODUCTION

Alzheimer's disease (AD) is a heterogeneous entity. This may result in differences in patient response to various drug treatments. One of the factors that might contribute to the heterogeneity of AD is the apolipoprotein E (ApoE) genotype. ApoE allele E4 is a well established risk factor for AD (Strittmatter et al., 1993; Lehtovirta et al., 1995). ApoE is a plasma protein involved in the transport of cholesterol and other lipids. ApoE is implicated in the growth and regeneration of nerves during development and following injury (Poirier, 1994). ApoE is present in senile plaques, neurofibrillary tangles, and cerebrovascular amyloid in AD. Different binding properties of ApoE isoforms to amyloid ß-protein (Aß) and tau protein also suggest that ApoE might be involved in the pathogenesis of AD (Strittmatter et al., 1994).

AD most commonly initially presents with memory loss followed by other cognitive dysfunctions. Recent magnetic resonance imaging (MRI) studies (Laakso et al., 1995) have indicated a pronounced decline in volume of the hippocampus as an early sign of AD. Here, we present data from MRI volumetric studies. First, we tested the hypothesis that AD patients carrying the E4 allele have a more severe hippocampal damage and more severe memory impairment than AD patients without E4. Second, we studied whether nondemented elderly subjects carrying the E4 allele show more severe hippocampal damage than those without E4. Third, we examined whether the response to an acute dosage of tetrahydroaminoacridine (THA) in AD patients was influenced by hippocampal volume.

SUBJECTS AND METHODS

1. We studied 26 patients fulfilling the NINCDS-ADRDA criteria of probable AD and 16 controls. The AD patients were in the early stage of the disease. The patients and controls underwent a thorough evaluation (Lehtovirta et al., 1995). The clinical severity was assessed using Mini-Mental Status examination (MMSE) and neuropsychological tests.
2. We measured hippocampal volumes in 32 healthy elderly individuals, 22 women, 10 men, mean age 69±6 years as described previously (Soininen et al., 1995a). Sixteen of the subjects fulfilled the criteria for age associated memory impairment (AAMI) and 16 had intact memory. None of the subjects fulfilled criteria for dementia.
3. We tested 17 patients fulfilling the criteria for probable AD in an acute THA trial. In this study, memory, visuoconstructive, executive and vigilance functions were assessed after administration of placebo and THA (per os one session 25 mg, one session 75 mg) (Riekkinen Jr et al, 1995).

Neuropsychological tests used in these studies have been reported earlier (Lehtovirta et al., 1995; Soininen et al., 1995a; Riekkinen Jr et al., 1995). Memory functions were examined with a list learning test, Heaton Visual Reproduction Test, and Benton visual memory test. The delayed recall was tested after a 30-minute delay.

The subjects were scanned with a 1.5 T Magnetom. The method has been described earlier in detail (Soininen et al., 1994; Laakso et al., 1995). We used volumes normalized for brain area in all statistical analyses.

The ApoE genotype was determined using DNA extraction and PCR amplification from venous EDTA-blood as described earlier (Lehtovirta et al., 1995).

RESULTS

Hippocampal volumes and ApoE in Alzheimer patients

The AD patients did not differ in age or sex from controls (Table I). The AD subgroups did not differ in age, sex, age at onset, education, or MMSE scores. ANOVA over the controls and AD subgroups showed a significant difference in the volumes of the right and left hippocampus ($p<0.0001$). The controls had larger hippocampi than AD subgroups. The AD 4/4 subjects displayed the most pronounced volume loss. Their right hippocampal volumer were significantly smaller (-54% of control) than all the other study groups (Table I).

The performance of AD subgroups on memory tests and other cognitive tests was significantly impaired compared to controls. The AD E4/4 patients had the lowest scores on delayed memory tests. The AD E4/4 group differed significantly from AD patients with E3/3 in delayed recognition of learned words (Duncan, $p<0.05$). Deficits in other cognitive domains were comparable between AD subgroups (Lehtovirta et al., 1995).

Hippocampal volumes and ApoE in nondemented elderly

In 32 nondemented elderly, we analyzed the MRI data for three groups: subjects with 2 E4 alleles (N=4), 1 E4 (N=10), and those without E4 (N=18). The groups did not differ in age, sex, education, or performance in an extensive battery of psychometric tests except Benton visual memory scores; the subjects without E4 allele had significantly higher scores than those with 1 E4 allele (ANOVA/Duncan, $p<0.05$) (Soininen et al., 1995a).

The hippocampal volumes did not differ between the study groups, but the volume difference between the right and left hippocampus differed significantly, ANOVA, $p<0.0001$. The Duncan post hoc analysis showed that the 2 E4 group differed from the two other groups, as did the 1 E4 group ($p<0.01$). The scatterplot of the right - left hippocampal volume difference showed that in all subjects without E4, the right hippocampus was larger than the left. In contrast, all subjects with 2 E4 alleles had their left hippocampus larger than the right.

Acute THA trial

This experiment was designed to investigate the hypothesis that marked hippocampal pathology may decrease the therapeutic effects of THA on memory functioning in AD. Eight patients performed better during treatment with THA 75 mg than during placebo or THA 25 mg. These responders performed better during baseline examination in tests assessing executive and declarative memory functions, and had higher MMSE scores and also larger hippocampi than nonresponders (Riekkinen Jr et al., 1995).

	Controls N=16	AD 4/4 N=5	AD 3/4 N=9	AD 3/3,3/2 N=12	P
W/M	10/6	3/2	4/5	5/7	
Age, y	70.2±4.7	65.0±11.4	71.6±7.6	68.5±8.7	
Age at onset, y	-	62.2±11.3	70.0±8.2	66.2±9.0	
MMSE	28.6±1.4 [a]	22.6±3.0	21.1±4.6	23.6±3.6	*
RH	3.71±0.11 [a]	1.69±0.55 [a]	2.41±0.50	2.53±0.56	*
LH	3.35±0.11 [a]	1.84± 0.41	2.04±0.49	2.20±0.70	*
Delayed memory					
List learning	10.0± 0.0 [a]	3.7±3.6 [b]	5.2±4.0	7.3±2.6	*
Heaton	11.8± 3.6 [a]	0.6±0.9	1.0±1.4	2.2±2.8	*

TABLE I. Clinical data, hippocampal volumes (RH, LH, cm^3) and memory scores in AD patients of different ApoE genotypes. Values are expressed as mean±S.D. ANOVA over the study groups: * $p<0.0001$. Duncan ($p<0.05$). [a] differs from all other groups; [b] AD 4/4 differs from controls and AD 3/3.

DISCUSSION

We hypothesized that AD patients carrying the E4 allele might have more prominent shrinkage of the hippocampus and a more severe memory loss than AD patients without E4 allele, despite equal global severity of dementia. Indeed, we found that the AD E4/4 patients had the most prominent volume loss in the hippocampus and also the lowest scores on tests assessing long-term verbal and visual memory. Otherwise, the profile of cognitive dysfunctions did not differ between AD subgroups with different ApoE genotypes.

Previous studies have indicated the role of medial temporal lobe structures, particularly of the hippocampus, in the processing of certain type of memory (Squire et al., 1992). In AD patients, reduced hippocampal volumes have also been reported to correlate with low scores in memory tests (Laakso et al., 1995).

Our data showed that nondemented elderly individuals carrying the E4 allele, particularly subjects homozygous for E4, may have minor damage in the hippocampus. Data from a population based study (Helkala et al, 1995) supported this finding and showed that elderly subjects without the E4 allele had better learning abilities than subjects carrying E4. Our finding suggests that identification of subjects with E4/4 genotype among the elderly suffering from memory impairment may help to find individuals at high risk for dementia and thus outline a possible target group for preventive procedures if they become available in the future.

The acute THA trial suggested that severe hippocampal atrophy may block the memory improving effect of THA. In that experiment, the small sample size did not allow the analysis of effects of ApoE genotype to patient response. However, it is possible that especially in patients carrying E4, the improving effect of THA on memory function is lacking due to severe hippocampal atrophy.

How ApoE mediates its action in AD is unknown. Increased accumulation of Aβ in brains of AD patients carrying the E4 allele (Schmechel et al., 1993) might be an explanation for the prominent atrophy of the hippocampus. Furthermore, Strittmatter et al. (1994) hypothesized that ApoE3 binds to tau protein, whereas ApoE4 does not bind to tau, and therefore indirectly promotes phosphorylation of tau and tangle formation.

Experimental studies suggest that ApoE is involved in synaptogenesis (Poirier, 1994). It has been proposed that cholesterol released during terminal breakdown is transported by the ApoE transport system to neurons undergoing reinnervation. Given the integral part of cholesterol and other lipoprotein transport in the brain in synaptogenesis, it is possible that AD patients differing in ApoE phenotype also have different capacities for synaptogenesis. During memory processing, strong synaptogenesis and reorganization normally take place in the hippocampus. Therefore, it is possible that ApoE isoforms influence synaptogenesis differently, causing differences in volume loss in AD patients with distinct ApoE phenotypes.

Recent data from two post mortem studies have suggested that AD patients carrying the E4 allele also have a more pronounced cholinergic deficit in the frontal cortex (Soininen et al., 1995b), temporal cortex and hippocampus (Poirier et al., 1994). ApoE may be of particular importance for the cholinergic system, which relies to a certain extent on the integrity of phospholipid homeostasis in neurons (Wurtman, 1992).

In conclusion, our data suggest that the ApoE E4 allele is associated with severe hippocampal atrophy in AD. Minor hippocampal changes can also be detected in nondemented elderly, particularly those with an E4/4 genotype. Moreover, we have found that AD patients who do not respond to THA treatment have significantly smaller hippocampi than THA-responders. These data suggest that the ApoE genotype may modify the severity of medial temporal lobe atrophy. It is of interest to study whether the ApoE genotype will also affect the response of AD patients to cholinesterase inhibitors. Thus, the ApoE polymorphism seems to be a major contributor to the heterogeneity of AD that may also have implications for therapeutic approaches in this disorder.

ACKNOWLEDGMENTS

The study was supported by the Medical Research Council of the Academy of Finland.

REFERENCES

Helkala EL, Koivisto K, Hänninen T et al. (1995): The association of apolipoprotein E with memory: a population based study. *Neurosci Lett* 1991:141-144.

Laakso MP, Soininen H, Partanen K et al. (1995): Volumes of hippocampus, amygdala and frontal lobes in the MRI-based diagnosis of early Alzheimer's disease: correlation with memory functions. *J Neural Transm* [P-D Sect] 9:73-86.

Lehtovirta M, Laakso MP, Soininen H et al. (1995): Volumes of hippocampus, amygdala, and frontal lobe in Alzheimer patients with different apolipoprotein E genotypes. *Neuroscience* 67:65-72.

Poirier J (1994): Apolipoprotein E in animal models of CNS injury and Alzheimer's disease. *Trends Neurosci* 17:525-530.

Riekkinen P Jr, Soininen H, Helkala E-L et al. (1995): Hippocampal atrophy, acute THA treatment and memory in Alzheimer's disease. *NeuroRepo*rt 6:1297-1300.

Schmechel DE, Saunders AM, Strittmatter WJ et al. (1993): Increased amyloid β-peptide deposition as a consequence of apolipoprotein E genotype in late-onset Alzheimer's disease. *Proc Natl Acad Sci USA* 90, 9649-9653.

Soininen H, Partanen K, Pitkänen A et al. (1995a): Decreased hippocampal volume asymmetry on MRI scans in nondemented elderly subjects carrying apolipoprotein E ε 4 allele. Neurology 45:391-392.

Soininen H, Partanen K, Pitkänen A et al. (1994): Volumetric MRI analysis of the amygdala and the hippocampus in subjects with age-associated memory impairment: correlation to visual and verbal memory. *Neurology* 44: 1660-1668.

Soininen H, Kosunen O, Helisalmi S et al. (1995b): A severe loss of choline acetyltransferase in the frontal cortex of patients carrying apolipoprotein E ε4 allele. *Neurosci Lett* 187:79-82.

Squire LR, Ojeman JG, Miezin FM et al. (1992): Activation of the hippocampus in normal humans: A functional anatomical study of memory. *Proc Natl Acad Sci USA* 89, 1837-1841.

Strittmatter WJ, Saunders AM, Schmechel D et al. (1993): Apolipoprotein E: High-avidity binding to β-amyloid and increased frequency of type 4 allele in late-onset familial Alzheimer's disease. *Proc Natl Acad Sci* USA 90, 1977-1981.

Strittmatter WJ, Weisgraber KH, Goedert M et al. (1994): Hypothesis: Microtubule instability and paired helical filament formation in the Alzheimer disease brain are related to apolipoprotein E genotype. *Exp Neurol* 125, 163-171.

Wurtman RJ (1992): Choline metabolism as a basis for the selective vulnerability of cholinergic neurons. *Trends Neurosci* 15: 117-122.

Alzheimer Disease: From Molecular Biology to Therapy
edited by R. Becker and E. Giacobini

GROUP PSYCHOTHERAPY

José Guimón
Department of Psychiatry, Geneva University Hospital,
Geneva, Switzerland

Elisabeth Basaguren
Fundación OMIE Bilbao, Spain

AIMS OF GROUPS IN PATIENTS WITH ALZHEIMER'S DISEASE

With patients who have different degrees of Alzheimer's disease (AD), the aim of group activities may range from simple " group work " to the most sophisticated dynamic group psychotherapy (Zilka, 1996). Groups which we have organized in Bilbao and Barcelona (Guimón, 1992) within the framework of institutions such as geriatric or psychogeriatric hospitals, old people's homes, hostels, etc., through strengthening the socialization of the patient, generally aim for an increase in the number of discharges, an improvement in aggressive behaviour, a better quality of life and sometimes, albeit rarely, for psychodynamic psychotherapy.

Groups which operate on an out-patient basis in day hospitals or centers, out-patient consultations etc., aim to achieve a better adjustment of the patient to his environment through various activities which provide advice, information and education. Sometimes techniques which allow the ventilation of emotions are used and sometimes, psychodynamic psychotherapy is also offered, as Zilka (1996) did in our department. Although patients with AD show large variations in their way of life (socio-economic conditions, previous successes, presence or absence of family, vulnerability to illnesses, etc.), group members help each other to complain less, to maintain relationships with one another and to reduce tensions.

TECHNICAL VARIANTS

There are techniques which perform "group work" which is oriented to carrying out various activities (music therapy, theatre or dance) and others which have a more psychoeducational aim.

As regards the approaches which involve more extensive psychotherapeutic aims, groups are, above all, based on open discussion. The main themes during the sessions are to do with physical complaints, idealization of the past, economic concerns, loneliness, fantasies and issues of rejection and lack of social recognition.

Psychotherapeutic programs often include varied activities combined with group sessions, techniques for the "development of the human potential", mutual help groups, etc.

Berland et al. (1979) brought together dynamically oriented groups with interpretation of the transference in a private home with retired patients; he underlines that enrichment of the personality continues throughout old age through elaboration of the experiences of death and loss. Johnson (1985), for his part, brought together groups which analyze transference by using presentations of some themes which regularly appeared during the sessions such as the closeness of death or somatic complaints.

Butler (1963, 1968) is really the author who has played the most active part in the development of "life review" techniques in elderly patients. Poulton and Strassberg (1986) pointed out that the use of reminiscences promotes the action of classical healing factors described by Yalom (1983) in group psychotherapy. Lescz et al. (1990) also conducted groups using active dynamic techniques to improve the self-esteem of elderly people by means of life review and he introduced approaches from psychology of the self and cognitive behaviour. It should be noted, however, that the use of these techniques requires good intellectual functioning.

As we have pointed out above, group activities are most effective when they are used within the framework of "milieu therapy" (Krassoievitch, 1993).

THE ROLE OF THE FACILITATOR

In groups which operate in institutions and, in general, in most educational groups based on different activities, the therapist plays an active role. He must provide comfort, explain, confront, desensitize, point out, negotiate, heal, decide. In exchange, in truly psychodynamic groups, particularly those which are dynamically oriented, he has a more passive role; he must be content to listen, to "reflect" and, rarely, to analyze. But this is rare in groups held with patients suffering from AD even in the early phases of the evolution.

The facilitator strives for improvement in social relations and to play down fears and dangers. He attempts, where possible, to "instill joy" into his groups. Many behaviours of a patient with AD, which outwardly appear to have no meaning, may be well understood if one regards the patient as much more frightened and furious than demented, and if one regards his reactions more as attempts to solve his problems than as evidence of his desire to create other problems.

The group facilitator is regarded as a father substitute, even if he is younger than the patients, and demented patients, thus, have a greater tendency to depend on him than younger patients.

The existence of groups helps the staff to think that they are performing a useful activity in an atmosphere which is very often filled with pessimism in therapeutic terms. Even maneuvers which aim to manipulate some "difficult" patients are, therefore, considered as methods which may be of use to them, and this enables the staff to get over their irritability due to the counter transference and, indirectly, to improve the patients' condition.

It can never be stressed enough how useful it is to pay very special attention to people looking after patients with AD as well as to families caring for them. A study (Coyne et al., 1993) stresses the risk of physical abuse faced by patients with AD on the part of those who care for them. The patients themselves often subject those who are looking after them to maltreatment as well. This all gives rise to a sordid, vicious circle which may affect normal care. For this reason it has been recommended that individuals working with this type of patient be supervised (a task which may sometimes be performed in a group setting) in order to help them to understand their aggressive reactions which are often due to counter-transference.

PSYCHODYNAMIC PROCESSES

In the treatment of demented patients some very particular defenses are superimposed on traditional defense mechanisms seen in all forms of psychotherapy and on specific mechanisms of group psychotherapies. The tendency to use denial is accentuated, especially with regard to the closeness of death. Sometimes it takes the form of selective memory loss concerning present or past suffering. It causes them to try to avoid listening to discourse which involves anxiety. Denial may also drive them into anti-phobic behaviour which may expose them to excessive danger. A large part of the rigidity found in the personality of people in the initial stages of dementia may also be interpreted more as a defense against threats than as the result of biological damage.

Very often the idealization of individuals, social conditions and past objects leads people in the initial stages of dementia to adopt regressive attitudes in order to try to once again benefit from supposed - though lost - advantages. In fact, a defensive tendency which is present at all stages of life and which is very often manifest in the dementing subject involves having recourse to some "secondary advantages" (special care, excessive length of time spent in hospital, etc.).

THERAPEUTIC RESULTS

In groups where there are elderly people who are relatively well preserved intellectually and in an out-patient setting, the advantages are obvious. Kubie points out that in his day center there were fewer instances of severe senile deterioration than expected, perhaps because the social environment stimulated and encouraged good social adjustment even when the study of the mental faculties revealed the presence of cerebral damage. Among the thousand patients who followed the program over the years, few of them had to be admitted to psychiatric hospitals. At the time of death the elderly patients who came to the center were shown to be more realistic and reacted much better than the professionals caring for them.

With patients who have AD in whom functional psychoses were additionally diagnosed the groups were shown to be more effective when the therapist acted in a more active and interventionist manner, and when the group discussions had particularly dealt with issues of everyday life than when they had dealt with psychopathology. Linden (1985), following the same line, directed a group in which he achieved an improvement in symptoms in 43% of patients (including cases of incontinence). This enabled patients to leave hospital three times more often than those in a control group.

Although patients whose condition is worse progress less with group techniques, some authors, taking into account their positive experiences, also recommend using these techniques. Thus, Wolk and Golbfard (1967) proposed weekly groups lasting an hour and a half for one year for 24 patients with non-organic problems and who had spent a large part of their life in a psychiatric hospital, and for 26 patients hospitalized because of cerebral damage. The condition of most patients improved, with those who had been in hospital for a long time progressing more-especially in terms of interpersonal relations, depression and anxiety. However, this study shows, as expected, that group work is more beneficial for elderly schizophrenics than for patients with Alzheimer's disease.

In our personal experience, in old people's homes where one is not striving for the discharge of patients the group may help patients to adjust to a protected existence. The mood and behaviour of patients suffering from mild brain damage improves considerably as a result of groups. Patients with serious brain damage obviously have much less improvement.

CONCLUSION

In an out-patient setting, group psychotherapy may be just as effective in patients with AD at the start of its course as in non-demented adults, both in groups formed only of individuals with AD, and in groups mixed with non-demented patients.

Group work is useful in institutions for several reasons: studying the participants means the staff know and judge the patients better; and the indispensable physical preparation of the patient which will enable him to take part in the groups is reflected in the improvement of his general condition. The staff find some individuals in a fairly good general condition but who, previously, had been ignored; they are better at listening to the patient's complaints and they receive information about their problems which helps motivate them and promotes professional satisfaction; the participation on the part of patients in group programs increases the interest of the families. Overall, a group improves the behaviour of individuals with truly serious brain damage to a relatively slight degree but, in contrast, results in more improvement in the behaviour of individuals with mild cerebral damage.

REFERENCES

Basaguren E et al. (1987): Experiencia grupoanalítica de terapia por el arte *Psiquis* 8:32-44.

Berland D and Poggi R (1979): Experience in group psychotherapy with the aged. *Int J Group Psychother* 1:87-108.

Butler RN (1963): The life review: An interpretation of reminiscence in the aged. *Psychiatry* 26:65-76.

Butler RN (1968): Towards a psychiatry of the life-cycle. In: *Aging and Modern Society,* Simon A and Epstein LJ, eds, Washington D.C.: APA, Psychiatric Research Report 23.

Coyne AC, Reichman WE and Bergit LJ (1993): The relationship between dementia and elder abuse. *Am J Psychoger* 4:643-647.

Guimón J (1970): La apraxia ideatoria. *Psiq y Psicol Med* 4:189-221.

Guimón J (1971): Apraxias y agnosias en las demencias degenarativas. In: *Apraxia y Agnosia,* Elexpuru Hermanos, eds., S.A.- Zamudio-Bilbao, pp. 28-68.

Guimón J (1992): Psicoterapia grupal: tendencias, indicaciones y eficacia. *Communicación psyquiátria* 17:271-300.

Johnson RI (1985): Experience in group psychotherapy with the elderly: A drama therapy approach. *Int J Group Psychother* 35:209-228.

Krassoievitch M (1993): Psicoterapia Geriátrica. *Fondo de Cultura Económica,* Mexico.

Lescz M, Feigenbaum E, Sadavoy I and Robinson A (1990): A men's group: Psychotherapy of elderly men. *Int J Group Psychother* 35:177-196.

Linden M (1985): Group psychotherapy with institutionalized senile women: Study in gerontologic human relations. *Int J Group Psychother* 3:150-3170.

Poulton J and Strassberg DS (1986): The therapeutic use of reminiscence. *Int J Group Psychother* 36:381-398.

Quinodoz D (1995): Psychanalyse, psychothérapie d'inspiration analytique : une aide pour les personnes âgées. *Méd et Hyg* 53:2223-2226.

Sadovoy J and Lescz M (1986): Congreso de 1984..

Wolk RL, Goldfarb AI (1967): The response to group psychotherapy of aged recent admissions compared with longterm mental hospital patients. *Amer J Psych* Vol 123(10):1251-1257.

Yalom ED (1983): *Patients Group Psychotherapy* New York:Basic Books.
Zilka E (1996): Expérience psychodynamique de gorupes de personnes âgées. *Méd et Hyg (In Press)*

Alzheimer Disease: From Molecular Biology to Therapy
edited by R. Becker and E. Giacobini

NON COGNITIVE SYMPTOMS IN ALZHEIMER'S DISEASE

Philippe H. Robert, Charles Henri Beau, Valérie Migneco, Valérie Aubin-Brunet and Guy Darcourt
Memory Center Cm DATA, Department of Psychiatry, UNSA
University of Nice Sophia Antipolis, France

In Alzheimer's disease (AD), non-cognitive symptoms, are particularly important because they are associated with caregiver distress, they increase the likelihood of institutionalization and may be associated with more rapid cognitive decline (Teri et al., 1990; Mortimer et al., 1992). The term non-cognitive has been chosen to distinguish these symptoms from those that are cognitive in origin (i.e. symptoms that arise directly from impairments in memory, language or visuospatial function). However, the distinction is not always clear (Rabins, 1994). In fact, blurred delimitation of the field is a central problem for investigation. As mentioned by Gilley (1993) in his review, previous studies have included behavioral manifestations of cognitive dysfunction, vegetative disturbances, psychiatric symptoms, personality disorders and activity of daily living. As a consequence of this variability in definitions, no single estimate can be made of the overall prevalence of disordered behavior in AD.

The term non-cognitive is used here to include disturbances in behavior, mood, perception and personality. The matter of this article is not to define once again non-cogntive symptoms but rather outlining other approaches allowing a better understanding of these symptoms and, consequently, a better way to treat them.

BEHAVIORAL DISORDERS IN FRONTAL DEMENTIA

Taking into account that the frontal lobes play a critical role in human behavior, multiple investigations of behavioral disorders have been performed in dementia involving frontal-subcortical dysfunction (Miller et al., 1991; Cummings, 1993; Mega and Cummings, 1994). This led to the description of three frontal-subcortical circuit-specific behavioral and neuropsychological syndromes. From a behavioral point of view, the dorsolateral prefrontal syndrome includes depression, anxiety, apathy or irritability. The orbitofrontal syndrome is characterized by personality changes, lability, euphoria, irritability, disinhibition and poor impulse control. Finally the frontal anterior cingulate syndrome includes apathy, diminished initiative and poverty of spontaneous speech.

These descriptions follow sometimes the overall psychiatric picture (anxiety, depression), and sometimes underline the importance of personality changes to consider single symptoms as a fully clinical dimension (irritability, apathy). This leads to a modification of the type of symptoms to be included in instruments designed to assess behavior in dementia. For example, the Neuropsychiatric Inventory Profile recently developed by Cummings et al. (1994) explores the following items; delusions, hallucinations, agitation/aggression, depression/dysphoria, anxiety, elation/euphoria, apathy/indifference, disinhibition, irritability/ lability, and aberrant motor behavior. Assessement of such a wide range of neuropsychiatric symptoms is different from assessment of activity of daily living and it makes it easier to understand the more and more frequent systematic evaluation of these dimensions in AD treatment trials.

RELATION BETWEEN COGNITIVE AND NON-COGNITIVE SYMPTOMATOLOGY

Although the deterioration of cognitive function is considered to be widespread in AD, cognitive frontal dysfunction has not been investigated in detail. A recent study (Binetti et al, 1996), demonstrated that executive dysfunction may be associated with early manifestation of AD. Another approach is to investigate relationships between frontal cognitive and non-cognitive impairment.

Emotional blunting has long been considered a core feature of schizophrenia. In 1978, Abrams and Taylor developed an original emotional blunting scale (EBS) in order to allow inter-rater agreement. Further factor analysis results (Berenbaum et al., 1987) indicated that blunting appears to consist of two independent components: lack of emotional expression and avolition. In fact, these dimensions have been described in various neuropsychiatric diseases involving frontal lobe dysfunction. We choose this description because it is frequently reported by elderly subjects, also by demented patients or their caregivers, to study the relation between cognitive and non-cognitive disturbances.

Population

In the memory center of Nice University, Department of Psychiatry, 70 elderly patients were evaluated using the emotional blunting scale (EBS). Only subjects with the highest (EBS > 15) and the lowest (EBS < 5) total score were included. The 55 selected patients were divided in two groups. The first one was composed by 43 patients with no diagnosis of dementia. This transnosographic group included patients with anxiety disorders (n = 7), affective disorders (n = 22), mild memory disorders (n = 7) and other psychiatric disorders (n = 7). The second group was composed by 12 demented subjects according to DSM IV criteria (1994).

Behavioral and Neuropsychological testing

Each subject underwent the following battery; EBS, MADRS depression scale (1979), Retardation scale (Wildlocher, 1981), Trail Making test A (1958) and a verbal fluency test including 3 formal and 3 semantic word fluency tasks.

Verbal fluency measures are commonly used by neuropsychologists. In this task, subjects have to name as many words as possible, beginning with a defined letter (formal fluency), or instances of a category (semantic fluency), in the order the instances occur to them (Spreen and Strauss, 1991). Patients with Alzheimer's dementia perform poorly in verbal fluency

	demented n = 12	non-demented n = 43	
Age	74.3 (4.5)	66.8 (5.8)	***
Educational level	8 (2.4)	9 (4)	
EBS	8.3 (8)	8.4 (8.4)	
MADRS	13.4 (7.2)	18.4 (9.6)	*
Retardation Scale	16 (11.2)	15 (9)	
MMSE	23.7 (2.8)	27.5 (2)	*
TMT A	103.9 (45)	72.3 (42)	
Formal Fluency	36 (15.7)	38 (16.8)	
Semantic fluency	38 (10)	45 (16.3)	
Verbal fluency / total	74 (22)	83 (30)	

TABLE 1, Behavioral and neuropsychological results in demented and non demented subjects. Mann-Whitney test: *P < 0,05,** P < 0,01,*** P < 0,001 tests compared with age and education-matched normal elderly controls (Monsch et al., 1992). Butters et al. (1987) attributed this difference to the deterioration of semantic knowledge (Martin and Fedio, 1983). Furthermore, poor performances are typically associated with frontal lobe damage (Milner, 1964; Allen et al., 1993; Ramier and Hecaen, 1970). This could be due to an impairment of strategies needed to generate a maximum of words in a given category in a limited time (Pasquier et al.,1995).

Results

Comparison between demented and non-demented subjects (Table 1) indicated no differences in verbal fluency performances. However, differences appeared when each group was divided according to EBS cutoff scores in blunted subjects (EBS>15) and non-blunted subjects (EBS<5). Within both groups, blunted subjects had lower performances at the verbal fluency test in comparison to non-blunted subjects (Fig. 1). These preliminary results confirmed that non-cognitive dimensions could be an important additional feature for subtyping different types of pathologies implicating cognitive disorders.

NEUROCHEMICAL AND ANATOMIC CORRELATES OF BEHAVIORAL DISORDERS IN DEMENTIA

The interest of therapeutic intervention on behavioral dimensions such as emotional blunting is emphasized by results of neurochemical studies. Palmer et al. (1988) demonstrated an excessive loss of anteriorly situated markers of the serotoninergic system in postmortem examinations of 17 AD patients with behavioral deficits. In the same line of evidence is the occurrence of psychotic symptoms in AD patients which is associated with significantly increased densities of senile plaques and neurofibrillary tangles in the frontal cortex (Zubenko et al., 1991). Furthermore, the same authors indicated a modest but generalized decrement in 5-HT and 5-HIAA in AD patients with either psychotic or depressive symptoms (Zubenko and Mossy, 1988). Brain imaging studies also provided evidence that behavioral changes in AD were associated with reduced metabolism in the frontal area (Chase et al., 1987; Kumar et al., 1990). In a recent positron-emission tomography study (Sultzer et al., 1995), non-cognitive symptoms were systematically evaluated in 21 AD patients. The authors reported a correlation between global cortical metabolic rate and two clusters of non-cognitive symptoms:agitation/disinhibition and anxiety/depression. Furthermore, also existed specific regional relationship between cortical hypometabolism and non-cognitive symptoms (e.g; agitation-disinhibition with frontal and temporal metabolic rate, behavioral

NON DEMENTED SUBJECTS EBS + (n = 15) EBS - (n = 28)

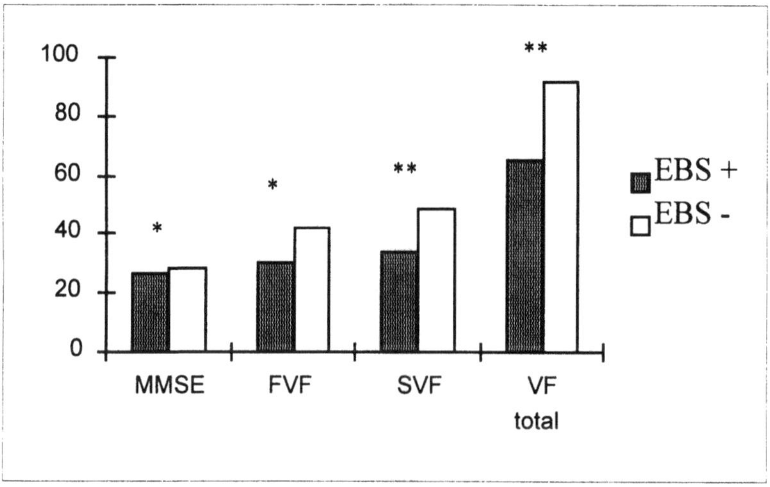

DEMENTED SUBJECTS EBS + (n = 4) EBS - (n = 8)

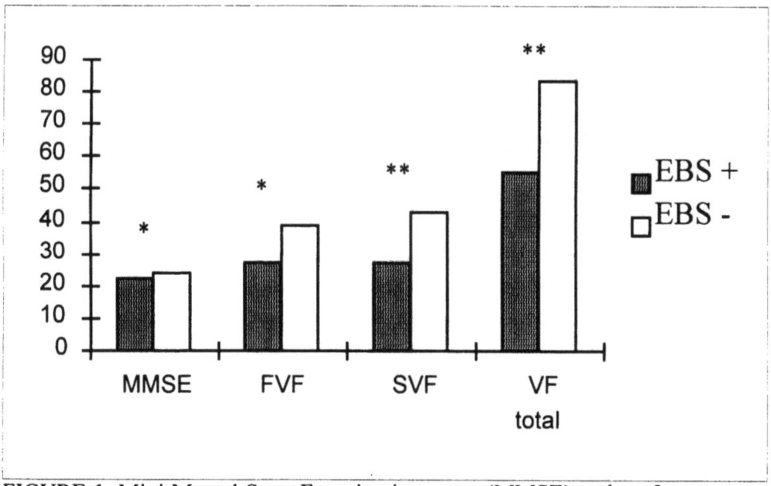

FIGURE 1, Mini Mental State Examination score (MMSE) and performances on verbal fluency test (FVF = formal verbal fluency, FVS = semantic verbal fluency, VF total = total score at all verbal fluency task) for non demented and demented patients divided according to emotional blunting scale (EBS) score (blunted group = EBS + = subjects with EBS > 15 / non blunted group EBS - = subjects with EBS < 5). Mann-Whitney test: * P < 0,05, ** P < 0,01, *** P < 0,001

retardation with individual regions in the frontal cortex, limbic system and subcortical gray matter).

These various findings suggest that frontal dysfunction is an important contributor to the expression of behavioral symptoms in AD. Behavioral symptoms in AD could be a direct expression of the degenerative brain changes in neurotransmitter system such as the serotoninergic. Thus, treatment of non-cognitive symptoms must be considered not only as an additional treatment but as a fundamental complementary part of the therapy.

REFERENCES

Abrams R and Taylor MA (1978): A rating scale for emotional blunting. *Am J Psychiatry* 135, 2:226-229.

Allen HA, Liddle PF and Frith CD (1993): Negative features, retrieval processes and verbal fluency in schizophrenia.*Brit J Psychiat* 163:769-775.

American Psychiatric Association (1994): Diagnostic and statistical manual of mental disorders, 4th Ed., *Amer Psych Assn* , Washington, DC

Berenbaum SA, Abrams R, Rosenberg S and Taylor MA (1987): The nature of emotional blunting : A factor-analytic study. *Psych Res* 20:57-67.

Binetti G, Magni E, Padovani A, Cappa SF, Bianchetti A and Trabucchi M (1996): Executive dysfunction in early Alzheimer's disease. *J Neuro, Neurosurg and Psych* 60:91-93.

Butters N, Granholm E, Salmon DP, Grant I and Wolfe J (1987): Episodic and semantic memory : a comparison of amnesic and demented patients. *J Clin Exp Neuropsychol* 9:479-497.

Chase TN, Burrows GH and Mohr E (1987): Cortical glucose utilization patterns in primary degenerative dementias of the anterior and posterior type. *Arch Gerontol Geriatr* 6:289-297.

Cummings JL (1993): Frontal-subcortical circuits and human behavior. *Arch Neurol* 50:873-880.

Cummings JL, Mega M, Gray K, Rosenberg-Thompson S., Carusi DA and Gornbein J (1994): The neuropsychiatric inventory: Comprehensive assessment of psychopathology in dementia. *Neurology* 44:2308-2314.

Gilley DW (1993): Behavioral and affective disturbances in Alzheimer's disease. *Arch Neurol* 48:112-130

Kumar A, Schapiro MB, Haxby JV et al. (1990): Cerebral metabolic and cognitive studies in dementia with frontal lobe behavioral features. *J Psychiatr Res* 24:97-109.

Martin A and Fedio P (1983): Word production and comprehension in Alzheimer's disease: the break down of semantic knowledge. *Brain Lang* 19:124-141.

Mega MS, Cummings JL (1994): Frontal-subcortical circuits and neuropsychiatric disorders. *J Neuropsych Clin Neurosci* 6:358-370.

Miller BL, Cummings JL, Villanueva-Meyer J et al. (1991): Frontal lobe degeneration:clinical, neuropsychological and SPECT characteristics. *Neurology* 41:1374-1382.

Milner B (1964): In: *The frontal granular cortex and behavior*. Warren JM and Akert K, eds., New-York, McGraw-Hill, pp. 313-331.

Monsch AU, Bondi MW, Butters N, Salmon DP, Katzman R and Thal LJ (1992): Comparisons of verbal fluency tasks in the detection of dementia of the Alzheimer type. *Arch Neurol* 49:1253-1258.

Montgomery SA, Asberg M (1979): A new depression scale designed to be sensitive to change. *Br J Psychiat* 134:382-389.

Mortimer JA, Ebbitt B, Jun SP et al. (1992): Predictors of cognitive and functional progression in patients with probable Alzheimer's disease. *Neurology* 42:1689-1696.

Palmer AM, Stratmann GC, Procter AW and Bowen DM (1988): Possible neurotransmitter basis of behavioral changes in Alzheimer's disease. *Ann Neurol* 23:616-620.

Pasquier F, Lebert F, Grymonprez L and Petit H (1995): Verbal fluency in dementia of frontal lobe type and dementia of Alzheimer type. *J Neurol Neurosurg Psychiat* 58 81-84.

Rabins PV (1994): Noncognitive symptoms in Alzheimer disease. In: *Alzheimer disease.* Terry RD, Katzman R, and Bick KL, eds.. New York: Raven Press, pp. 419-429

Ramier AM and Hecaen H (1970): Rôle respectif des atteintes frontales et de la latéralisation lésionnelle dans les déficits de la fluence verbale.Revue *Neurologique* 123:17-22.

Spreen O and Strauss E (1991): A compendium of neuropsychological tests. New York: Oxford University Press.

Sultzer D, Mahler ME, Mandelkern MA, Cummings JL, Van Gorp WG, Hinkin CH and Berisford MA (1995): The relationship between psychiatric symptoms and regional cortical metabolism in Alzheimer's disease: *J Neuropsych Clin Neurosci* 7:476-484.

Teri L, Hughes JP, Larson EB (1990): Cognitive deterioration in Alzheimer's disease:behavioral and health factors. *J Gerontol* 45:58-63.

Wildlöcher D (1981): L'échelle de ralentissement dépressif. Fondements théoriques et premières applications. psychologie *Médicale* 13B:53-60.

Zubenko GS and Mossy J (1988): Major depression in primary dementia: clinical and neuropathologic correlates. *Arch Neurol* 45:1182-1186.

Zubenko GS, Mossy J, Martinez AJ, Rao G, Claassen D, Rosen J, Kopp U (1991): Neuropathologic and neurochemical correlates of psychosis in primary dementia. *Arch Neurol* 48:619-624.

Alzheimer Disease: From Molecular Biology to Therapy
edited by R. Becker and E. Giacobini

DEPRESSION AND ALZHEIMER'S DISEASE

Carl-Gerhard Gottfries
Göteborg University, Institute of Clinical Neuroscience, Department of Psychiatry and Neurochemistry, Mölndal, Sweden

INTRODUCTION

Alzheimer's disease or Alzheimer-type dementia is a heterogeneous disorder which must be divided into at least an early onset form, Alzheimer's disease (AD), and a late onset form, senile dementia of the Alzheimer type (SDAT). AD seems to be a more cortical disorder, while SDAT also includes damage to white matter tissue (Svennerholm and Gottfries, 1994). This is of importance when discussing noncognitive disorders in AD/SDAT.

PREVALENCE

Depressive symptoms, commonly seen in vascular dementia, are also seen in Alzheimer-type dementia. The prevalence figures vary with the methodology for identifying depression and the definition of depressive disorders used. Different types of mood disorders can be distinguished in syndromes seen concomitant with dementia. Major depressive disorders seem to occur in 15-20% of AD patients (see Folstein and Bylsma, 1994). In a study of 56 patients with probable AD, 75% suffered from anxiety and 56% from slowed thinking; sadness was present in 47%, suicidal thoughts in 20% and 15% were considered to have major depressive disorders.

Anxiety, irritability and restlessness are symptoms often seen together with depressed mood. Depressive disorders in AD patients have a symptomatology that differs from the major depressive disorders seen in younger and nondemented patients. Depressive symptoms have been reported in between 0 and 87% of AD patients.

From clinical experience it is also clear that depressed mood is a symptom seen more frequently in mild and moderate dementia than in severe dementia. The explanation may be that severely demented patients do not express depressive symptoms in a way that is readily identified by rating scales.

ETIOLOGICAL ASPECTS

It is clear that psychosocial factors are of importance for the development of depression in patients with AD/SDAT. In mild and moderate forms of AD/SDAT the patients are aware of their disability. In fact, depression is common not only among the patients, but also among their caregivers. Another factor of importance with regard to the etiology of depression in the elderly is the neurochemical changes that occur in the brains of Alzheimer-afflicted patients. It was shown in 1983 (Gottfries et al.) that not only the cholinergic system but also the monoaminergic systems are disturbed in patients with Alzheimer-type dementia. Significantly reduced concentrations of 5-hydroxytryptamine (5-HT), 5-hydroxyindoleacetic acid (5-HIAA), noradrenaline (NA), dopamine (DA) and homovanillic acid (HVA) were reported in discrete areas of the brain. Interestingly, 3-methoxy-4-hydroxyphenylglycol (HMPG), the end metabolite of NA metabolism, was significantly increased in the caudate nucleus, whereas it was decreased in the hippocampus. Findings of changes of the monoamines in Alzheimer-afflicted brains are summarized in Figure 1 (see Gottfries, 1988).

In studies by Förstl et al. (1992), neuropathological changes in the locus coeruleus and substantia nigra were studied in 52 patients with Alzheimer-type dementia. Neuronal counts in the locus coeruleus were significantly lower in patients with AD/SDAT and depression than in a subgroup of patients without depression. Depression contributed significantly to the variance of neuronal counts in the locus coeruleus, even when covarying for gender, age at onset, cognitive impairment and cortical Alzheimer pathology. In the depressed Alzheimer-demented patients, cognitive impairment was less obvious and the loss of cells in the nucleus basalis of Meynert was also less pronounced. The increased concentrations of HMPG in some discrete brain areas in Alzheimer-afflicted patients concomitant with reduced NA concentrations may indicate a higher turnover rate in this system in Alzheimer-afflicted patients.

The monoamine metabolites in the cerebrospinal fluid (CSF) have also been studied. A study by Blennow et al., (1991) included 123 patients with AD/SDAT and 57 healthy controls. The AD/SDAT group showed significantly lower mean levels of HVA and 5-HIAA than the control group.

In studies of GBR 12935-binding and imipramine-binding (markers for presynaptic terminals in DA and 5-HT systems) reduced numbers of DA and 5 HT nerve terminals have been shown in discrete areas of Alzheimer-afflicted brains (Gottfries, 1990).

Within psychiatry there are advanced hypotheses about monoamines. The serotonin and NA systems, in particular, are thought to be of importance for the control of mood and emotional disturbances in the human brain. Thus, the data reported above indicate that Alzheimer-afflicted patients may have lowered thresholds for depressive disorders.

It is well known that in depression there is increased activity in the hypothalamic-pituitary-adrenal (HPA) axis, indicating an increased release of cortisol. However, it has also been shown that 40-60% of patients with dementia disorders have increased HPA axis activity. A study by Gottfries et al. (1994) showed that in the total group of dementias there were significant correlations between the severity of dementia and the post-dexamethasone (DST) levels. The frequency of pathological DST also correlated significantly with the severity of dementia. The percentage of pathological DST was lowest in the AD group (40%). It was somewhat higher in the group of vascular dementias (49%) and still higher in the SDAT group (54%). When the relationship between post-DST cortisol levels and rating scores (GBS ratings) was analyzed, significant correlations were found mainly in the vascular group. Intellectual impairment, anxiety, fear-panic and restlessness correlated significantly with post-DST cortisol levels in this group. In the SDAT group, impairment of motor performance, intellectual impairment, confusion and depressed mood correlated with post-DST cortisol levels. In the AD group (n=34), there were no significant correlations between the post-DST cortisol levels and the clinical symptoms. Raadsheer et al. (1995) summarized their findings of HPA axis activity in patients with dementia. Their data indicate that HPA axis activity is only moderately increased in AD, possibly as a result of the hippocampal damage that diminishes hippocampal inhibition of HPA axis activity. Their data also indicate that in dementia, in which increased HPA axis activity has been reported, the prevalence of depression increased. This is true for AD, Cushing syndrome and depression. Corticotropin-releasing hormone (CRH) might be causal in the development of depression. Taken together, the data suggest that hyperactivity in the HPA axis may contribute to the development of depression rather than to the pathogenesis of AD/SDAT.

It is clear that Alzheimer patients suffer from increased psychosocial stress. Neurochemical changes in the brain, especially monoamines, may lower the threshold for depression. Neuroendocrine disturbances caused by the dementia process may further contribute to the development of depression.

TREATMENT

In a Nordic multicentre study (Nyth and Gottfries, 1990) the clinical efficacy of citalopram, a selective serotonin reuptake blocker (SSRI drug), was investigated in 98 patients with moderate AD/SDAT or vascular dementia using a combined double-blind and open technique with placebo and citalopram. Analyses were made for each diagnosis after four weeks of double-blind treatment. Patients with AD/SDAT treated with citalopram showed significant improvement in emotional bluntness, confusion, irritability, anxiety, fear-panic, depressed mood and restlessness. These

							CORTEX				
	Pons	Hypothal	Caudat	Putam	Thalam	Hippoc	Front	Temp	Cing	Hippoc	Unspec
DA	↓	↓	↓	↓	↓					↓	
HVA			↓	↓			↓	↓	↑		
GBR 12935 binding				↓							
NA		↓	↓	↓		↓	↓	↓			↓
HMPG			↑			↓					↓
DBH						↓	↓	↓			
5-HT		↓	↓			↓	↓	↓			↓
5-HIAA		↓				↓	↓	↓			↓
Imipramine binding				↓			↓	↓	↓		

FIGURE 1. Changes in the monoamines in the human brain of patients withAlzheimer-type dementia (AD/SDAT). DA = dopamine, HVA = homovanillic acid, GBR 12935 binding = ligands to DA uptake sites, NA = noradrenaline, HMPG, 3-methoxy-4-hydroxyphenylglycol, DBH = dopamine-β-hydroxylase, 5-HT = 5-hydroxytryptamine, imipramine binding = ligands to 5-HT uptake sites

improvements were not found after treatment with placebo. There was no significant improvement in patients with vascular dementia; however, this group was small in number. No improvement was recorded in motor performance or cognitive functioning. Citalopram provoked few and comparatively mild side-effects in this group of patients. Another study (Balldin et al., 1988) showed that the use of SSRI drugs also reduced HPA axis activity in patients with dementia disorders.

REFERENCES

Balldin J, Gottfries CG, Karlsson I, Lindstedt G, Långström B and Svennerholm L (1988): Relationship between DST and the serotonergic system. Results from treatments with two 5-HT reuptake blockers in dementia disorders. *Int J Geriatr Psych* 3:17-26.

Blennow K, Wallin A, Gottfries CG, Lekman A, Karlsson I, Skoog I and Svennerholm L (1991): Significance of decreased lumbar CSF levels of HVA and 5-HIAA in Alzheimer's disease. *Neurobiol Aging* 13:107-113.

Folstein MF and Bylsma FW (1994). In: *Alzheimer's disease*, Terry RD, Catsman R and Bick KL, eds. New York: Raven Press, pp. 27-40.

Förstl H, Burns A, Luthert P, Cairns N, Lantos P and Levy R (1992): Clinical and neuropathological correlates of depression in Alzheimer's disease. *Psychol Med* 22:877-884.

Gottfries CG (1988): Alzheimer's disease. A critical review. *Compr Gerontol* C 2:47-62.

Gottfries CG (1990): Disturbance of the 5-hydroxytryptamine metabolism in brains from patients with Alzheimer's dementia. *J Neural Transm* [Suppl] 30:33-43.

Gottfries CG, Adolfsson R, Aquilonius SM, Carlsson A, Eckernäs SÅ, Nordberg A, Oreland L, Svennerholm L, Wiberg Å and Winblad B (1983): Biochemical changes in dementia disorders of Alzheimer type (AD/SDAT). *Neurobiol Aging* 4:261-271.

Gottfries CG, Balldin J, Blennow K, Bråne G, Karlsson I, Regland B and Wallin A (1994): Regulation of the hypothalamic-pituitary-adrenal axis in dementia disorders. *Ann NY Acad Sci* 746:336-344.

Nyth AL and Gottfries CG (1990): The clinical efficacy of citalopram in treatment of emotional disturbances in dementia disorders. A Nordic multicentre study. *Br J Psychiatry* 157:894-901.

Raadsheer FC, Kamphorst W, Purba JS, Ravid R, Tilders FJH, Swaab DF (1995): Hyperactivity of the hypothalamus-pituitary-adrenal axis may contribute to depression rather than to dementia. In: *Neurogerontology and neurogeriatrics*. Cabelos R, Frey H, Nishimura T and Winblad B. Vol 1. Barcelonea, Spain: Prous Science, pp. 51-60.

Svennerholm L and Gottfries CG (1994) Membrane lipids selectively diminished in Alzheimer brains suggest synapse loss as primary event in early onset form (Type I) and demyelination in late onset form (Type II). *J Neurochemistry* 62:1039-1047.

Alzheimer Disease: From Molecular Biology to Therapy
edited by R. Becker and E. Giacobini

PSYCHOMOTOR THERAPY AND ALZHEIMER'S DISEASE

Jacques Richard, Philippe Bovier, Jean-Philippe Bocksberger
University Clinic of Geriatric Psychiatry, 1225, Geneva

INTRODUCTION

The psychomotor therapy lays within a specific axis of observation, which rules in selected circumstances the data of the individual psychomotor organization. This field of observation is defined by several elements.

Spontaneous action tell us about the way the individual apprehends environment, space and the world of objects. It speaks about one's own capability to undertake an activity, to organize and to realize it, to pursue and to complete it, and to comply with the necessity of the milieu. It is a sign about the body's expression when in relation to other people.

The mastery of the body's immobility and of its static and dynamic equilibrium tells about its adaptability. The same can be said about the coordination and the dissociation of the body's elements when at rest, initiating or holding a movement.

The clever way of manipulating objects or imitating gestures and postures informs about the perception the individual has of his own body and of that of the other people.

All these facts determine the subject's bearings, his reactions, his relational net or the distance he establishes between him and the others. Observation of these facts allow to categorize the conception he has of his body schema and the image he does of it. These facts are contained in the variation of the muscle tone, which is directly connected with affectivity.

Space and time are the factors which basically condition psychical plasticity. The individual's availability, his power of attention, of memorization, of programmation, of inhibition, his operational capability, his aptitude to mentalize and to shift from the field of reality to the imaginary one, are related to the perception of space and the notion of duration. Cognition, motivation and emotion can therefore show through the motor realization.

Such a particular field of observation supports the action of the therapist in psychomotricity. In his own specific way to approach the other's body, the therapist in psychomotricity set it in situation and interacts with it. During this meeting the therapist seeks to alleviate a suffering that is bodily felt and expressed. He tries to promote a better integration of the individual into his environment. He has in view an educational, re-educational and especially psychotherapeutic benefit.

The Alzheimer's disease was swiftly given a position among the numerous application's areas of the psychomotor therapy. From the very beginning of the affection, indeed, the sufferer endures obvious modifications of his motor behavior. He has, for a short time, the tools (language, praxia, gnosia, etc.) to describe the discomfort that rose from the progressive alterations of the cognitive functions in his central nervous system. We can notice that these cognitive functions may longer keep their operationility when put into function together with the psychomotor therapist. This fact has contributed to reconsider the correlations that were established between cerebral lesions and the behaviors they permit. Later in the evolution, the cognitive functions express themselves into the motor realization only.

THE BACKGROUND OF THE PSYCHOMOTOR THERAPIST

The psychomotor therapist refers to various means, (Richard, 1995) such as the psychotherapeutic relaxation of J. de Ajuriaguerra (De Saugy, 1981), eutony (Alexander, 1985), relational kinesitherapy, movement or dance therapy (Bernstein, 1984), musicotherapy (Harber, 1975), rhythmic (Jacques-Dalcroze 1909), etc.

The way the patient starts and controls his own motricity - by himself or with some exterior help - provides a necessary information to the psychomotor taking in charge. The focus is the corporeal style. The therapist in psychomotricity finds access to the body's experience, that in its turn will open the way to the intimate relational system of the patient, which is more or less conscious and open to the world. Poverty, excess, ritualization of the motor expressions are as many components of the evaluation of the possible or effective therapeutic impact.

The attention given to the corporeal experience that is involved in the variations of muscle tone driven in the relaxation, is an original element of the psychomotor approach. The clever way the patient becomes aware of the new mode of being that this relaxation unveils in himself, shows the significance, the value and the disorders of his relational system. Here the therapist has in hand a transitional object to access to a dialogue that, by the means of muscle tone, indicates the investment's quality of the relations mediated by the body.

Some patients can bodily experiment changes of muscle tone, but show an impenetrable body to the glance, to attention and to introspection. Others are far too open to relaxation, and particularly eagerly demanding on the relational and affective levels. To put oneself to action instead of being acted upon constitutes one of the adaptive conditions to the environment. Anything giving evidence for it will ease the approach of the movement's dynamics and will also contribute to the possible interplay between the corporeal expression and the degree of current affective mobility.

Several corporeal domains can be reached: the *felt body*, the *moved body*, the *dialogued body*.

The felt body is the one of the perceptive afferences, of the sensations and their conjugated integrations. This level constitutes the starting point of becoming aware of the body.

The moved body qualifies the state of emotions that permeates the organism.

The dialogued body (spoken, thought and represented) describes the subject's capability of being in relation with himself and with others, either on the verbal or fantasmatic level.

At times, desequilibrium can appear between these different domains. The motor expression accounts for it. The psychomotor therapy's role will be to equilibrate and to harmonize the relations of these various levels, having in view a better adaptation of the individual to his milieu.

PSYCHOMOTOR CONTEXT IN ALZHEIMER'S DISEASE

In Alzheimer disease, some motor behaviors appear which are linked to the evolution of the affection. They constitute a ground or a motor context. However the therapist's action exerts itself beyond these motor behaviors. In this modified motor context, his focus remains the physiognomy of the movements, attitudes, gestures and postures, estimating the influence of various situational or environmental factors.

These motor behaviors, in Alzheimer's Disease, can be connected to 3 kinds of functional modifications:

1. Some of these motor behaviors evoke organization's modes which goes back into the ontogenesis, to evolutional steps of the individual. They appeared as well as natural articulations of the motor development, i.e. changes related to the maturation and the functional realization. They realize, structure and surpass themselves in order to acquire some kind of balance. They will contribute to the estimating of the level the patient will adapt to.
2. Other kinds of motor behaviors seem to correspond into the ontogenesis, to structurations which disappear, in order to allow the function's willful realization. On the one hand, they evoke very primitive behaviors that are inscribed within anatomical structures. On the other hand, they evoke some maturation's steps of the motor development that involve both the affectivity and the personality of the individual.
3. Finally, other motor behaviors are neo-behaviours that own only remote and questionable references to the individual's development stages. If not conditioned by some apragmatism linked to the patient's style of life, their presence implies a low leveled readaptation.

Therefore, from earlier descriptions (Richard et al, 1988; Bocksberger et Richard 1995), the following facts can be brought together.

A voluntary muscle slackness can possibly not happen. Or it may happen with difficulties. It may happen irregularly. In may vary according to the patient's attention. In which cases, it will be question of *«paratonie»*.

As regards the *opposition's hypertonia* it can show during the flexion and the extension movements of member's part. Basically it is an apparently willful spontaneous contraction of the antagonist muscles of the movement one is seeking to produce. It is irregular. It can show more obviously on the upper members than on the inferior members, or vice-versa.

The *anticipation of passive movements* constitutes, so to speak, the negative side of oppositionism. The patient moves in such a way as to precede the imposed passive movement. Sometime it can be followed by the *motor iteration.*The *lessening of passivity* can be clearly seen examining the dangling of the various members segments.

The *maintenance of postures* or *catalepsy* bears the spontaneous attitude, and especially the provoked ones.

An abrupt and fixed look, that can be paralleled to a wavering look, may intervene while visually exploring around. In most cases, such a glance belongs to the *BALINT degraded form of psychic paralysis.*

According to circumstances, *tactile and visual oral reflexes* can be provoked in approaching some long and sharp object to the mouth, or something spherical, or something flat. There can be added a *sign of cardinal points* by André-Thomas.

The sign of the *hand accompanied to the face*, consists in the difficulty the examiner has to lead the hand of the patient towards the patient's own face. This movement may be blocked at about 10 cm from the patient's face whereas the examiner can easily enough touch with his hand the patient's face, and whereas the latter spontaneously touches it.

We won't dwell upon the *reflexes of forced grasping* (tactile and proprioceptive), the *plantar reflex*, *magnetization or withdrawal phenomenon's*, nor on the *syncinesis* or the *stereotypies*. We will but quote the *POTZL phenomenon* (paradoxical contracting at the quadriceps during the LASEGUE maneuver) the *sole supporting reaction*, the *impossibility to bend the knee while holding the foot extension*, the *difficulty to drop a ball ready to be thrown and to throw it away*, the *striking to the ground* and the *trampling for walk.*

Here is the patient experimenting a metamorphosis of his body. The corporeal attitude changes. Gestures get altered, slowed down, after getting difficult. Sensibilities and senses are being modified. The harmony of the whole is put into question.

PSYCHOMOTOR THERAPY AND ALZHEIMER DISEASE

In Alzheimer's Disease, the psychomotor therapy is successful first by the simple fact that, like some other treatment modes, it allows to put into working the physical and the psychic functions of what can and must work. From this point of view, it contributes to partly restore the patient's relationship, in effacing some syncinetic behaviors and a certain number of motor stereotypies. Psychomotor therapy, can also prevent the occurring of paratonic phenomena, of opposition hypertonia and of archaic reflexes. It also delays their emergence at middle term within the known evolution of the Alzheimer's disease.

More specifically, the impact of the psychomotor therapy has often been proved through the way the patient endures the disease. It provides a way to take an interest not only in the cognitive aspects of the disease - that has focused till recently every trial of taking in charge - but also to find an access to the poorly investigated affectivity of this kind of patient, and to find the means to propose an efficient help.

The means of evaluating that we won't develop here, are based upon what is observable of the patient's body through psychomotor concepts, such as muscle tone (in the body's space, in time, in its relationship with emotivity), the quality of breathing, the setting into action, the rhythm of the action, the corporeal use of space or the capacities of equilibration, coordination and dissociation of the various corporeal parts. The body's representation shows through the discourse held about the body, in the knowledge about the body schema, so as in the united, linked, affirmed and invested image of the body.

REFERENCES

Alexander G (1985): Eutony: the holistic discovery of the total person. Felix Morrow New York.

Bocksberger Jph, Richard J (1995): Neurologie et psychomotricité. In: *La thérapie psychomotrice,* Masson, Paris pp 82-90.

Bernstein PL (1984): Theoretical approaches in Dance Movement Therapy. Kendall Hunt Publishing Company.

Harber G (1975): Grundlage des musiktherapie un Musikpsychologie. Gustav Fischer Verlag, Stuttgart.

Jacques-Dalcroze (1909): L'éducation par le rythme. Rythme, 7, Genève.

Saugy D (De) (1981): La relaxation selon J. de Ajuriaguerra. Ses origines, son utilisation. Médecine et Hygiène, Genève.

Richard J, Rubio L (1995): La thérapie psychomotrice. Masson, Paris.

Richard J, Constantinidis J, Bouras C (1988). La maladie d'Alzheimer. PUF, Paris.

Alzheimer Disease: From Molecular Biology to Therapy
edited by R. Becker and E. Giacobini

BEHAVIORAL TECHNIQUES FOR TREATMENT OF PATIENTS WITH ALZHEIMER'S DISEASE

Linda Teri
Department of Psychiatry and Behavioral Science
University of Washington, Seattle, WA

Behavioral techniques have been employed in the treatment of patients with Alzheimer's disease (AD) in a variety of ways for a variety of problems. This paper presents two systematic approaches to training caregivers in behavioral strategies designed to reduce patient behavioral disturbance. Both treatments have been employed in clinical practice for over ten years and are being studied as part of separate controlled clinical trials in which they are being compared to either psychosocial or medical treatment alternatives. Information on the clinical strategies employed, the study design, and preliminary data, when available, is provided here.

The rationale behind a behavioral approach to the treatment of behavioral problems in patients with AD has been discussed in detail elsewhere (Teri, 1986; Teri and Uomoto, 1986; Teri and Uomoto, 1991; Teri and Logsdon, 1990). In summary, behavioral problems are viewed as a series of behaviors which are initiated and maintained by person-environment interactions that can be observed and modified. For patients with AD, many of the critical person-environment interactions revolve around the patient-caregiver dyad. Thus, intervention seeks to alter the dyadic interaction in order to decrease problem behaviors. Treatment is individualized and systematic, focusing on current observable interactions of direct relevance to the problem under consideration.

Similar to other caregiver education programs, behavioral treatment provides basic education about AD and information about community and family resources in order to assist the caregiver with caregiving responsibilities. More uniquely, behavioral treatment involves teaching caregivers methods of behavior observation and change, identifying and developing strategies to maximize patient function, and teaching effective problem-solving skills for day-to-day difficulties in patient care.

Teaching methods of behavior observation and change involves introducing the caregiver to the importance of behavioral observation and analysis and teaching them how to identify individual behavior problems and the factors that lead up to and maintain problems. Once the chain of problem occurrence and response is understood, caregivers are guided through a systematic approach to change that problem.

Depression

Depression and dementia often coexist (Teri and Wagner, 1992; Wragg and Jeste, 1988). Depression-related problems such as increased tearfulness, dysphoric mood, loss of interest in previously enjoyable activities, sleep and appetite disturbances, fatigue, and expressions of worthlessness, guilt, or self-reproach have been found associated to higher rates of patient behavioral and functional impairment, and caregiver distress, burden and depression (Weiner et al., 1994; Pearson et al., 1989; Drinka et al., 1987). Thus, depression in dementia adversely effects patients and caregivers alike.

The Seattle Protocol (Teri, 1994; Teri et al., 1994; Teri and Uomoto, 1991) is a behavioral treatment program that consists of nine weekly sessions of one hour duration. Patients and caregivers are (a) provided with education about AD and the rationale for behavioral intervention; (b) taught methods of behavior change; (c) given strategies for identifying and increasing patient pleasant activities; (d) instructed in methods to help understand and maximize the patient's remaining cognitive and functional abilities; (e) taught effective problem-solving techniques for the day-to-day difficulties of patient care, especially those related to depression behaviors; (f) given aid with caregiving responsibilities; and (g) provided with plans for maintaining and generalizing treatment gains, once treatment ends.

In addition to being used in clinical practice, this treatment approach has been investigated in a controlled clinical trial in which patient-caregiver pairs were randomly assigned to one of three treatment conditions or a wait list control. Assessments were obtained by interviewers blind to subject assignment at pre, post, 6 and 12 month follow-up. Measures included standardized depression measures such as the Hamilton Depression Rating Scale (HDRS; Hamilton, 1967) and the Cornell Scale for Depression in Dementia (CSDD; Alexopoulos et al., 1988); and indices of caregiver distress such as the Burden Inventory (BI; Zarit et al., 1980) and the Beck Depression Inventory (BDI; Beck et al., 1961). Preliminary results suggested significant levels of improvement in patient depression after treatment, as measured by these standardized measures of depression. Furthermore, significant reduction in caregiver depression was also obtained. Patients and caregivers in the behavioral treatment condition improved significantly more than did those in the waiting list control condition (Teri 1994).

Agitation

Agitation ranks as one of the most often cited reasons for nursing admission and the use of medication or restraint in these settings (Colerick and George,

1986; Cohen et al., 1993). Agitation, verbal and nonverbal aggression, wandering, physical violence and restlessness are just a few of the many terms used to indicate the array of problems typically considered "agitation" (Cohen-Mansfield et al., 1989).

The focus of behavioral treatment is to train caregivers in behavior analysis and management in order to successfully reduce patient agitation. Strategies are developed for identifying and confronting behavioral disturbances that are associated with the agitation that interfere with engaging in pleasant activities or otherwise cause conflict between the patient, caregiver, and others. Using the skills of behavior observation and analysis taught and reinforced throughout each session, caregivers are taught strategies for changing these problems and improving patient care. (This strategy of problem identification and treatment has been called the ABC's of behavior change and is detailed in a video training program by the author, entitled "Managing and Understanding Behavior Problems in Alzheimer's Disease and Related Disorders," (Teri, 1990)). Caregivers are also provided with education and training in communication skills, realistic expectations, identifying and increasing pleasant events for patients, and strategies for maintenance of treatment gains.

As part of a masked placebo controlled trial, this behavioral treatment is offered over a 16 week period. Patient-caregiver dyads are randomly assigned to one of four treatment arms: Haldoperidol or Trazodone (two commonly prescribed medications for agitation), placebo, or behavior management. Those assigned to medication or placebo receive a weekly dose titration for 8 weeks (beginning at .5 mg Haldoperidol or 50 mg Trazodone, or placebo with a maximum allowable dose of 3 mg Haldoperidol or 300 mg Trazodone) followed by an 8-week maintenance phase. Dosing guidelines are extensive and provide the clinician with information about how to increase, decrease, or stabilize dosing.

Assessments are collected at baseline, midpoint, post treatment, 3, 6, and 12 months follow-up. They include measures of patient behavioral disturbance, such as the Cohen-Mansfield Agitation Inventory (CMAI; Cohen-Mansfield et al., 1989); the Revised Memory & Behavior Problem Checklist (RMBPC; Teri et al., 1992), and the CERAD-Behavioral Rating Scale for Dementia (BRSD, Tairot et al., 1991), as well as a measure of caregiver burden, the Screen for Caregiver Burden (SCB; Vitaliano et al., 1991).

Currently underway at 22 sites across the United States, outcome data is not yet available on this trial. Thus far, 144 patient-caregivers have been enrolled.

CONCLUSION

This paper presented two behavioral treatment programs for patients with dementia. One focused on depression, one on agitation. In both behavioral protocols, intervention is very structured and involves a gradually increasing behavior analytic approach to patient care. Each strategy taught and developed is integrated into a structured and well thought out intervention aimed at the particular complaint or constellation of complaints that seems most relevant for a given patient – caregiver dyad. The caregiver, who is often a family member, is active in treatment and is taught skills to reduce behavioral problems and enhance patient care. Few contraindications exist, therefore behavioral intervention can be offered instead of or in conjunction with other interventions. However, caregivers, and patients to a lesser extent, must be willing and able to devote the time and effort necessary for successful intervention. For patients without involved caregivers, behavioral intervention may pose unique problems in implementation. For those with involved careproviders, behavioral strategies appear to be a viable approach to the reduction of depression and agitation in patients with AD.

The ultimate utility of these interventions await further study. Additional investigation of the depression program in other settings with patients of more severe impairment and comparisons to commonly employed antidepressants would be worthwhile. Future studies of the agitation program might involve training of institutionalized staff and involvement of long-term care patients. For both approaches, further research and clinical application is needed.

ACKNOWLEDGMENTS

This study was funded in part by the Seattle Protocol; and a grant from the National Institute of Mental Health MH4326 (Teri, Logsdon, Uomoto, and McCurry). This study also is funded by a grant from the National Institution of Aging AG10483. (Teri, Raskind, Logsdon, Peskind, Weiner, Whitehouse, Grundman, Hill, and Thal).

REFERENCES

Alexopoulos GS, Abrams RC, Young RC and Shamoian CA (1988): Cornell scale for depression in dementia. *Bio Psychia* 23:271-284.

Beck AT, Ward C, Mendelson M, Mock J and Erbaugh J (1961): An inventory for measuring depression. *Arch Gen Psychia* 4:561-571.

Cohen CA, Gold D, Shulman K and Wortley J (1993): Factors determining the decision to institutionalize dementing individuals: A prospective study. *Gerontologist* 33:714-720.

Cohen-Mansfield J, Marx MS and Rosenthal AS (1989): A description of agitation in a nursing home. *J Gerontol* 44:77-84.

Colerick E and George LK (1986): Predictors of institutionalization among caregivers of Alzheimer's patients, *J Amer Geriat Soc* 34:493-498.

Drinka JK, Smith JC and Drinka PJ (1987): Correlates of depression and burden for informal caregivers of patients in a geriatrics referral clinic. *J Amer Geriat Soc* 35:522-525.

Hamilton M (1967): Development of a rating scale for primary depressive illness. *Brit J Soc Clin Psychol* 6:278-296

Pearson J, Teri L, Reifler B and Raskind M (1989): Functional status and cognitive impairment in Alzheimer's disease patients with and without depression. *J Amer Geriat Soc* 37:1117-1121.

Pearson JL, Teri L, Wagner A, Truax P and Logsdon R (1993): The relationship of problem behaviors in dementia patients to the depression and burden of caregiving spouses. *Amer J Alzh Dis Rel Disord & Res* 8:15-22.

Tairot P, Blazina L, Mack J and Patterson M (1991): CERAD Behavior Rating Scale for Dementia administration manual. Unpublished manual available from CERAD, (716)274-7508.

Teri L (1986): Treating depression in Alzheimer's Disease: Teaching the caregiver behavioral strategies. Presented to American Psychological Association, Washington DC

Teri L (1990): Managing and Understanding Behavior Problems in Alzheimer's Disease and Related Disorders. Training program with video tapes and written manual, Seattle, Washington.

Teri L (1994): Behavioral treatment of depression in patients with dementia. *Alz Dis Assoc Disord* 8:66-74.

Teri L and Logsdon R (1990): Assessment and management of behavioral disturbances in Alzheimer's disease patients. *Comp Ther* 16:36-42.

Teri L and Uomoto J (1986): Treatment of depression in Alzheimer's disease: Helping caregivers to help themselves and their patients. Presented to Gerontological Society of America, Illinois.

Teri L and Uomoto J (1991): Reducing excess disability in dementia patients: Training caregivers to manage patient depression. *Clin Geront* 10:49-63.

Teri L and Wagner A (1992): Alzheimer's disease and depression. *J Consult Clin Psychol* 3:379-391.

Teri L, Logsdon R, Wagner A and Uomoto J (1994): The caregiver role in behavioral treatment of depression in dementia patients. In: *Stress Effects on Family Caregivers of Alzheimer's Patients*, Light E, Lebowitz B and Niederehe G, eds. New York: Springer Press, pp. 185-204.

Teri L, Truax P, Logsdon R, Uomoto J, Zarit S and Vitaliano PP (1992): Assessment of behavioral problems in dementia: The Revised Memory and Behavior Problems Checklist. *Psychol Aging* 7:622-631.

Vitaliano PP, Russo J, Young HM, Becker J and Maiuro RD (1991): The Screen for Caregiver Burden. *Gerontologist* 31:76-83.

Weiner MF, Edland SD and Luszczynska H (1994): Prevalence and incidence of major depression in Alzheimer's disease. *Amer J Psychia* 151:1006-1009.

Wragg RE and Jeste DV (1988): Neuroleptics and alternative treatments: management of behavioral symptoms and psychosis in Alzheimer's disease and related conditions. *Psychiatr Clin North Am* 11:195-213.

Zarit SH, Reever KE and Bach-Peterson J (1980): Relatives of the impaired elderly: Correlates of feelings of burden. *Gerontologist* 6:649-655.

Hamilton M (1967). Development of a rating scale for primary depressive illness. [illegible]

[illegible]

Alzheimer Disease: From Molecular Biology to Therapy
edited by R. Becker and E. Giacobini

TREATMENT OF PSYCHOSOCIAL DISTURBANCES

Jean-Marie Léger, Jean-Pierre Clément, Sandrine Paulin
Department of Psychiatry, University of Limoges,
Hospital Center Esquirol, Limoges, France

INTRODUCTION

Dementia can be characterized by an acquired, progressive, more or less irreversible intellectual impairment. Cummings and Benson (1983) have medically defined it as «an acquired and progressive disturbance of the cognitive functions concerning different parts of mental capacity, and resulting from a cerebral dysfunction». Nevertheless, these acquired neuropsychological deficiencies must be sufficient to disturb relationships with others and limit the self-sufficiency (Orgogozo and Auriacombe, 1995).

In DSM IV (APA, 1994), dementia disorder is defined by the development of multiple cognitive deficits (memory impairment and one (or more) of the following cognitive disturbances: aphasia, apraxia, agnosia or disturbances in executive functioning (abstracting, planning, sequencing, initiating and organizing complex activities). These cognitive deficits imply significant impairment in occupational and social functioning and represent a significant decline from a previous level of functioning. The impairment of personality is not listed any more in comparison in DSM IV with DSM III-R and, against actual recommendations (Rebok and Folstein, 1993), psychiatric and behavioral disorders are completely absent from these criteria.

Yet, we should recognize that the care of demented patients, whether at home with family help or in hospital or a nursing home with medical assistance is, above all, the treatment of behavioral disturbances. If these disturbances continue they can disrupt harmony between patient and family and diminish tolerance of the latter.

DEFINITION OF PSYCHOSOCIAL DISTURBANCES

Psychosocial disturbances should be seen as abnormal behavior, which may have an impact on caregivers. They are certainly neither cognitive, nor conative disorders, but resulting from some of the affective disorders, delusional disorders and/or delirium, from expressions of anxiety and personality changes. In this view, when there is dementia, at whatever stage, it is more logical to have a dimensional model in analyzing symptomatology, because all diagnostic categories would be interested. The two main dimensions (requiring to be supervised by care) are psychomotor behavior and behavior in relationships. But they are often linked.

Concerning psychomotor behavior of the demented patient, main deviations which have effects on caregivers are :

-disorganized agitation
-major restlessness --or on the contrary - major apathy
-wandering with no goal, sometimes mislaying things and, too often, it is misinterpreted as running away
-pacing, when patients go, back and forth, up and down
-repetitive tumbles, sometimes resulting from previous disturbed behavior
-maladjusted vocal behavior -unceasing shouting, screaming, and howling
-boisterousness in the night
-and finally some major stereotypes

Disorders of behavior in relationships are :

-hostile attitudes and violent movements
-angry, insulting, scatological and malicious talk
-attitudes of opposition and indifference (blunted affect)
-psychical excitement with logorrhea, singing, playing, or even lustful gestures
-impulsive behavior
-inconsistency in eating («allotriophagies», Richard et al, 1989) and hygiene
-unremitting quarrelsome exchanges
-continuous search for consideration and reassurance
-frequent hyperemotive reactions
-eventually, in early stages of dementia, addictive behaviors particularly alcoholism, spending money extravagantly and thoughtless decision-making
-and finally delusional states, in which whether the demented person is convinced he is being persecuted and that somebody wishes to inflict loss on him, or he will deny some fact or the person of some close relative, such as his spouse (non-recognition of the spouse's syndrome), with possibly a turbulent or aggressive behavior.

ETIOLOGIES OF PSYCHOSOCIAL DISTURBANCES

Beside memory impairment and its resulting effect on instrumental functions and reasoning, explanations of these psychosocial disturbances are often linked with changes in physical or affective environment. It may concern moving from one house to another or take place inside the home, changes in close relationships (the coming or going of a family member, the changing of intervening carers), loss of personal property, main caregiver in a state of exhaustion, indeed, unjust blame or abuse of the demented elderly (Coyne et al,1993).

More, when language is too impaired, the demented subject is all the more likely to manifest by behavior, because «dementalization» of the person is important and because the demented person is in contact with his environment only by way of perception. So, some behaviors perceived as maladjusted, by family and caregivers, are often the expression of anxious and/or depressive disorders. They have the worth of appeal to these others, and we must not forget that there are often the warning signals of underlying somatic symptoms, particularly unnoticed orthopedic problem, urinary retention, fecal dysfunction, hypoglycemia, drug withdrawal or different pains.

In the severe stage of the disease, these psychomotor behaviors, in particular pacing, are perhaps adaptation strategies against the precarious restrictive situation of the demented patient by anxiolytic autostimulation or, in the case of shouting, by autostimulation to fulfill relational and sensorial deficiencies. Agitation has also been explained as supplying an insufficient pare-excitations system by stimulating the failing vigilance.

Nevertheless, neurochemical alterations after degenerative process, affecting not only cholinergic pathway but also noradrenergic, dopaminergic, serotoninergic and peptidergic ones (Lussier and Stip, 1995), could explain some behavioral disturbances. Particularly, serotoninergic dysfunction could explain some symptoms associated with dementia, like mood disorders or sleep disorders (Prinz et al, 1982 ; Julien, 1985). So, catecholaminergic depletion could express by effect on attention and awakening mechanisms (Julien, 1985). In this biological view, impact of medications correctly taken or not by the patient are also not negligible.

Finally, we may outline that psychosocial disturbances, by decreasing tolerance of family circle or caregiver, could induce reactions of fear, of anxiety, excessive auto and heteromedication, reactions of rejection, indeed aggressiveness, and inopportune and maladjusted decisions.

TREATMENT OF PSYCHOSOCIAL DISTURBANCES

What solutions could we suggest, by relatives request, in order to reduce or remove these psychosocial disturbances ?

Firstly, we must not forget that the appearance of a new abnormal behavior invites the question: is it a biological, affective or social curable etiology?

Drugs

Then, psychotropic drugs, in a lot of cases permit to raise a preliminary solution to posed problems. Each drug category remains, of course, indicated if a psychiatric condition is diagnosed (anxiety, depression, delusion, delirium or dementia). Nevertheless, it is a difficult way of process because dementia can mask the usual expression of these disorders, even if more and more specific scales to assess symptoms in dementia have been proposed (Alexopoulos et al, 1988; Sunderland et al, 1988). There are few studies reporting improvement with antidepressive medications, not only on depressive symptoms but also on behavioral disturbances.

Lebert et al. (1994a) used trazodone in 13 patients with a dementia of Alzheimer type during 10 weeks at 75 mg per day and observed decreasing irritability, anxiety, restlessness and an improvement of mood.

With trazodone alone, then with L-tryptophane, Greenwald et al. (1986) showed in one demented patient that shouting and violence stopped, considering these symptoms as an unusual feature of depression and they reminded that impulsiveness, aggressiveness and compulsions are presentations characterized by serotoninergic cerebral dysfunction.

Lebert et al. (1994b) also used fluoxetine 20 mg per day in 10 patients during 8-week period with the same conclusions: there was a postdrug reduction of emotional lability, irritability, anxiety and fear-panic, without change in cognition.

Nyth and Gottfries (1990) reported results of a multicenter clinical study in 98 demented patients, assessing the effects of citalopram, a new SSRI with a dose of 10 to 30 mg per day, during 4 weeks using a combined double-blind and open technique with placebo. They showed a significant improvement of «affective» disorders (anxiety, depression, irritability, emotional bluntness), without significant improvement on cognitive impairment, suggesting the concept of «emotion stabilizing drug».

So, besides benzodiazepines prescribed for anxiety but which have detrimental adverse effects with the elderly demented, like falls because of myorelaxation, fixation memory disorders or risk of delirium; besides neuroleptics prescribed for agitation and delusion but which can also induce extrapyramidal disorders, it appears more and more interesting, in a dimensional way, to consider what could be the place of drugs known to be antidepressive to treat psychosocial disorders in dementia.

Relational cares

But with the drug treatment, it is essential to recommend other therapeutic solutions, depending on whether subject has to live in institutions or at home.

1. Care in nursing homes

These are based on team work with a perfect understanding of interdisciplinarity principles (Léger et al, 1995). Objectives are first a subject's well adjustment to the ward, which needs to be a safety place of social life. Then, medical team has to carry the most possible tolerant attitude, highly regarded to show to the elderly their best self-image and, therefore, to produce a good relation with increasing narcissism and self-identity. It has to fight with efficacy against dependence and sensorial desafferentation finding balance between motherly attitudes and excessive stimulation. These institutions must integrate psychotherapic support (described afterwards). They also include family support by interviews, because, if a return to home is planned, it is sometimes joined with resumption of previous difficulties. It is yet important to explain all aspects of this family affair by an improvement of partners communication.

2. Care in the home maintenance background

A care network is important which needs different medical and social partners, generally working around the general practitioner.

Firstly, there are different home service deliveries: nurses watching for drugs observance and clinical state of patient, helping with washing and dressing; duties for supplying meals at home or housework.

But this is also to receive these patients in psychogeriatric day-hospital, where they benefit from different specific care: stimulation programs of cognitive functions, rehabilitation of everyday life know-hows, resocialization and psychological support. In addition to the objective is to relieve family circle and to leave in the home the demented patient, psychogeriatric day-hospitals offers an adapted care to the demented persons to restore their quality of life and family's.

The usual techniques are specifically stimulation exercises of the memory, which takes vigilance, perceptive abilities, motivation and affectivity into account and which insist on span, intentional attention, sensory stimulation, organization of informations with rationality or by mental imagery, spatial and temporal registration and associative recruitment. Other methods stimulate skills to recognize and to do activities of daily living. Many projects also have a resocialization goal. Group speeking sessions have the objective to revive language, rehabilitate recollections and stimulate communication and discussion. Gymnastic staffs also exist to hold the body schema, by relaxation and bodily care, and lastly, sessions of expression and of creativity. Care programs are given individually or within small groups. They integrate psychotherapy sessions.

Psychotherapy

Individual or group psychotherapy, and rather by the association of the two even if language impairment is a restricting factor, are possible and need to be studied, valorized and applied insofar as this is all along the sessions, whom setting is different from adult psychotherapy, that very old or more recent anxiety and fears could be resolved (Clément and Léger, 1995).

Firstly, psychological treatment needs a relational empathic attitude with the demented elderly person and his family circle. It is allowed to approach detrimental misbehavior in a reactive dimension, taking different collected data into account (biography, the recent past, previous personality, other concomitant disorders, actual treatment, context of the crisis and quality of actual relationships of family).

Another goal of psychotherapic action is to restore skills and potentiality of the demented patient. Therefore this psychological support is the more often practiced after a crisis period.

New technical methods begin to settle. They lean on the concept of dementalization of reflexive thinking which takes its course to figurative thinking, disrupted thinking and to primitive thinking (Péruchon and Le Gouès, 1991; Grosclaude, 1995). This psychotherapy requires a well-containing action (to avoid dispersal and distortion of representations, of affects and of emotions and to replace a satisfactory communication between disturbed family and the others) and a back-up action which in his regression process, allows the elderly to have at his disposal toward a support point, hooking of reinvestments in a kind of «helping alliance».

Sometimes, family also gains from this support, but here still, we must avoid that the caregiver appropriates the patient, who has to stay in a well defined therapeutic setting.

Space of thinking in psychotherapy requires that therapist must introduce into his field of consideration, more than the conscient and the unconscient, but also the non-conscient, the unintentional (Clément and Léger, 1995).

The ability to be psychotherapist of demented patients involves the capability to support a psychism cut off from its thinking apparatus (Le Gouès, 1985). This kind of psychocological therapy is yet often in a difficult way facing with the problem of an empty thinking in dementia or by this released primary thinking, which is nevertheless produced by the memory: this material with whom the self has structure, this ability to build mental connections. The affect, mark of a secondary thinking, which disappears with the destruction of personality, is present but rough, without gradation. Thing representation is preserved, but word representation is lost. The objective of psychotherapy is to restore psychism, to start spontaneous products of the person, and of his residual mental activity. It is based on a restitution action and on a reconnection of cleaved fragments and traces, and finally on a research of the subject where he is in reduction (Grosclaude, 1987).

Psychotherapeutic abilities depend on stage of dementia. So, if we really want to change the process in its evolution and reduce psychosocial disturbances, we may adapt setting and rules of the psychotherapeutic action in accordance with the remaining ability of the patient to be able to invest the space of thinking. But we also may never let the patient alone in it and know returning to the «corporeality» to proceed towards mental productions by the touch and the relaxation training.

CONCLUSION

The treatment of psychosocial disturbances of the demented person is therefore necessarily a multidisciplinary approach, taking into account the person and of his relatives. It is essentially based on new applications of existing biological therapy and psychotherapy. This treatment, as every trial to understand dementia, is still in the research phase, which keeps the hope to give not only palliative care, but also curative care, and perhaps preventative solutions.

Finally, pharmacological and psychological approaches will be viewed as complementary and in the future, studies of psychological treatments need to be improved in number of ways. (Orrell and Woods, 1996).

REFERENCES

Alexopoulos GS, Abrams RC, Young RC, Shamoian CA (1988): Cornell scale for depression in dementia. *Biol Psychiatry*, 23 : 271-284.

American Psychiatric Association (1994): Diagnostic and Statistical Manual for Mental Disorders. Fourth Ed (DSM IV). American Psychiatric Press, Washington DC.

Clément JP and Léger JM (1995 : Prolégomènes à une psychothérapie des démences. *Psychologie Médicale*, 27, sp 3 : 150-155.

Coyne AC, Reichman WE, Berbig LJ (1993): The relationship between dementia and elder abuse. *Am J Psychiat*, 150, 4 : 643-646.

Cummings JL and Benson DF (1983): Dementia: a clinical approach. Butterworths, Boston.

Greenwald BS, Marin DB, Silverman SM (1986): Serotoninergic treatment of screaming and banging in dementia. *Lancet*, ii:1464-1465.

Grosclaude M (1987): Le dément sénile: un sujet perdu, un sujet (re)trouvable ? *Psychologie Médicale*, 19, 8:1267-9.

Grosclaude M (1995): Premier Colloque Universitaire Européen de Gérontologie «Psychothérapies des démences», Strasbourg, 14-15 juin 1994. *Psychologie Médicale*, 27, sp. 3 et 4.

Julien RM (1985): A primer of drug action. RC Atkinson, G Lindzey and RF Thompson. eds, 4th Ed, New-York: Freeman & Company, pp. 238-243.

Lebert F, Pasquier F, Petit H(1994a): Behavioural effects of trazodone in Alzheimer 's disease. *J Clin Psychiatry* 55:536-538.

Lebert F, Pasquier F, Petit H(1994b): Behavioural effects of fluoxetine in dementia of Alzheimer type. *Int J Geriat Psychiatry* 9,7:590-591.

Le Gouès G (1985): La vie mentale tardive face à la réécriture de la perte. *Psychologie Médicale* 17, 8:1121-1124.

Léger JM, Clément JP, Tessier JF (1995): Stratégies spécifiques chez le sujet âgé. In : *Thérapeutique Psychiatrique*, JL Senon, D Sechter and D Richard, eds. Hermann, Collection Science et Pratique Médicales, pp. 899-917.

Lussier I and Stip E (1995): Relations entre les systèmes neurochimiquement définis et la mémoire : problèmes posés par la démence de type Alzheimer. *J Psychiat Neurosci* 20, 1:49-66.

Nyth AL and Gottfries CG (1990): The clinical efficacy of citalopram in treatment of emotional disturbance in dementia disorders. A nordic multicenter study. *Br J Psychiatry* 157:894-901.

Orgogozo JM and Auriacombe S (1995): Syndrome démentiel. Encycl Méd Chir (Paris-France), *Neurologie*, 17-023-A-30, 5 p.

Orrell M and Woods B (1996): Tacrine and psychological therapies in dementia - No contest ? *Int. J. Geriat. Psychiatry* 11:189-192.

Péruchon M and Le Gouès G (1991): Les processus de pensée dans la maladie d'Alzheimer. Approche psychanalytique. *Bull Psychol* tome XLIV, 398:11-14.

Prinz PN, Vitaliano PP, Vitiello MV et al. (1982): Sleep, EEG and mental function changes in senile dementia of the Alzheimer's type. *Neurobiol Aging* 3:363-370.

Rebok GW and Folstein MF (1993): Dementia. *J Neuropsy Clin Neurosci* 5:265-76.

Richard J, Fortini K, Droz P (1989): Les allotriophagies et l'élaboration de la connaissance médicale. In: *Troubles des conduites alimentaires*, Confrontations Psychiatriques n°31, Ed. Médicales Spécia, Paris, 319-334.

Sunderland T, Alterman IS, Young D et al. (1988): A new scale for the assessment of depressed mood in demented patients. *Am J Psychiatry* 145:955-959.

Alzheimer Disease: From Molecular Biology to Therapy
edited by R. Becker and E. Giacobini

SEROTONINERGIC SYMPTOMATOLOGY IN DEMENTIA

Valérie M. Aubin-Brunet, Charles H. Beau,
Geneviéve Asso, Philippe H. Robert and Guy Darcourt
Memory Center Cm DATA, Department of Psychiatry
UNSA University of Nice Sophia Antipolis, France.

INTRODUCTION

In adults, serotonin disturbances have been implicated in general depression (see Gjerris et al., 1987). In 1970, Van Praag et al. demonstrated a decrease in serotonin turnover in patients with endogenous depression, due to a probenecid induced effect on the accumulation of its metabolite, 5-HIAA, in cerebrospinal fluid (CSF). In 1976, Asberg, observing the bimodal 5-HIAA distribution in the CSF, suggested a relationship between low 5-HIAA levels and a high incidence of suicide attempts and great aggressiveness. The lowest levels of 5-HIAA are found following violent suicide attempts (Asberg et al., 1984, Brown et al., 1982 and Brown and Goodwin, 1986). Many studies have demonstrated close correlations between low 5-HIAA CSF levels and insomnia (Banki and Arato, 1983; Hartmann and Greewald, 1984); aggressiveness (Coccaro, 1989; Kruesi et al., 1990); impulsiveness (Zuckerman, 1986; Plutchik and Van Praag; 1989); irritability (Linnoila et al., 1983; Lopez-Ibor, 1988; Van Praag et al., 1988, 1990); pain (Basbaum and Fields, 1984); alcoholism (Roy et al., 1990) and eating disorders (Brewerton et al., 1990). Animal models have identified a common dimension to these behavioral disturbances, i.e. intolerance of waiting and an inability to defer a reaction (Soubrié, 1986).

Little research has been done on 5HT neurotransmission disorders with significant clinical consequences in elderly patients. Type 5HT1 and 5HT2 receptor levels tend to decrease with age (Marcusson et al., 1984b). 5HT1 and 5HT2 binding decreases by 30 to 50% in the frontal cortex and by 70% in the temporal cortex with age (Timiras et al., 1983), data confirmed by studies using C11 labeled methylspiperone and positron tomography (see McEntee and Crook, 1991).

- Sleep disorders
- Eating disorders
- Pain sensitivity
- Irritability
- Mood swings
- Impatience
- Aggressiveness
- Anxiety
- Suicidal ideas
- Intolerance of isolation

TABLE I: The 10-item Serotonin Rating Scale (SRS) (Aubin et al., 1993)

SEROTONIN RATING SCALE

We studied clinical symptoms traditionally associated with serotonin, in subjects over 60, using a transnosographic approach. This led to the construction of the SRS (Serotonin Rating Scale: Aubin et al., 1993), a clinical tool to assess a wide range of behaviors related to serotonin deficiency.

The scale is quantitative and continuous from normal to abnormal. Based on an initial factor analysis, a 10-item dimensional scale (Table I) has been developed and validated in 175 patients over the age of 60 hospitalized in psychiatric wards.

Each item is scored from 0 to 6. Preliminary studies established the validity and reliability of SRS and indicated a cutoff score of 20/60. Subjects with scores above 20 might have a serotonin deficiency (Aubin-Brunet, 1993). A factor analysis showed an homogeneous factorial structure, highly selective items, and a distribution into four factors (Table II).

	Factor 1 "loss control"	Factor 2 of "anxiety"	Factor 3 "depression"	Factor 4 "addiction"
% of variance	44.7	13.3	9.4	8.5
Mood swings	.93			
Aggressiveness	.90			
Irritability	.87			
Impatience	.65			
Sleep disorders		.93		
Anxiety		.59		
Suicidal ideas			.94	
Intolerance of isolation			.69	
Eating disorders				.76
Pain sensitivity				.74

TABLE II: Factorial analysis on the SRS 10 items: four-factor solution after oblique rotation

The principal factors were “loss of control", followed by "anxiety", "depression" and "addiction". These subgroups of symptoms are clinically significant and agree with data in the literature (Apter et al.,1990).

SEROTONINERGIC SYMPTOMATOLOGY IN DEMENTIA

The aim of this study was to evaluate "serotoninergic symptomatology" in patients with dementia.

Population and methods:
Forty patients (15 men, 25 women) consulting the memory center of Nice University Psychiatric Department were assessed. All were over 60 (mean age = 74.9 ± 7.9). They all had dementia according to DSM IV criteria. The Mini Mental State Examination (MMSE: Folstein et al., 1975) had to be between 15 and 25 (mean score = 20.7 ± 3.2). Each subject underwent the following behavioral assessment: Montgomery and Asberg Depression Rating Scale (MADRS, 1979), Anxiety scale (Covi and Lipman, 1984) and the Serotonin Rating Scale. The Kolmogonov-Smirnov test showed that all the data were normally distributed and that a parametric test could be used (Pearson test).

RESULTS

Of the 40 patients assessed, 11(27.5%) had moderate to severe symptoms of anxiety (total score >6 on the Covi Anxiety Scale). According to the MADRS total score, depression was identified in 9 (23%) patients at a cutoff score of 20/60, and in 17 (43.5%) patients with a cutoff of 17/60. According to the SRS total score, serotonin deficiency was identified in 17 (43.5%) patients (cutoff 20/60); while 9/17 had an isolated deficiency (MADRS<20, COVI<6).

Table III shows the relationships between SRS and MMSE, MADRS, COVI and age. Scores on the Covi Anxiety Scale were strongly correlated with the SRS total score ($r=0.55$; $p< 0.0002$). A weaker correlation was found between the MADRS total score and the SRS total score ($r=0.49$; $p< 0.001$).

Furthermore, according to the factorial scores, the "loss of control" factor correlated with COVI ($r=0.31$; $p< 0.05$) but not with MADRS ($r= 0.217$, $p< 0.185$). "Anxiety" correlated strongly with COVI ($r=0.67$; $p<0.0001$) and

	AGE	MMS	MADRS	COVI
SRS total score			**	***
F. “loss of control”				*
F. “anxiety”			***	***
F. “depression”			***	**

TABLE III: Correlations of SRS and factorial scores with MMS, MADRS, COVI and age (Pearson). * = $p<0.05$; ** = $p<0.001$; *** = $p<0.0001$

MADRS ($r = 0.65$; $p < 0.0001$). As expected, the "depression" factorial score correlated with MADRS ($r = 0.583$, $p < .0001$) and COVI ($r = 0.464$, $p < 0.002$).

DISCUSSION

This population of elderly patients with dementia confirms the frequency of depressive symptoms (23% with MADRS cutoff score of 20/60) and anxious symptoms (27.5% according to COVI anxiety scale), the latter being less widely described.

The SRS was designed to identify presumed serotonin deficiency symptoms other than anxious and depressive syndromes. This clinical dimension appeared to be present in a large number of the elderly subjects with dementia (42.5%). In a previous study of 130 elderly non-demented subjects, this dimension was present but at a lower frequency (33%). Clinically, serotonin deficiency can lead to behavioral disturbances (loss of control, appetite disorders, sleep disorders) and affect disturbances (anxiety, suicidal ideas, intolerance of seclusion).

Serotonin deficiency, as assessed by the SRS, is independent of age and the severity of cognitive impairment (assessed by MMSE). On the other hand, it is strongly correlated to anxiety (assessed by COVI) and weakly to depression (assessed by MADRS).

The main SRS factor "loss of control" (including mood swing, aggressiveness, irritability and impatience) was less strongly correlated to anxiety than was the total score ($r = 0.31$, $p<0.05$). This factor was not correlated to depression ($r = 0.21$, $p<0.18$). These results suggest that symptoms of serotonin deficiency are independent of the depressive dimension.

These data may contribute to modifying the treatment of behavioral disorders in dementia. In the case of patients with depressive or anxious disorders, test for specific serotonin-linked symptoms could be an additional factor in the choice of antidepressant. This choice of a serotoninergic antidepressant could also be helpful when behavioral disorders (aggressiveness, impatience, irritability) are absent, or even with depressive and/or anxious symptoms. This may be an alternative to neuroleptics which are classically used in these cases and are often poorly tolerated.

REFERENCES:

American Psychiatric Association (1994): Diagnostic and statistical manual of mental disorder (DSM-IV). Washington DC, APA (ed.).

Apter A, Van Praag HM, and Plutchik R et al.(1990): Interrelationships among anxiety, aggression, impulsivity and mood: a serotoninergically linked cluster? *Psychiatry Res.* 32:191-199.

Asberg M, Traskman L and Thoren P (1976): 5-HIAA in the cerebrospinal fluid: A biochemical suicide predictor? *Arch Gen Psychiatry*, 33:1193-1197.

Asberg M, Betilsson L, Martensson B, Scalia-Tomba GP, Ythoren P, Traskman L (1984): CSF Monoamine Metabolites in Melancholia. *Acta Psychiatr Scand* 69: 201-219.

Aubin V, Jouvent R, Wildöcher D, Darcourt G (1993): Modélisation hypothétique du déficit sérotoninergique chez le sujet agé. *L'Encéphale* XIX:37-46

Aubin-Brunet V (1993): Echelle d'évaluation clinique du deficit sérotoninergique chez les personnes agées. *L'Encéphale* XIX: 413-416

Banki CM and Arato M (1983): Amine metabolites, neuroendocrine findings, and personality dimension as correlates of suicidal behavior. *Psychiatry Research* 10: 253-261.

Basbaum AI, Fields HL (1984): Endogenous pain controls systems. Brainstem spinal pathways and endorphin circuits. *Ann Rev Neurosci* 7: 309-338.

Brewerton N TD, Brandt HA, Lessem MD, Murphy DL, Jimerson DC (1990): Serotonin and eating disorders. In: *Serotonin in Major Psychiatric Disorders* . Coccaro EF and Murphy DL, eds., Progress in Psychiatric Series, Inc. Washington DC, American Psychiatric Press, pp 153-184.

Brown GL, Ebert MH, Goyer PF et al. (1982): Aggression, suicide and serotonin: relationship to CSF amine metabolites. *Am. J. Psychiatry* 139:741-746.

Brown GL and Goodwin FK (1986): Human aggression and suicide. *Suicide and Life-Threatening Behavior* 16:141-161.

Coccaro EF (1989): Central serotonin and impulsive aggression. *Br. J. Psychiatry* 155 Suppl.8:52-62.

Covi L, Lipman RS (1984): Primary depression or primary anxiety? A possible psychometric approach to a diagnostic dilemma. *Clin Neuro Pharmacol.* 7 Suppl. 1:924-925

Folstein MF, Folstein SE, McHugh PR (1975): Mini Mental Scale. A practical method for grading the cognitive state of patients for the clinicians. J. *Psychiatry Res.* 12:189-198

Gjerris A, Sorensein AS, Rafaelsen et al. (1987): 5HT and 5HIAA in Cerebrospinal Fluid in Depression. *J Affective Disord* 12:13-22.

Hartmann E, Greewald D (1984): Tryptophan and human sleep: an analysis of 43 studies. In: *Progress in Tryptophan and Serotonin research.* Schlossberger HG, Kochen W, Linzen B et al. eds., Berlin, Walter de Gruyter, pp 297-304

Kruesi MJP, Rapoport JL, Hamburger S et al. (1990): Cerebrospinal fluid monoamine metabolites, aggression, and impulsivity in disruptive behavior disorders of children and adolescents. *Arch. Gen. Psychiatry* 47: 419-426.

Linnoila M, Virkkunen M, Scheinin M, et al. (1983): Low cerebral fluid 5-hydroxy-indoleacetic acid concentration differentiates impulsive from non-impulsive violent behaviour. *Life Sci.* 33: 2609-2614.

Lopez- Ibor JJ (1988): The involvement of serotonin in psychiatric disorders and behaviour. *Br J Psychiat* 153 (suppl. 3): 26-39.

Marcusson J, Morgon D, Winblad B, Finch C (1984b): Serotonin-2 binding sites in human frontal cortex and hippocampus. Selective loss of S2 sites with age. *Brain Res* 311:51-6

McEntee WJ, Crook TH (1991): Serotonin, memory and the aging brain. *Psychopharm* 103:143-9

Montgomery SA and Asberg M (1979): A new depression scale designed to be sensitive to change. *Br. J. Psychiatry* 134: 382-389

Plutchik R and Van Praag HM (1989): The measurement of suicidality, aggressivity and impulsivity. *Prog. Neuro-Psychopharmacol. & Biol. Psychiat.* 13: S23-S34

Roy A, Virkkunen M and Linnoila M (1990): Serotonin in suicide, violence and alcoholism. In: *Serotonin in Major Psychiatric Disorders.* Coccaro EF and Murphy DL, eds., Progress in Psychiatrie Series, Inc. Washington DC. American Psychiatric Press, pp 185-208.

Soubrié P (1986): Reconciling the role of central serotonin neurons in human and animal behavior. *Behav. and Brain Sci.* 9:319-364.

Timiras P, Cole G, Croteau M, Hudson D, Miller C, Segall L P (1983): Changes in brain serotonin with aging and modification through precursor availability. In: *Aging: Aging brain and ergot alkaloids.* A. Agnoli et al., eds. Vol. 23, New-York, Raven Press, pp 23-35

Van Praag HM, Korf J and Puite J (1970): 5-Hydroxyindolacetic acid levels in the cerebrospinal fluid of depressive patients treated with probenecid. *Nature* 225:1259 - 1260

Van Praag HM, Kahn R, Asnis GM et al. (1988): Beyond nosology in Biological Psychiatry . 5-HT disturbances in Mood, Aggression and Anxiety Disorders. In: *New Concept in Depression,* Briley M., Fillion G. eds, London, McMillian Press, pp 96-119

Van Praag HM, Asnis GM, Brown SL and Korn ML (1990) Monoamines and abnormal behavior: a multi-dimensional perspective. *Br J Psychia* 157: 723-34

Zuckerman M (1986): Serotonin, impulsivity, and emotionality. *Behav. Br Sci.* 9:348-349.

Part V

SOCIAL ISSUE IN ALZHEIMER DISEASE

Alzheimer Disease: From Molecular Biology to Therapy
edited by R. Becker and E. Giacobini

INFORMED CONSENT AND ALZHEIMER DISEASE RESEARCH: INSTITUTIONAL REVIEW BOARD POLICIES AND PRACTICES

Theodore R. LeBlang and Jean L. Kirchner
Department of Medical Humanities
Southern Illinois University School of Medicine, Springfield, IL

INTRODUCTION

It is the nature of Alzheimer Disease (AD) that patients will suffer impaired memory, failure of the ability to reason, and loss of competency (Marson et al., 1994). Thus the question of whether or not AD patients may grant informed consent to participation in dementia research is a compelling consideration for clinical investigators as well as Institutional Review Boards (IRBs).

Because there are a variety of incentives to conduct research involving AD subjects, IRBs function to balance and accommodate two important interests–facilitating research and protecting human subjects (Resau, 1995; Popp and Moore, 1994). In this latter context, IRBs are responsible for making decisions regarding whether, and to what extent, research involving AD subjects may be undertaken and precisely how informed consent will be obtained from AD subjects or their legally authorized representatives.

To evaluate whether or not there are uniform policies and practices pursuant to which IRBs make such decisions, the authors conducted a survey of IRBs in 1995. The results of this survey offer useful insights into IRB decision-making standards. This article summarizes the findings of the IRB survey and compares these findings with customary practices reported in the literature.

IRB SURVEY

Utilizing the 1995 Membership Directory of the Applied Research Ethics National Association, the authors forwarded a survey to persons identified as IRB chairpersons or administrators. Two hundred fourteen surveys were sent to individuals at one hundred eighty-nine institutions. The number of surveys mailed exceeded the number of specific institutions surveyed because, in some instances, an IRB chair and IRB administrator at the same institution received a survey. Each survey was assigned a number and survey recipients were informed that responses would remain confidential.

Ninety-three responses were received from eighty-nine institutions, indicating that for some institutions, both an IRB chair and administrator elected to respond independently. Of the ninety-three responses, many IRBs indicated that they did not evaluate protocols involving AD research. Only thirty-nine responses were considered substantially complete. These were utilized to generate the data upon which the following discussion is based.

COMPETENCE OF THE AD SUBJECT

The issue of competency assessment in dementia research has been the focus of considerable attention. Investigators often face the issue of whether or not AD patients are competent to grant consent to participate in research. In this regard, commentators have observed that there is uncertainty regarding procedures for evaluating the validity of consent as well as for obtaining informed consent (Marson et al., 1995). Thus, IRBs were asked if they required investigators to determine whether AD subjects had decisional capacity (competence) before obtaining informed consent for participation in research. Of thirty-nine respondents, thirty-four indicated that investigators were required to make such a determination. Of the five that responded in the negative, two indicated that AD patients were presumed to lack competence, two referenced use of consent forms for next-of-kin or legally authorized individuals, and one indicated that this determination is "usually part of screening criteria."

From the group of thirty-four IRBs that require investigators to first determine whether or not AD subjects are competent, sixteen responded that the investigator must include an explanation in the protocol describing the method for making this determination. Seventeen had no such requirement and one stated that investigators are "sometimes" required to provide an explanation.

NATURE OF AD RESEARCH PERMITTED

IRBs also were queried about the types of research in which AD subjects would be permitted to participate. Four research categories were the subject of inquiry: (A) research in which there *is* potential therapeutic benefit for the AD subject and there is *no more* than minimal risk; (B) research in which there *is* potential therapeutic benefit for the AD subject and there is *more than* minimal risk; (C) research in which there is *no* potential therapeutic benefit to the AD subject (but potential benefit to others) and there is *no more* than minimal risk; and (D) research in which there is *no* potential therapeutic benefit to the AD subject (but potential benefit to others) and there is *more than* minimal risk. IRBs also were asked: [1] whether they would permit an AD subject to consent to participate in each of the above categories of research if the subject were competent; and [2] whether they would permit a legally authorized representative for the AD subject to consent to the subject's participation in each of the above categories of research, if the subject lacked competence. The following table (see below) summarizes responses of IRBs that specifically answered the question "yes" or "no." Data from IRBs that answered "not applicable" or with a narrative comment are not reflected in the table.

When there is potential for therapeutic benefit and there is *no more* than minimal risk (Category A), IRBs were unanimous in permitting AD subjects to participate, regardless of competence. When research had the potential to be therapeutic and there was *more than* minimal risk (Category B), between 87 percent and 88 percent of IRBs would allow AD subjects to participate, again regardless of competence. When the research offered *no* therapeutic benefit (but potential benefit to others) (Category C), and there was *no more* than minimal risk, 82 percent of IRBs would allow AD subjects who lacked competence to participate, and 88 percent of IRBs would allow competent AD subjects to participate. Thus, in research categories A, B, and C, there was a clear consensus regarding participation in research.

Of interest is the finding that when research offered *no* therapeutic benefit to AD subjects (but potential benefit to others) (Category D), and there was *more than* minimal risk, IRBs generally were unwilling to permit AD subjects to participate. Thirty-four percent of IRBs would permit competent AD subjects to participate in such research and twenty-three percent of IRBs would permit participation by AD subjects who lacked competence. These IRB responses are generally consistent with views expressed by commentators who have considered this issue (High et al., 1994; Kapp, 1994; Moore, 1994).

PROXY DECISION-MAKING FOR AD SUBJECTS

The involvement of family surrogates as proxy decision-makers for AD subjects is widespread (High et al., 1994; Kapp, 1994; Resau, 1995). Regardless of legal status, these individuals typically grant informed consent for participation in research by AD subjects who lack competence (Sachs, 1994).

Competency of AD Subject	**Category of Research**							
	Category A		Category B		Category C		Category D	
	Yes	No	Yes	No	Yes	No	Yes	No
[1]	31	0	27	4	28	4	10	19
[2]	36	0	29	4	27	6	7	23

Under federal regulations, when a research subject is not competent to give informed consent, consent must be obtained from the subject's legally authorized representative. A legally authorized representative is defined as "an individual or judicial or other body authorized under applicable law to consent on behalf of a prospective subject to the subject's participation in the procedure(s) involved in the research" (45 CFR 46.102(d)). Yet, federal regulations provide no other guidance for identifying a legally authorized representative (Moore, 1994). Thus, the current model for obtaining informed consent focuses on proxy consent from a family member as the legally authorized representative, *along with* the AD subject's assent (Sachs et al., 1994). This practice is widely accepted because it is believed that participation of family surrogates involves individuals who are most concerned about the welfare of AD subjects, thus justifying presumptive decision-making authority (High et al., 1994; Resau, 1995).

Survey responses from IRBs reflected similar thinking. In situations where an AD subject was presumed or determined to lack competence, IRBs were asked whether they would permit a family member or non-family member to serve as a legally authorized representative for the AD subject, to give informed consent for the AD subject's participation in research. Thirty IRBs responded affirmatively and seven responded negatively. Of the seven IRBs that would refuse, responses indicated that consent must be given by a court-appointed guardian, a designated health care agent under a written power of attorney, a health care surrogate functioning under state law, or a combination of these individuals.

Of the thirty IRBs that would allow consent by either a family member or a non-family member, eleven would permit either to grant consent and nineteen would restrict this prerogative to family members. Thus, a substantial majority of IRBs interpret the federal definition of "legally authorized representative" to include family members and/or non-family members. This IRB practice echoes the model of proxy consent broadly described in the literature and typically utilized in practice.

ASSENT OF THE AD SUBJECT

The traditional practice of obtaining informed consent from AD subjects who lack competence involves the use of proxy consent from a legally authorized representative along with the AD subject's assent (Sachs, 1994). In some situations, where the competence of the AD subject is questionable, informed consent to participation in research may be obtained from both the proxy *and* the subject (High et al., 1994). IRB responses generally paralleled these views.

Of the thirty IRBs that would permit a family member or non-family member to serve as a legally authorized representative, nineteen indicated they also required the investigator to obtain the AD subject's written assent; nine did not require it; one indicated that it would depend on the state of incapacity;

and one responded "not applicable." Of the nine IRBs that do not *require* written assent, two indicated that oral assent is used instead, one indicated that written assent is desired, and one indicated that it depends on the subject's mental capacity.

PROXY DECISION-MAKING STANDARDS

There is much discussion in the literature regarding whether proxy decision-makers should apply the best interests standard (what is most beneficial to the subject) or the substituted judgment standard (what the subject would decide if competent) when consenting to participation in research. Generally, if the AD subject's values and preferences are known, use of the substituted judgment standard is preferred (Sachs, 1994). When such information is unavailable, the best interests standard typically is utilized (Kapp, 1994). In practice, however, it appears that both standards are used, particularly when a legally authorized representative has limited knowledge of the AD subject's values and preferences (Kapp, 1994; Sachs, 1994). Of the thirty IRBs that permit a family member or non-family member to serve as a legally authorized representative, four require the proxy to make decisions based on the best interests standard, four require use of the substituted judgment standard, and fifteen allow either or both standards to be applied. Six either have no policy or have not addressed the issue and one recommends, but does not require, use of the best interests standard.

Given the usefulness of information regarding an AD subject's personal values and preferences, IRBs were asked if they required investigators to determine whether legally authorized representatives had such knowledge before informed consent was obtained. Of the thirty IRBs that permit a family member or non-family member to serve as a legally authorized representative, ten require the investigator to make such a determination before informed consent is obtained.

A related consideration involves conflicts of interest between proxy decision-makers and AD subjects (Sachs, 1994). Thus, IRBs were asked whether efforts were made to address this problem. Of the thirty IRBs that permit a family member or non-family member to serve as a legally authorized representative, seven require the investigator to determine whether the proxy has a real or potential financial, emotional, or other conflict with the subject. These IRBs require the investigator to identify a different legally authorized representative under such circumstances. Of the remaining IRBs, which responded that the investigator is not required to make such a determination or that the policy has not been addressed, three indicated such a determination is desirable.

CONCLUSION

Since enactment of the National Research Act in 1974, public policy favoring protection of human research subjects has been clear. IRBs have occupied an essential role in this context. Among other things, IRBs must ensure that informed consent is obtained from research subjects or legally authorized representatives and that it is documented in accordance with federal regulations.

The results of the IRB survey discussed herein offer useful insights regarding IRB policies and practices for addressing issues that arise in the context of research involving AD subjects. It is reasonable to conclude, based upon these data, that IRBs are fulfilling their role of affording protection to this vulnerable population while, at the same time, endeavoring to facilitate important research regarding diagnosis and treatment of this pernicious disease.

ACKNOWLEDGMENTS

The authors thank Carol Herndon for providing information used to facilitate IRB survey distribution and Sarah Peters for careful manuscript preparation.

REFERENCES

High DM, Whitehouse PJ, Post SG and Berg L (1994): Guidelines for addressing ethical and legal issues in Alzheimer Disease research. *Alzheimer Disease and Associated Disorders* 8 (Supp. 4):66.

Kapp MB (1994): Proxy decision making in Alzheimer Disease research: durable powers of attorney, guardianship, and other alternatives. *Alzheimer Disease and Associated Disorders* 8 (Supp. 4):28.

Marson DC, Ingram KK, Cody HA and Harrell LE (1995): Assessing the competency of patients with Alzheimer's Disease under different legal standards: a prototype instrument. *Arch Neurol* 52:949.

Marson DC, Schmitt FA, Ingram KK and Harrell LE (1994): Determining the competency of Alzheimer patients to consent to treatment and research. *Alzheimer Disease and Associated Disorders* 8 (Supp. 4):5.

Moore DL (1994): An IRB member's perspective on access to innovative therapy. *Albany L Rev* 57:559.

Popp AJ and Moore DL (1994): Institutional review board evaluation of neuroscience protocols involving human subjects. *Surg Neurol* 41:162.

Resau LS (1995): Obtaining informed consent in Alzheimer's research. J *Neuroscience Nursing* 27(1):57.

Sachs GA (1994): Advance consent for dementia research. *Alzheimer Disease and Associated Disorders* 8 (Supp. 4):19.

Sachs GA, Stocking CB, Stern R, Cox DM, Hougham G and Sachs RS (1994): Ethical aspects of dementia research: informed consent and proxy consent. *Clinical Research* 42:403.

Alzheimer Disease: From Molecular Biology to Therapy
edited by R. Becker and E. Giacobini

LEGAL ISSUES IN ALZHEIMER DISEASE RESEARCH IN FRANCE

Alain Garay
Paris, France

INTRODUCTION

Therapy testing on man is as old as medicine itself. In honor of this, Article 15 of the Code of Medical Ethics, in no uncertain terms, requires doctors to "ensure the regularity of ... biomedical research." Alzheimer's disease, a form of senile dementia and a biological enigma affecting more than 400,000 persons in France, remains a challenge to experimental medicine (*LaRevue du Practicien*, 1995). For French specialists, neurologists and gerontologists, THA (Tetrahydroaminoacrinine or tacrine) appeared to be a possible therapy, leading to an administrative marketing permit in April, 1994 (*La Presse médicale*, 1995). At present, the "best candidates" for tacrine are notably "those who are in a position to be included in research protocols for more efficient and tolerated medicine" (Derouesne, (1994); *LaRevue du Practicien*, 1994).

All publications concerning the legal and scientific aspects of therapy testing stressed the need for the consent of the patient who is both subject and object of the experiments (Woodward, 1979; Bouvenet and Eschwege, 1994). These numerous articles filled a real need: the respect due to volunteers involved in research programs, the protection of their dignity, their independence, their rights, their needs and their personal benefit. They constitute the basis upon which the different legislations, regulations and key ideas are developed which precede any ethics on experimentation (Roy, 1986).

PATIENT CONSENT: A BASIS FOR ETHICS ON EXPERIMENTATION

For what reasons is it necessary to request that consent be included in therapy testing? From a utilitarian viewpoint, a primary argument underlines the necessity to make the patient collaborate with the results of the research. From a legal view point, other arguments insist to the fact that consent is the expression of a medical contract concluded by both the doctor and patient. It corresponds to a mutual agreement, in this case, a therapy-testing contract.

Finally, the last type of argument is established based on an ethical viewpoint. As a fundamental expression of freedom, the basis of therapy-testing is the result of a choice: to accept or refuse to participate in a scientific experiment.

Historically, these are the reasonings upon which the judgment rendered in Nuremberg were based, in the case United States V. Karl Brandt et al., classified by the Nuremberg Code, which raised the principle of voluntary consent of a human being to all biomedical research (Nuremberg Code , 1949). The declarations made in Helsinki (1964), and by the World Medical Assembly of Manila (1981), also signed by the World Health Organization and the Council of International Medical Science Organization accepted this ethical requirement. A distinction was made between therapy research which comes within the scope of medical care, and non-therapy research with no application to the patient. In the latter case, it concerns an act with a purely scientific outcome. In a decision dated October 8, 1984, the National Committee of Ethics clearly defined the limits of this distinction. The first situation concerns testing on a patient or subject at risk, the effects of inducing an individual for a result that is immediate or foreseeable. The second hypothesis concerns testing on a patient which does not produce any direct or personal benefit (Bouvier, 1986).

What must be understood by research with a direct therapeutic outcome ("therapeutic experimentation")? According to a survey carried out in France by the Council of State, entitled "Sciences of life - From Ethics to Law", published in 1988, "tests must be distinguished from technical innovations, used by doctors in the patient's immediate interest, which would be a matter of medical ethics or common law." This official study concluded that "tests are only ethically and legally acceptable if a certain number of requirements are met," including respect for the general principle of free and informed consent (*La Documentation française*, 1988).

However, according to certain individuals, Professor Jean-Marie Auby for example, this can include a "therapeutic act having an element of innovation". In an attempt to save the patient or ease his pain, a medical, surgical, pharmaceutical, or other technique could be administered possessing an element of innovation. This technique might, for example, not be officially recognized (medicine), or even currently in use. This act is essentially of a therapeutic nature. It aims primarily at helping the patient and, as a secondary characteristic, has/is, an experimental aspect. The tests could result in conclusions which would be of benefit to other patients, and the patient himself (*Litec*, 1993).

This definition appears to be justified, especially in light of the circular concerning the mode of prescription for COGNEX™ (*La Presse médicale*, 1995) in public health establishments (unidentified). Indeed, such a legal and administrative disposition constitutes an important innovation in the French system of issuing pharmaceutical substances.

CONSENT AS CONDITION FOR EXPERIMENTATION ON HUMAN BEINGS

The law N° 88-1138, dated December 20, 1988, concerning the protection of individuals who volunteer for biomedical research has filled in gaps in the legal system (Code of Public Health, 1990). This disposition applies to "tests or experiments organized and practiced on human beings with the aim of developing biological and medical knowledge". However, this law also covers " therapeutic" tests and experiments, that is to say treatment of patients in the broadest sense (Assemblé Nationale). Furthermore, this text covers tests organized by a pharmaceutical establishment in order to obtain a product marketing permit.

The consent of the patient, who is protected by the 1988 law, is a fundamental requirement, traditionally formulated in relation to acts on the human body and which are not expressly authorized by law. This requirement applies especially when the medical act has an experimental nature (Montador, 1971; Beaudoin and Parizeau, 1987) Sabatini, 1990; Duprat, 1982; Thouvenin, 1985, 1988; Fagot-Largeault, 1994).

1. Conditions of Consent (Article 209-9 of the law dated December 20, 1988)

1.1. Consent must be free, that is to say voluntary, unforced, without mental or physical coercion. It must not be obtained through manipulation or psychological pressure. Consent would be invalid if the patient were to be deceived with respect to the direct benefit or scientific nature of the research, or with respect to risks and inconveniences, etc. (Maurain and Viala, 1985). The European Commission of Human Rights has clearly stipulated that « medical treatment of an experimental nature, conducted without the consent of the patient can, in certain circumstances, be considered as a violation of article 3 » (European Commission of Human Rights Comm, 1983). In another ruling, it considered that « forced medical intervention, even if it is of minimal importance, is a breach of an individual's rights to respect of private life » (Article 8 § 1) (Bick (1979).

1.2. Consent must be informed. The informed patient must have understood the full import of common medical information. The patient must have in his possession all pertinent elements in order for him to make a decision based on the specifications of the test and his own personal references, such as values and life style.

1.3. It must be stated. Article L 209-9 stipulates that a written document be given to the patient or, if this is not possible, certified by a third party who is totally independent of the investigator and the promoter. Actual written consent, apart from its legal role, thus follows ethical and educational aims.

2. The Official Form of the Consent

According to the terms of Article L 209-9, the investigator, or the doctor who represents him, must inform the patient of the five following elements:

- the purpose of the research, the methods to be used (Parizeau, 1988), and its duration (scientific and therapeutic interests);
- expected benefits;
- foreseeable limits and risks, including those involved should the research be stopped before its completion;
- the opinion of the consulting committee for the protection of persons undergoing biomedical research (Article L 209-2) (Conditions of Consent, 1988);
- the patient must be informed that he can refuse to give his consent or withdraw it at any moment without running any liability.

3. Consent of Incapable Individuals

Article L 209-10 outlines legal and administrative solutions based on case law in areas of medical acts.

3.1. Unemancipated: Consent will be given by both parents or by one in the event of the spouse's death, or the parent with custody of the child, should the parents be divorced.

3.2. Minors or majors under warship. The guardian will express his consent, however the law makes provision for two situations:

- when research with a direct therapeutic benefit presents no serious risks, the guardians approval is sufficient;
- for other cases, the guardian cannot act without authorization from other family members or the juvenile welfare judge.

3.3. Minors or majors under guardianship capable of expressing their own will: The law accepts that their consent should be obtained and outlines that their refusal or their withdrawal of consent must not be overruled.

4. The Law Penalizes the Absence of Consent

If these conditions are not respected, the person who has this test carried out or carries out this test himself, without having requested consent, can be penalized (imprisonment for six months to three years and a fine or 12,000 - 200,000 French Francs, or one of these two penalties only) (Labbé, 1995).

The release of a medicine on the market implies that its quality, efficiency and safety have been proved. This means that in France the Autorisation de Mise sur le Marché (Product Marketing Permit) is not given until the benefit/risk report has been declared overall in favor of the patient. This scientific and administrative process, which takes between 7 and 12 years and is expensive, entails respect for the obligatory requirement to test, that is strictly regulated by law.

This disposition is included in the public health policy which emphasizes more and more the health market. « The benefit of the patient therefore appears (in the research) as a by-product of therapeutic testing (Jasmin, 1991).

REFERENCES

Article L 209-9 (1988): Before proceeding with any research on human beings, all investigators are obligated to submit a project to this committee. In: *Conditions of Consent*, December 20.

Assemblée Nationale (Charles M): This analysis is based on the interpretation of the law of the House of Representatives according to whom the law embraces "research (diagnostic, therapeutic and preventative research) including the examination of medical substances, as well as research linked with all new surgical technics".

Beaudoin J-L and Parizeau M-H (1987): Legal and Ethic Reflections Concerning Consent to Medical Treatment, *Medicine-Science* 3: 8-12.

Bick C (1983): Concerning European case law. Medical Research and Human Rights, the European approach. *JPC* I:3719.

Bouvenet G, and Eschwege E (1994): Therapy Testing and Informed Consent - Methodological Bases, Legal and Ethical Aspects. *La Revue du Practicien* Vol. 44:2767-9.

Bouvier F (1986): An opinion relating to "Ethical Problems Raised by the Testing of New Treatments on Men" In: *Consent to Therapy: Realities and Perspectives*, *J C P I*:3249.

Code of Public Health (1990): The application decree N° 90-872 dated September 27, 1990, introduced in the regulations section a book II b entitled "Protection of Individuals Volunteering for Biomedical Research".

COGNEX™ (1994): This document signed on September 9, 1994 by the directors of social security of hospitals and health, outlines the mode of prescriptions and the pharmacological oversight of COGNEX, one of the first medicines with a degree of efficiency in treating Alzheimer's disease.

Commission, (1979): X versus Austria. December 13, N° 8278-78.

Derouesne C, (1994): Which kind of patients with Alzheimer's disease can be treated and how? Reply to V. Lemaire's note In: *Le Concours Médical*, September 17, pp. 2392-2393.

Duprat J-P (1982): Information and Consent of the Patient in the Case of Tests of Medicine on Humans. *RTDSS*:369.

European Convention of Human Rights Commission (1983): establishes the prohibition to submitting of an individual to inhumane or degrading treatment. *Article 3* March 2, N° 9974/82: D.R. 32:292.

Fagot-Largeault A (1994): Informed Consent, *Medicine and Law*, N° 6, May-June, pp. 55-56.

Jasmin C (1991): Bioethics and Therapeutic Research: Informed Consent, *Médecine-Science* May, N°5, Vol.7: p. 473.

La Documentation française (1988): Documentary notes and studies. -5, N°.4855:23.

La Presse médicale (1995): This substance, commercialized under the tradename COGNEX was previously only available in the United States. February 4, N° 5, pp. 285.

La Revue du Practicien (1994): Coverage Forms for Alzheimer's Patients. September 26, pp. 17-20.

La Revue du Practicien (1995): This neurological degenerative complaint is characterized notably by the appearance of problems of memory, personality and language. December 18, pp. 7-70.

Labbée X (1995): Medical Experimentation, *Médecine et Droit* N° 11:10-15.

Litec (1993): Biomedical Research on Man: Medical and Hospital-Related Law. 5-, Fasc. 34:3.

Maurain C and Viala G (1985): On the difficulty of a right to therapeutic information: *Legal Limits of Therapeutic Information. JPC* I:3203.

Montador J (1971): The absence of Consent and Hospital Public Service's Responsibility, *RTDSS:* 180-191.

Nuremberg Code (1949):."It is absolutely necessary to obtain the voluntary consent of the patient ...For the duration of the tests, the volunteer patient will have the right to decide to stop the tests if these cause him mental or physical discomfort and if, in any other manner, the continuation of the tests seems impossible to him". In: *United States vs Karl Brandt et al.*

Parizeau M H (1988): Randomization, double blinds, placebos or reference medicine In: *The Concept of Consent to Human Experimentation, doctorate of philosophy thesis, University Paris XII*

Roy D J (1986): Medical and Research Practices; North-American Perspectives on Consent. *Medicine and Hygiene* 44:2014-2017.

Sabatini J (1990): From Absence of Consent to Tape of a Person. *Rev Prat.* 40:935-936.

Thouvenin D (1985): Human Experimentation: on the Concept of Informed Consent, *Technical Culture*, November, N° 15:109.

Thouvenin D (1988): The Reference to the Care Contract in Experimentation on Humans, In: *Medical Ethics and Human Rights,* Actes Sud-Inserm:123-146.

Woodward W (1979): Informed consent of volunteers: a direct measurement of comprehension and retention of information, *Clinical research* 27:248.

Alzheimer Disease: From Molecular Biology to Therapy
edited by R. Becker and E. Giacobini

ETHICAL PROBLEMS IN THERAPEUTIC RESEARCH IN ALZHEIMER'S DISEASE PATIENTS

Stéphanie Thibault, Lucette Lacomblez and Christian Derouesné
Department of Neurology, Salpêtrière Hospital, Paris, FRANCE.

INTRODUCTION

Ethical thought in therapeutic research is relatively new, especially in dementia. In France, the protection of patients participating in biomedical research is regulated since 1988 by the Huriet law, revised in 1991, 1993 and 1994. But this law does not take into consideration the specific problem of dementia. In the United States, ethical thought began in 1978, after the disclosure of two public scandals. The first one was the administration of cancerous cells to dying subjects without their approval. The second one was the famous Tukegee Experiment in which penicillin treatment was deliberately not given to the patients who had been followed for syphilis since 1930 in the study of the disease's natural history, a study which went on until 1972. Furthermore, these patients were all black Americans. The repercussion of these scandals was the organization of a National Committee for the protection of human rights in biomedical research, under the hospice of the American Association of Physicians.

The publication of ethical considerations concerning demented persons dates only from 1989. In this pathology, biomedical research raises the question of informed consent because of the incapacity for some AD patients to decide for themselves about their participation in clinical research, due to their memory, cognitive and decision making deficits, restricting their insight and autonomy.

Our reflection should be focused on questions about therapeutic research in AD patients, but these specific questions bring us back to general concerns on biomedical research.

WHAT ARE THE TERMS OF THE PROBLEM?

The absence of demonstrated animal models for AD restricts the assessment of possible effectiveness of an antidementia drug only in AD patients. However, because of the frequency and the severity of the disease, we desperately need to advance in scientific knowledge to develop effective treatment. For AD subjects included in a clinical research protocol, the deal is very different from the usual medical agreement in which the physician is engaged to do what he can for his patient's benefit. The agreement with an investigator is a social one, that is to say the obligation of results which only concern a better understanding of the disease or the demonstration of the activity of a new compound from which future patients could benefit. These aims point out the contradictions between scientific requirements and the obligation for clinicians to fully protect the rights and welfare of vulnerable patients. Even if, in France, the Huriet law gives a legal framework, it does not solve all the problems.

IS THERE A SPECIFICITY FOR THERAPEUTIC RESEARCH TOWARDS OTHER BIOMEDICAL RESEARCH?

If the Huriet law makes the distinction between research with and without individual benefit, it does not specify the characteristics of these two kinds of research. At the moment, only the open trials require the qualification of trial with direct individual benefit. But, from a scientific point of view, there is a widespread agreement that testing in placebo-controlled, double-blind, parallel-groups study design is required to assess the efficacy of an antidementia drug. This is difficult to accept in the context of any individual direct benefit because a quarter to a half of the subjects are assigned to the placebo group, and therefore without any expectant therapeutic effect. Nevertheless, we feel better today because some studies have shown a significant decrease of the progression of the disease for patients taking a placebo, in comparison with those taking no treatment at all. Several explanations exist for this placebo effect. First, the psychometric evaluation improvement could be due to the experience of the situation of test. Second, the patients included in a clinical research protocol are especially selected and not representative of the entire demented population. Finally, in a trial, the active medical and psychological caring of the subjects seems to be sufficient to explain the placebo effect and to influence the course of the disease.

However, is this argument sufficient to consider a placebo-controlled study design as research with direct individual benefit? The answer is no, and the distinction between therapeutic research, its methodology and the biomedical research is unclear. That is why we have to analyze the principles of research as they have been stated by the American College of Physicians in the United States and by the Huriet law in France.

Six general bioethical principles for clinical and behavioral research were published in 1992 by the American College of Physicians. They are non-malfeasance (the duty to do "no harm"), beneficence (the duty to promote the good of the patient), autonomy (the right of the patient to self-determination), confidentiality (the respect for the patient's control over his/her information), veracity (truth telling) and justice (the fairness of the distribution of goods and services).

NON-MALFEASANCE

Therapeutic research must imply minimal burden and risks for the patient. Then, research that involves potential risks and no direct benefit for the patient may only be justified if the anticipated knowledge proves that it is of vital importance for the understanding of the disease in the future, and that the specific research protocol is therefore reasonably likely to generate such knowledge.

However, different problems exist. The toxicity of the drug is often not completely known, even after the trials of phase I and II. This implies that the Promoter must take out a special insurance and as stated in the Huriet law, the ban of the inclusion of patients who do not benefit from National Health Insurance (article L. 209-16).

Tacrine has been approved for use in AD. So the inclusion of patients in a clinical research protocol raises the problem of the confirmation of the non-efficiency of that drug. However, the cognitive benefit with Tacrine appears modest and limited to a minority of AD patients. Furthermore, frequent monitoring for hepatotoxicity and medical evaluation of tolerance is recommended for the first eighteen weeks of therapy. So, in our opinion, and because we clearly need more effective and less toxic drugs, it is not necessary to prove the absence of response to Tacrine before the inclusion of a patient in a clinical research protocol.

Another specific problem appears with long-duration trials. Trial periods of up to two years are increasingly used in studies assessing the effect of drugs in slowing the rate of progression of AD. Theoretically, it commands us to carry the trial through to a successful conclusion, whatever the progression of the disease is for an individual patient. In the case of withdrawal, the patient must be evaluated at the supposed date of the end of the study. This implies the impossibility for a patient who cognitively decreased to take any other treatment or to be included in another study from which he could benefit. This position is very difficult to argue and the solution could be a methodological one. Before the beginning of the trial, the criteria accepted for the maximum allowed decrease of the cognitive functions should be defined. The analysis could then be done with something like survival curves. This could also avoid focusing the analysis only on the patients who continue the study and who may represent a special population with a relatively slow progression of the disease.

BENEFICENCE

The distinction between therapeutic research with and without direct individual benefit is not clear for placebo-controlled studies designed. The principle of beneficence conflicts in some cases with the social need for scientific progress by means of human experimentation. In fact, therapeutic research is often coupled with another larger study, such as the determination of the apolipoprotein E genotype, the effects of the drug on cerebral biochemistry or on the level of metabolites in cerebrospinal fluid. The position of this kind of research is questionable because, at the moment, there is absolutely no direct benefit for the participating subjects. The justification of this research is only based on the fact that we may have the possibility in the future to predict the drug efficacy according to these parameters. So it is very reasonable to consider that this research has no direct individual benefit and has to be limited to those who could clearly help in the advance of scientific knowledge of the human being. Qualified researchers and research sites are necessary to limit the potential risks for the subjects.

Some Ethic Review Committees mainly oriented in the rights of the persons and not in the methodological considerations, judge that their role is to ensure that the patients have been correctly informed about the potential risks and burdens of the study, refusing those with high risks. In our opinion, clinical research with no chance of scientific advance because of wrong hypothesis and/or incorrect methodology is not ethical. This position has been applied to the Salpêtrière Hospital since the creation of the Ethic Review Committee and is therefore in conformity with the Huriet law.

AUTONOMY

Autonomy is the right of the patient to self-determination or the obligation of physicians to inform their patients as adequately as possible about the burdens, the risks and the benefits of the treatment.

Autonomy raises the question of informed consent and the patient's competency. These important problems in dementia are discussed in another chapter.

CONFIDENTIALITY

Confidentiality simply prescribes some administrative solutions like setting medical records in a key-locked cupboard and the banning of disclosing any information, respecting professional secrecy. However, we have to keep in mind that a patient signing the informed consent form accepts the monitoring of the trial and the access for the Promoter to the source data. And even if the personnel is bound by professional secrecy, at present there is no professional code of ethics and deontology for them.

VERACITY

Veracity means the obligation of truth telling which prescribes us to inform the patient of his/her diagnosis. Indeed, how can a patient really make a decision with full knowledge of the facts if he does not know the nature of the disease and its risks? It is, unquestionably, a delicate point. The Huriet law makes provision for some information not to be disclosed when the diagnosis can not be told "in the best interests of an ill person" (article L. 209-9). But, this must be exceptional, and in such cases the protocol must specify the reason. We think that telling the truth to the patient and his family, as it is commonly done in the United States, is a demonstration of respect and the best way to consider the patient as a responsible adult and to reduce the dramatization of the diagnosis. In reality, the problem is to do with the relationship between the investigator, his patient and the family. Nonetheless, telling the truth could sometimes cause mental stress such as anxiety or depression. As a consequence, the patients may lose confidence in their physician, which in the end could harm them.

JUSTICE

The question of justice in France is very different to that in the United States, where the problem of minorities is strong. Nevertheless, a few questions have been raised. If we consider that only biomedical research, and particularly therapeutic trials, should help us to advance in the treatment of diseases like AD, we must notice that there is an important inequality in the repatriation of patients included in trials. A contradiction exists between the necessity to widen the patient's recruitment and the required conditions for research sites. Furthermore, at University we do not learn the attitude of investigator which is very different from usual medical practice. In classic medical caring, the clinician has to do what he can for his patient, with his possibilities of action. On the contrary, in clinical research, the question is, "Is the drug active or not, safe or not?"

The trials are presently more and more burdensome, with the necessity of restricting the patient population in order to control confounding variables related to concurrent diseases and concomitant medications, as well as the severity of the disease or the non-institutionalization of the patients. Because AD can affect both women and men of any social class, level of education, ethnic group, race or economic status, AD research must reflect this diversity. But, because of easier recruitment it is likely that most subjects who participate in AD research protocols are white, middle-class, well-educated, and non-institutionalized, suffering from only mild to moderate impairments. Under-representation of minorities may deprive these groups of potential benefits of the research, just as disproportionate inclusion may overburden other groups. Even so, patients might not belong to ethnic minorities because of the linguistic problems for the validity of the tests.

The most difficult problem is the one of phase III therapeutic studies. During the development of a new drug, phase III must theoretically include patients representative of the entire AD population. But, because of marketing reasons, the development does not distinguish phase II from phase III. The selection of the patients and the methodology are identical. Such an attitude results in the problem of a very well known example: tacrine has been accepted as an antidementia drug, but as for its current use, we have no idea about its interactions with other medications nor with concomitant diseases.

CONFLICTS OF INTEREST

The patient's best interests can compete with the clinician's or the family's interests. For the practician who chooses the patients for the trials, conflicts of interest are financial but are also scientific and ethical.

Financial conflicts of interest concern the relationships between the investigators and the pharmaceutical industry. Commercial factors must not influence clinical decision making. For example, a researcher must not exert any pressure on the patient to participate in the "well paid" research protocol conducted by that physician. To date, guidelines for physicians have dealt with financial relationships between clinicians and the pharmaceutical industry.

Ethical conflicts of interest could exist between the interests of a patient to be included in a trial and the trial's interests. To eliminate this difficulty, strategies include the practice of assigning recruitment responsibilities to practicians who are knowledgeable of the research protocol but are not directly involved in providing clinical care to the subject. The decision maker is then different from the investigator.

Scientific conflicts of interest appear when, for example, an investigator keeps in a long-term therapeutic research protocol a patient who has cognitively decreased, favoring the protocol at the expense of the patient. Another example is the inclusion of exams not directly in relation with the trial but of a pure scientific interest.

For the family, a financial conflict of interest could exist when the patients have to pay for classic medical care and not for their participation in a clinical research protocol. Theoretically, the family has to act in the best interests of the AD subject. But there is sometimes a conflict between the patient's best interests estimated by the surrogate, and the decision the patient would have made if competent with the problem of the previously discussed informed consent. When a patient is cognitively impaired as a result of dementia, the informed consent of a legal representative must be obtained. But we have to keep in mind that no person can give "consent" on behalf of another, they can only give their approval to the procedure. The investigators have then to pay scrupulous attention to detect behavioral changes in the patient who may communicate his refusal. In other situations, the surrogates disapprove the participation in a clinical research protocol, even if the participation in the trial is in the best interest of the patient, because of the burdens caused by the monitoring requirements. For example it is when they have to come twice a month to the hospital for the patient's evaluation. Their obligation is also to oversee the compliance of the treatment and noting the potential adverse events, and the expected efficiency on cognitive and/or behavioral impairments. To encourage their participation, the research protocols should take into consideration the surrogate's responsibilities more.

CONCLUSION

All these ethical questions need to be discussed between Promoters, Investigators and the Institutional Review Board, according to the aim of the therapeutic trial: slowing the progression of the disease or relieving symptoms, maximizing functions, comfort and dignity.

With the rapid advance of AD research, greater and more complex problems undoubtedly lie ahead, including those that will occur in the event that presymptomatic genetic testing for AD becomes available. Ethical and methodological thought in biomedical research in AD should be developed and taught University School of Medicine.

ACKNOWLEDGMENTS

We thank Nancy Winchester for revising the manuscript.

REFERENCES

American College of Physicians (1989): Cognitively impaired subjects. *Ann Intern Med,* 111: 843-848.

Berghmans RLP and Ter Meulen RHJ (1995): Ethical issues in research with dementia patients. *Int J Geriatr Psychiatry,* 10: 647-652.

Binstock RH, Post SG and Whitehouse PJ (1992): Dementia and aging: ethics, value, and policy choice. Binstock RH, Post SG and Whitehouse PJ, eds. Baltimore: The Johns Hopkins University Press.

High DM (1994): Ethical and legal issues in Alzheimer's disease research. *Alzheimer Disease and Related Disorders,* 8, suppl 4.

Post SG (1995): The moral challenge of Alzheimer disease. Baltimore: The Johns Hopkins University Press.

Stanley B (1985): Geriatric psychiatry: ethical and legal issues. Washington: Am Psychiatric Press.

patients who may communicate his refusal [illegible] the investigator [illegible]

Alzheimer Disease: From Molecular Biology to Therapy
edited by R. Becker and E. Giacobini

MAKING CONSENT WORK IN ALZHEIMER DISEASE RESEARCH

George J. Agich
Departments of Medical Humanities and Psychiatry,
Southern Illinois University School of Medicine

Informed consent has been the veritable ethical gold standard for research with human subjects at least since the *Nuremberg Code* (1949), but even the *Nuremberg Code* includes other requirements such as protection of subjects from physical or mental suffering and injury, the requirement of a qualified researcher using an appropriate research design, and freedom of subjects to withdraw from the experiment at anytime. Similarly, although informed consent is central in recent research ethics, it is balanced by other principles. For example, *The Belmont Report* (1979) endorsed three cardinal ethical principles for research involving human subjects: respect for persons, beneficence, and justice. Respect for persons involves both protection of human subjects as well as respect for autonomous choice, which is the basis for informed consent. Thus, although informed consent is important, it needs to be viewed within a wider ethical framework. Awareness of this wider framework is certainly evident in discussions of Alzheimer disease (AD) research ethics (Dubler, 1984; High, 1992; High et al., 1994; High and Doole, 1995), but the focus of attention has nonetheless been on the issue of informed consent. The recent literature has focused on proxy consent and advance directives for research, rather than actual consent by the AD subject. The question addressed in this paper is whether there is a role for actual consent by AD subjects and how can such consent be made to work.

Lack of attention to the practical problems associated with actually securing informed consent in AD subjects is arguably due to two factors. First, an assumed connection between AD and decisional incapacity has led researchers to focus on proxy consent and advance directives for research (Kapp, 1994; Sachs, 1994). Obviously, if the subject is assumed to lack a capacity to consent because of the diagnosis of AD, then some other mechanism has to be employed in order to secure consent. Second, AD researchers are reported to believe that protection of subjects, rather than respecting autonomy, is the central purpose of informed consent (Kapp, 1994).

Respecting autonomy as such appears far too abstract and theoretical a purpose to command much attention. Respecting autonomy, however, is a basic meaning of respect for persons, which the *Belmont Report* (1979) established as one of the basic ethical principles regulating research with human subjects, so unless AD researchers can come to terms with the meaning of autonomy, the issue of informed consent in AD subjects is likely to remain unresolved.

Because persons afflicted with AD are on a trajectory involving declining cognitive capacities, it is assumed that they have a declining capacity for decision making. However, there are no compelling accounts of the direct relationship between cognitive capacity and decisional capacity. Some cognitive capacity is clearly required for any decision making, so that at the extreme cognitive incapacity naturally involves decisional incapacity. Before the extreme is reached, however, it is unclear what the relationship involves. The uncertainty regarding the relationship between cognitive and decisional capacity is partly tied up with the fact that decisional capacity is not a general capacity, but a specific and instrumental capacity that also admits of degrees of ability. For example, one can be capable of making some decisions, but not others; and, one can be more or less capable of making some decisions at certain times and under certain circumstances, but not at other times or under other circumstances. In the case of consent to participate in research, it matters significantly whether the research project is readily understandable. It matters whether the subject's mode of participation involves being monitored, receiving medications, having blood drawn, and so on. It matters whether the research involves risks and the specific kind and degree of risks. And, finally, it matters whether the project is therapeutic or experimental in nature. These and other considerations are critically relevant for defining the standard of consent that is ethically required for participation of AD subjects in research.

As James Drane (1984) has argued, a sliding-scale of standards for informed consent to therapeutic interventions is both practically and ethically defensible. When interventions have high probabilities of benefit, low risk, and the medical circumstances are clear, the decision making standard that it is reasonable to impose upon a patient is actually quite minimal. However, if no therapeutic goal can reasonably be expected and if risks are significant, then a higher decision making standard is needed before a subject should be allowed to participate. In the first case, simple assent (implicit or explicit) and awareness is appropriate, whereas in the second case a reflective and critical understanding coupled with a judgment that the risks are reasonable in light of the patient's own beliefs and values is required. Following this approach would lead investigators and IRBs to reexamine the function of autonomy in informed consent.

Informed consent is a complex psychological, social, ethical, and legal concept that requires reanalysis if autonomy is to be respected in AD research. Reanalysis is important, because the purpose of informed consent has tended to be marginalized in discussions over proxy consent, advance consent for research, and determining decision making capacity. Making informed consent work in AD research requires a substantive framework for understanding autonomy, one that is robust concrete and practical. We need an account of autonomy that is actually present in the everyday life of AD subjects themselves. Articulating such a framework (Agich, 1990, 1993, 1994, 1995) is beyond the scope of this essay, but the implications of a theory of *actual autonomy* for informed consent in AD research can be summarized.

AD research subjects can be viewed along a spectrum; at one end are early or probable AD subjects, in which symptoms of memory loss and confusion are present, but there is no significant impairment of decision making, and at the other end are subjects with AD dementia, who clearly lack decisional capacity. At the severe end of the spectrum, the use of proxy consent is obviously necessary, but involving proxies in the decision to participate in research does not guarantee that autonomy is respected. Decision making standards widely in use for proxies are problematic from the point of view of respecting actual autonomy. The substituted judgment standard requires that explicit choices and articulated values of the patient be personally known to the surrogate, but those choices and values are not necessarily the patient's *present* choices or values. In fact, the best interest standard is invoked whenever there is no knowledge of the patient's own preferences or values, so clearly the patient's own autonomy is hardly at stake. Other approaches such as advance directives for research and the use of techniques such as narrative ethics, discourse ethics, or values history (Kapp, 1994; High, 1992; Post et al., 1994) seem to share an underlying commitment regarding respect for autonomy, namely, that decisions made for an incapacitated subject not only should be consistent with that subject's own values, but are ethically justified on the basis of those values. (Annas and Glantz, 1986; Cassel, 1988). However, dementia, other serious cognitive disabilities, and psychological changes associated with AD call into question who the subject actually is. Such alterations are an ethically relevant aspect of respect for autonomy (Dresser, 1994); respect for autonomy requires that we respect the present and actual autonomy, not just the person's prior decision making or some idealized concept of autonomy (Agich, 1993; 1995). Actual autonomy always involves identifications, that is, a formed set of beliefs, dispositions, habits, and values that define who the individual is. A formed identity, however, is always in the process of unfolding and is never completely finished. That is why the use of advance directives for research, values histories, and narrative ethics need to be approached cautiously. Inevitably, they look to the past rather than present manifestations of who the person autonomously is. All of these approaches share a rather idealized notion of autonomy. Autonomy, however, is an everyday and not a specialized concept confined only to informed consent situations. So, the question for AD research is whether another approach can be developed to identify and respect the *actual* autonomy of AD subjects.

Attention to the subject's daily expressions of likes and dislikes, of what is important or not important for the subject is a critical first step in respecting an AD subject's autonomy. Rather than focusing on a formal process of consent, assessment of the subject's everyday decisions might be a more reliable way to respect autonomy. Everyday decisions, however, are not at all like the standard understanding of the decision to consent or refuse to consent (Agich, 1995). Standard understandings of informed consent involve the assumption of a deliberative process in which a subject is given pertinent information about the research, its risks, benefits, and alternatives and is able to weigh this information and to decide based on the subject's own beliefs and values. This standard understanding, however, assumes a relatively high level of cognitive functioning as well as presumes a model of rationality that does not seriously account even for normal psychological factors affecting decision making.

Focusing on actual autonomy would require researchers and IRBs alike to devise procedures for identifying the patient's preferences and beliefs as currently expressed in everyday settings. When proxy decision makers are not themselves firsthand familiar with the subject's daily life, they can only express abstract preferences and beliefs. Under these circumstances, it would be appropriate for researchers and IRBs to rely more heavily on day-to-day caregivers and to elicit from them their understanding of the patient's acceptance or rejection of the subject's research participation, the awareness and understanding of the subject regarding the nature and purpose of the research, and the emotional burden that participation in research might impose on an impaired and vulnerable AD subject.

Thus, consent to participate in research, like consent for treatment, can rely on a sliding scale of standards depending on the medical or research circumstances, i.e., the degree of risk and the certainty regarding knowledge of outcome, that the subject is being asked to consider. Research protocols that have a therapeutic goal such as delaying the onset of symptoms which do not also carry with them more than minimal risk would thus require only *agreement* on the part of the subject. Proxy consent would not add anything to the subject's agreement; the subject's willingness alone would de controlling. Analogously, for research involving more than minimal risk and with no direct benefit to the subject himself, disagreement or expression of unwilling to participate in the research activities would afford an ethically sound basis for refusal. Under these circumstances, proxy consent would not be adequate to allow a subject to participate.

At least one commentator (Kapp 1994) has asserted that there exits a firm consensus that the chief role for a proxy decision maker in a research setting is to protect the incapacitated potential subject from harm. However, this transmutes respect for autonomy which is the principle underlying informed consent into a principle of nonmaleficence, namely, avoidance of harm (Annas and Glantz, 1986). Respect for autonomy, however, requires that the patient's own choice should be respected, not simply that harm be avoided. From this point of view, the use of a proxy consent procedures is best understood as a way to ensure that an incapacitated subject's beliefs and values do, in fact, underlie decision making, but only if the subject's own present preferences agree with the proxy's decision. Hence, one obvious requirement for proxy decision making is that the proxy be intimately familiar with the subject's daily existence. That is why family members are often perceived as the best proxies, because under most circumstances they are reasonably assumed to know and share the subject's everyday preferences, desires, likes, and dislikes.

For research with AD patients, it may be important that refusal to participate be based on current expressions of disagreement, discomfort, or dislike by the AD subject in the present. These everyday preferences are unlikely to be known by a process of formal consent giving involving reading and signing consent forms. Such standard ways of getting informed consent misses what is most important about respecting actual autonomy, namely, the present expressions of the subject regarding participation. The problems with standard ways of thinking about informed consent is that the information flow is mostly in the direction of the subject and from the researcher, however, the ethically significant aspect of consent is the acceptance or agreement of the subject without which the participation would not be authorized. Hence, investigators and IRBs need to think more critically about what can be an authorizing action or what counts as an authorizing intention in AD subjects.

REFERENCES

Agich GJ (1990): Reassessing autonomy in long-term care. *Hastings Center Report* 20 (6):12-17.

Agich GJ (1993): *Autonomy and Long-Term Care*. New York and Oxford, Oxford University Press.

Agich GJ (1994): Autonomy in Alzheimer disease. In: *Alzheimer Disease: Therapeutic Strategies*, Giacobini E and Becker R, eds. Boston: Birkhäuser, pp. 464-469.

Agich GJ (1995): Actual autonomy and long-term care decision making. In: *Long-Term Care Decisions: Ethical and Conceptual Dimensions*, McCullough LB and Wilson NL, eds. Baltimore, MD: John Hopkins University Press, pp. 113-36.

Annas GJ, Glantz LH (1986): Rules for research in nursing homes. *N Engl J Med* 315:1157-8.

Cassel CK (1988): Ethical issues in the conduct of research in long term care. *Gerontologist* 28 (Suppl.):90-96.

Drane J (1984): Competency to give an informed consent. *JAMA* 282:925-7.

Dresser R (1994): Missing persons: legal perceptions of incompetent patients. *Rutgers Law Review* 46 (2):609-719.

Dubler NN (1984): The ethics of research. *Generations*. 9:18-21.

High DM, and Doole MM (1995): Ethical and legal issues in conducting research involving elderly subjects. *Behavioral Sciences and the Law* 13:319-35.

High DM, Whitehouse PJ, Post SG, Berg L (1994): Guidelines for addressing ethical and legal issues in Alzheimer's disease research: a position paper. *Alzheimer Disease and Associated Disorders* 8 (Suppl. 4):66-74.

High DM (1992): Research with Alzheimer's disease subjects: informed consent and proxy decision making. *Journal of the American Geriatric Society* 40:950-957.

Kapp MB (1994): Proxy decision making in Alzheimer's disease research: durable power of attorney, guardianship, and other alternatives. *Alzheimer's Disease and Associated Disorders* 8 (Suppl. 4):28-37.

National Commission for the Protection of Human Subjects of Biomedical and Behavioral Research (1979): *The Belmont Report.* Washington, DC: U S Government Printing Office.

Nuremberg Code (1949): *Trials of War Criminals Before the Nuremberg Military Tribunals Under Control Council Law No. 10* (Vol.2). Washington, DC: U S Government Printing Office.

Post SG, Ripich DN, Whitehouse PJ (1994): Discourse ethics: research, dementia, and communication. *Alzheimer's Disease and Associated Disorders* 8 (Suppl. 4): 58-65.

Sachs GA (1994): Advanced Consent for dementia research. *Alzheimer's Disease and Associated Disorders* 8 (Suppl. 4) 19-27.

Alzheimer Disease: From Molecular Biology to Therapy
edited by R. Becker and E. Giacobini

INDUSTRY PERSPECTIVES ON THE MARKETING OF ANTI-ALZHEIMER DISEASE THERAPY

Robert van der Mark
Sandoz Pharma, Basel, Switzerland

THE ROLE OF PHARMACEUTICAL MARKETING

Most pharmaceutical companies originated in the late 18th or 19th century as chemical and later as dye-stuff producers. The typical activities of those involved in marketing in those early years was simply to find people to buy whatever the production department was producing at the time. As the industry moved into the 20th century the modern era of drug design and development began, although the bulk of active medicines were still galenicals.

The role of the industry has changed and the current climate is moving towards one in which the pharmaceutical company focuses on research into diseases and the development of drugs as part of an integrated health care package. In addition to medicines, such packages may include components like the development and provision of diagnostic tools and patient support services, such as toll-free telephone hot-lines and nurses to help patients and their families with problems that the disease brings. More and more pharmaceutical companies now see their activities as part of disease management rather than simply selling medicines. Consequently, the boundary between the clinician delivering care and the pharmaceutical industry is becoming more blurred as the pharmaceutical industry is beginning to see itself as the clinician's partner in the provision of health care.

Over the last decade or so, there has been an increasing focus on containing the cost of health care and on the need for economic justifications for the pricing and reimbursement of health-care interventions such as medicines. Most companies have specialist economists working closely with the medical and marketing divisions of their development teams.

Perhaps the most important central role of the Marketing Department is to identify the needs of particular patient groups and doctors. Marketing Departments interview patients, caregivers, physicians and payers in most large countries in order to find out what is needed and what the perceived attitudes are about the illness and its management. These interviews are carried out either with individuals, or in small groups, where it may be possible to obtain extra benefit from the group interactions that occur. Market research allows the accumulation of large data bases containing such information as the number of prescriptions that are written for particular medicines, for what indications they are prescribed, what specialty is responsible for diagnosis and/or prescribing, how patterns differ from country to country and within countries. The result of the research is to inform the Marketing Department of the needs and current treatment practices in a particular disease area and how this differs, if at all, from one customer to another. This is used to make recommendations to the Research and Development team (R & D).

Activities most commonly associated with the Marketing Department are mainly concerned with assembling and disseminating information about the new medicine. Typical activities include satellite symposia and trade exhibits at congresses, or the materials to support basic messages about the product for sales representative to use in the doctor's office. Today, these materials often go beyond the product itself and extend to such things as diagnostic tools, and patient/carer information and educational material.

THE HISTORY OF TACRINE

There was a resurgence in interest in Alzheimer's disease(AD) in the 1980s, led by developments in the USA of new diagnostic guidelines – NINCDS-ADRDA in 1984 (McKhann et al.), and the elaboration of the cholinergic hypothesis for the loss of cognitive ability and memory and the proposition that cholinergic activation provided a therapeutic approach to the symptomatic treatment of AD (Perry, 1986). In 1986, Summers et al. published the results of a study which suggested that the cholinergic approach might provide miraculous benefit to patients with AD. Parke-Davis acquired the rights to tacrine and agreed with the FDA on the use of the cognitive tests developed at Mount Sinai, together with a global rating scale. Although those involved in Alzheimer research were aware of this, the vast majority of clinicians responsible for treating patients with the disease were quite unaware of these developments.

When the results of the tacrine trials were published (Farlow et al., 1992), most general practitioners, geriatricians, neurologists and psychiatrists were unimpressed by the small benefits offered by the drug (in part, possibly, due to inadequate study design and low doses) and there was general disappointment that tacrine was not the 'miracle drug' as had been suggested. Meanwhile, research had moved on and there were some real breakthroughs made in research on plaques and tangles.

Those responsible for medication budgets were hesitant to make tacrine freely available saying that it gave little benefit for considerable cost.

Nevertheless, most HMOs in the USA eventually added the drug to their formularies. Today the use of tacrine is very small as a function of the prevalence of AD. This is despite the fact that there are reports of patients who have done well for several years (Wilcock et al., 1994) and a more balanced medical opinion in general.

MARKETING CHALLENGES

In the absence of a cure, marketing of an anti-Alzheimer drug provides many challenges. First, the general population is mostly quite unaware of the early signs and symptoms of the disease. Many general practitioners still believe that senility is an unavoidable sign of advancing years. Because there is no effective treatment for AD, many general practitioners are unwilling to consider AD as a possible diagnosis. This is despite the fact that it may become easier for a family to cope with a difficult relative if they can be told that the abnormal behaviour is due to a specific illness. Early AD, i.e. the stage at which acetylcholinesterase inhibitors are most likely to be effective, is therefore probably considerably under-diagnosed.

Second, there is the challenge of how to assess a patient's response to treatment. General practitioners cannot be expected to carry out sophisticated assessments of cognitive function at regular intervals. Linked with this is the issue of what is a realistic expectation of efficacy. How should the doctor judge who should be prescribed treatment and for how long? It has been suggested that the benefits of treatment are likely to be best appreciated by caregivers. Caregivers may therefore be in the best position to judge whether the drug is actually offering any benefits to the patient.

Leading experts in the AD field will have a collective responsibility to offer guidance on questions such as these to medical professionals at the front-line of AD management. Pharmaceutical companies can help facilitate such debate where possible.

EXPECTATIONS OF TREATMENTS FOR ALZHEIMER'S DISEASE

The grounds upon which an AD will be judged vary according to who does the judging. For example, regulatory authorities probably would want to know the probability levels of changes on the ADAS-Cog and CIBIC-Plus scales. Physicians responsible for the management of patients with AD may have little understanding of these scales. For them, a simple global assessment could be sufficient. Patients and caregivers will base their judgment compared to how the patient was before. Often a small improvement or less worsening of the symptoms will be seen as a major benefit. At the other end of the spectrum, basic research will probably want to see the establishment of a link between treatment and the disease process itself, and until this is established, they may regard symptomatic therapy with suspicion.

Payers, on the other hand, think about benefit in terms of cost. Unless a clear societal cost reduction in the overall care of individual patients can be demonstrated, it is likely that the availability of treatment will be limited. However, political pressure to have certain health care interventions available can be very effective. It must be borne in mind that not all benefits can be measured in monetary terms.

Despite frequent media announcements of breakthroughs in AD research, it is unlikely that there will be any miracle drugs or cures within the next 7 years or so, the length of time a patient diagnosed with AD today might expect to live. What then is realistic to expect of non-curative Alzheimer's treatment? To answer this we need to keep in mind a number of complicated issues. First, the syndrome that we know as AD probably has many causes. Second, in the absence of reliable biological markers, the symptoms are used as a measure of the disease severity, its progression and the effects of treatment. Furthermore, clinical studies assessing symptoms over only six months will not provide evidence about long-term outcome. However, by the same token, neither will it be possible (on the basis of six-month studies) to exclude the possibility that acetylcholinesterase inhibitors do provide long term benefit or indeed that they can delay the progress of the disease. Indeed, there are reports of long-term benefit (Wilcock et al, 1994; Knopman et al., 1996), as well as encouraging preclinical data on APP processing (Lahiri et al., 1994; Mori et al., 1995). Therefore the best that any non-curative therapy is likely to offer as monotherapy is that some patients will have less worsening of symptoms than they would without the drug. The duration of this symptomatic effect is important as well as whether progress of the disease itself is affected.

HEALTH ECONOMICS

Social attitudes and values towards health care and its funding are undergoing rapid change at the moment. This creates a major challenge for those concerned with the delivery of care in AD, where early costs tend to be indirect or hidden, and institutionalization costs are difficult to estimate accurately. In many countries, institutionalization appears to be more dependent upon such factors as the financial status of the family or awareness of the availability of services than on the patient's cognitive or behavioural status.

Once again, 6-month placebo-controlled studies will not provide hard facts for health economic analyses. Neither will monetary considerations reflect the true value to families of having grandfather more like his old self for another year or so. The price at which a new Alzheimer's medicine is offered for sale, therefore, depends very much on such things as cost of the disease, who benefits and what is the true value of treatment in both economic and social terms.

NEW DOCTRINE VS OLD HABITS

Whenever the first new type of medicine becomes available in a particular therapeutic area, there is the challenge of altering long-established habits in order to enable use of the new medicine to become established. Some countries are conspicuously slow to adapt to particular therapeutic trends. In AD, this also holds true, and although the reluctance of some countries to adopt new trends represents a challenge, it also represents an opportunity, should it prove possible to change old habits. Many general practitioners see the early symptoms of AD as normal signs of aging and regard only the late stage of the illness as AD. In many countries, early symptoms are widely believed to be of mainly vascular origin.

The clinical trials programs that most multinational companies undertake today have been developed to meet the regulatory requirements of most of the large countries in the world. The licenses and labelling that are issued for these drugs will therefore be essentially the same world-wide, so that the indication will be listed as Alzheimer's, and not as vascular dementia or age-related senility. The challenge for the Marketing Department will be to persuade doctors to diagnose AD early enough in those countries where the habit is to diagnose dementias e.g., 'age-related cerebrovascular insufficiency'.

CONCLUSIONS

The future for the Marketing Departments of pharmaceutical companies planning to introduce new anti-Alzheimer medicines will present many challenges, not least of which is the relative lack of knowledge about the disease, its signs, symptoms and diagnosis. There will be many old prejudices and habits to overcome, and it will be necessary to communicate benefit to patients and their caregivers in a way that realistically describes the value of treatment. Nevertheless, the pharmaceutical industry has changed over the last decade and the prospects are good for the move towards a collaborative approach to management between the medical profession and industry, away from the straightforward selling of medicines that was historically the principle role of the Marketing Department.

REFERENCES

Farlow M, Gracon SI, Hershey LA, Lewis KW, Sadowsky CH, Dolan-Ureno J and the Tacrine Study Group (1992): A controlled trial of tacrine in Alzheimer's disease. *JAMA* 268:2523–2529.

Knopman D, Schneider L, Davis K et al. (1996): Long-term tacrine(Cognex) treatment effects on nursing home placement and mortality. *Neurology* (In Press).

Lahiri DK, Lewis S, Farlow MR (1994): Tacrine alters the secretion of the beta-amyloid precursor protein in cell lines. *J Neurosci Res* 37:777–787.

McKhann G, Drachman D, Folstein M et al. (1984): Clinical diagnosis of Alzheimer's disease: report of the NINCDS-ADRDA Work Group under the auspices of Department of Health and Human Services Task Force on Alzheimer's Disease. *Neurology* 34:9939–944.

Mori F, Lai C-C, Fusi F, Giaccobini E (1995): Cholinesterase inhibitors increase secretion of APPs in rat brain cortex. *NeuroReport 6:633–636.*

Perry EK (1986): The cholinergic hypothesis – ten years on. *Br Med Bull* 42:63–69.

Summers WK, Majovski LV, Marsh GM, Tachiki K, Kling A. (1986): Oral tetrahydroamino-acridine in long-term treatment of senile dementia, Alzheimer's type. *N Engl J Med* 327:1253–1259.

Wilcock GK, Scott M, Pearsall T (1994): Long-term use of tacrine. *Lancet* 343:294.

Alzheimer Disease: From Molecular Biology to Therapy
edited by R. Becker and E. Giacobini

CODEM: A LONGITUDINAL STUDY ON ALZHEIMER DISEASE COSTS

Marco Trabucchi, Karin M. Ghisla and Angelo Bianchetti
Alzheimer's Unit, "S.Cuore Fatebenefratelli" Hospital and Geriatric Research Group, Brescia, Italy

INTRODUCTION

In Italy, an Alzheimer disease (AD) prevalence of 2.6% in over 60 has been demonstrated; this prevalence leads to a total of 283,000 people affected by the disease (Rocca et al., 1993). The costs of resources required to provide care for AD patients have a considerable impact on the economy of families and of the health system. The overall burden is made up of various components: direct costs for formal care, indirect costs, and costs for research and education (Max, 1993). Since detailed information on this problem do not exist for Italy (Trabucchi et al., 1995), aim of the study is to measure direct and indirect costs in a sample of hundred community dwelling AD patients followed for one year.

METHODS

The study is based on information from 103 AD patients (25 males and 78 females; mean age 77.7±7.3) and their caregivers consecutively referred to the Alzheimer's Unit in the "S. Cuore Fatebenefratelli" Hospital (Brescia, Italy) from July, 1994 to September, 1995.

Inclusion criteria were: a) diagnosis of Alzheimer Disease probable according to NICDS-ADRDA criteria (McKhann et al. 1984), and b) living at home at the time of inclusion in the study.

Patients consecutively admitted to the Alzheimer Unit or to the Day Hospital from July, 1994 were asked to participate at the study. The sample was balanced for Clinical Dementia Rating scale level to obtain a comparable proportion of patients in each of the three levels of the scale (0.5-1: questionable or mild dementia; 2: moderate dementia; 3: severe dementia) (Hughes et al., 1982). Patients were evaluated at baseline and after 6 and 12 months to assess cognitive and functional status, behavioral symptoms, concurrent diseases.

Cognitive function was assessed using the Mini Mental State Examination (MMSE) (Folstein et al. 1975); functional status using the Activity of Daily Living (ADL) scale (Katz et al., 1963) and the Instrumental Activity of Daily Living (IADL) Scale (Lawton and Brody 1969). Depressive symptoms were evaluated using the Geriatric Depression Scale (Yesavage et al., 1983). Behavioral symptoms were assessed using the Cohen-Mansfield Agitation Inventory (Cohen-Mansfield et al., 1986). Concurrent diseases were assessed using the Greenfield Index (Greenfield et al., 1987).

The patient's families were evaluated at baseline in order to obtain information on structure, socio-economic status, employment and house modifications. A social worker visited the caregivers every week for the first three months and every two weeks for the following nine months, collecting data on all direct and indirect costs (hours spent for caregiving, drugs and aids use, physician visits, examinations, hospitalization or day services use, other type of health or social services). All the costs are calculated in Italian lira (for the transformation in US $ we estimated 1 $ = 1,600 Italian lira). For the determination of caregiving costs, we used information obtained from national contract for domestic workers and mean costs sustained by public health services in northern Italy (hour costs for domestic workers: $ 6.25; hr costs for surveillance: $4.90; hr costs for patient caring: $7.25; hr costs for nursing activities: $18.70). The protocol of the study was approved by the local Ethical Committee.

The paper is based on the information collected at baseline and during the first three months of longitudinal observation.

RESULTS

At the baseline, the mean patients' MMSE score was 10.2±7.5 (10.7% of the sample had a MMSE higher than 21, 23.3% from 15 to 20, 28.1% from 8 to 14 and 37.9% 7 or less). 34.0% of the patients had a CDR level of 0.5-1, 25.2% a CDR level of 2 and 40.8% a CDR level of 3. 26.2% of the AD patients were totally independent for the ADL, 32.0% lost 1 or 2 functions and 41.8% lost three or more functions (the mean functions lost were 2.4±2.0).

3% of principal caregivers were spouses, 45.7% children, 16.5% other relatives and 4.9% non-family members. 35.4% of the patients had to stop their profession for dementia, and 20.4% of principal caregivers reduced or stopped their profession for caring. 35.0% of the sample had to modify the house to make it more suitable for AD patient (the mean expenditure was $7,900 US).

The mean weekly drug expenditure at baseline was 10.8±10.7 US $0.77±4.1 US $ were spent for nootropics drugs, 1.48±1.9 for neuroleptics or benzodiazepines, and 2.2±4.3 for antidepressants (TABLE I).

	mean±SD
nootropics	0.77±4.1
benzodiazepines or neuroleptics	1.48±1.9
antidepressant drugs	2.2±4.3
cardiovascular disorders' drugs	2.8±3.4
pulmonary disorders' drugs	0.03±0.2
gastrointestinal disorders' drugs	1.3±3.8
vitamins	0.83±3.4
endocrine disorders' drugs	0.18±1.1
non-steroidal antinflammatory drugs	0.14±0.9
genitourinary disorders' drugs	0.1±1.3
all drugs	*10.8±10.7*

TABLE I. Mean weekly drug expenditure for AD patients at baseline (in US $)

Caregiver	*Caring activities*	*Hours/ week*	*Estimated costs*
Relatives	instrumental activities of daily living §	18.3	114.5
	activities of daily living *	5.4	39.1
	surveillance	81.0	400.1
	nursing activities °	0.1	1.7
Paid providers	instrumental activities of daily living §	2.4	14.8
	activities of daily living *	0.5	3.8
	surveillance	3.6	17.9
	nursing activities °	0.1	2.4
Volunteers	instrumental activities of daily living §	0.3	0.1
	activities of daily living *	0.0	0.0
	surveillance	0.0	0.0
	nursing activities °	0.0	0.0
Public services	instrumental activities of daily living §	0.1	0.4
	activities of daily living *	0.3	2.2
	surveillance	0.1	0.1
	nursing activities °	0.2	3.0

TABLE II. Mean weekly informal and formal caregiving hours and relative costs (in US $) based on first three months longitudinal observations. § activities include: housekeeping, cooking, transportation; * activities include: bathing, grooming, eating, mobility; ° activities include: drugs monitoring, medications.

Data based on the first three months of longitudinal observation show that 109.3±62.1 hours/week were spent caring for the patients: 20.6±14.7 for housework, 6.3±8.0 for care of the person, 85.5±48.7 for patient supervision

Independent variables	*b*	*Beta*	*p (T)*	R^2
part a				
MMSE	-4.5	-0.38	0.03	0.15
part b				
MMSE	-23.57	-0.38	0.04	0.15

TABLE III. Variables independent predictors of hours weekly spent for care by primary caregiver (*part a*) and of weekly costs for care (calculated in US $) (*part b*) based on first three months longitudinal observations only in patients with MMSE higher than 8 (n=64). b value represents the number of hours (*part a*) or the US $ (*part b*) weekly spent for a one-unit change of independent variables. Regression analysis using stepwise deletion of reduntand variables. Variables not in the equation: ADL, CDR, IADL

(90% of this time is provided by informal caregivers); nurse care was provided above all by public services (0.4$\pm$1.3 hours/week). The weekly total costs for patients' care is 585.2$\pm$330.3 US $ (see TABLE II).

The relation between clinical variables and costs for care is not linear; higher costs were found in AD patients in the moderate stage of dementia. A multivariate analysis indicate that MMSE score is the best independent predictor of cost for care in patients in a mild to moderate stage of the disease (MMSE higher than 8). For a one-point change at MMSE the weekly cost for care varies of 23.57 US $ (see TABLE III).

CONCLUSIONS

This study offers the first data based on longitudinal observation on direct and indirect costs of AD in Italy.

Preliminary analysis based on cross-sectional information shows that caregiver services are provided predominantly by family. Indirect costs (loss of employment, house modifications) are substantial. Nearly fifty percent of total drug expenditure is due to nootropics, antidepressant and psychotropics use.

Data on the first three months longitudinal observation show that the relation between clinical variables and costs for care is not linear; in particular higher costs were found in AD patients in the moderate stage of the disease. A multivariate analysis indicate that MMSE is the best independent predictor of cost for care in patients with a MMSE higher than 8.

These data may be helpful in the cost-benefit analysis of pharmacological and non-pharmacological therapeutic strategies.

ACKNOWLEDGMENTS

The study is supported by a grant from Bayer S.p.A. (Milan, Italy). We acknowledge G.Fattore, M.C. Cavallo and E. Vendramini (Ce.RGAS. Bocconi University, Milan) for their helpful suggestions in the preparation of the protocol, and Mrs. P.Brochetti for secretariat assistance.

REFERENCES

Cohen-Mansfield J (1986): Agitated behavior in the elderly: II. Preliminary results in the cognitively deteriorated. *J Am Geriatr Soc* 34:722-727.

Folstein MF, Folstein SE, McHug PR (1975): Mini-Mental State: a practical method for grading the cognitive state of patients for the clinician. *J Psychiatr Res* 12:189-198.

Greenfield S, Blano DM, Elashoff RM (1987): Development and testing of a new index of comorbidity. *Clin Res* 35:346A.

Hughes CP, Berg L, Danziger WL et al. (1982): A new clinical scale for the staging of dementia. *Br J Psychiat* 140:566-572.

Katz S, Ford AB, Moskowitz RW, Jackson BA, Jaffee MW (1963): The index of ADL: a standardized measure of biological and psychosocial function. *J Am Med Assoc* 185: 914-919.

Lawton MP, Brody E. (1969): Assessment of older people: self maintaining and instrumental activities of daily living. *Gerontologist* 9:179-186.

Max W (1993): The economic impact of Alzheimer's disease. *Neurology* 43(supp.4):S6-S10.

McKhann G, Drachman D, Folstein M, Katzman R, Price D, Stadlan EM (1984): Clinical diagnosis of Alzheimer's disease. *Neurology* 34:939-944.

Rocca Wa, Hofman A et al. (1993): Frequency and distribution of Alzheimer's disease in Europe: a collaborative study of 1980-1990 prevalence findings. *Annals Neurol* 30:9.

Trabucchi M, Govoni S, Bianchetti A (1995): Socio-economic aspects of Alzheimer's disease treatment. In: *Alzheimer disease: Therapeutic Strategies*. Giacobini E, Becker R (eds). Birkhauser, Boston, pp 459-463.

Yesavage JA, Brink TL, Rose TL, Andrey M (1983): Development and validation of a geriatric depression screening scale. *J Psychiatr Res* 17:37-49.

Alzheimer Disease: From Molecular Biology to Therapy
edited by R. Becker and E. Giacobini

HEALTH ECONOMIC ASPECTS OF ALZHEIMER'S DISEASE AND THE IMPLICATIONS FOR DRUG DEVELOPMENT AND PRICING

M. Hardens
Benefit B.V., Schipholweg 9f, 2316 XB Leiden, The Netherlands

CONTEXT

Among chronic medical disorders senile dementia has been the focus of major epidemiological studies. Jorm et al. (1988, 1987) identified 47 prevalence studies performed between 1945 and 1985. This exceptional interest can be explained both by the social and medical impact of the disease and by the methodological problems associated with its definition.

Dementia is associated with more than 60 conditions, the two most common being Alzheimer's disease (AD) and vascular dementia (Skoog et al. 1992). Whether AD represents two or more distinct conditions distinguished by the age at onset is a matter of substantial debate and research (Schoenberg, 1986). For this reason, some investigators restrict the designation "Alzheimer's Disease" to cases with onset below the age of 65 years (with a more rapidly progressive course), and use the terms "senile dementia of the Alzheimer's type" (SDAT), for those cases with onset at age 65 or above. The term "dementia of the Alzheimer's type" (DAT) is used in this text to refer to this condition regardless of age of onset, unless otherwise specified.

Disease prevalence figures vary among different countries and studies. Mas et al. (1987) concluded in their meta-analysis of all available studies in 1987, that the prevalence of senile dementia of the Alzheimer's type ranged from 1% to 5.8% for 65 years of age or older. Other studies describe the prevalence at a much higher level for the same age range, between 10% and 12% (Evans et al., 1989; Rocca et al., 1986; Pfeffer et al., 1987). Evans et al. (1989) extrapolated these figures at a national level yielding 2.88 million Alzheimer patients in the US in 1989. With the lengthening of life expectancy, projections for the year 2050 led to an estimated number of patients with SDAT ranging from 7.5 million to 14.3 million in the US alone. Similarly, based on population projections up to the year 2040 in West Germany, a recent study estimated that the number of patients with severe and moderate dementia will rise by 50% between 1990 and 2040 (Hafner and Loffler, 1991).

Prevalence varies also as a function of age: Evans et al. (1989) estimated the prevalence of SDAT at 3.0% among the 65-74 years old, at 18.7% among the 75-84 years old and at 47.2% among the 85 years and older.

Other epidemiological sources found a constant doubling of prevalence rates every 5 years. Furthermore, a prevalence study of dementia in Europe described no major geographical differences in the prevalence of Alzheimer's disease across countries when controlling for age and sex (Hofman et al., 1991).

The overall European prevalences for the 5 year age groups, from 60 to 94 years, were 1.0%, 1.4%, 4.1%, 5.7%, 13%, 26.6%, and 32.2%, respectively. Some studies report that AD is more common in women than in men, others do not report differences. There have been no racial differences reported to date. Considering all these epidemiological features, the public health impact of AD is very likely to increase with the increasing longevity of the population.

SEVERITY OF DEMENTIA

It is not possible to deduce the overall severity of dementia from one single aspect (Teunisse, 1991). Demented patients have cognitive impairment, disability in daily life and often behavioral disturbance. Some of the important clinical features of dementia are the impairment in short-term and long-term memory, abstract thinking and judgment, and disturbances of the cortical function and personality changes. In addition, patients are progressively unable to wash or dress themselves or perform basic tasks. Signs of disturbed behavior are often expressed as aggressiveness and non-social behavior. Few studies have directly analyzed the relationship between cognitive functioning and behavior and functioning in daily life. Their findings suggest that only a moderate association exists. However, as a result of both the cognitive and non-cognitive aspects of disease, the Quality of Life of patients and their relatives is negatively affected.

CARING FOR ALZHEIMER PATIENTS

Caring for patients with AD is particularly demanding and requires compassion and skills that go beyond the range of sophisticated and effective therapies characterizing modern medical practice. Several studies suggest that most families of Alzheimer's patients experience a great deal of emotional stress throughout their relative's illness (Morrissey et al., 1990; Zarit et al., 1985). Because of the social dimension of the disease, these studies have widely described the needs for patient care as well as the care structure. The impact of the disease depends mainly on social and cultural backgrounds.

Thus, the burden for the patient's family and for society also varies accordingly.

In France, the majority of patients suffering from AD live at home, and only 12% of patients are estimated to be institutionalized (Jouan-Flahault et al., 1989). According to this study, patients are more likely to live with their spouse (45%), children (23%), alone (20%) or with others (12%). These patients often need care all day long, depending on their level of severity. Estimates of the time spent by family members or relatives for the care of the demented elderly have been reported: a severely demented elder receives an average of 8.06 hours a day of care, whereas a mild and moderate case receives about 3.2 hours a day (Hu et al., 1986). Families are often supported in this painful task by social workers who provide nursing, prepare meals and help the patient in his daily activities. These services are often paid directly by families or included in social coverage schemes. However, a lot of variation exists among different countries.

For all these reasons, institutionalized patients are not necessarily the most severe cases (McLennan et al., 1984). The reason for institutionalization is strongly related to the family's ability to take care of the patients' needs, and surprisingly does not directly reflect the underlying severity of the disease.

THE ECONOMIC IMPACT OF ALZHEIMER'S DISEASE

The high prevalence, the degenerative component of the disease as well as its social and familial implications suggests that the economic impact of the disease may be dramatic for society, health insurance systems, and patient's and their families. As an illustration, Schoenberg (1986) indicates that patients with chronic neurologic diseases occupy more than 50% of the nursing home beds in the USA.

Two studies performed in the US concerning the total cost of the disease are of particular interest. Huang et al. (1988) gave the following figures for senile dementia:

- <u>Total direct costs:</u> $13,260 billion, $6,360 billion of which represent direct medical costs, $2,560 billion nursing home care costs, and $4,340 billion social services. These costs included short-term hospitalizations, physician's services, drug consumption and other medical supplies and expenditures.
- <u>Total indirect costs:</u> $74,630 billion, $31,460 billion of which represent community home care and $43,170, premature deaths and work loss. These costs included the time costs imputed for community home care as well as the loss of the lifetime productive value of human capital and the subjective values of loss of life as a result of dementia. They did not include the time spent in visiting nursing homes or accompanying the demented elderly to receive additional physician services.

The second US study, a pilot by Hu et al. (1986) explored the relationship between illness severity and cost structure, while taking into account only direct medical costs and social services. The average cost per year of an Alzheimer patient was found to be about $11,735, ranging from $14,812 for a severe type, to $6,515 for a mild type.

A British prospective study assessing the nursing home and hospital use of patients with Alzheimer's-type dementia showed that the median length of stay in nursing homes was 2.75 years, over ten times the average length of stay for all diagnoses in the UK. Nursing home charges were then estimated to be between $4,3 and $6,4 million ($35,000 - $52,000/ patient) Gray and Fenn, 1993).

Comparing these figures with the total cost of myocardial infarction and the cost of schizophrenia, this disease is certainly one of the most dramatic medical and economic challenges that our society will face in the coming years (Rice et al., 1993; Hartunian et al., 1980; Gunderson and Mosher, 1975).

DRUG DEVELOPMENT

Drug therapies for AD can generally be classified into 4 types: preventive, therapeutic, symptomatic and palliative. Preventive treatments are those that prevent the individual from developing the disease in the first place. Results from a recent study indicate that estrogen may play a role in reducing the risk of AD in some women (Cotton, 1994). The mechanism of this effect of estrogen is unknown. No other preventive therapies for AD have been identified as yet.

A therapeutic treatment improves and perhaps even reverses the course of AD. This specifically means reducing the symptoms, slowing the disease progression or increasing survival time. A symptomatic therapy would treat the symptoms of the disease, but not alter the actual course. This might mean improving the patient's cognitive or functional abilities or reducing aggressive behaviors. A palliative therapy serves only to make the patient comfortable.

Because of the uncertain etiology and pathogenesis of AD, it is difficult to assess the effectiveness of a drug in terms of therapeutic, symptomatic or palliative capabilities. In fact, the three treatment goals are difficult to separate for this disease. All the drugs commonly used in the treatment and management of patients with AD have been evaluated in terms of their association with improvement on various instruments that assess cognition, memory and behavior. One of the most commonly used instruments, the Alzheimer's Disease Assessment Scale (ADAS), measures all the symptoms common to patients with AD, including cognitive impairment, memory, orientation and behavioral impairment. The clinical relevance of small, but statistically significant improvements in these scales is currently a matter of much debate. This presents a significant difficulty for the development of new drugs for treating AD. The introduction of tacrine has clearly showed that

those responsible for running the health care systems will only be prepared to reimburse such treatments if the manufacturers of such products can prove a clear, clinically relevant advantage in day-to day practice. As it is difficult to quantify any advantages of these treatments it will be even harder to argue that the price of such drugs is justified and thus that the treatment is cost-effective.

The quantification of the economic impact of new treatments poses less methodological difficulties but is far from simple. The present generation of drug therapies seem to improve the cognitive abilities in patients with AD. However, functional status continues to decline and although the therapies may slow progression of Alzheimer's symptoms, they do not prevent or reverse the disease. From an economic perspective the possible benefits of these therapies include:

- a delay in institutionalization of the patient;
- a reduction in the number of hours of informal caregiving required in the short term;
- an improvement in Quality of Life for patients and caregivers.

The economic evaluation of drug therapies under development for AD presents still many challenges as the benefits of treatment affect indirect costs through reduced caregiver hours and improved Quality of Life for both patient and caregiver. The quantification of informal caregiving hours is controversial and fraught with methodological problems.

In order to advance the field, instruments need to be developed that are sensitive to the specific circumstances of patients with AD and their caregivers. This is of major importance because only when the true value of new drugs for the treatment of AD can be scientifically documented will our societies be prepared to pay their high acquisition cost.

REFERENCES

Cotton P (1994): Constellation of risks and processes seen in search for Alzheimer's clues. *JAMA* 271:89-91.

Evans et al. (1989): Prevalence of Alzheimer's Disease in a community population of older persons. *JAMA* 262:2551-2556.

Gray A, Fenn P (1993): Alzheimer's Disease: the burden of illness in England. *Health Trends* 25:31-37.

Gunderson JG, Mosher LR (1975): The cost of Schizophrenia. *Am J Psychiat* 132:901-906.

Hafner H, Loffler W (1991): Trends in the number of patients with senile dementia and need for nursing care in the upcoming 50 years - a demographic projection based on epidemiologic data for former West Germany. *Off-Gesundheitswes* 53:681-686.

Hartunian NS, Smart CN, Thompson MS (1980): The incidence and economic costs of major health impairments: a comparative analysis of cancer, motor vehicle injuries, heart disease, and stroke. Lexington, Mass. *Lexington Books*.

Hofman A, Rocca WA, Brayne C et al. (1991): The prevalence of dementia in Europe: a collaborative study of 1980-1990 findings. Eurodem Prevalence Research Group, *Int J Epidemiol* 20:736-748.

Hu TW, Huang LF, Cartwright WS (1986): Evaluation of the costs of caring for the senile demented elderly: a pilot study. *Gerontol Soc Amer* 26:158-163.

Huang LF, Cartwright WS, Hu TW (1988): The economic cost of senile dementia in the United States. *Public Health Report* (103):3-7.

Jorm AF, Korten AE, Henderson AS (1987): The prevalence of dementia: a quantitative integration of the literature. *Acta Psychiatr Scand* 76:465-479.

Jorm AF, Korten AE, Jacomb PA (1988): Projected increases in the number of dementia cases for 29 developed countries: application for a new method for making projections. *Acta Psychiatr Scand* 78:493-500.

Jouan-Flahault C, Seroussi MC, Colvez A (1989): Absence de liaison entré démence sénile et âge parental. Enquête cas-témoins en Haute-Normandie. *Revue Epidem et Santé Publ* 37:73-75.

Mas JL, Alperovitch A, Derouesné C (1987): Epidémiologie de la démence de type Alzheimer. *Revue Neurologique* 3:161-171.

McLennan WJ, Isles Fe, McDougalls S et al. (1984): Medical and social factors influencing admission to residential care. *Brit Med J* 288:701-703.

Morrissey E, Becker J, Rubert M (1990): Coping resources and depression in the caregiving spouses of Alzheimer's patients. *Brit J Med Psych* 63:161-171.

Pfeffer RI, Afifi AA, Chance JM (1987): Prevalence of Alzheimer's Disease in a retirement community. *Am J Epidemiol* 125:420-436.

Rice DR, Fox PJ, Max W et al. (1993): The economic burden of Alzheimer's Disease care. *Health Aff* 12:165-176.

Rocca WA, Amaducci LA, Schoenberg BS (1986): Epidemilogy of clinically-diagnosed Alzheimer's Disease. *Ann Neurol* 19:415-424.

Schoenberg BS (1986): Epidemiology of dementia. *Neurol Clins* (4).

Skoog I, Nilsson L, Palmertz B et al. (1992): A population-based study of dementia in 85 year olds. *N Engl J Med* 28:153-158.

Teunisse S, Derix M, van Crevel H (1991): Assessing the severity of dementia. Patient and Caregiver. *Arch Neurol* 48:274-277.

Zarit SH, Orr NK, Zarit JM (1985): *The hidden victims of Alzheimer's Disease*. New York, New York Univ Press.

Alzheimer Disease: From Molecular Biology to Therapy
edited by R. Becker and E. Giacobini

INTERNATIONAL WORKING GROUP FOR THE HARMONIZATION OF DEMENTIA DRUG GUIDELINES: A PROGRESS REPORT

Peter J. Whitehouse
Department of Neurology, Case Western Reserve University School of Medicine, Cleveland, OH

Alzheimer's disease(AD) and related disorders are public health and social problems of growing magnitude in all countries, but particularly in those with high proportion of older citizens. Worldwide major research and development efforts in academia, industry and government are bringing the rapid development in the fields of molecular neurobiology and biosystems science to focus on interventions improving the quality of life of patients with dementia. Regulatory guidelines in the United States, Canada, Europe and Japan have been proposed which serve some essential features butdiffer on important aspects such as trial design and outcome measures. To date, few multicentered trials in dementiahave involved sites in different countries outside Europe because of the practical difficulties of executing the same protocol across national boundaries.

In 1994, we started a process to promote harmonization of activities in the major pharmaceutical markets. After an initial planning meeting held in Minnesota in July 1995, we sent a survey outlining options for content and processes of future meetings to interested parties in academia, industry and government. We formed the International Working Group for Harmonization of Dementia Drug Guidelines (IWGHDDG) with the mission to assist efficient development of drugs to improve the quality of life of persons affected by dementia. We established two essential goals: first to study and harmonize regulatory guidelines developed in the major markets of the world, and second to facilitate the development of multinational, multisite protocols.

After the initial meeting and survey, we agreed that we would form subgroups to examine critical issues identified by participants. A list of the subgroups and chairpersons is shown in Table I. The principle focus of these groups is Phase III studies of drugs to improve the cognitive symptoms including outcome measures and design issues. We created other groups to study protocols for prevention, slowing progression of disease and treatment

of noncognitive symptoms. Yet other groups focus on broader issues such as ethical and cultural concerns.

This preplanning allowed the development of an agenda for a meeting in London held over two days in September, 1995. At this meeting all the working groups met having prepared position papers prior to the meeting. The position papers reviewed the literature, the existing guidelines, recommended areas for harmonization and additional research. In July of 1996, we will meet at the International Alzheimer's Conference in Osaka to finalize the working position papers that will then be published along with a consensus paper in Alzheimer's Disease and Associated Disorders International Journal.

Although membership of the harmonization process includes academics, regulators and industry, industry provides financial support almost exclusively. A list of membership, the working subgroups and other written material concerning the harmonization process can be found on the harmonization page (http://weatherhead.cwru.edu/harmon/).

Activities of the harmonization group as a whole are being coordinated with regional harmonization events. For example, at the end of April, 1996, a conference organized by the Canadian consortium of academic centers called C5R will hold a meeting with regulators in Canada to move forward the process of their guidelines being accepted officially. In July, 1996 after the Osaka Conference, we will hold a meeting in Japan organized by the Ministry of Health and Welfare, Japanese Pharmaceutical Manufacturer's Association, pharmaceutical companies and academic centers. This meeting will advance the development of Japanese guidelines in both vascular dementia and AD. At the London meeting, Japanese representatives announced that these guidelines would be developed. We expect initial drafts by the time of the meeting in July. We are planning a meeting on the use of data from trials in Europe perhaps to coincide with the next meeting of the International Conference on Harmonization (ICH) in Brussels.

A number of co-sponsoring organizations have joined our process including the World Federation of Neurology Research Group on Dementia, the International Psychogeriatrics Association, and with Alzheimer's Disease International. We meet in conjunction with the International Alzheimer's Conference which will meet next in July, 1998, in Amsterdam. Sponsorship from the World Health Organization has been granted although it is officially pending. The International Pharmaceutical Manufacturer's Association has also participated. This group is the secretariat for the ICH. Although we have no formal connections to ICH, we are working under their conceptual umbrella. ICH is a group of principally regulators; it is not focusing on specific diseases but is working on the more general process of making pre-clinical/clinical submissions more uniform around the world. Sponsorship from other organizations is currently under consideration (Table II).

QUALITY OF LIFE SUBGROUP ACTIVITIES

To illustrate the mechanism by which the subgroups are working, we will highlight the activities of the Quality of Life subgroup chaired by Dr. Jean-Marc Orgogozo and Dr. Peter J. Whitehouse. As with all the subgroups, the committee has the goal to review current guidelines concerning the role of quality of life in drug development, existing assessment instruments and research on quality of life in dementia. We are considering the role of quality of life measurement in Phase III and IV studies. We will make recommendations concerning future research and areas in need of harmonization. As with all the subgroups, a planning document was prepared in preparation for the London meeting that was circulated to members of the subcommittee. We discussed this in small groups and then presented to the group as a whole. We developed a tentative definition of quality of life, adapted from the World Health Organization process, "the integration of self-perceptions, the satisfactory cognitive functioning, personal activities, psychological well-being and social interactions." The group has developed some tentative conclusions 1) quality of life is not specifically considered in the guidelines; 2) there is increasing interest in the assessment of quality of life in dementia both for caregivers and patients. Using both self-rating and observer measures; 3) more work on the definition of quality of life is needed and stronger links to pharmaco-economic analyses, for example, cost/utility analysis will be required. We believe that although data on quality of life should be collected in both Phase III and Phase IV studies, approval should not be based on quality of life measures. We do expect the pricing decisions would be influenced by the ability of the medication to demonstrate an effect on quality of life of the patient or the caregiver. The quality of life group will post its draft reports on the World Wide Web and invites comments from members of the harmonization process as well as other interested parties.

FUTURE PLANS

In the future, we plan to advance the process of harmonization through meetings and electronic networking. We will work with regulatory bodies, including FDA, European Medicines Evaluation Agency, Koseisho and the Canadian government to provide advice as needed for the development of guidelines. We expect to be able to comment on the 6th draft of the European Union Guidelines which may become the penultimate version. We plan to use the World Wide Web to maintain our collaborations and educational activities. Publication of the Alzheimer's Disease and Associated Disorders International Journal supplemental issue is expected in the Fall of 1996. All those interested in joining the group should contact Peter J. Whitehouse at (e-mail address: pjw3@po.cwru.edu) or another member of the steering committee.

TABLE I.

International working dementia drug guidelines

Group of harmonization of working group facilitators

ADL's	Serge Gauthier
Clinical Global Measures	Barry Reisberg
Cultural Issues	Luigi Amaducci
Diagnostic Criteria	Zaven Khachaturian
Ethical Issues	Stephen Post
Length of Trials	Raymond Levy
Objective Psychometric Tests	Steven Ferris
Pharmacoeconomics	Bengt Winblad
Prevention Protocols	Leon Thal
Protocols to Demonstrate Slowing Progression of Disease	Martin Rossor
Protocols for Treating Noncognitive Symptoms	Akira Homma
Quality of Life	Co-chairs: Peter Whitehouse and Jean-Marc Orgogozo
Translation	Howard Feldman

TABLE II.

International Working Group on Harmonization of Dementia Drug Guidelines

Affiliated Organizations

World Federation of Neurology Dementia Research Group
International Psychogeriatric Association
Alzheimer Disease International
International Alzheimer Conference
World Health Organization (officially pending)
International Federation of Pharmaceutical Manufacturers Association (secretariat for International Conference of Harmonization)

Alzheimer Disease: From Molecular Biology to Therapy
edited by R. Becker and E. Giacobini

DEVELOPING SAFE AND EFFECTIVE ANTIDEMENTIA DRUGS

Paul D. Leber
Division of Neuropharmacological Drug Products
ODE I, CDER, Food and Drug Administration

INTRODUCTION

Within the United States, most activities related to the development, clinical testing, and marketing of new drug products are subject to the provisions and requirements of the Federal Food, Drug and Cosmetic Act. The Act is enforced by the Food and Drug Administration [FDA], an administrative agency within the Executive Branch of America's Federal Government.

The Act is intended to ensure the quality and reliability of the armamentarium. Toward this end, the Act requires that all legally marketed new drugs be the subject of an approved new drug application [NDA]. To gain approval of an NDA, a sponsor must compile and submit to the FDA all information and evidence specified by the Act as necessary to show that a drug product is in compliance with the requirements of the Act. It is the responsibility of the FDA to review each NDA and to approve it, unless, upon review, the agency determines that the reports submitted fail to show that the drug product involved meets the Act's requirements.

The Act instructs the FDA to disapprove an application, if, among other findings:

- the NDA does not include reports of adequate kinds and numbers of tests to show whether or not the product will be safe for use under the conditions of use recommended in its proposed labeling, or
- the results of such tests show the product to be unsafe for use or fail to show that the product is safe for use as recommended , or
- there is inadequate information available to make a determination that the drug will be safe for use, or
- there is a lack of substantial evidence to show that the drug will have the effects its sponsor claims it will have under the conditions of use described in its proposed labeling, or

- based upon a fair evaluation of all material facts, the labeling under which the sponsor proposes to market the product is false or misleading in any particular.

Although the words of the Act make plain the importance Congress attached to ensuring the safety of marketed drugs, the Act's language is not so obvious as to Congressional intent where the standard of drug efficacy is concerned. Historians of drug regulation opine that this can be explained by the fact that those who devised the effectiveness standard were mindful that experts could reasonably and responsibly disagree about the strength of evidence needed to support a conclusion that a particular drug product had been shown to be effective. Accordingly, Congress set the effectiveness evidence standard at a substantial, and not at a preponderant, level. Those who drafted the effectiveness requirement, however, clearly intended to limit the kind of evidence that experts could consider in reaching a determination of effectiveness. Thus, the definition of substantial evidence given in the Act is explicit in its demand that evidence of effectiveness be 1) derived from adequate and well controlled clinical investigations and 2) be of a kind and quality that would allow an expert in the management of the disease or condition for which the investigational treatment is intended, to conclude responsibly, that the drug will have the effect claimed for it under the conditions of use described in the labeling proposed by its sponsor.

The Act provides insufficient information to determine from it words alone precisely how drug products are to be evaluated. The practical day to day enforcement of the requirements of the Act, therefore, turns on rules and policies promulgated by the FDA [e.g., 21 CFR 312: Investigational New Drugs) and 21 CFR 314: New Drug Applications). Although these address issues common to all drug classes (e.g., the kind of experimental designs and control conditions that would be acceptable for use in an adequate and well controlled clinical investigation), they provide little guidance as to therapeutic class specific matters. Although this might seem to grant sponsors considerable advantage, it has the opposite effect if a sponsor learns only after completing a development effort that the agency finds it incomplete or inappropriate. While such an outcome is unlikely if sponsors maintain an ongoing dialog with agency staff, both the regulator and regulated face a quandary when a product under development is the first member of a new therapeutic class.

Precisely such a situation arose in 1986. In the aftermath of a report in the New England Journal of Medicine describing the successful use of tacrine in the management of patients with dementia, commercial interest in marketing products with an antidementia indications increased substantially. At the instruction of Frank Young, then the Commissioner of Food and Drugs, the author, with advice and help of several members of FDA's Peripheral and Central Nervous System [PCNS] Advisory Committee, initiated an effort to develop guidance on antidementia drug development for the regulated

industry. The goal was to provide a clear description of the clinical investigations that would, by design, be accepted as a source of substantial evidence for an unmodified antidementia claim. A draft of the Guidelines was made available for comment in 1990. In February of 1992, the PCNS AC met to discuss the draft and the comments that had been received concerning it.

The principles enumerated in the draft guidelines were largely endorsed by the PCNS AC; the guidelines have not been formally adopted by the agency, however. Nevertheless, with minor exceptions, the guidance they offer is essentially identical to that a sponsor seeking advice from the Division of Neuropharmacological Drug Products would receive today in regard to the development of an antidementia drug product.

This is not to be taken as an implication that the draft guidelines are a final product or a conclusion that they cannot be improved further. To the contrary, the draft guidelines address, as several observers have noted, only a very narrow part of the spectrum of deficits and disabilities that are concomitants of the dementia syndrome. The narrowness of their scope, however, is intentional; the goal was not to delineate every possible antidementia related claim, but to provide sponsors with a clear statement of the nature of the evidence needed for an unmodified antidementia claim.

The guidance offered reflects the view that any drug product granted an unmodified antidementia claim must be shown to have a clinically meaningful effect on the core cognitive deficits that characterize dementia. To demonstrate that it possesses these attributes, it must be shown, in more than one randomized controlled trial, to have effects that can be detected on two different kinds of outcome assessments. First, its effects on the cognitive deficits of dementia must be demonstrated on a validated, performance based measure, for example, the cognitive subpart of the Alzheimer's Disease Assessment Scale (ADAS-Cog). Second, the effect must be detected on a measure that is deemed capable of assessing a change of clinical importance. This second component of the dual assessment requirement has proved controversial. The original draft guidelines recommended a clinician's global assessment as the means to ensure that an effect produced was clinically meaningful. The argument made was that if a clinician could detect an effect based entirely on information gained during an interview with a patient, the effect would have to be of a magnitude that would be clinically important. It is critical to this argument that the physician base his/her assessment on information gained at interview from the patient, otherwise, the clinician's assessment might not reflect an observable change in clinical state, but the opinion of a caregiver aware of side effects, etc.

While the author believes that what is now called the CIBIC (Clinician's Interview based Impression of Change) is still the best means to serve the aims he intended, the community of experts is clearly willing to rely on a less demanding test of clinical effect. Accordingly, the agency now accepts an assessment obtained by a clinician at an interview conducted with and/or relying upon information derived from family members, significant others, and/ or caregivers (the so called CIBIC Plus) as a means to satisfy the second component of its compound outcome requirement. An effect detected on a validated measure of activities of daily living would also satisfy the requirement.

Other trial design issues are not controversial. Evidence of an antidementia effect should be obtained in randomized, parallel design trials of at least 3 months, and preferably 6 months, duration. Because premature subject discontinuations are common in such studies, every effort should be made to obtain information on the status of drop-outs at the point in time at which they would have been assessed had they remained in the study. This information is of value in assessing the potential of drop-outs to bias estimates of treatment effect.

Because of the potential for carryover effects, trial designs that attempt to enrich the sample randomized by pre-exposing study candidates to the treatment being investigated in a pre-randomization phase are strongly discouraged.

When a difference favoring a drug is found under the clinical trial conditions enumerated, however, a question arises invariably as to whether the difference derives from a symptomatic effect on performance or one on the disease process itself. How best to distinguish these possibilities has long been of interest.

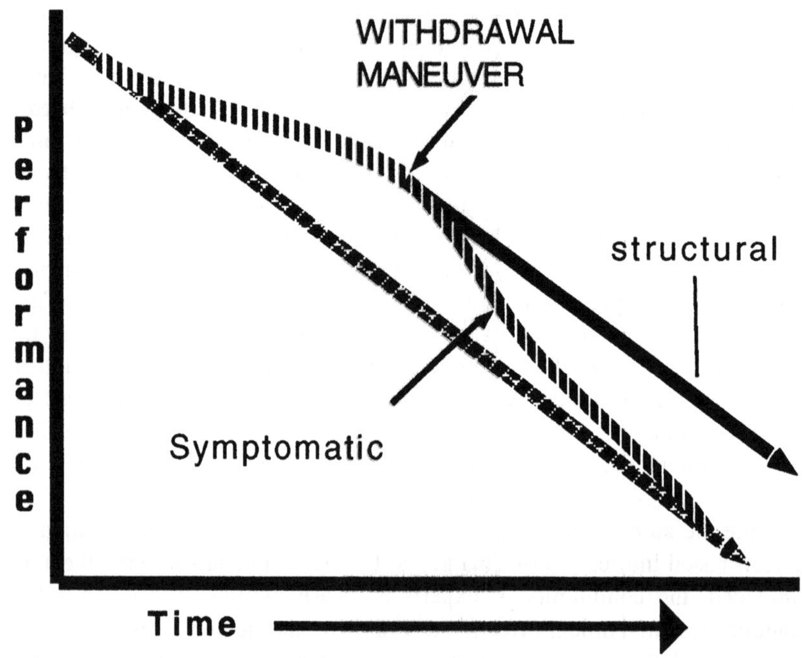

One approach, pictured above, involves re-randomization. At the completion of a randomized controlled trial, members of the group of patients assigned to active treatment are re-randomized to receive either placebo or the active treatment to which they were initially randomized. If the active treatment group re-randomized to placebo fails to maintain the advantage it had over the group receiving placebo from the time of initial randomization, the effects of the drug cannot be viewed as having had a lasting effect and the product must be deemed to have a symptomatic effect. If, however, the gains relative to the placebo group are maintained, the drug must have had an effect on the course of the illness.

This investigational maneuver, however, has always been somewhat unappealing for it requires patients to withdraw from a treatment they believe has proven useful. An alternative research design I have proposed that is based on delayed randomization, avoids this problem.

In the design I propose, patients are followed for an interval on placebo. At some point, a proportion of the cohort is randomized to active treatment; after a suitable delay, the remainder of the cohort is randomized to the identical active treatment regimen. If the treatment employed produces only symptomatic effects, the delay in treatment initiation should not affect the magnitude of treatment response, and, the performance of the subset of the

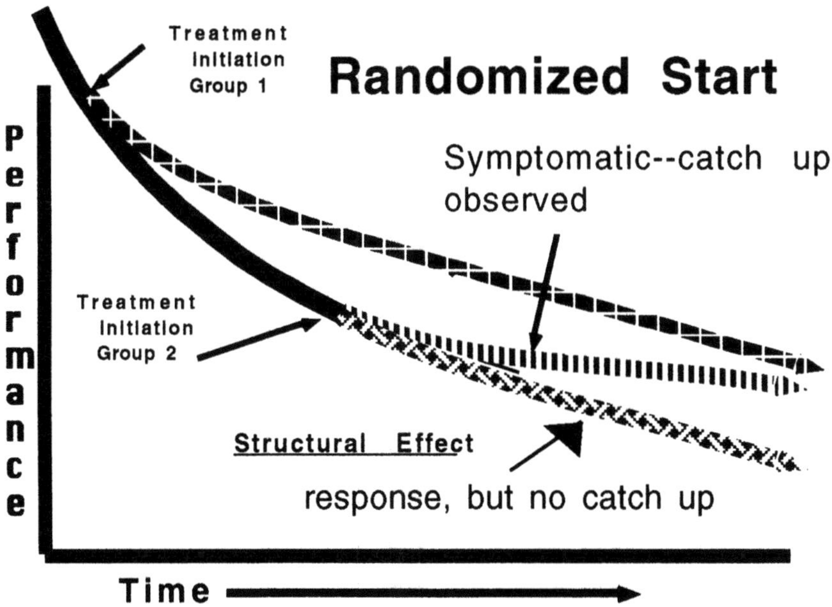

cohort assigned to treatment after the delay will "catch up" to those assigned to active treatment without delay. If, however, the group randomized with delay fails to achieve parity with the group randomized without delay, one must conclude that the drug had an effect on the progression of the dementing process. The advantage of the strategy lies in the fact that no patient presumed to be doing well because of treatment is required to be withdrawn from it.

Finally, a word about safety. Knowledge of a drug's risks comes from clinical experience. While small samples may provide insight into the common untoward concomitants of a drug's use, risks of importance to the public health cannot be excluded without extensive experience. Unfortunately, the time required to gain this experience imposes delays that undermine the goal of accelerating the pace with which new products are made available for marketing. Once the effectiveness of a drug is documented in a controlled trial, therefore, efforts to gain experience with the product should be expanded rapidly. Large simple trials, treatment protocols, are encouraged as long as they do not interfere with the completion of the controlled trials required to demonstrate effectiveness.

NOTICE: Dr. Leber is with the Food and Drug Administration, Rockville, Maryland. However, this article was written by Dr. Leber in his private capacity. No official support or endorsement by the Food and Drug Administration is intended or should be inferred.

Alzheimer Disease: From Molecular Biology to Therapy
edited by R. Becker and E. Giacobini

A JAPANESE PERSPECTIVE ON THE WORK OF THE INTERNATIONAL GROUP ON HARMONIZATION OF DRUG GUIDELINES

Akira Homma
Department of Psychiatry, Tokyo Metropolitan Institute of Gerontology,
35-2 Sakaecho, Itabashiku, Tokyo, 173 Japan

INTRODUCTION

It was reported that the elderly with age-associated dementia would comprise 6.8% of the aged population in Japan in 1995 (Ohtsuka, 1991). Approximately 70 to 75 percent of those are living in the community and their families care for them (Karasawa, 1988). Approximately half of the demented elderly living in the community are still mildly demented. Thus, effective antidementia drugs to prevent progression of dementia have been actually required by the relatives or families caring for the demented elderly in Japan. Up to now, a variety of antidementia drugs have been developed in Japan. However, recent trials of some acetylcholinesterase inhibitors failed to show satisfactory results. In this chapter, practical issues to be discussed in conducting drug trials and the necessity to develop infrastructures for the drug trials before harmonizing drug guidelines in Japan will be briefly described.

ISSUES TO BE DISCUSSED IN DEVELOPING ANTIDEMENTIA DRUGS

A variety of drugs have been developed for patients with Alzheimer type dementia (ATD) in Japan as shown in the Table I. Most of them are still in the phase II stage. Generally, in Japan, four cases including randomly assigned two cases with placebo and two cases with an active drug are allocated as a set in double-blind placebo-controlled trials to compare an active drug and placebo. Usually, at least, a single set is provided to each site. Also, a single investigator is recommended to complete a set including four cases. However, a set including four cases has not been always completed by a single investigator. It seems no doubt that such a way of the trial influence the results of the trial, in particular, the judgment of the clinical evaluation.

1. *ACh inhibitors*	
• E2020 (Eisai)	PII
• NIK-247 (Nikken Kagaku)	PIII
• ENA-713 (Sandoz)	PII
• TAK-147 (Takeda)	PII
2. *ACh releasers*	
• T-588 (Toyama Kagaku)	PII
3. *M1 agonists*	
• YM-796 (Yamanouchi)	PII
• AF-102B (Snow Brand)	PII
4. *Neuropeptides*	
• TRH-SR (Takeda) : for VD	PII
• Ebiratide (Hoechst)	PII
• JTP-4819 (Japan Tobacco) :PPCE inhibitor	PII
5. *Others*	
• DM-9384 (Daiichi):Nootropics	PII
• S-5810 (Shionogi):BDZ inverse agonist	PII
• Nimodipine (Bayer): for VD	PIII

TABLE I. Current antidementia drugs being developed in Japan. VD: vascular dementia, BDZ: benzodiazepine, PII: Phase II study, PIII: Phase III study

Diagnostic criteria of dementia including NINCDS-ADRDA, DSM-IV, and ICD-10 are quite commonly used in the routine clinical situations in Japan. However, reports on inter-rater reliability or validity for the criteria are quite sparse in Japan, except for the results by Kitamura et al. (1985) reporting satisfactory inter-rater agreements of DSM-III and a conventional classification of severity of dementia between three psychiatrists. In the trials for ATD in Japan, psychometric tests such as ADAS-Cog are not always administered by psychologists. In some sites, they are administered by unexperienced physicians. Heterogeneity of the administrator may bias the results of the trials. The other issue on the administration of performance-based tests which has to be discussed is the time for the administration of the test. For instance, in the results to develop a Japanese version of ADAS-Cog, an average administration time was increased with age of the subjects (Homma et al., 1992). Although no data is available that the administration time will bias the results of the trials, it might be worthwhile to be considered as a subject for the discussion.

Figure 1 shows a difference in the assessment of the global improvement rating (GIR) between experienced and inexperienced investigators in a certain phase III study in Japan. There was an apparent difference in the GIR by experienced investigators between placebo and an active drug, while no marked difference was found in inexperienced investigators. Figure 2 shows some difference in the assessment of the GIR between completed sets and

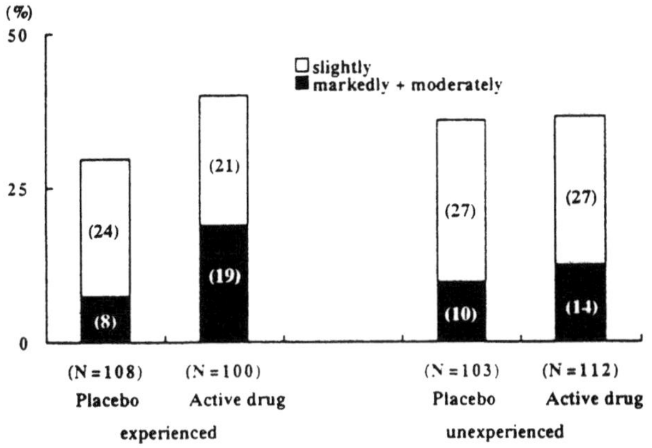

FIGURE 1. Difference of the global improvement rating by experienced and unexperienced investigators.

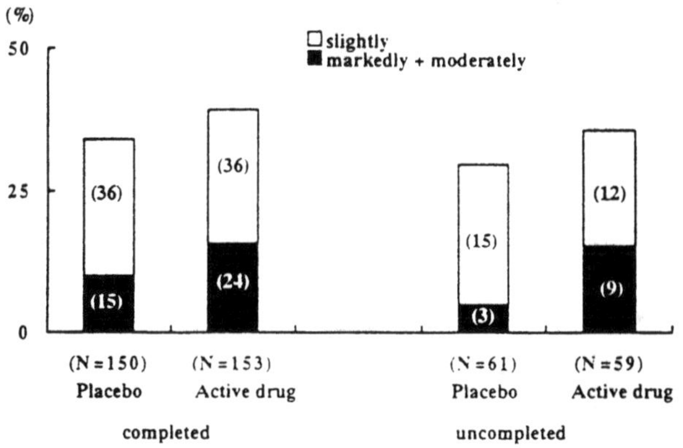

FIGURE 2. Difference of the global improvement rating by completed sets of four cases and uncompleted sets.

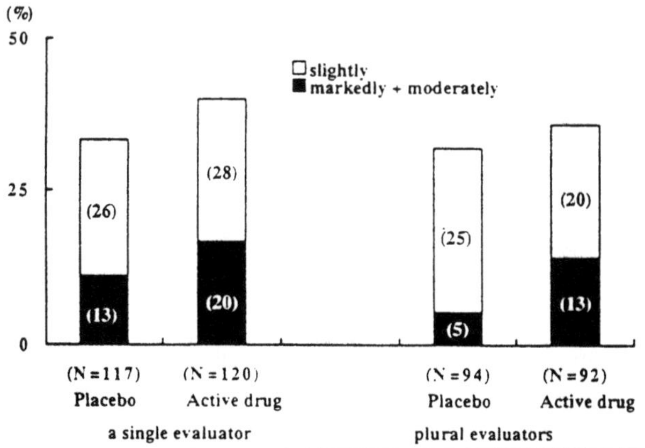

FIGURE 3. Difference of the global improvement rating in the completed sets of four cases by a single investigator and more than two investigators.

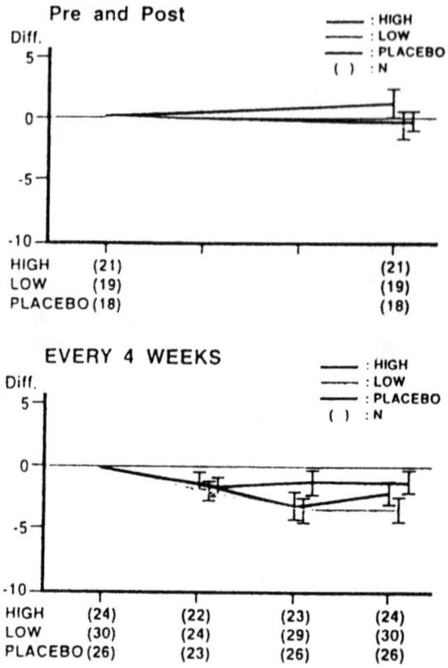

FIGURE 4. Difference of ADAS-J Cog scores by frequency of administration. The upper figure shows the scores of ADAS-J Cog administered before and after the trial. The lower figure shows the scores of ADAS-J Cog administered every four weeks during the trial period of three months.

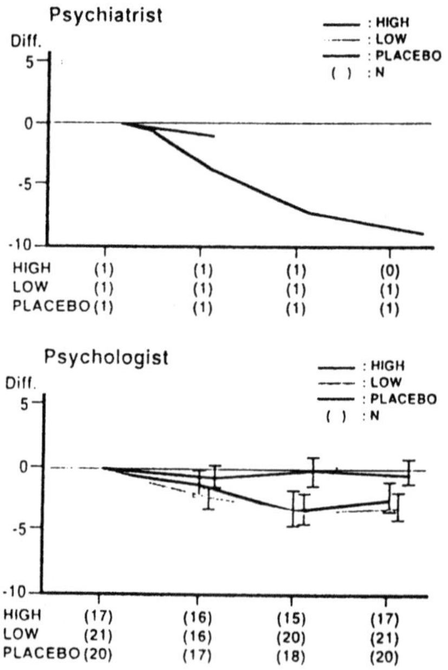

FIGURE 5. Difference of ADAS-J Cog scores by an administrator. The upper figure shows the scores of ADAS-J Cog administered by physicians. The lower figure shows the scores of ADAS-J Cog administered by psychologists.

uncompleted sets. The difference seems reasonable. Figure 3 indicates a difference of the GIR in the sets completed by a single and plural investigators. Again some difference may be indicated. Figure 4 shows the difference of ADAS-Cog score by frequency of the administration of the test in the other double-blind placebo-controlled trial. The upper figure indicates the change of the ADAS-Cog scores obtained only before and after the three months trial. Differences among the three doses are not so clear. However, the lower figure shows the changes of the ADAS-Cog scores administered as indicated in the protocol, that is, every four weeks. The differences among three doses become more clear. Also, Figure 5 shows the change of the ADAS-Cog scores by the type of administrator. Though the sample size is very small, the upper figure indicates the results by psychiatrists. In these sites, psychiatrists administered the ADAS-Cog because psychologists were not available during the trial period. The lower figure shows the results by psychologists. Changes of the ADAS-Cog scores in the patients treated with placebo seem quite different in both figures.

CONCLUSION

These data strongly support the necessity of an appropriate infrastructure with an official guideline of drug trials for ATD patients in Japan. In particular, recognition of trials as research and training of the standardized interview procedures seem most essential in conducting the trials for AD patients in Japan. An outline of the guideline is to be available until the end of this July, when the Fourth Congress of Alzheimer's Disease and Associated Disorders is to be held in Osaka, Japan. This will be a good time to start to construct suitable infrastructures for the appropriate drug trials in Japan.

REFERENCES

Homma A. Fukuzawa K. Tsukada Y. Ishii T. Hasegawa K and Mohs R (1992): Development of a Japanese version of Alzheimer's Disease Assessment Scale (ADAS). *Jap J Geriat Psychiat* 6:647-655. (in Japanese)

Karasawa A (1988): Prevalence of age-associated dementia. *Treatment* 70:638-642. (in Japanese)

Kitamura T. Maruyama S. Ohtuska T. Shimonaka Y and Nakazato K (1985): Inter-rater reliability of DSM-III and Karasawa's classification of senility. *Gerontopsychiatry* 2:774-777. (in Japanese)

Ohtsuka T (1991): Report on the care and the care system for patients with dementia. Tokyo: Foundation of Longevity Sciences. pp.89.

uncompleted sets. The difference seems reasonable. Figure 3 indicates the difference of the GIR in the sets completed as a single and plural investigation. Again some difference may be indicated. Figure 4 shows the difference of ADAS-Cog score by frequency of the administration of the test in the other double blind placebo controlled trial. The upper figure shows the change of the ADAS-Cog score obtained [illegible] and after the three months trial. Difference among the [illegible]. However, it is not clear how [illegible] ADAS-Cog [illegible] and [illegible] three times become more correct. Also, Figure 5 shows the change of the ADAS-Cog scores by [illegible] the sample size is very small [illegible]

[illegible]

Alzheimer Disease: From Molecular Biology to Therapy
edited by R. Becker and E. Giacobini

THE RONALD AND NANCY REAGAN RESEARCH INSTITUTE OF THE ALZHEIMER'S ASSOCIATION

Z. S. Khachaturian and T. S. Radebaugh

BACKGROUND AND RATIONALE

Current efforts toward developing treatments to delay the symptoms of AD will not be adequate to overcome the demographic forces which are increasing the total number of affected individuals. The prevalence of AD increases nearly exponentially with age, that is, the percentage of the population affected doubles every decade beyond age 65. By delaying the onset of disabling symptoms for all age groups by five to ten years, it would be possible to reduce the number of people affected by half and cut the cost of care significantly.

In twenty years, attempts to postpone institutionalization will be too little and too late. It is imperative that we begin now to expand research aimed at finding more effective treatment and better behavioral management approaches. We must accelerate the process of drug discovery and the development of strategies for the prevention of AD. There is an urgent need to place into action specific plans to achieve the long range goals of the Alzheimer's Association of reducing the **duration** of illness, decreasing the **numbers** of persons affected by AD, and lowering the **cost** of long term care. To achieve these overall objectives, the Alzheimer's Association created the ***Ronald and Nancy Reagan Research Institute.***

The Alzheimer's Association, by establishing the Reagan Institute, launched a bold new initiative to rally the leadership of academia, the pharmaceutical industry, family support groups, and the American people behind a plan of integrated effort aimed at accelerating the discovery of treatments. The specific aim of the Institute is to implement some of the goals of the ***1994-1998 Strategic Plan of the Alzheimer's Association***. The Institute will mediate cooperation and coordination among all groups with an interest in AD research. It will expand research aimed at discovering therapeutic targets and agents and facilitate the development of treatment strategies targeted at maintaining the independent functioning of AD patients for an additional five, ten, or fifteen years. It will also help to garner the necessary resources for the systematic expansion of research programs.

The Association, through the Institute, expects to forge an alliance and a new "social contract" between several partners but particularly between the family care providers and scientists to solve the problem of AD. In the present climate of worldwide economic austerity, the Alzheimer's Association is taking charge and providing the leadership, through the Reagan Institute, for mobilizing this nation's scientific resources to accelerate the discovery of treatments to delay the onset of AD. Progress toward the goal of discovering treatments could be substantially accelerated by the elimination of some historical barriers and administrative, legal, and philosophical traditions. There is an urgent need to break such barriers and to form alliances among such groups as the government, academia, biotechnology and pharmaceutical companies, and private organizations. As resources become scarce and the scientific ventures become financially risky, it makes sense to pool the resources of all interested parties. The future success of this venture will rely on such joint ventures. At the present while national priorities for domestic programs are being reevaluated and debated, it is an opportune moment in history to initiate a venture and place into practice the idea of private-public working partnership to solve a major public health problem, a concept long promoted by President Ronald Reagan.

The potential success of a joint venture between partners such as families, scientists and industry is virtually assured because all the essential elements are at hand; the ultimate goal is well defined, there are many good ideas, the necessary human capital is available, and the critical infrastructures are in place. The scientific community is confident that the objectives of this plan can be achieved within a reasonable time frame. Clearly, no one can predict precisely when and where a major scientific breakthrough will occur or how fast we will be able to achieve this goal, but we must start on the long and difficult road to this goal now. Even modest progress would provide much welcome relief to the families of AD patients.

MISSION OF THE REAGAN INSTITUTE

The mission of the ***Ronald and Nancy Reagan Research Institute*** is to conduct, support and promote research on finding the means to delay the onset of symptoms, to prevent and ultimately to cure AD. The Institute will expand the ongoing research support program of the Alzheimer's Association to: accelerate the discovery of risk factors; early detection and treatments; development of strategies targeted at maintaining the independent functioning of AD patients for an additional five, ten, or fifteen years. The scientific agenda of the Institute is driven by three goals:

- ***To find the cause(s) of Alzheimer's disease*** - spanning from its biological underpinnings to the impact of behavioral and social factors on disease progression.
- ***To maintain independent functioning for an additional 5 to 10 years*** - through the development of safe, effective pharmacological and behavioral approaches as treatment.
- ***To prevent Alzheimer's disease*** - by discovering selective risk factors and their interactions with genetic and epigenetic factors.

Discover the Cure for Alzheimer's Disease

One of the most important tasks of the ***Ronald and Nancy Reagan Research Institute*** is to conduct, support and promote research on finding the means to delay the onset of symptoms, to prevent and ultimately to cure AD

Develop Innovative Scientific Programs

The second critical task of the Institute is to develop unique and innovative programs that are designed;

a- to increase the scientific and technical knowledge necessary to fulfill the mission of the Institute,

b- to facilitate the transfer of technology and knowledge,

c- to create synergism among research laboratories by forming collaborative international working groups,

d- foster the establishment of strategic alliances between academia, industry, financial institutions and the Association, and

e- to leverage the scientific talents and resources of academia, industry and the government to function more efficiently and effectively in the process of discovery and development of treatments. The trans-university drug discovery teams, the formation of scientific working groups, the International Society for Alzheimer's Disease Research, and the organization of research planning workshops are only some of the initial programs of the Reagan Institute.

Develop a National Fund for Research on Alzheimer's Disease

The third, but perhaps more pragmatic, responsibility of the Institute is to provide the Alzheimer's Association attractive targets for large endowments. The relationship between the scientific programs of the Institute and the Association's success in fund raising are closely linked in a symbiotic relationship. The Institute's scientific programs are important mechanisms or tools for developing a major fund raising campaign.

Forge Alliances Between Scientists and Families

For the fourth undertaking, the Institute will create programs and offer the means for strengthening the bond between the scientific community and AD family members thus enlarging the Alzheimer's Association's constituency. The

Institute will forge new working partnerships between scientists and strengthen the allegiance between the families and the scientists with an interest in solving the problem of AD. The ability of the Alzheimer's Association to attract large major gifts for biomedical research to some extent depends on the vitality of its scientific programs. To build diverse and attractive scientific programs, the Association needs a strong affiliation with the medical and scientific world.

Support Research

The fifth task of the Institute is to foster the continuation and the expansion of the Alzheimer's Association's long standing grants and awards program to support scientific research. The restructuring of the Association's program of research support is work in progress. The responsibility of the Institute in this area is to provide technical advice and serve a consultative role in the effort to streamline the process of grants review and administration.

FUNCTIONAL STRUCTURE OF THE INSTITUTE

The Institute will have a medical and scientific advisory committee representing leaders from academia, medicine, science, health care, industry (pharmaceutical, biotechnology), law, finance and families. The **functional structure** of the ***Ronald and Nancy Reagan Research Institute*** will have two operational units to administer new scientific initiatives of the Alzheimer's Association.

A - Program of Discovering Treatments, Early Markers & Risk Factors

The Drug Discovery Program, the Discovery of Early Markers Program, and the Discovery of Risk Factors Program, are some of the initiatives that are being planned by this program. The formation of the drug discovery teams as part of the Drug Discovery Program is one of the most important initial activities of this program.

B - Program of Technology Transfer

This program will organize research planning workshops, support the formation of scientific work-groups, publish position papers and help shape public opinions/policies on medical/scientific issues. The sponsorship of the formation of the International Society for Alzheimer's Disease Research is one of the activities of this program. It will focus on forging international alliances and establishing linkages between scientists, family members, and organizations around the world with an interest in AD.

DRUG DISCOVERY PROGRAM

The **Drug Discovery Program** is a component of the **Program of Discovering Treatments, Early Markers & Risk Factors.** The **Drug Discovery Program,** along with the other programs, is designed to achieve the objective of accelerating the discovery of a cure and raising support for research.

Program Structure

The **Drug Discovery Program** will be composed of multiple collaborative scientific teams with representatives from the very best scientific laboratories of academic, medical, pharmaceutical, biotech, government and private institutions around the world. The teams will be organized as trans-institution working groups by projects where each participant's knowledge, expertise, perspective, techniques, and skills will bring a unique but synergistic contribution to the solution of the scientific problem at hand. The working groups or teams, representing the world's best and brightest investigators, will be formed around a scientific theme or problem concerning targets for treatment, discovery of new molecules or solving problems related to drug targeting or delivery systems. Each group will select its leader and determine its own scientific thrust on the basis of the group members' interests, unique technical strengths and the priorities of the commercial partner(s) of that group.

Program Aims

The objective of the **Drug Discovery Program** is to establish joint ventures between academia and industry by creating working partnerships between academia based pre-clinical investigators and those with a stronger interest in development, based in industry. This program will facilitate the transfer of knowledge and technologies derived from basic science laboratories to those interested in the practical applications to develop products. The aim is to increase the diversity, quantity and quality of ideas that enter the pipeline of products being developed for the treatment of AD. The Institute will create research and development alliances between the drug discovery teams and commercial partners from the pharmaceutical industry, biotechnology companies, and/or venture capital groups. The principle is to find the best fit between the technical/scientific know-how and the resources of the various parties. The **Drug Discovery Program** is intended to provide an even-handed opportunity to determine the relative merits of **all differing** ideas on therapeutic approaches. It is structured to insure a level playing field to test the scientific and/or the commercial viability of novel strategies to treatments, no matter how unconventional or contrary to the views of the prevailing scientific orthodoxy these ideas might be. The program is to be eclectic and ecumenical; it is not intended to become the private domain of any single interest.

GUIDING PRINCIPLES FOR THE FORMATION OF **DRUG DISCOVERY TEAMS:**

The **Drug Discovery Teams** will be organized on the basis of the following criteria:

a- commitment and dedication to discovering treatments for AD,
b- a history of innovative thinking and productive research,
c- willingness to work in collaboration with other members of the Institute,
d- readiness to be part of joint research ventures with pharmaceutical or biotechnology partners,
e- acceptance of the principle that a portion of the grant or royalty derived income will be returned to the Institute for reinvestment into the revolving research fund of the Institute, and
f- an idea and research plan for which funding can be secured.

The creation and viability of a team will be determined by two factors: the scientific merit of the problem area chosen, and the willingness of a donor or partner to subsidize the research. The science will dictate the formation and termination of work groups. This will provide the maximum flexibility to the Institute.

INTERNATIONAL SOCIETY FOR ALZHEIMER'S DISEASE RESEARCH

The ***Ronald and Nancy Reagan Research Institute*** is the headquarters for **The International Society for Alzheimer's Disease Research (Society).** The **Society** fulfills several important needs of the growing world community of scientists interested in AD research.

The mission of the **Society** is to accelerate the pace of research aimed at discovering the cause(s) of Alzheimer's disease, early and accurate detection methods, effective treatments, and ways to care for patients and families with dignity.

The purpose of the **Society** is to create a broad based forum for rapid communication of scientific and technical information, and is organized to achieve the following goals:

- promote research on AD, encourage international cooperative studies and cross-disciplinary research
- foster communication among investigators and accelerate the transfer of technology,
- disseminate scientific information,
- organize local, regional, national and international research workshops or conferences,

- form work groups to make recommendations for public policy,
- serve as a local, regional, national, and international scientific resource to voluntary health organizations, private foundations, governments, and industry,
- create alliances between family members of AD patients and the world scientific community.

The **Society** is a multi-disciplinary scientific organization. The membership will include all relevant professions, specialties and disciplines ranging from structural biology, protein chemistry, neurology, psychiatry, neuropathology, molecular biology and genetics, clinical and neuropsychology, neuroepidemiology, nursing, social work, health care economics, and the social scientists investigating health services as a route to improving patient care.

The **Society** encompasses all areas of research concerned with the neurobiology of healthy brain aging, the biology of the dementias, diagnosis, etiology, treatment, care and management of AD. The members, which may number nearly 3000, can serve as reviewers of grant applications for local, national or international research funding organizations, and advisors or spokespersons on scientific matters to the public media, policy formulators and legislators.

To accelerate the pace of major discoveries in AD, it is critical to improve communication of research findings. Information exchange between scientists must be rapid and frequent. Members of the **Society** will be offered access to a professionally designed and maintained World Wide Web for Alzheimer's research to keep members informed about meetings and conferences, funding opportunities, titles and references of major publications, abstracts of funded projects, clinical trials planned or in progress and recruitment information, and other relevant items. This Web site will provide efficient access to information such as imaging data, scientific literature, and resources. Membership in the **Society** will include a subscription to a journal, central to the scientific interests of the particular member, to ensure that the members' research interests and information needs are covered. The **Society** will also establish an international speaker's bureau aimed at facilitating rapid dissemination and translation of research results for the general public.

The **Society** is envisioned as a new, and vital partner in the global effort to conquer AD. It will (1) promote additional funding for research, (2) work to strengthen the bond between family members and scientists, and (3) facilitate the rapid dissemination of research findings to scientists and people around the world.

- form workgroups to make recommendations for public policy.
- serve as a local, regional, national, and international scientific resource to community health organizations, private foundations, government, and industry;
- create alliances between family members of AD patients and the world scientific community.

The Society [illegible] multidisciplinary [illegible] membership will include [illegible] relevant [illegible] disciplines ranging from [illegible] biology, [illegible] neurology, psychiatry, neuropathology, molecular biology and genetics, clinical and neuropsychology, neuroepidemiology, [illegible] health care economics, and the social sciences [illegible]

[illegible]

SUBJECT INDEX

AUTHOR INDEX

MIX
Papier aus verantwortungsvollen Quellen
Paper from responsible sources
FSC® C105338

If you have any concerns about our products,
you can contact us on
ProductSafety@springernature.com

In case Publisher is established outside the EU,
the EU authorized representative is:
Springer Nature Customer Service Center GmbH
Europaplatz 3, 69115 Heidelberg, Germany

Printed by Libri Plureos GmbH
in Hamburg, Germany